Springer Handbook
of Electronic and Photonic Materials

Safa Kasap, Peter Capper (Eds.)

Springer Handbook of Electronic and Photonic Materials
Organization of the Handbook

Each chapter has a concise summary that provides a general overview of the subject in the chapter in a clear language. The chapters begin at fundamentals and build up towards advanced concepts and applications. Emphasis is on physical concepts rather than extensive mathematical derivations. Each chapter is full of clear color illustrations that convey the concepts and make the subject matter enjoyable to read and understand. Examples in the chapters have practical applications. Chapters also have numerous extremely useful tables that summarize equations, experimental techniques, and most importantly, properties of various materials. The chapters have been divided into five parts. Each part has chapters that form a coherent treatment of a given area. For example,

Part A contains chapters starting from basic concepts and build up to up-to-date knowledge in a logical easy to follow sequence. Part A would be equivalent to a graduate level treatise that starts from basic structural properties to go onto electrical, dielectric, optical, and magnetic properties. Each chapter starts by assuming someone who has completed a degree in physics, chemistry, engineering, or materials science.

Part A Fundamental Properties
2 Electrical Conduction in Metals and Semiconductors
3 Optical Properties of Electronic Materials: Fundamentals and Characterization
4 Magnetic Properties of Electronic Materials
5 Defects in Monocrystalline Silicon
6 Diffusion in Semiconductors
7 Photoconductivity in Materials Research
8 Electronic Properties of Semiconductor Interfaces
9 Charge Transport in Disordered Materials
10 Dielectric Response
11 Ionic Conduction and Applications

Part B provides a clear overview of bulk and single-crystal growth, growth techniques (epitaxial crystal growth: LPE, MOVPE, MBE), and the structural, chemical, electrical and thermal characterization of materials. Silicon and II–VI compounds and semiconductors are especially emphasized.

Part B Growth and Characterization
12 Bulk Crystal Growth-Methods and Materials
13 Single-Crystal Silicon: Growth and Properties
14 Epitaxial Crystal Growth: Methods and Materials
15 Narrow-Bandgap II–VI Semiconductors: Growth
16 Wide-Bandgap II–VI Semiconductors: Growth and Properties
17 Structural Characterization
18 Surface Chemical Analysis
19 Thermal Properties and Thermal Analysis: Fundamentals, Experimental Techniques and Applications
20 Electrical Characterization of Semiconductor Materials and Devices

Part C covers specific materials such as crystalline Si, microcrystalline Si, GaAs, high-temperature semiconductors, amorphous semiconductors, ferroelectric materials, and thin and thick films.

Part C Materials for Electronics
21 Single-Crystal Silicon: Electrical and Optical Properties
22 Silicon-Germanium: Properties, Growth and Applications
23 Gallium Arsenide
24 High-Temperature Electronic Materials: Silicon Carbide and Diamond
25 Amorphous Semiconductors: Structure, Optical, and Electrical Properties
26 Amorphous and Microcrystalline Silicon
27 Ferroelectric Materials
28 Dielectric Materials for Microelectronics
29 Thin Films
30 Thick Films

Part D examines materials that have applications in optoelectronics and photonics. It covers some of the state-of-the-art developments in optoelectronic materials, and covers III–V Ternaries, III–Nitrides, II–VI compounds, quantum wells, photonic crystals, glasses for photonics, non-linear photonic glasses, nonlinear organic, and luminescent materials.

Part E provides a survey on novel materials and applications such as information recording devices (CD, video, DVD) as well as phase-change optical recording. The chapters also include applications such as solar cells, sensors, photoconductors, and carbon nanotubes. Both ends of the spectrum from research to applications are represented in chapters on molecular electronics and packaging materials.

Part D Materials for Optoelectronics and Photonics
31 III–V Ternary and Quaternary Compounds
32 Group III Nitrides
33 Electron Transport within the III–V Nitride Semiconductors, GaN, AlN, and InN: A Monte Carlo Analysis
34 II–IV Semiconductors for Optoelectronics: CdS, CdSe, CdTe
35 Doping Aspects of Zn-Based Wide-Band-Gap Semiconductors
36 II–VI Narrow-Bandgap Semiconductors for Optoelectronics
37 Optoelectronic Devices and Materials
38 Liquid Crystals
39 Organic Photoconductors
40 Luminescent Materials
41 Nano-Engineered Tunable Photonic Crystals in the Near-IR and Visible Electromagnetic Spectrum
42 Quantum Wells, Superlattices, and Band-Gap Engineering
43 Glasses for Photonic Integration
44 Optical Nonlinearity in Photonic Glasses
45 Nonlinear Optoelectronic Materials

Part E Novel Materials and Selected Applications
46 Solar Cells and Photovoltaics
47 Silicon on Mechanically Flexible Substrates for Large-Area Electronics
48 Photoconductors for X-Ray Image Detectors
49 Phase-Change Optical Recording
50 Carbon Nanotubes and Bucky Materials
51 Magnetic Information-Storage Materials
52 High-Temperature Superconductors
53 Molecular Electronics
54 Organic Materials for Chemical Sensing
55 Packaging Materials

Glossary of Defining Terms There is a glossary of *Defining Terms* at the end of the handbook that covers important terms that are used throughout the handbook. The terms have been defined to be clear and understandable by an average reader not directly working in the field.

使 用 说 明

1.《电子与光子材料手册》原版为一册,分为A、B、C、D、E五部分。考虑到各部分内容相对独立完整,为使用方便,影印版按部分分为5册。

2.各册在页脚重新编排页码,该页码对应中文目录。保留了原书页眉及页码,其页码对应原书目录及主题索引。

3.各册均有完整5册书的内容简介。

4.作者及其联系方式、缩略语表各册均完整呈现。

5.名词术语表、主题索引安排在第5册。

6.文前页基本采用中英文对照形式,方便读者快速浏览。

材料科学与工程图书工作室
联系电话　0451-86412421
　　　　　0451-86414559
邮　　箱　yh_bj@yahoo.com.cn
　　　　　xuyaying81823@gmail.com
　　　　　zhxh6414559@yahoo.com.cn

Springer 手册精选系列

电子与光子材料手册

光电子学与光子材料

【第4册】

Springer
Handbook of
Electronic
and Photonic
Materials

〔加拿大〕Safa Kasap
〔英　国〕Peter Capper 主编

（影印版）

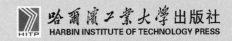
哈尔滨工业大学出版社
HARBIN INSTITUTE OF TECHNOLOGY PRESS

黑版贸审字08-2012-031号

Reprint from English language edition:
Springer Handbook of Electronic and Photonic Materials
by Safa Kasap and Peter Capper
Copyright © 2007 Springer US
Springer US is a part of Springer Science+Business Media
All Rights Reserved

This reprint has been authorized by Springer Science & Business Media for distribution in China Mainland only and not for export there from.

图书在版编目（CIP）数据

电子与光子材料手册. 第4册，光电子学与光子材料=Handbook of Electronic and Photonic Materials Ⅳ Materials for Optoelectronics and Photonics：英文／（加）卡萨普（Kasap S.），（英）卡珀（Capper P.）主编. —影印本. —哈尔滨：哈尔滨工业大学出版社，2013.1

（Springer手册精选系列）
ISBN 978-7-5603-3763-0

Ⅰ.①电…Ⅱ.①卡…②卡…Ⅲ.①电子材料-手册-英文②光学材料-手册-英文③光电子学-手册-英文Ⅳ.①TN04-62②TB34-62③TN201-62

中国版本图书馆CIP数据核字（2012）第189740号

责任编辑	杨　桦　许雅莹
出版发行	哈尔滨工业大学出版社
社　　址	哈尔滨市南岗区复华四道街10号 邮编150006
传　　真	0451-86414749
网　　址	http://hitpress.hit.edu.cn
印　　刷	哈尔滨市石桥印务有限公司
开　　本	787mm×960mm 1/16 印张 24.75
版　　次	2013年1月第1版　2013年1月第1次印刷
书　　号	ISBN 978-7-5603-3763-0
定　　价	78.00元

（如因印刷质量问题影响阅读，我社负责调换）

序 言

本书的编辑、作者、出版人都将庆祝这本卓著书籍的出版,这对于电子与光子材料领域的工作者也将是无法衡量的好消息。从以往编辑的系列手册看,我认为本书的出版是值得的,坚持出版这样一本书也是必要的。本书之所以显得特别重要,是因为它在这个领域,内容覆盖范围广泛,涉及的方法也是当今的最新研究进展。在这样一个迅速发展的领域,这是一个相当大的挑战,它已经赢得了人们的敬意。

早期的手册和百科全书也都注重阐述半导体材料的发展趋势,而且必须覆盖半导体材料广泛的研究范围和所涉及的现象。这是可以理解的,原因在于半导体材料在电子领域中的主导地位。但没有多少人有足够的勇气预测未来的发展趋势。1992年,Mahajan和Kimerling在其《简明半导体材料百科全书和相关技术》一书的引言中做了尝试,并且预测未来的挑战将是纳米电子领域、低位错密度的III-V族衬底技术、半绝缘III-V族衬底技术、III-V族图形外延技术、替换电介质和硅接触技术、离子注入和扩散技术的发展。这些预测或多或少地成为了现实,但是这也同样说明做出这样的预测是多么的困难。

十年前没有多少人会想到III族氮化物在这本书中将成为重要的部分。与制备相关的问题是,作为高熔点材料,在受欢迎的能在光谱蓝端作光发射器的材料中它们的熔点并不高。这是一个很有意思的话题,至少与解决早期光谱红端的固体激光器工作寿命短的问题一样有趣。总地说来,光电子学和光子学在前十年中已经呈现出一些令人瞩目的研究进展,这些在本书中得到了体现,范围从可见光发光器件材料到红外线材料。书中Part D的内容范围很宽,包括III-V族和II-VI族光电子材料和能带隙工程,以及光子玻璃、液态晶体、有机光电导体和光子晶体的新领域。整个部分反映了材料的光产生、工艺、光传输和光探测,包括所有用光取代电子的必要内容。

在电子材料这一章(Part C)探讨了硅的进展。毋庸置疑地是,硅是占据了电子功能和电子电路整个范围主导地位的材料,包括新电介质和其他关于缩减电路和器件的几何尺寸以实现更高密度的封装方面的内容,以及其他书很少涉及的领域,薄膜、高温电子材料、非晶和微晶材料。增加硅使用寿命的新技术成果(包括硅/锗合金)在书中也有介绍,并且又一次提出了同样问题,即,预测硅过时时间是否过于超前!铁电体——一类与硅非常有效结合的材料同样也出现在书中。

Part E章节中（新型材料和选择性的应用）使用了一些极好的新方法开辟了新领地。我们大都知道且频繁使用信息记录器件，但是很少知道，涉及器件使用的材料或原理，比如说CD、视频、DVD等。本书介绍了磁信息存储材料，同样介绍了相变光记录材料，使我们充分与当前发展步伐保持同步更新。该章也同样介绍了太阳能电池、传感器、光导体和碳纳米管的应用，这样大量的工作也体现出编写内容汲取到了世界范围的广度。本章各节中的分子电子和封装材料从研究到应用都得到了呈现。

  本书的突出优点在于它的内容覆盖了从基础科学（Part A）到材料的制备、特性（Part B）再到材料的应用（Part C~E）。实际上，书中介绍了涉及的所有材料的广泛应用，这就是本书为什么将会实用的原因之一。就像我之前提及的那样，我们之中没有多少人能够成功地预测未来的发展方向和趋势，在未来十年占领这个领域的主导地位。但是，本书教给我们关于材料的基本性能，可用它们去满足将来的需要。我热切地把这本书推荐给你们。

Prof. Arthur Willoughby
Materials Research Group,
University of Southampton,
UK

前　言

不同学科各种各样的手册，例如电子工程、电子学、生物医学工程、材料科学等手册被广大学生、教师、专业人员很好地使用着，大部分的图书馆也都藏有这些手册。这类手册一般包含许多章（至少50章）内容，在已确定的学科内覆盖广泛的课题；学科选材和论述水平吸引着本科生、研究生、研究员，乃至专业工程人员；最新课题提供广泛的信息，这对该领域所有初学者和研究人员是非常有帮助的；每隔几年，就会有增加新内容的新版本更新之前的版本。

电子和光子材料领域没有类似手册的出版，我们出版这本《电子与光子材料手册》的想法是源自于对手册的需求。它广泛覆盖当今材料领域内的课题，在工程学、材料科学、物理学和化学中都有需要。电子和光子材料真正是一门跨学科的学问，它包含了一些传统的学科，如材料科学、电子工程、化学工程、机械工程、物理学和化学。不难发现，机械工程人员对电子封装实施研究，而电子工程人员对半导体特性进行测量。只有很少的几所大学创建了电子材料或光子材料系。一般来说，电子材料作为一个"学科"是以研究组或跨学科的活动出现在"学院"中。有人可能会对此有异议，因为它事实上是一个跨学科领域，非常需要既包括基础学科又要有最新课题介绍的手册，这就是出版本手册的原因。

本手册是一部关于电子和光子材料的综合论述专著，每一章都是由该领域的专家编写的。本手册针对于大学四年级学生或研究生、研究人员和工作在电子、光电子、光子材料领域的专业人员。书中提供了必要的背景知识和内容广泛的更新知识。每一章都有对内容的一个介绍，并且有许多清晰的说明和大量参考文献。清晰的解释和说明使手册对所有层次的研究者有很大的帮助。所有的章节内容都尽可能独立。既有基础又有前沿的章节内容将吸引不同背景的读者。本手册特别重要的一个特点就是跨学科。例如，将会有这样一些读者，其背景（第一学历）是学化学工程的，工作在半导体工艺线上，而想要学习半导体物理的基础知识；第一学历是物理学的另外一些读者需要尽快更新材料科学的新概念，例如，液相外延等。只要可能，本手册尽量避免采用复杂的数学公式，论述将以半定量的形式给出。手册给出了名词术语表（Glossary of Defining Terms），可为读者提供术语定义的快速查找——这对跨学科工具书来说是必须的。

编者非常感激所有作者们卓越的贡献和相互合作,以及在不同阶段对撰写这本手册的奉献。真诚地感谢Springer Boston的Greg Franklin在文献整理以及手册出版的漫长的工作中给予的支持和帮助。Dr.Werner Skolaut在Springer Heidelberg非常熟练地处理了无数个出版问题,涉及审稿、绘图、书稿的编写和校样的修改,我们真诚地感谢他和他所做出的工作——使得手册能够吸引读者。他是我们见过的最有奉献精神和有效率的编者。

感谢Arthur Willoughby教授的诸多建设性意见使得本手册更加完善。他在材料科学杂志(Journal of Materials Science)积累了非常丰富的编辑经验;电子材料这一章在书中起着重要作用,不仅仅是选取章节,而且还要适应读者需要。

最后,编者感谢所有的成员(Marian, Samuel and Tomas; and Nicollette)在全部工作中的支持和付出的特别耐心。

Dr. Peter Capper
Materials Team Leader,
SELEX Sensors and Airborne Systems,
Southampton, UK

Prof. Safa Kasap
Professor and Canada Research Chair,
Electrical Engineering Department,
University of Saskatchewan,
Canada

Foreword

The Editors, Authors, and Publisher are to be congratulated on this distinguished volume, which will be an invaluable source of information to all workers in the area of electronic and photonic materials. Having made contributions to earlier handbooks, I am well aware of the considerable, and sustained work that is necessary to produce a volume of this kind. This particular handbook, however, is distinguished by its breadth of coverage in the field, and the way in which it discusses the very latest developments. In such a rapidly moving field, this is a considerable challenge, and it has been met admirably.

Previous handbooks and encyclopaedia have tended to concentrate on semiconducting materials, for the understandable reason of their dominance in the electronics field, and the wide range of semiconducting materials and phenomena that must be covered. Few have been courageous enough to predict future trends, but in 1992 Mahajan and Kimerling attempted this in the Introduction to their Concise Encyclopaedia of Semiconducting Materials and Related Technologies (Pergamon), and foresaw future challenges in the areas of nanoelectronics, low dislocation-density III-V substrates, semi-insulating III-V substrates, patterned epitaxy of III-Vs, alternative dielectrics and contacts for silicon technology, and developments in ion-implantation and diffusion. To a greater or lesser extent, all of these have been proved to be true, but it illustrates how difficult it is to make such a prediction.

Not many people would have thought, a decade ago, that the III-nitrides would occupy an important position in this book. As high melting point materials, with the associated growth problems, they were not high on the list of favourites for light emitters at the blue end of the spectrum! The story is a fascinating one – at least as interesting as the solution to the problem of the short working life of early solid-state lasers at the red end of the spectrum. Optoelectronics and photonics, in general, have seen one of the most spectacular advances over the last decade, and this is fully reflected in the book, ranging from visible light emitters, to infra-red materials. The book covers a wide range of work in Part D, including III-V and II-VI optoelectronic materials and band-gap engineering, as well as photonic glasses, liquid crystals, organic photoconductors, and the new area of photonic crystals. The whole Part reflects materials for light generation, processing, transmission and detection – all the essential elements for using light instead of electrons.

In the Materials for Electronics part (Part C) the book charts the progress in silicon – overwhelmingly the dominant material for a whole range of electronic functions and circuitry – including new dielectrics and other issues associated with shrinking geometry of circuits and devices to produce ever higher packing densities. It also includes areas rarely covered in other books – thick films, high-temperature electronic materials, amorphous and microcrystalline materials. The existing developments that extend the life of silicon technology, including silicon/germanium alloys, appear too, and raise the question again as to whether the predicted timetable for the demise of silicon has again been declared too early!! Ferroelectrics – a class of materials used so effectively in conjunction with silicon – certainly deserve to be here.

Prof. Arthur Willoughby
Materials Research Group,
University of Southampton,
UK

The chapters in Part E (Novel Materials and Selected Applications), break new ground in a number of admirable ways. Most of us are aware of, and frequently use, information recording devices such as CDs, videos, DVDs etc., but few are aware of the materials, or principles, involved. This book describes magnetic information storage materials, as well as phase-change optical recording, keeping us fully up-to-date with recent developments. The chapters also include applications such as solar cells, sensors, photoconductors, and carbon nanotubes, on which such a huge volume of work is presently being pursued worldwide. Both ends of the spectrum from research to applications are represented in chapters on molecular electronics and packaging materials.

A particular strength of this book is that it ranges from the fundamental science (Part A) through growth and characterisation of the materials (Part B) to applications (Parts C–E). Virtually all the materials covered here have a wide range of applications, which is one of the reasons why this book is going to be so useful. As I indicated before, few of us will be successful in predicting the future direction and trends, occupying the high-ground in this field in the coming decade, but this book teaches us the basic principles of materials, and leaves it to us to adapt these to the needs of tomorrow. I commend it to you most warmly.

Preface

Other handbooks in various disciplines such as electrical engineering, electronics, biomedical engineering, materials science, etc. are currently available and well used by numerous students, instructors and professionals. Most libraries have these handbook sets and each contains numerous (at least 50) chapters that cover a wide spectrum of topics within each well-defined discipline. The subject and the level of coverage appeal to both undergraduate and postgraduate students and researchers as well as to practicing professionals. The advanced topics follow introductory topics and provide ample information that is useful to all, beginners and researchers, in the field. Every few years, a new edition is brought out to update the coverage and include new topics.

There has been no similar handbook in electronic and photonic materials, and the present Springer Handbook of Electronic and Photonic Materials (SHEPM) idea grew out of a need for a handbook that covers a wide spectrum of topics in materials that today's engineers, material scientists, physicists, and chemists need. Electronic and photonic materials is a truly interdisciplinary subject that encompasses a number of traditional disciplines such as materials science, electrical engineering, chemical engineering, mechanical engineering, physics and chemistry. It is not unusual to find a mechanical engineering faculty carrying out research on electronic packaging and electrical engineers carrying out characterization measurements on semiconductors. There are only a few established university departments in electronic or photonic materials. In general, electronic materials as a "discipline" appears as a research group or as an interdisciplinary activity within a "college". One could argue that, because of the very fact that it is such an interdisciplinary field, there is a greater need to have a handbook that covers not only fundamental topics but also advanced topics; hence the present handbook.

This handbook is a comprehensive treatise on electronic and photonic materials with each chapter written by experts in the field. The handbook is aimed at senior undergraduate and graduate students, researchers and professionals working in the area of electronic, optoelectronic and photonic materials. The chapters provide the necessary background and up-to-date knowledge in a wide range of topics. Each chapter has an introduction to the topic, many clear illustrations and numerous references. Clear explanations and illustrations make the handbook useful to all levels of researchers. All chapters are as self-contained as possible. There are both fundamental and advanced chapters to appeal to readers with different backgrounds. This is particularly important for this handbook since the subject matter is highly interdisciplinary. For example, there will be readers with a background (first degree) in chemical engineering and working on semiconductor processing who need to learn the fundamentals of semiconductors physics. Someone with a first degree in physics would need to quickly update himself on materials science concepts such as liquid phase epitaxy and so on. Difficult mathematics has been avoided and, whenever possible, the explanations have been given semiquantitatively. There is a *"Glossary of Defining Terms"* at the end of the handbook, which can serve to quickly find the definition of a term – a very necessary feature in an interdisciplinary handbook.

Dr. Peter Capper
Materials Team Leader, SELEX Sensors and Airborne Systems, Southampton, UK

Prof. Safa Kasap
Professor and Canada Research Chair, Electrical Engineering Department, University of Saskatchewan, Canada

The editors are very grateful to all the authors for their excellent contributions and for their cooperation in delivering their manuscripts and in the various stages of production of this handbook. Sincere thanks go to Greg Franklin at Springer Boston for all his support and help throughout the long period of commissioning, acquiring the contributions and the production of the handbook. Dr. Werner Skolaut at Springer Heidelberg has very skillfully handled the myriad production issues involved in copy-editing, figure redrawing and proof preparation and correction and our sincere thanks go to him also for all his hard work in making the handbook attractive to read. He is the most dedicated and efficient editor we have come across.

It is a pleasure to thank Professor Arthur Willoughby for his many helpful suggestions that made this a better handbook. His wealth of experience as editor of the Journal of Materials Science: Materials in Electronics played an important role not only in selecting chapters but also in finding the right authors.

Finally, the editors wish to thank all the members of our families (Marian, Samuel and Thomas; and Nicollette) for their support and particularly their endurance during the entire project.

Peter Capper and Safa Kasap
Editors

List of Authors

Martin Abkowitz
1198 Gatestone Circle
Webster, NY 14580, USA
e-mail: *mabkowitz@mailaps.org*,
abkowitz@chem.chem.rochester.edu

Sadao Adachi
Gunma University
Department of Electronic Engineering,
Faculty of Engineering
Kiryu-shi 376-8515
Gunma, Japan
e-mail: *adachi@el.gunma-u.ac.jp*

Alfred Adams
University of Surrey
Advanced Technology Institute
Guildford, Surrey, GU2 7XH,
Surrey, UK
e-mail: *alf.adams@surrey.ac.uk*

Guy J. Adriaenssens
University of Leuven
Laboratorium voor Halfgeleiderfysica
Celestijnenlaan 200D
B-3001 Leuven, Belgium
e-mail: *guy.adri@fys.kuleuven.ac.be*

Wilfried von Ammon
Siltronic AG
Research and Development
Johannes Hess Strasse 24
84489 Burghausen, Germany
e-mail: *wilfried.ammon@siltronic.com*

Peter Ashburn
University of Southampton
School of Electronics and Computer Science
Southampton, SO17 1BJ, UK
e-mail: *pa@ecs.soton.ac.uk*

Mark Auslender
Ben-Gurion University of the Negev Beer Sheva
Department of Electrical
and Computer Engineering
P.O.Box 653
Beer Sheva 84105, Israel
e-mail: *marka@ee.bgu.ac.il*

Darren M. Bagnall
University of Southampton
School of Electronics and Computer Science
Southampton, SO17 1BJ, UK
e-mail: *dmb@ecs.soton.ac.uk*

Ian M. Baker
SELEX Sensors and Airborne Systems Infrared Ltd.
Southampton, Hampshire SO15 0EG, UK
e-mail: *ian.m.baker@selex-sas.com*

Sergei Baranovskii
Philipps University Marburg
Department of Physics
Renthof 5
35032 Marburg, Germany
e-mail: *baranovs@staff.uni-marburg.de*

Mark Baxendale
Queen Mary, University of London
Department of Physics
Mile End Road
London, E1 4NS, UK
e-mail: *m.baxendale@qmul.ac.uk*

Mohammed L. Benkhedir
University of Leuven
Laboratorium voor Halfgeleiderfysica
Celestijnenlaan 200D
B-3001 Leuven, Belgium
e-mail: *MohammedLoufti.Benkhedir
@fys.kuleuven.ac.be*

Monica Brinza
University of Leuven
Laboratorium voor Halfgeleiderfysica
Celestijnenlaan 200D
B-3001 Leuven, Belgium
e-mail: *monica.brinza@fys.kuleuven.ac.be*

Paul D. Brown
University of Nottingham
School of Mechanical, Materials and
Manufacturing Engineering
University Park
Nottingham, NG7 2RD, UK
e-mail: *paul.brown@nottingham.ac.uk*

Mike Brozel
University of Glasgow
Department of Physics and Astronomy
Kelvin Building
Glasgow, G12 8QQ, UK
e-mail: *mikebrozel@beeb.net*

Lukasz Brzozowski
University of Toronto
Sunnybrook and Women's Research Institute,
Imaging Research/
Department of Medical Biophysics
Research Building, 2075 Bayview Avenue
Toronto, ON, M4N 3M5, Canada
e-mail: *lukbroz@sten.sunnybrook.utoronto.ca*

Peter Capper
SELEX Sensors and Airborne Systems Infrared Ltd.
Materials Team Leader
Millbrook Industrial Estate, PO Box 217
Southampton, Hampshire SO15 0EG, UK
e-mail: *pete.capper@selex-sas.com*

Larry Comstock
San Jose State University
6574 Crystal Springs Drive
San Jose, CA 95120, USA
e-mail: *Comstock@email.sjsu.edu*

Ray DeCorby
University of Alberta
Department of Electrical
and Computer Engineering
7th Floor, 9107-116 Street N.W.
Edmonton, Alberta T6G 2V4, Canada
e-mail: *rdecorby@trlabs.ca*

M. Jamal Deen
McMaster University
Department of Electrical
and Computer Engineering (CRL 226)
1280 Main Street West
Hamilton, ON L8S 4K1, Canada
e-mail: *jamal@mcmaster.ca*

Leonard Dissado
The University of Leicester
Department of Engineering
University Road
Leicester, LE1 7RH, UK
e-mail: *lad4@le.ac.uk*

David Dunmur
University of Southampton
School of Chemistry
Southampton, SO17 1BJ, UK
e-mail: *d.a.dunmur@soton.ac.uk*

Lester F. Eastman
Cornell University
Department of Electrical
and Computer Engineering
425 Phillips Hall
Ithaca, NY 14853, USA
e-mail: *lfe2@cornell.edu*

Andy Edgar
Victoria University
School of Chemical and Physical Sciences SCPS
Kelburn Parade/PO Box 600
Wellington, New Zealand
e-mail: *Andy.Edgar@vuw.ac.nz*

Brian E. Foutz
Cadence Design Systems
1701 North Street, Bldg 257-3
Endicott, NY 13760, USA
e-mail: *foutz@cadence.com*

Mark Fox
University of Sheffield
Department of Physics and Astronomy
Hicks Building, Hounsefield Road
Sheffield, S3 7RH, UK
e-mail: *mark.fox@shef.ac.uk*

Darrel Frear
RF and Power Packaging Technology Development,
Freescale Semiconductor
2100 East Elliot Road
Tempe, AZ 85284, USA
e-mail: *darrel.frear@freescale.com*

Milan Friesel
Chalmers University of Technology
Department of Physics
Fysikgränd 3
41296 Göteborg, Sweden
e-mail: *friesel@chalmers.se*

Jacek Gieraltowski
Université de Bretagne Occidentale
6 Avenue Le Gorgeu, BP: 809
29285 Brest Cedex, France
e-mail: *Jacek.Gieraltowski@univ-brest.fr*

Yinyan Gong
Columbia University
Department of Applied Physics
and Applied Mathematics
500 W. 120th St.
New York, NY 10027, USA
e-mail: *yg2002@columbia.edu*

Robert D. Gould[†]
Keele University
Thin Films Laboratory, Department of Physics,
School of Chemistry and Physics
Keele, Staffordshire ST5 5BG, UK

Shlomo Hava
Ben-Gurion University of the Negev Beer Sheva
Department of Electrical
and Computer Engineering
P.O. Box 653
Beer Sheva 84105, Israel
e-mail: *hava@ee.bgu.ac.il*

Colin Humphreys
University of Cambridge
Department of Materials Science and Metallurgy
Pembroke Street
Cambridge, CB2 3!Z, UK
e-mail: *colin.Humphreys@msm.cam.ac.uk*

Stuart Irvine
University of Wales, Bangor
Department of Chemistry
Gwynedd, LL57 2UW, UK
e-mail: *sjc.irvine@bangor.ac.uk*

Minoru Isshiki
Tohoku University
Institute of Multidisciplinary Research
for Advanced Materials
1-1, Katahira, 2 chome, Aobaku
Sendai, 980-8577, Japan
e-mail: *isshiki@tagen.tohoku.ac.jp*

Robert Johanson
University of Saskatchewan
Department of Electrical Engineering
57 Campus Drive
Saskatoon, SK S7N 5A9, Canada
e-mail: *johanson@engr.usask.ca*

Tim Joyce
University of Liverpool
Functional Materials Research Centre,
Department of Engineering
Brownlow Hill
Liverpool, L69 3BX, UK
e-mail: *t.joyce@liv.ac.uk*

M. Zahangir Kabir
Concordia University
Department of Electrical and Computer Engineering
Montreal, Quebec S7N5A9, Canada
e-mail: *kabir@encs.concordia.ca*

Safa Kasap
University of Saskatchewan
Department of Electrical Engineering
57 Campus Drive
Saskatoon, SK S7N 5A9, Canada
e-mail: *safa.kasap@usask.ca*

Alexander Kolobov
National Institute of Advanced
Industrial Science and Technology
Center for Applied Near-Field Optics Research
1-1-1 Higashi, Tsukuba
Ibaraki, 305-8562, Japan
e-mail: *a.kolobov@aist.go.jp*

Cyril Koughia
University of Saskatchewan
Department of Electrical Engineering
57 Campus Drive
Saskatoon, SK S7N 5A9, Canada
e-mail: *kik486@mail.usask.ca*

Igor L. Kuskovsky
Queens College, City University of New York (CUNY)
Department of Physics
65-30 Kissena Blvd.
Flushing, NY 11367, USA
e-mail: *igor_kuskovsky@qc.edu*

Geoffrey Luckhurst
University of Southampton
School of Chemistry
Southampton, SO17 1BJ, UK
e-mail: *g.r.luckhurst@soton.ac.uk*

Akihisa Matsuda
Tokyo University of Science
Research Institute for Science and Technology
2641 Yamazaki, Noda-shi
Chiba, 278-8510, Japan
e-mail: *amatsuda@rs.noda.tus.ac.jp*,
a.matsuda@aist.go.jp

Naomi Matsuura
Sunnybrook Health Sciences Centre
Department of Medical Biophysics,
Imaging Research
2075 Bayview Avenue
Toronto, ON M4N 3M5, Canada
e-mail: *matsuura@sri.utoronto.ca*

Kazuo Morigaki
University of Tokyo
C-305, Wakabadai 2-12, Inagi
Tokyo, 206-0824, Japan
e-mail: *k.morigaki@yacht.ocn.ne.jp*

Hadis Morkoç
Virginia Commonwealth University
Department of Electrical
and Computer Engineering
601 W. Main St., Box 843072
Richmond, VA 23284-3068, USA
e-mail: *hmorkoc@vcu.edu*

Winfried Mönch
Universität Duisburg-Essen
Lotharstraße 1
47048 Duisburg, Germany
e-mail: *w.moench@uni-duisburg.de*

Arokia Nathan
University of Waterloo
Department of Electrical
and Computer Engineering
200 University Avenue W.
Waterloo, Ontario N2L 3G1, Canada
e-mail: *anathan@uwaterloo.ca*

Gertrude F. Neumark
Columbia University
Department of Applied Physics
and Applied Mathematics
500W 120th St., MC 4701
New York, NY 10027, USA
e-mail: *gfn1@columbia.edu*

Stephen K. O'Leary
University of Regina
Faculty of Engineering
3737 Wascana Parkway
Regina, SK S4S 0A2, Canada
e-mail: *stephen.oleary@uregina.ca*

Chisato Ogihara
Yamaguchi University
Department of Applied Science
2-16-1 Tokiwadai
Ube, 755-8611, Japan
e-mail: *ogihara@yamaguchi-u.ac.jp*

Fabien Pascal
Université Montpellier 2/CEM2-cc084
Centre d'Electronique
et de Microoptoélectronique de Montpellier
Place E. Bataillon
34095 Montpellier, France
e-mail: *pascal@cem2.univ-montp2.fr*

Michael Petty
University of Durham
Department School of Engineering
South Road
Durham, DH1 3LE, UK
e-mail: *m.c.petty@durham.ac.uk*

Asim Kumar Ray
Queen Mary, University of London
Department of Materials
Mile End Road
London, E1 4NS, UK
e-mail: *a.k.ray@qmul.ac.uk*

John Rowlands
University of Toronto
Department of Medical Biophysics
Sunnybrook and Women's College
Health Sciences Centre
S656-2075 Bayview Avenue
Toronto, ON M4N 3M5, Canada
e-mail: *john.rowlands@sri.utoronto.ca*

Oleg Rubel
Philipps University Marburg
Department of Physics
and Material Sciences Center
Renthof 5
35032 Marburg, Germany
e-mail: *oleg.rubel@physik.uni-marburg.de*

Harry Ruda
University of Toronto
Materials Science and Engineering,
Electrical and Computer Engineering
170 College Street
Toronto, M5S 3E4, Canada
e-mail: *ruda@ecf.utoronto.ca*

Edward Sargent
University of Toronto
Department of Electrical
and Computer Engineering
ECE, 10 King's College Road
Toronto, M5S 3G4, Canada
e-mail: *ted.sargent@utoronto.ca*

Peyman Servati
Ignis Innovation Inc.
55 Culpepper Dr.
Waterloo, Ontario N2L 5K8, Canada
e-mail: *pservati@uwaterloo.ca*

Derek Shaw
Hull University
Hull, HU6 7RX, UK
e-mail: *DerekShaw1@compuserve.com*

Fumio Shimura
Shizuoka Institute of Science and Technology
Department of Materials and Life Science
2200-2 Toyosawa
Fukuroi, Shizuoka 437-8555, Japan
e-mail: *shimura@ms.sist.ac.jp*

Michael Shur
Renssellaer Polytechnic Institute
Department of Electrical, Computer,
and Systems Engineering
CII 9017, RPI, 110 8th Street
Troy, NY 12180, USA
e-mail: *shurm@rpi.edu*

Jai Singh
Charles Darwin University
School of Engineering and Logistics,
Faculty of Technology, B-41
Ellengowan Drive
Darwin, NT 0909, Australia
e-mail: *jai.singh@cdu.edu.au*

Tim Smeeton
Sharp Laboratories of Europe
Edmund Halley Road, Oxford Science Park
Oxford, OX4 4GB, UK
e-mail: *tim.smeeton@sharp.co.uk*

Boris Straumal
Russian Academy of Sciences
Institute of Sold State Physics
Institutskii prospect 15
Chernogolovka, 142432, Russia
e-mail: *straumal@issp.ac.ru*

Stephen Sweeney
University of Surrey
Advanced Technology Institute
Guildford, Surrey GU2 7XH, UK
e-mail: *s.sweeney@surrey.ac.uk*

David Sykes
Loughborough Surface Analysis Ltd.
PO Box 5016, Unit FC, Holywell Park, Ashby Road
Loughborough, LE11 3WS, UK
e-mail: *d.e.sykes@lsaltd.co.uk*

Keiji Tanaka
Hokkaido University
Department of Applied Physics,
Graduate School of Engineering
Kita-ku, N13 W8
Sapporo, 060-8628, Japan
e-mail: *keiji@eng.hokudai.ac.jp*

Charbel Tannous
Université de Bretagne Occidentale
LMB, CNRS FRE 2697
6 Avenue Le Gorgeu, BP: 809
29285 Brest Cedex, France
e-mail: *tannous@univ-brest.fr*

Ali Teke
Balikesir University
Department of Physics, Faculty of Art and Science
Balikesir, 10100, Turkey
e-mail: *ateke@balikesir.edu.tr*

Junji Tominaga
National Institute of Advanced Industrial
Science and Technology, AIST
Center for Applied Near-Field Optics Research,
CAN-FOR
Tsukuba Central 4 1-1-1 Higashi
Tsukuba, 3.5-8562, Japan
e-mail: *j-tomonaga@aist.go.jp*

Dan Tonchev
University of Saskatchewan
Department of Electrical Engineering
57 Campus Drive
Saskatoon, SK S7N 5A9, Canada
e-mail: *dan.tonchev@usask.ca*

Harry L. Tuller
Massachusetts Institute of Technology
Department of Materials Science and Engineering,
Crystal Physics and Electroceramics Laboratory
77 Massachusetts Avenue
Cambridge, MA 02139, USA
e-mail: *tuller@mit.edu*

Qamar-ul Wahab
Linköping University
Department of Physics,
Chemistry, and Biology (IFM)
SE-581 83 Linköping, Sweden
e-mail: *quw@ifm.liu.se*

Robert M. Wallace
University of Texas at Dallas
Department of Electrical Engineering
M.S. EC 33, P.O.Box 830688
Richardson, TX 75083, USA
e-mail: *rmwallace@utdallas.edu*

Jifeng Wang
Tohoku University
Institute of Multidisciplinary Research
for Advanced Materials
1-1, Katahira, 2 Chome, Aobaku
Sendai, 980-8577, Japan
e-mail: *wang@tagen.tohoku.ac.jp*

David S. Weiss
NexPress Solutions, Inc.
2600 Manitou Road
Rochester, NY 14653-4180, USA
e-mail: *David_Weiss@Nexpress.com*

Rainer Wesche
Swiss Federal Institute of Technology
Centre de Recherches en Physique des Plasmas
CRPP (c/o Paul Scherrer Institute), WMHA/C31,
Villigen PS
Lausanne, CH-5232, Switzerland
e-mail: *rainer.wesche@psi.ch*

Roger Whatmore
Tyndall National Institute
Lee Maltings, Cork , Ireland
e-mail: *roger.whatmore@tyndall.ie*

Neil White
University of Southampton
School of Electronics and Computer Science
Mountbatten Building
Highfield, Southampton SO17 1BJ, UK
e-mail: *nmw@ecs.soton.ac.uk*

Magnus Willander
University of Gothenburg
Department of Physics
SE-412 96 Göteborg, Sweden
e-mail: *mwi@fy.chalmers.se*

Jan Willekens
University of Leuven
Laboratorium voor Halfgeleiderfysica
Celestijnenlaan 200D
B-3001 Leuven, Belgium
e-mail: *jan.willekens@kc.kuleuven.ac.be*

Acknowledgements

D.33 Electron Transport Within the III–V Nitride Semiconductors, GaN, AlN, and InN: A Monte Carlo Analysis
by Brian E. Foutz, Stephen K. O'Leary, Michael Shur, Lester F. Eastman

Financial support from the Office of Naval Research and the Natural Sciences and Engineering Research Council of Canada is gratefully acknowledged. The use of equipment granted from the Canada Foundation for Innovation, and equipment loaned from the Canadian Microelectronics Corporation, is also acknowledged.

D.36 II–VI Narrow-Bandgap Semiconductors for Optoelectronics
by Ian M. Baker

The author wishes to express his gratitude to Mike Kinch of DRS Technologies, Kadri Vural and Jose Arias of Rockwell/Boeing and Marion Reine and coworkers at BAE SYSTEMS, Lexington for supplying material for this chapter and valuable advice. Also the advice and support from my technical colleagues, particularly: Peter Capper, Chris Maxey, Chris Jones and Les Hipwood, and my management here at SELEX Infrared, particularly Graham Hall. Thanks also to my wife, Lesley, for help with the English.

D.37 Optoelectronic Devices and Materials
by Stephen Sweeney, Alfred Adams

It is a pleasure to acknowledge the many people with whom the authors have worked with over the years. In particular, we would like to thank the staff and students, past and present, at the University of Surrey for their wide-ranging contributions to this work. We would also like to thank the editor, Safa Kasap, for his support and encouragement in preparing this chapter. On a personal note, SJS would like to thank his wife for her support whilst writing this chapter.

D.44 Optical Nonlinearity in Photonic Glasses
by Keiji Tanaka

The author would like to thank his students, K. Sugawara and N. Minamikawa, for preparing illustrations and giving comments.

目 录

缩略语

Part D 光电子材料与光子材料

31 Ⅲ-Ⅴ族三元与四元化合物 3
- 31.1 Ⅲ-Ⅴ族三元与四元化合物的介绍 3
- 31.2 插值模型 4
- 31.3 结构参数 5
- 31.4 机械、弹性和晶格振动特性 7
- 31.5 热能特性 9
- 31.6 能带参数 11
- 31.7 光学特性 16
- 31.8 载流子传输特性 18
- 参考文献 19

32 Ⅲ族氮化物 21
- 32.1 氮化物的晶体结构 23
- 32.2 氮化物的晶格常数 24
- 32.3 氮化物的机械性能 25
- 32.4 氮化物的热学性能 29
- 32.5 氮化物的电学特性 34
- 32.6 氮化物的光学特性 45
- 32.7 氮合金特性 59
- 32.8 总结与结论 62
- 参考文献 63

33 Ⅲ-Ⅴ族氮化物半导体GaN，AlN和InN中电子传输特性：蒙特卡罗分析 73
- 33.1 半导体中的电子传输与蒙特卡罗仿真近似 74

33.2 体纤维锌矿GaN，AlN和InN的稳态与瞬态电子传输特性 78

33.3 Ⅲ-V族氮化物半导体中电子传输特性：回顾 90

33.4 结论 94

参考文献 94

34 光电子的Ⅱ-Ⅵ族半导体：CdS,CdSe,CdTe 97

34.1 背景 97

34.2 太阳能电池 97

34.3 辐射探测器 102

34.4 结论 108

参考文献 108

35 锌基宽禁带半导体掺杂 111

35.1 ZnSe 111

35.2 ZnBeSe材料 116

35.3 ZnO 117

参考文献 119

36 光电子的窄带隙Ⅱ-Ⅵ族半导体 123

36.1 应用与传感器设计 126

36.2 HgCdTe与相关合金的光导检测 128

36.3 SPRITE检测 132

36.4 相关合金的光导检测 134

36.5 HgCdTe光导检测结论 135

36.6 HgCdTe的光伏器件 135

36.7 Ⅱ-Ⅵ族半导体的发射器件 150

36.8 HgTe-CdTe简化维度势能 151

参考文献 151

37 光电子器件与材料 155

37.1 光电子器件的介绍 156

37.2 发光二极管与半导体激光器 158

37.3 单模激光器……172

37.4 光学放大器……174

37.5 调制器……175

37.6 光探测器……179

37.7 结论……182

参考文献……183

38 液态晶体……185

38.1 液态晶体的介绍……185

38.2 液态晶体的基本物理学……192

38.3 液态晶体器件……199

38.4 显示材料……208

参考文献……217

39 有机光导体……221

39.1 卡尔逊与静电复印术……222

39.2 操作方法与临界材料特性……224

39.3 OPC特性描述……233

39.4 OPC体系结构和组成……235

39.5 光接收器构成……244

39.6 总结……245

参考文献……246

40 荧光材料……251

40.1 发光中心……253

40.2 晶格交替……255

40.3 热刺激发光……257

40.4 光（光学的-）刺激发光……258

40.5 实验技术——光致发光……259

40.6 应用……260

40.7 典型荧光粉……263

参考文献……263

41 近红外纳米可调光子晶体与可见电磁光谱ㅤ265
41.1 PC综述ㅤ266
41.2 静态PCs传统结构方法论ㅤ269
41.3 可调PCsㅤ279
41.4 总结与结论ㅤ282
参考文献ㅤ283

42 量子阱、超晶格和带隙工程ㅤ289
42.1 带隙工程与量子限制原理ㅤ290
42.2 量子局限结构光电子特性ㅤ292
42.3 发射器ㅤ300
42.4 检测器ㅤ302
42.5 调制器ㅤ304
42.6 未来发展方向ㅤ305
42.7 结论ㅤ306
参考文献ㅤ306

43 光子集成玻璃ㅤ309
43.1 光子材料玻璃的主要贡献ㅤ310
43.2 光学集成玻璃ㅤ318
43.3 集成光源激光玻璃ㅤ321
43.4 总结ㅤ325
参考文献ㅤ327

44 光子玻璃的光学非线性ㅤ331
44.1 均质玻璃的三级非线性ㅤ332
44.2 极化玻璃的二级非线性ㅤ337
44.3 粒子嵌入式系统ㅤ338
44.4 光诱导现象ㅤ339
44.5 总结ㅤ340
参考文献ㅤ340

45 非线性光电子材料 343
45.1 背景 343
45.2 照明因变量折射率与非线性品质因数（FOM）... 345
45.3 体与多量子阱（MQW）无机晶体半导体 348
45.4 有机材料 352
45.5 纳米晶体 355
45.6 其他非线性材料 356
45.7 结论 357
参考文献 357

Contents

List of Abbreviations

Part D Materials for Optoelectronics and Photonics

31 III–V Ternary and Quaternary Compounds ... 735
- 31.1 Introduction to III–V Ternary and Quaternary Compounds ... 735
- 31.2 Interpolation Scheme ... 736
- 31.3 Structural Parameters ... 737
- 31.4 Mechanical, Elastic and Lattice Vibronic Properties ... 739
- 31.5 Thermal Properties ... 741
- 31.6 Energy Band Parameters ... 743
- 31.7 Optical Properties ... 748
- 31.8 Carrier Transport Properties ... 750
- References ... 751

32 Group III Nitrides ... 753
- 32.1 Crystal Structures of Nitrides ... 755
- 32.2 Lattice Parameters of Nitrides ... 756
- 32.3 Mechanical Properties of Nitrides ... 757
- 32.4 Thermal Properties of Nitrides ... 761
- 32.5 Electrical Properties of Nitrides ... 766
- 32.6 Optical Properties of Nitrides ... 777
- 32.7 Properties of Nitride Alloys ... 791
- 32.8 Summary and Conclusions ... 794
- References ... 795

33 Electron Transport Within the III–V Nitride Semiconductors, GaN, AlN, and InN: A Monte Carlo Analysis ... 805
- 33.1 Electron Transport Within Semiconductors and the Monte Carlo Simulation Approach ... 806
- 33.2 Steady-State and Transient Electron Transport Within Bulk Wurtzite GaN, AlN, and InN ... 810
- 33.3 Electron Transport Within III–V Nitride Semiconductors: A Review ... 822
- 33.4 Conclusions ... 826
- References ... 826

34 II–IV Semiconductors for Optoelectronics: CdS, CdSe, CdTe ... 829
- 34.1 Background ... 829
- 34.2 Solar Cells ... 829
- 34.3 Radiation Detectors ... 834
- 34.4 Conclusions ... 840
- References ... 840

35	**Doping Aspects of Zn-Based Wide-Band-Gap Semiconductors**	843
35.1	ZnSe	843
35.2	ZnBeSe	848
35.3	ZnO	849
	References	851

36	**II–VI Narrow-Bandgap Semiconductors for Optoelectronics**	855
36.1	Applications and Sensor Design	858
36.2	Photoconductive Detectors in HgCdTe and Related Alloys	860
36.3	SPRITE Detectors	864
36.4	Photoconductive Detectors in Closely Related Alloys	866
36.5	Conclusions on Photoconductive HgCdTe Detectors	867
36.6	Photovoltaic Devices in HgCdTe	867
36.7	Emission Devices in II–VI Semiconductors	882
36.8	Potential for Reduced-Dimensionality HgTe–CdTe	883
	References	883

37	**Optoelectronic Devices and Materials**	887
37.1	Introduction to Optoelectronic Devices	888
37.2	Light-Emitting Diodes and Semiconductor Lasers	890
37.3	Single-Mode Lasers	904
37.4	Optical Amplifiers	906
37.5	Modulators	907
37.6	Photodetectors	911
37.7	Conclusions	914
	References	915

38	**Liquid Crystals**	917
38.1	Introduction to Liquid Crystals	917
38.2	The Basic Physics of Liquid Crystals	924
38.3	Liquid-Crystal Devices	931
38.4	Materials for Displays	940
	References	949

39	**Organic Photoconductors**	953
39.1	Chester Carlson and Xerography	954
39.2	Operational Considerations and Critical Materials Properties	956
39.3	OPC Characterization	965
39.4	OPC Architecture and Composition	967
39.5	Photoreceptor Fabrication	976
39.6	Summary	977
	References	978

40	**Luminescent Materials**	983
40.1	Luminescent Centres	985
40.2	Interaction with the Lattice	987
40.3	Thermally Stimulated Luminescence	989
40.4	Optically (Photo-)Stimulated Luminescence	990

40.5	Experimental Techniques – Photoluminescence	991
40.6	Applications	992
40.7	Representative Phosphors	995
	References	995

41 Nano-Engineered Tunable Photonic Crystals in the Near-IR and Visible Electromagnetic Spectrum — 997

41.1	PC Overview	998
41.2	Traditional Fabrication Methodologies for Static PCs	1001
41.3	Tunable PCs	1011
41.4	Summary and Conclusions	1014
	References	1015

42 Quantum Wells, Superlattices, and Band-Gap Engineering — 1021

42.1	Principles of Band-Gap Engineering and Quantum Confinement	1022
42.2	Optoelectronic Properties of Quantum-Confined Structures	1024
42.3	Emitters	1032
42.4	Detectors	1034
42.5	Modulators	1036
42.6	Future Directions	1037
42.7	Conclusions	1038
	References	1038

43 Glasses for Photonic Integration — 1041

43.1	Main Attributes of Glasses as Photonic Materials	1042
43.2	Glasses for Integrated Optics	1050
43.3	Laser Glasses for Integrated Light Sources	1053
43.4	Summary	1057
	References	1059

44 Optical Nonlinearity in Photonic Glasses — 1063

44.1	Third-Order Nonlinearity in Homogeneous Glass	1064
44.2	Second-Order Nonlinearity in Poled Glass	1069
44.3	Particle-Embedded Systems	1070
44.4	Photoinduced Phenomena	1071
44.5	Summary	1072
	References	1072

45 Nonlinear Optoelectronic Materials — 1075

45.1	Background	1075
45.2	Illumination-Dependent Refractive Index and Nonlinear Figures of Merit (FOM)	1077
45.3	Bulk and Multi-Quantum-Well (MQW) Inorganic Crystalline Semiconductors	1080
45.4	Organic Materials	1084
45.5	Nanocrystals	1087
45.6	Other Nonlinear Materials	1088
45.7	Conclusions	1089
	References	1089

List of Abbreviations

2DEG	two-dimensional electron gas	CuPc	copper phthalocyanine
		CuTTBPc	tetra-tert-butyl phthalocyanine
		CV	chemical vapor

A

		CVD	chemical vapor deposition
AC	alternating current	CVT	chemical vapor transport
ACCUFET	accumulation-mode MOSFET	CZ	Czochralski
ACRT	accelerated crucible rotation technique	CZT	cadmium zinc telluride
AEM	analytical electron microscopes		
AES	Auger electron spectroscopy		

D

AFM	atomic force microscopy	
ALD	atomic-layer deposition	
ALE	atomic-layer epitaxy	DA	Drude approximation
AMA	active matrix array	DAG	direct alloy growth
AMFPI	active matrix flat-panel imaging	DBP	dual-beam photoconductivity
AMOLED	amorphous organic light-emitting diode	DC	direct current
APD	avalanche photodiode	DCPBH	double-channel planar buried heterostructure

B

		DET	diethyl telluride
		DFB	distributed feedback
b.c.c.	body-centered cubic	DH	double heterostructure
BEEM	ballistic-electron-emission microscopy	DIL	dual-in-line
BEP	beam effective pressure	DIPTe	diisopropyltellurium
BH	buried-heterostructure	DLC	diamond-like carbon
BH	Brooks–Herring	DLHJ	double-layer heterojunction
BJT	bipolar junction transistor	DLTS	deep level transient spectroscopy
BTEX	m-xylene	DMCd	dimethyl cadmium
BZ	Brillouin zone	DMF	dimethylformamide
		DMOSFET	double-diffused MOSFET
		DMS	dilute magnetic semiconductors

C

		DMSO	dimethylsulfoxide
		DMZn	dimethylzinc
CAIBE	chemically assisted ion beam etching	DOS	density of states
CB	conduction band	DQE	detective quantum efficiency
CBE	chemical beam epitaxy	DSIMS	dynamic secondary ion mass spectrometry
CBED	convergent beam electron diffraction	DTBSe	ditertiarybutylselenide
CC	constant current	DUT	device under test
CCD	charge-coupled device	DVD	digital versatile disk
CCZ	continuous-charging Czochralski	DWDM	dense wavelength-division multiplexing
CFLPE	container-free liquid phase epitaxy	DXD	double-crystal X-ray diffraction
CKR	cross Kelvin resistor		
CL	cathodoluminescence		

E

CMOS	complementary metal-oxide-semiconductor	
CNR	carrier-to-noise ratio	
COP	crystal-originated particle	EBIC	electron beam induced conductivity
CP	charge pumping	ED	electrodeposition
CPM	constant-photocurrent method	EDFA	erbium-doped fiber amplifier
CR	computed radiography	EELS	electron energy loss spectroscopy
CR-DLTS	computed radiography deep level transient spectroscopy	EFG	film-fed growth
		EHP	electron–hole pairs
CRA	cast recrystallize anneal	ELO	epitaxial lateral overgrowth
CTE	coefficient of thermal expansion	ELOG	epitaxial layer overgrowth
CTO	chromium(III) trioxalate	EM	electromagnetic
		EMA	effective media approximation

ENDOR	electron–nuclear double resonance		IFIGS	interface-induced gap states
EPD	etch pit density		IFTOF	interrupted field time-of-flight
EPR	electron paramagnetic resonance		IGBT	insulated gate bipolar transistor
ESR	electron spin resonance spectroscopy		IMP	interdiffused multilayer process
EXAFS	extended X-ray absorption fine structure		IPEYS	internal photoemission yield spectroscopy
			IR	infrared
			ITO	indium-tin-oxide

F

J

FCA	free-carrier absorption		JBS	junction barrier Schottky
f.c.c.	face-centered cubic		JFET	junction field-effect transistors
FET	field effect transistor		JO	Judd–Ofelt
FIB	focused ion beam			
FM	Frank–van der Merwe			

K

FPA	focal plane arrays		KCR	Kelvin contact resistance
FPD	flow pattern defect		KKR	Kramers–Kronig relation
FTIR	Fourier transform infrared		KLN	$K_3Li_2Nb_5O_{12}$
FWHM	full-width at half-maximum		KTPO	$KTiOPO_4$
FZ	floating zone			

G

L

GDA	generalized Drude approximation		LB	Langmuir–Blodgett
GDMS	glow discharge mass spectrometry		LD	laser diodes
GDOES	glow discharge optical emission spectroscopy		LD	lucky drift
			LDD	lightly doped drain
GF	gradient freeze		LEC	liquid-encapsulated Czochralski
GMR	giant magnetoresistance		LED	light-emitting diodes
GOI	gate oxide integrity		LEIS	low-energy ion scattering
GRIN	graded refractive index		LEL	lower explosive limit
GSMBE	gas-source molecular beam epitaxy		LF	low-frequency
GTO	gate turn-off		LLS	laser light scattering
			LMA	law of mass action

H

HAADF	high-angle annular dark field		LO	longitudinal optical
HB	horizontal Bridgman		LPE	liquid phase epitaxy
HBT	hetero-junction bipolar transistor		LSTD	laser light scattering tomography defect
HDC	horizontal directional solidification crystallization		LVM	localized vibrational mode

M

HEMT	high electron mobility transistor		MBE	molecular beam epitaxy
HF	high-frequency		MCCZ	magnetic field applied continuous Czochralski
HOD	highly oriented diamond			
HOLZ	high-order Laue zone		MCT	mercury cadmium telluride
HPc	phthalocyanine		MCZ	magnetic field applied Czochralski
HPHT	high-pressure high-temperature		MD	molecular dynamics
HRXRD	high-resolution X-ray diffraction		MEED	medium-energy electron diffraction
HTCVD	high-temperature CVD		MEM	micro-electromechanical systems
HVDC	high-voltage DC		MESFET	metal-semiconductor field-effect transistor
HWE	hot-wall epitaxy		MFC	mass flow controllers
			MIGS	metal-induced gap states

I

IC	integrated circuit		ML	monolayer
ICTS	isothermal capacitance transient spectroscopy		MLHJ	multilayer heterojunction
			MOCVD	metal-organic chemical vapor deposition
IDE	interdigitated electrodes		MODFET	modulation-doped field effect transistor

MOMBE	metalorganic molecular beam epitaxy	PL	photoluminescence
MOS	metal/oxide/semiconductor	PM	particulate matter
MOSFET	metal/oxide/semiconductor field effect transistor	PMMA	poly(methyl-methacrylate)
		POT	poly(n-octyl)thiophene
MOVPE	metalorganic vapor phase epitaxy	ppb	parts per billion
MPc	metallophthalocyanine	ppm	parts per million
MPC	modulated photoconductivity	PPS	polyphenylsulfide
MPCVD	microwave plasma chemical deposition	PPY	polypyrrole
MQW	multiple quantum well	PQT-12	poly[5,5'-bis(3-alkyl-2-thienyl)-2,2'-bithiophene]
MR	magnetoresistivity		
MS	metal–semiconductor	PRT	platinum resistance thermometers
MSRD	mean-square relative displacement	PSt	polystyrene
MTF	modulation transfer function	PTC	positive temperature coefficient
MWIR	medium-wavelength infrared	PTIS	photothermal ionisation spectroscopy
		PTS	1,1-dioxo-2-(4-methylphenyl)-6-phenyl-4-(dicyanomethylidene)thiopyran

N

		PTV	polythienylene vinylene
NDR	negative differential resistance	PV	photovoltaic
NEA	negative electron affinity	PVD	physical vapor transport
NeXT	nonthermal energy exploration telescope	PVDF	polyvinylidene fluoride
NMOS	n-type-channel metal–oxide–semiconductor	PVK	polyvinylcarbazole
NMP	N-methylpyrrolidone	PVT	physical vapor transport
NMR	nuclear magnetic resonance	PZT	lead zirconate titanate
NNH	nearest-neighbor hopping		

Q

NSA	naphthalene-1,5-disulfonic acid		
NTC	negative temperature coefficient	QA	quench anneal
NTD	neutron transmutation doping	QCL	quantum cascade laser
		QCSE	quantum-confined Stark effect

O

		QD	quantum dot
OLED	organic light-emitting diode	QHE	quantum Hall effect
OSF	oxidation-induced stacking fault	QW	quantum well
OSL	optically stimulated luminescence		
OZM	overlap zone melting		

R

		RAIRS	reflection adsorption infrared spectroscopy

P

		RBS	Rutherford backscattering
PAE	power added efficiency	RCLED	resonant-cavity light-emitting diode
PAni	polyaniline	RDF	radial distribution function
pBN	pyrolytic boron nitride	RDS	reflection difference spectroscopy
Pc	phthalocyanine	RE	rare earth
PC	photoconductive	RENS	resolution near-field structure
PCA	principal component analysis	RF	radio frequency
PCB	printed circuit board	RG	recombination–generation
PDMA	poly(methylmethacrylate)/poly(decyl methacrylate)	RH	relative humidity
		RHEED	reflection high-energy electron diffraction
PDP	plasma display panels	RIE	reactive-ion etching
PDS	photothermal deflection spectroscopy	RIU	refractive index units
PE	polysilicon emitter	RTA	rapid thermal annealing
PE BJT	polysilicon emitter bipolar junction transistor	RTD	resistance temperature devices
		RTS	random telegraph signal
PECVD	plasma-enhanced chemical vapor deposition		

S

PEN	polyethylene naphthalate		
PES	photoemission spectroscopy		
PET	positron emission tomography	SA	self-assembly
pHEMT	pseudomorphic HEMT	SAM	self-assembled monolayers

SAW	surface acoustic wave	TMA	trimethyl-aluminum
SAXS	small-angle X-ray scattering	TMG	trimethyl-gallium
SCH	separate confinement heterojunction	TMI	trimethyl-indium
SCVT	seeded chemical vapor transport	TMSb	trimethylantimony
SE	spontaneous emission	TO	transverse optical
SEM	scanning electron microscope	TOF	time of flight
SIMS	secondary ion mass spectrometry	ToFSIMS	time of flight SIMS
SIPBH	semi-insulating planar buried heterostructure	TPC	transient photoconductivity
		TPV	thermophotovoltaic
SIT	static induction transistors	TSC	thermally stimulated current
SK	Stranski–Krastanov	TSL	thermally stimulated luminescence
SNR	signal-to-noise ratio		

U

SO	small outline		
SOA	semiconductor optical amplifier		
SOC	system-on-a-chip	ULSI	ultra-large-scale integration
SOFC	solid oxide fuel cells	UMOSFET	U-shaped-trench MOSFET
SOI	silicon-on-insulator	UPS	uninterrupted power systems
SP	screen printing	UV	ultraviolet
SPECT	single-photon emission computed tomography		

V

SPR	surface plasmon resonance		
SPVT	seeded physical vapor transport	VAP	valence-alternation pairs
SQW	single quantum wells	VB	valence band
SSIMS	static secondary ion mass spectrometry	VCSEL	vertical-cavity surface-emitting laser
SSPC	steady-state photoconductivity	VCZ	vapor-pressure-controlled Czochralski
SSR	solid-state recrystallisation	VD	vapor deposition
SSRM	scanning spreading resistance microscopy	VFE	vector flow epitaxy
STHM	sublimation traveling heater method	VFET	vacuum field-effect transistor
SVP	saturated vapor pressure	VGF	vertical gradient freeze
SWIR	short-wavelength infrared	VIS	visible
		VOC	volatile organic compounds

T

		VPE	vapor phase epitaxy
		VRH	variable-range hopping
TAB	tab automated bonding	VUVG	vertical unseeded vapor growth
TBA	tertiarybutylarsine	VW	Volmer–Weber
TBP	tertiarybutylphosphine		

W

TCE	thermal coefficient of expansion		
TCNQ	tetracyanoquinodimethane		
TCR	temperature coefficient of resistance	WDX	wavelength dispersive X-ray
TCRI	temperature coefficient of refractive index	WXI	wide-band X-ray imager

X

TDCM	time-domain charge measurement		
TE	transverse electric		
TED	transient enhanced diffusion	XAFS	X-ray absorption fine-structure
TED	transmission electron diffraction	XANES	X-ray absorption near-edge structure
TEGa	triethylgallium	XEBIT	X-ray-sensitive electron-beam image tube
TEM	transmission electron microscope	XPS	X-ray photon spectroscopy
TEN	triethylamine	XRD	X-ray diffraction
TFT	thin-film transistors	XRSP	X-ray storage phosphor
THM	traveling heater method		

Y

TL	thermoluminescence		
TLHJ	triple-layer graded heterojunction	YSZ	yttrium-stabilized zirconia
TLM	transmission line measurement		
TM	transverse magnetic		

Part D Materials for Optoelectronics and Photonics

31 III–V Ternary and Quaternary Compounds
Sadao Adachi, Gunma, Japan

32 Group III Nitrides
Ali Teke, Balikesir, Turkey
Hadis Morkoç, Richmond, USA

33 Electron Transport Within the III–V Nitride Semiconductors, GaN, AlN, and InN: A Monte Carlo Analysis
Brian E. Foutz, Endicott, USA
Stephen K. O'Leary, Regina, Canada
Michael Shur, Troy, USA
Lester F. Eastman, Ithaca, USA

34 II–IV Semiconductors for Optoelectronics: CdS, CdSe, CdTe
Jifeng Wang, Sendai, Japan
Minoru Isshiki, Sendai, Japan

35 Doping Aspects of Zn-Based Wide-Band-Gap Semiconductors
Gertrude F. Neumark, New York, USA
Yinyan Gong, New York, USA
Igor L. Kuskovsky, Flushing, USA

36 II–VI Narrow-Bandgap Semiconductors for Optoelectronics
Ian M. Baker, Southampton, UK

37 Optoelectronic Devices and Materials
Stephen Sweeney, Guildford, UK
Alfred Adams, Surrey, UK

38 Liquid Crystals
David Dunmur, Southampton, UK
Geoffrey Luckhurst, Southampton, UK

39 Organic Photoconductors
David S. Weiss, Rochester, USA
Martin Abkowitz, Webster, USA

40 Luminescent Materials
Andy Edgar, Wellington, New Zealand

41 Nano-Engineered Tunable Photonic Crystals in the Near-IR and Visible Electromagnetic Spectrum
Harry Ruda, Toronto, Canada
Naomi Matsuura, Toronto, Canada

42 Quantum Wells, Superlattices, and Band-Gap Engineering
Mark Fox, Sheffield, UK

43 Glasses for Photonic Integration
Ray DeCorby, Edmonton, Canada

44 Optical Nonlinearity in Photonic Glasses
Keiji Tanaka, Sapporo, Japan

45 Nonlinear Optoelectronic Materials
Lukasz Brzozowski, Toronto, ON, Canada
Edward Sargent, Toronto, Canada

31. III–V Ternary and Quaternary Compounds

III–V ternary and quaternary alloy systems are potentially of great importance for many high-speed electronic and optoelectronic devices, because they provide a natural means of tuning the magnitude of forbidden gaps so as to optimize and widen the applications of such semiconductor devices. Literature on the fundamental properties of these material systems is growing rapidly. Even though the basic semiconductor alloy concepts are understood at this time, some practical and device parameters in these material systems have been hampered by a lack of definite knowledge of many material parameters and properties.

This chapter attempts to summarize, in graphical and tabular forms, most of the important theoretical and experimental data on the III–V ternary and quaternary alloy parameters and properties. They can be classified into six groups: (1) Structural parameters; (2) Mechanical, elastic, and lattice vibronic properties; (3) Thermal properties; (4) Energy band parameters; (5) Optical properties, and; (6) Carrier transport properties. The III–V ternary and quaternary alloys considered here are those of Group III (Al, Ga, In) and V (N, P, As, Sb) atoms. The model used in some cases is based on an interpolation scheme and, therefore, requires that data on the material parameters for the related binaries (AlN, AlP, GaN, GaP, etc.) are known. These data have been taken mainly from the Landolt-Börnstein collection, Vol. III/41, and from the *Handbook on Physical Properties of Semiconductors Volume 2: III–V Compound Semiconductors*, published by Springer in 2004. The material parameters and properties derived here are used with wide success to obtain the general properties of these alloy semiconductors.

31.1	Introduction to III–V Ternary and Quaternary Compounds	735
31.2	Interpolation Scheme	736
31.3	Structural Parameters	737
	31.3.1 Lattice Parameters and Lattice-Matching Conditions Between III–V Quaternaries and Binary Substrates	737
	31.3.2 Molecular and Crystal Densities	737
31.4	Mechanical, Elastic and Lattice Vibronic Properties	739
	31.4.1 Microhardness	739
	31.4.2 Elastic Constants and Related Moduli	739
	31.4.3 Long-Wavelength Phonons	739
31.5	Thermal Properties	741
	31.5.1 Specific Heat and Debye Temperature	741
	31.5.2 Thermal Expansion Coefficient	741
	31.5.3 Thermal Conductivity	741
31.6	Energy Band Parameters	743
	31.6.1 Bandgap Energy	743
	31.6.2 Carrier Effective Mass	744
	31.6.3 Deformation Potential	746
31.7	Optical Properties	748
	31.7.1 The Reststrahlen Region	748
	31.7.2 The Interband Transition Region	749
31.8	Carrier Transport Properties	750
References		751

31.1 Introduction to III–V Ternary and Quaternary Compounds

III–V semiconducting compound alloys are widely used as materials for optoelectronic devices such as light-emitting diodes, laser diodes and photodetectors, as well as for electronic transport devices such as field effect transistors, high electron mobility transistors and heterojunction bipolar transistors. In a ternary alloy, the bandgap energy E_g and the lattice parameter a are generally both functions of a single composition parameter, so they cannot be selected independently. In quaternary alloys, on the other hand, the two composition parameters allow E_g and a to be selected independently, within the constraints of a given alloy–substrate system. Even though the basic semiconductor alloy concepts are understood at this time, the determination of some practical device parameters has been hampered by a lack of definite knowledge of many material parameters. This chapter provides data on the fundamental material properties of III–V ternary and quaternary alloys. The model used here is based on an interpolation scheme and thus requires that values of the material parameters for the related endpoint binaries are known. We therefore begin with the constituent binaries and gradually move on to alloys. The phenomenon of spontaneous ordering in semiconductor alloys, which can be categorized as a self-organized process, is observed to occur spontaneously during the epitaxial growth of certain alloys, and results in modifications to their structural, electronic and optical properties. This topic is omitted from the coverage [31.1].

31.2 Interpolation Scheme

The electronic energy band parameters of III–V compound alloys and their dependence on alloy composition are very important device parameters, and so they have received considerable attention in the past. Investigations of many device parameters have, however, been hampered by a lack of definite knowledge of various material parameters. This necessitates the use of some kind of interpolation scheme. Although the interpolation scheme is still open to experimental verification, it can provide more useful and reliable material parameters over the entire range of alloy composition [31.2].

If one uses the linear interpolation scheme, the ternary parameter T can be derived from the binary parameters (B) by

$$T_{A_xB_{1-x}C} = xB_{AC} + (1-x)B_{BC} \equiv a + bx \quad (31.1)$$

for an alloy of the form $A_xB_{1-x}C$, where $a \equiv B_{BC}$ and $b \equiv B_{AC} - B_{BC}$. Some material parameters, however, deviate significantly from the linear relation (31.1), and exhibit an approximately quadratic dependence on the mole fraction x. The ternary material parameter in such a case can be very efficiently approximated by the relationship

$$T_{A_xB_{1-x}C} = xB_{AC} + (1-x)B_{BC} + C_{A-B}x(1-x)$$
$$\equiv a + bx + cx^2, \quad (31.2)$$

where $a \equiv B_{BC}$ and $b \equiv B_{AC} - B_{BC} + C_{A-B}$, and $c \equiv -C_{A-B}$. The parameter c is called the bowing or nonlinear parameter.

The quaternary material $A_xB_{1-x}C_yD_{1-y}$ is thought to be constructed from four binaries: AC, AD, BC, and BD. If one uses the linear interpolation scheme, the quaternary parameter Q can be derived from the Bs by

$$Q(x,y) = xyB_{AC} + x(1-y)B_{AD} + (1-x)yB_{BC}$$
$$+ (1-x)(1-y)B_{BD}. \quad (31.3)$$

If one of the four binary parameters (e.g., B_{AD}) is lacking, Q can be estimated from

$$Q(x,y) = xB_{AC} + (y-x)B_{BC} + (1-y)B_{BD}. \quad (31.4)$$

The quaternary material $A_xB_yC_{1-x-y}D$ is thought to be constructed from three binaries: AD, BD, and CD. The corresponding linear interpolation is given by

$$Q(x,y) = xB_{AD} + yB_{BD} + (1-x-y)B_{CD}. \quad (31.5)$$

If the material parameter can be given by a specific expression owing to some physical basis, it is natural to consider that the interpolation scheme may also obey this expression. The static dielectric constant ε_s is just the case that follows the Clausius–Mosotti relation. Then, the interpolation expression for the $A_xB_{1-x}C_yD_{1-y}$ quaternary, for example, has the form

$$\frac{\varepsilon_s(x,y) - 1}{\varepsilon_s(x,y) - 2} = xy\frac{\varepsilon_s(AC) - 1}{\varepsilon_s(AC) - 2} + x(1-y)\frac{\varepsilon_s(AD) - 1}{\varepsilon_s(AD) - 2}$$
$$+ (1-x)y\frac{\varepsilon_s(BC) - 1}{\varepsilon_s(BC) - 2}$$
$$+ (1-x)(1-y)\frac{\varepsilon_s(BD) - 1}{\varepsilon_s(BD) - 2}. \quad (31.6)$$

When bowing from the anion sublattice disorder is independent of the disorder in the cation sublattice, the interpolation scheme is written by incorporating these cation and anion bowing parameters into the linear interpolation scheme as

$$Q(x,y) = xyB_{AC} + x(1-y)B_{AD} + (1-x)yB_{BC}$$
$$+ (1-x)(1-y)B_{BD} + C_{A-B}x(1-x)$$
$$+ C_{C-D}y(1-y) \quad (31.7)$$

for the $A_xB_{1-x}C_yD_{1-y}$ quaternary, or

$$Q(x,y) = xB_{AD} + yB_{BD} + (1-x-y)B_{CD}$$
$$+ C_{A-B-C}xy(1-x-y) \quad (31.8)$$

for the $A_xB_yC_{1-x-y}D$ quaternary.

If relationships for the ternary parameters Ts are available, the quaternary parameter Q can be expressed either as ($A_xB_{1-x}C_yD_{1-y}$)

$$Q(x, y) = \frac{x(1-x)[yT_{ABC}(x) + (1-y)T_{ABD}(x)]}{x(1-x) + y(1-y)}$$
$$+ \frac{y(1-y)[xT_{ACD}(y) + (1-x)T_{BCD}(y)]}{x(1-x) + y(1-y)}, \quad (31.9)$$

or ($A_xB_yC_{1-x-y}D$)

$$Q(x, y) = \frac{xyT_{ABD}(u) + y(1-x-y)T_{BCD}(v)}{xy + y(1-x-y) + x(1-x-y)}$$
$$+ \frac{x(1-x-y)T_{ACD}(w)}{xy + y(1-x-y) + x(1-x-y)} \quad (31.10)$$

with

$$u = (1-x-y)/2, \quad v = (2-x-2y)/2,$$
$$w = (2-2x-y)/2. \quad (31.11)$$

31.3 Structural Parameters

31.3.1 Lattice Parameters and Lattice-Matching Conditions Between III–V Quaternaries and Binary Substrates

The lattice parameter a (c) is known to obey Vegard's law well, i.e., to vary linearly with composition. Thus, the lattice parameter for a III–V ternary can be simply obtained from (31.1) using the binary data listed in Table 31.1 [31.3, 4]. Introducing the lattice parameters in Table 31.1 into (31.3) [(31.5)], one can also obtain the lattice-matching conditions for $A_{1-x}B_xC_yD_{1-y}$ ($A_xB_yC_{1-x-y}D$) quaternaries on various III–V binary substrates (GaAs, GaSb, InP and InAs). These results are summarized in Tables 31.2, 31.3, 31.4 and 31.5.

31.3.2 Molecular and Crystal Densities

The molecular density d_M can be obtained via

$$d_M = \frac{4}{a^3} \quad (31.12)$$

for zinc blende-type materials, and

$$d_M = \frac{4}{a_{\text{eff}}^3} \quad (31.13)$$

for wurtzite-type materials, where a_{eff} is an effective cubic lattice parameter defined by

$$a_{\text{eff}} = \left(\sqrt{3}a^2c\right)^{1/3}. \quad (31.14)$$

The X-ray crystal density g can be simply written, using d_M, as

$$g = \frac{Md_M}{N_A}, \quad (31.15)$$

Table 31.1 Lattice parameters a and c and crystal density g for some III–V binaries at 300 K

Binary	Zinc blende	Wurtzite		g (g/cm^{-3})
	a (Å)	a (Å)	c (Å)	
AlN	–	3.112	4.982	3.258
AlP	5.4635	–	–	2.3604
AlAs	5.661 39	–	–	3.7302
AlSb	6.1355	–	–	4.2775
α-GaN	–	3.1896	5.1855	6.0865
β-GaN	4.52	–	–	6.02
GaP	5.4508	–	–	4.1299
GaAs	5.653 30	–	–	5.3175
GaSb	6.095 93	–	–	5.6146
InN	–	3.548	5.760	6.813
InP	5.8690	–	–	4.7902
InAs	6.0583	–	–	5.6678
InSb	6.479 37	–	–	5.7768

Table 31.2 Lattice-matching conditions for some III–V quaternaries of type $A_xB_{1-x}C_yD_{1-y}$ at 300 K. $x = \frac{A_0 + B_0 y}{C_0 + D_0 y}$

Quaternary	Substrate	A_0	B_0	C_0	D_0	Remark
$Ga_xIn_{1-x}P_yAs_{1-y}$	GaAs	0.4050	−0.1893	0.4050	0.0132	$0 \leq y \leq 1.0$
	InP	0.1893	−0.1893	0.4050	0.0132	$0 \leq y \leq 1.0$
$Al_xIn_{1-x}P_yAs_{1-y}$	GaAs	0.4050	−0.1893	0.3969	0.0086	$0.04 \leq y \leq 1.0$
	InP	0.1893	−0.1893	0.3969	0.0086	$0 \leq y \leq 1.0$

where M is the molecular weight and $N_A = 6.022 \times 10^{23}$ mole^{-1} is the Avogadro constant. We list g for some III–V binaries in Table 31.1. Alloy values of d_M and g can be accurately obtained using Vegard's law, i.e., (31.1), (31.3), and (31.5).

Table 31.3 Lattice-matching conditions for some III–V quaternaries of type $A_xB_{1-x}C_yD_{1-y}$ at 300 K. $y = \frac{A_0 + B_0 x}{C_0 + D_0 x}$

Quaternary	Substrate	A_0	B_0	C_0	D_0	Remark
$Al_xGa_{1-x}P_yAs_{1-y}$	GaAs	0	0.0081	0.2025	−0.0046	$0 \leq x \leq 1.0$
$Al_xGa_{1-x}As_ySb_{1-y}$	GaSb	0	0.0396	0.4426	0.0315	$0 \leq x \leq 1.0$
	InP	0.2269	0.0396	0.4426	0.0315	$0 \leq x \leq 1.0$
	InAs	0.0376	0.0396	0.4426	0.0315	$0 \leq x \leq 1.0$
$Al_xGa_{1-x}P_ySb_{1-y}$	GaAs	0.4426	0.0396	0.6451	0.0269	$0 \leq x \leq 1.0$
	GaSb	0	0.0396	0.6451	0.0269	$0 \leq x \leq 1.0$
	InP	0.2269	0.0396	0.6451	0.0269	$0 \leq x \leq 1.0$
	InAs	0.0376	0.0396	0.6451	0.0269	$0 \leq x \leq 1.0$
$Ga_xIn_{1-x}As_ySb_{1-y}$	GaSb	0.3834	−0.3834	0.4211	0.0216	$0 \leq x \leq 1.0$
	InP	0.6104	−0.3834	0.4211	0.0216	$0.47 \leq x \leq 1.0$
	InAs	0.4211	−0.3834	0.4211	0.0216	$0 \leq x \leq 1.0$
$Ga_xIn_{1-x}P_ySb_{1-y}$	GaAs	0.8261	−0.3834	0.6104	0.0348	$0.52 \leq x \leq 1.0$
	GaSb	0.3834	−0.3834	0.6104	0.0348	$0 \leq x \leq 1.0$
	InP	0.6104	−0.3834	0.6104	0.0348	$0 \leq x \leq 1.0$
	InAs	0.4211	−0.3834	0.6104	0.0348	$0 \leq x \leq 1.0$
$Al_xIn_{1-x}As_ySb_{1-y}$	GaSb	0.3834	−0.3439	0.4211	0.0530	$0 \leq x \leq 1.0$
	InP	0.6104	−0.3439	0.4211	0.0530	$0.48 \leq x \leq 1.0$
	InAs	0.4211	−0.3439	0.4211	0.0530	$0 \leq x \leq 1.0$
$Al_xIn_{1-x}P_ySb_{1-y}$	GaAs	0.8261	−0.3439	0.6104	0.0616	$0.53 \leq x \leq 1.0$
	GaSb	0.3834	−0.3439	0.6104	0.0616	$0 \leq x \leq 1.0$
	InP	0.6104	−0.3439	0.6104	0.0616	$0 \leq x \leq 1.0$
	InAs	0.4211	−0.3439	0.6104	0.0616	$0 \leq x \leq 1.0$

Table 31.4 Lattice-matching conditions for some III–V quaternaries of type $A_xB_yC_{1-x-y}D$ at 300 K. $y = A_0 + B_0 x$

Quaternary	Substrate	A_0	B_0	Remark
$Al_xGa_yIn_{1-x-y}P$	GaAs	0.5158	−0.9696	$0 \leq x \leq 0.53$
$Al_xGa_yIn_{1-x-y}As$	InP	0.4674	−0.9800	$0 \leq x \leq 0.48$

Table 31.5 Lattice-matching conditions for some III–V quaternaries of type $AB_xC_yD_{1-x-y}$ at 300 K. $x = A_0 + B_0 y$

Quaternary	Substrate	A_0	B_0	Remark
$AlP_xAs_ySb_{1-x-y}$	GaAs	0.7176	−0.7055	$0 \leq y \leq 0.96$
	InP	0.3966	−0.7055	$0 \leq y \leq 0.56$
	InAs	0.1149	−0.7055	$0 \leq y \leq 0.16$
$GaP_xAs_ySb_{1-x-y}$	GaAs	0.6861	−0.6861	$0 \leq y \leq 1.0$
	InP	0.3518	−0.6861	$0 \leq y \leq 0.51$
	InAs	0.0583	−0.6861	$0 \leq y \leq 0.085$
$InP_xAs_ySb_{1-x-y}$	GaSb	0.6282	−0.6899	$0 \leq y \leq 0.911$
	InAs	0.6899	−0.6899	$0 \leq y \leq 1.0$

31.4 Mechanical, Elastic and Lattice Vibronic Properties

31.4.1 Microhardness

The hardness test has been used for a long time as a simple means of characterizing the mechanical behavior of solids. The Knoop hardness H_P for $Ga_xIn_{1-x}P_yAs_{1-y}$ lattice-matched to InP has been reported [31.5], and is found to increase gradually from $520\,kg/mm^2$ for $y = 0$ ($Ga_{0.47}In_{0.53}As$) to $380\,kg/mm^2$ for $y = 1.0$ (InP). It has also been reported that the microhardness in $Al_xGa_{1-x}N$ thin film slightly decreases with increasing AlN composition x [31.6].

31.4.2 Elastic Constants and Related Moduli

Although the elastic properties of the III–V binaries have been studied extensively, little is known about their alloys. Recent studies, however, suggested that the elastic properties of the alloys can be obtained, to a good approximation, by averaging the binary endpoint values [31.7, 8]. We have, therefore, listed in Tables 31.6 and 31.7 the elastic stiffness (C_{ij}) and compliance constants (S_{ij}) for some III–V binaries with zinc blende and wurtzite structures, respectively. Table 31.8 also summarizes the functional expressions for the bulk modulus B_u, Young's modulus Y, and Poisson's ratio P. Note that Y and P are not isotropic, even in the cubic zinc blende lattice.

31.4.3 Long-Wavelength Phonons

The atoms of a crystal can be visualized as being joined by harmonic springs, and the crystal dynamics can be analyzed in terms of a linear combination of $3N$ normal modes of vibration (N is the number of different types of atoms; different in terms of mass or ordering in space). In alloys, the nature of the lattice optical spectrum depends on the difference between the quantities representing the lattice vibronic properties of the components. If these quantities are similar, then the optical response of an alloy is similar to the response of a crystal with the quantities averaged over the composition (one-mode behavior). In one-mode systems, such as most I–VII alloys, a single set of long-wavelength optical modes appears, as schematically shown in Fig. 31.1. When the parameters differ strongly, the response of a system is more complex; the spectrum contains a num-

Table 31.6 Elastic stiffness (C_{ij}) and compliance constants (S_{ij}) for some cubic III–V binaries at 300 K

Binary	C_{ij} (10^{11} dyn/cm^2)			S_{ij} (10^{-12} cm^2/dyn)		
	C_{11}	C_{12}	C_{44}	S_{11}	S_{12}	S_{44}
AlP	15.0*	6.42*	6.11*	0.897*	−0.269*	1.64*
AlAs	11.93	5.72	5.72	1.216	−0.394	1.748
AlSb	8.769	4.341	4.076	1.697	−0.5618	2.453
β-GaN	29.1*	14.8*	15.8*	0.523*	−0.176*	0.633*
GaP	14.050	6.203	7.033	0.9756	−0.2988	1.422
GaAs	11.88	5.38	5.94	1.173	−0.366	1.684
GaSb	8.838	4.027	4.320	1.583	−0.4955	2.315
InP	10.22	5.73	4.42	1.639	−0.589	2.26
InAs	8.329	4.526	3.959	1.945	−0.6847	2.526
InSb	6.608	3.531	3.027	2.410	−0.8395	3.304

* Theoretical

Table 31.7 Elastic stiffness (C_{ij}) and compliance constants (S_{ij}) for some wurtzite III–V binaries at 300 K

Binary	C_{ij} (10^{11} dyn/cm^2)						S_{ij} (10^{-12} cm^2/dyn)					
	C_{11}	C_{12}	C_{13}	C_{33}	C_{44}	C_{66}^{*1}	S_{11}	S_{12}	S_{13}	S_{33}	S_{44}	S_{66}^{*2}
AlN	41.0	14.0	10.0	39.0	12.0	13.5	0.285	−0.085	−0.051	0.283	0.833	0.740
α-GaN	37.3	14.1	8.0	38.7	9.4	11.6	0.320	−0.112	−0.043	0.276	1.06	0.864
InN	19.0	10.4	12.1	18.2	0.99	4.3	0.957	−0.206	−0.499	1.21	10.1	2.33

*1 $C_{66} = 1/2(C_{11} - C_{12})$, *2 $S_{66} = 2(S_{11} - S_{12})$

Table 31.8 Functional expressions for the bulk modulus B_u, Young's modulus Y, and Poisson's ratio P in semiconductors with zinc blende (ZB) and wurtzite (W) structures

Parameter	Structure	Expression	Remark
B_u	ZB	$(C_{11} + 2C_{12})/3$	
	W	$[(C_{11} + C_{12})C_{33} - 2C_{13}^2]/(C_{11} + C_{12} + 2C_{33} - 4C_{13})$	
Y	ZB	$1/S_{11}$	(100), [001]
		$1/(S_{11} - S/2)$	(100), [011]
		$1/S_{11}$	(110), [001]
		$1/(S_{11} - 2S/3)$	(110), [111]
		$1/(S_{11} - S/2)$	(111)
	W	$1/S_{11}$	$c \perp l$
		$1/S_{33}$	$c \parallel l$
P	ZB	$-S_{12}/S_{11}$	(100), $m = [010]$, $n = [001]$
		$-(S_{12} + S/2)/(S_{11} - S/2)$	(100), $m = [011]$, $n = [0\bar{1}1]$
		$-S_{12}/S_{11}$	(110), $m = [001]$, $n = [1\bar{1}0]$
		$-(S_{12} + S/3)/(S_{11} - 2S/3)$	(110), $m = [1\bar{1}1]$, $n = [1\bar{1}\bar{2}]$
		$-(S_{12} + S/6)/(S_{11} - S/2)$	(111)
	W	$(1/2)[1 - (Y/3B_u)]$	$c \perp l, c \parallel l$

$S = S_{11} - S_{12} - (S_{44}/2)$; $l =$ directional vector; $m =$ direction for a longitudinal stress; $n =$ direction for a transverse strain ($n \perp m$)

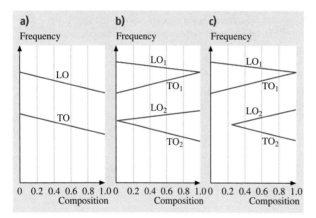

Fig. 31.1a–c Three different types of long-wavelength phonon mode behavior in ternary alloys: (**a**) one-mode; (**b**) two-mode; and (**c**) one-two-mode

cal modes with frequencies characteristic of each end member and strengths that are roughly proportional to the respective concentrations.

As seen in Table 31.9, the long-wavelength optical phonons in III–V ternaries exhibit either one-mode or two-mode behavior, or more rigorously, three different types of mode behavior: one-mode, two-mode, and one-two-mode behaviors. The one-two-mode system exhibits a single mode over only a part of the composition range, with two modes observed over the remaining range of compositions.

In a quaternary alloy of the $A_xB_{1-x}C_yD_{1-y}$ type, there are four kinds of unit cells: AC, AD, BC, and BD. On the other hand, in the $A_xB_yC_{1-x-y}D$ type there are three kinds of unit cells: AD, BD, and CD. We can, thus, expect four-mode or three-

ber of bands, each of which corresponds to one of the components, and it has an intensity governed by its content in the alloy ("multimode" behavior). For example, a two-mode system exhibits two distinct sets of opti-

Table 31.9 Behavior of the long-wavelength optical modes in III–V ternary and quaternary alloys

Behavior	Alloy
One mode	AlGaN(LO), AlInN, GaInN, AlAsSb
Two mode	AlGaN(TO), AlGaP, AlGaAs, AlGaSb, AlInAs, AlInSb, GaInP, GaInAs, GaNAs, GaPAs, GaPSb
One–two mode	AlInP, GaInSb, InAsSb
Three mode	AlGaAsSb, GaInAsSb, AlGaInP, AlGaInAs, InPAsSb
Four mode	GaInPSb, GaInPAs

mode behavior of the long-wavelength optical modes in such quaternary alloys ([31.9]; Table 31.9). However, the $Ga_xIn_{1-x}As_ySb_{1-y}$ quaternary showed three-mode behavior with GaAs, InSb and mixed InAs/GaAs characteristics [31.10]. The $Ga_xIn_{1-x}As_ySb_{1-y}$ quaternary was also reported to show two-mode or three-mode behavior, depending on the alloy composition [31.11].

The long-wavelength optical phonon behavior in the $Al_xGa_{1-x}As$ ternary has been studied both theoretically and experimentally. These studies suggest that the optical phonons in $Al_xGa_{1-x}As$ exhibit the two-mode behavior over the whole composition range. Thus, the $Al_xGa_{1-x}As$ system has two couples of the transverse optical (TO) and longitudinal optical (LO) modes; one is the GaAs-like mode and the other is the AlAs-like mode. Each phonon frequency can be expressed as [31.12]

- TO (GaAs): $268 - 14x$ cm^{-1},
- LO (GaAs): $292 - 38x$ cm^{-1},
- TO (AlAs): $358 + 4x$ cm^{-1},
- LO (AlAs): $358 + 71x - 26x^2$ cm^{-1}.

It is observed that only the AlAs-like LO mode shows a weak nonlinearity with respect to the alloy composition x.

31.5 Thermal Properties

31.5.1 Specific Heat and Debye Temperature

Since alloying has no significant effect on elastic properties, it appears that using the linear interpolation scheme for alloys can provide generally acceptable specific heat values (C). In fact, it has been reported that the C values for InP_xAs_{1-x} [31.13] and $Al_xGa_{1-x}As$ [31.14] vary fairly linearly with alloy composition x. It has also been shown [31.12] that the Debye temperature θ_D for alloys shows very weak nonlinearity with composition. From these facts, one can suppose that the linear interpolation scheme may provide generally acceptable C and θ_D values for III–V semiconductor alloys. We have, therefore, listed in Table 31.10 the III–V binary endpoint values for C and θ_D at $T = 300$ K. Using these values, the linearly interpolated C value for $Al_xGa_{1-x}As$ can be obtained from $C(x) = 0.424x + 0.327(1-x) = 0.327 + 0.097x$ (J/gK).

Table 31.10 Specific heat C and Debye temperature θ_D for some III–V binaries at 300 K

Binary	C (J/gK)	θ_D (K)	α_{th} (10^{-6} K^{-1})
AlN	0.728	988	3.042 ($\perp c$), 2.227 ($\parallel c$)
AlP	0.727	687	
AlAs	0.424	450	4.28
AlSb	0.326[*1]	370[*1]	4.2
α-GaN	0.42	821	5.0 ($\perp c$), 4.5 ($\parallel c$)
GaP	0.313	493[*2]	4.89
GaAs	0.327	370	6.03
GaSb	0.344[*1]	240[*1]	6.35
InN	2.274	674	3.830 ($\perp c$), 2.751 ($\parallel c$)
InP	0.322	420[*1]	4.56
InAs	0.352	280[*1]	≈ 5.0
InSb	0.350[*1]	161[*1]	5.04

[*1] At 273 K, [*2] at 150 K

31.5.2 Thermal Expansion Coefficient

The linear thermal expansion coefficient α_{th} is usually measured by measuring the temperature dependence of the lattice parameter. The composition dependence of α_{th} has been measured for many semiconductor alloys, including $Ga_xIn_{1-x}P$ [31.15] and GaP_xAs_{1-x} [31.16]. These studies indicate that the α_{th} value varies almost linearly with composition. This suggests that the thermal expansion coefficient can be accurately estimated using linear interpolation. In fact, we plot in Fig. 31.2 the 300 K value of α_{th} as a function of x for the $Al_xGa_{1-x}As$ ternary. By using the least-squares fit procedure, we obtain the linear relationship between α_{th} and x as $\alpha_{th}(x) = 6.01 - 1.74x$ (10^{-6} K^{-1}). This expression is almost the same as that obtained using the linear interpolation expression: $\alpha_{th}(x) = 4.28x + 6.03(1-x) = 6.03 - 1.75x$ (10^{-6} K^{-1}). The binary endpoint values of α_{th} are listed in Table 31.10.

31.5.3 Thermal Conductivity

The lattice thermal conductivity κ, or the thermal resistivity $W = 1/\kappa$, results mainly from interactions between phonons and from the scattering of phonons by crystalline imperfections. It is important to point out that when large numbers of foreign atoms are added to the host lattice, as in alloying, the thermal conductivity may

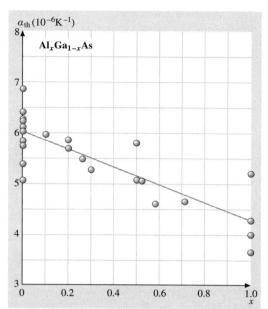

Fig. 31.2 Thermal expansion coefficient α_{th} as a function of x for the $Al_xGa_{1-x}As$ ternary at $T = 300$ K. The experimental data are gathered from various sources. The *solid line* is linearly interpolated between the AlAs and GaAs values

Table 31.11 Thermal resistivity values W for some III–V binaries at 300 K. Several cation and anion bowing parameters used for the calculation of alloy values are also listed in the last column

Binary	W (cmK/W)	C_{A-B} (cmK/W)
AlN	0.31*[1]	
AlP	1.11	
AlAs	1.10	
AlSb	1.75	
α-GaN	0.51*[1]	$C_{Al-Ga} = 32$
GaP	1.30	$C_{Ga-In} = 72$
GaAs	2.22	$C_{P-As} = 25$
GaSb	2.78	$C_{As-Sb} = 90$
InN	2.22*[2]	
InP	1.47	
InAs	3.33	
InSb	5.41–6.06	

*[1] Heat flow parallel to the basal plane, *[2] ceramics

decrease significantly. Experimental data on various alloy semiconductors, in fact, exhibit strong nonlinearity with respect to the alloy composition. Such a composition dependence can be successfully explained by using the quadratic expression of (31.2) or (31.6) [31.17].

In Fig. 31.3 we compare the results calculated from (31.2) [(31.7)] to the experimental data for $Al_xGa_{1-x}As$, $Al_xGa_{1-x}N$ and $Ga_xIn_{1-x}As_yP_{1-y}$/InP alloys. The binary W values used in these calculations are taken from Table 31.11. The corresponding nonlin-

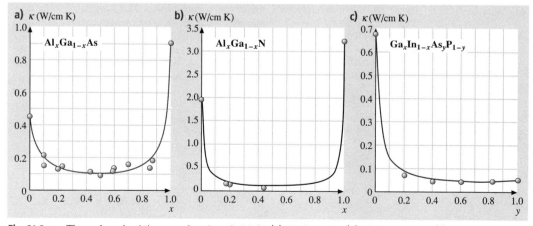

Fig. 31.3a–c Thermal conductivity κ as a function of $x(y)$ for (**a**) $Al_xGa_{1-x}As$, (**b**) $Al_xGa_{1-x}N$, and (**c**) $Ga_xIn_{1-x}As_yP_{1-y}$ lattice-matched to InP at $T = 300$ K. The experimental data (*solid circles*) are gathered from various sources. The *solid lines* represent the results calculated from (31.2) and (31.6) using the binary endpoint values and nonlinear parameters in Table 31.11

ear parameters C_{A-B} are also listed in Table 31.11. The agreement between the calculated and experimental data is excellent. By applying the present model, it is possible to estimate the κ (or W) values of experimentally unknown III–V alloy systems, such as GaAs$_x$Sb$_{1-x}$ and Al$_x$Ga$_y$In$_{1-x-y}$As.

31.6 Energy Band Parameters

31.6.1 Bandgap Energy

Lowest Direct and Lowest Indirect Band Gaps

The bandgap energies of III–V ternaries usually deviate from the simple linear relation of (31.1) and have an approximately quadratic dependence on the alloy composition x. Table 31.12 summarizes the lowest direct gap energy E_0 and the lowest indirect gap energies E_g^X and E_g^L for some III–V binaries of interest here. The corresponding nonlinear parameters C_{A-B} are listed in Table 31.13 [31.18]. Note that the E_g^X and E_g^L transitions correspond to those from the highest valence band at the Γ point to the lowest conduction band near X ($\Gamma_8 \to X_6$) or near L ($\Gamma_8 \to L_6$), respectively. The E_0 transitions take place at the Γ point ($\Gamma_8 \to \Gamma_6$).

Figure 31.4 plots the values of E_0 and E_g^X as a function of alloy composition x for the Ga$_x$In$_{1-x}$P ternary at $T = 300$ K. The solid lines are obtained by introducing the numerical values from Tables 31.12 and 31.13 into (31.2). These curves provide the direct-indirect crossover composition at $x \approx 0.7$. Figure 31.5 also shows the variation in composition of E_0 in the Ga$_x$In$_{1-x}$As, InAs$_x$Sb$_{1-x}$ and Ga$_x$In$_{1-x}$Sb ternaries. It is understood from Table 31.13 that the bowing parameters for the bandgap energies of III–V ternaries are negative or very small, implying a downward bowing or a linear interpolation to within experimen-

Table 31.12 Band-gap energies, E_0, E_g^X and E_g^L, for some III–V binaries at 300 K. ZB = zinc blende

Binary	E_0 (eV)	E_g^X (eV)	E_g^L (eV)
AlN	6.2	–	–
AlN (ZB)	5.1	5.34	9.8*
AlP	3.91	2.48	3.30
AlAs	3.01	2.15	2.37
AlSb	2.27	1.615	2.211
α-GaN	3.420	–	–
β-GaN	3.231	4.2*	5.5*
GaP	2.76	2.261	2.63
GaAs	1.43	1.91	1.72
GaSb	0.72	1.05	0.76
InN	0.7–1.1	–	–
InP	1.35	2.21	2.05
InAs	0.359	1.37	1.07
InSb	0.17	1.63	0.93

* Theoretical

Table 31.13 Bowing parameters used in the calculation of E_0, E_g^X and E_g^L for some III–V ternaries. * W = wurtzite; ZB = zinc blende

Ternary	Bowing parameter C_{A-B} (eV)		
	E_0	E_g^X	E_g^L
(Al,Ga)N (W)	−1.0	–	–
(Al,Ga)N (ZB)	0	−0.61	−0.80
(Al,In)N (W)	−16 + 9.1x	–	–
(Al,In)N (ZB)	−16 + 9.1x		
(Ga,In)N (W)	−3.0		
(Ga,In)N (ZB)	−3.0	−0.38	
(Al,Ga)P	0	−0.13	
(Al,In)P	−0.24	−0.38	
(Ga,In)P	−0.65	−0.18	−0.43
(Al,Ga)As	−0.37	−0.245	−0.055
(Al,In)As	−0.70	0	
(Ga,In)As	−0.477	−1.4	−0.33
(Al,Ga)Sb	−0.47	0	−0.55
(Al,In)Sb	−0.43		
(Ga,In)Sb	−0.415	−0.33	−0.4
Al(P,As)	−0.22	−0.22	−0.22
Al(P,Sb)	−2.7	−2.7	−2.7
Al(As,Sb)	−0.8	−0.28	−0.28
Ga(N,P) (ZB)	−3.9		
Ga(N,As) (ZB)	−120.4 + 100x		
Ga(P,As)	−0.19	−0.24	−0.16
Ga(P,Sb)	−2.7	−2.7	−2.7
Ga(As,Sb)	−1.43	−1.2	−1.2
In(N,P) (ZB)	−15		
In(N,As) (ZB)	−4.22		
In(P,As)	−0.10	−0.27	−0.27
In(P,Sb)	−1.9	−1.9	−1.9
In(As,Sb)	−0.67	−0.6	−0.6

* In those case where no value is listed, linear variation should be assumed

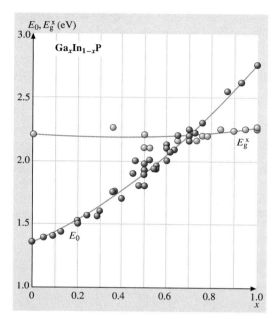

Fig. 31.4 Variation of the lowest direct gap (E_0) and lowest indirect gap energies (E_g^X) in the $Ga_xIn_{1-x}P$ ternary at $T = 300\,K$. The experimental data are gathered from various sources. The *solid lines* are calculated from (31.2) using the binary endpoint values and bowing parameters in Tables 31.12 and 31.13

tal uncertainty (Figs. 31.4, 31.5). It should be noted that nitrogen incorporation into (In,Ga)(P,As) results in a giant bandgap bowing of the host lattice for increasing nitrogen concentration [31.19]. We also summarize in Table 31.14 the expressions for the E_0 gap energy of some III–V quaternaries as a function of alloy composition.

Higher-Lying Band Gaps

The important optical transition energies observed at energies higher than E_0 are labeled E_1 and E_2. We summarize in Table 31.15 the higher-lying bandgap energies E_1 and E_2 for some III–V binaries. The corresponding bowing parameters for these gaps are listed in Table 31.16.

31.6.2 Carrier Effective Mass

Electron Effective Mass

Since the carrier effective mass is strongly connected with the carrier mobility, it is known to be one of the most important device parameters. Effective masses can be measured by a variety of techniques, such as the Shubnikov-de Haas effect, magnetophonon resonance, cyclotron resonance, and interband magneto-optical effects. We list in Table 31.17 the electron effective mass (m_e^Γ) at the Γ-conduction band and the density of states (m_e^α) and conductivity masses (m_c^α) at the X-conduction and L-conduction bands of some III–V binaries. We also list in Table 31.18 the bowing parameters used when calculating the electron effective mass m_e^Γ for some III–V

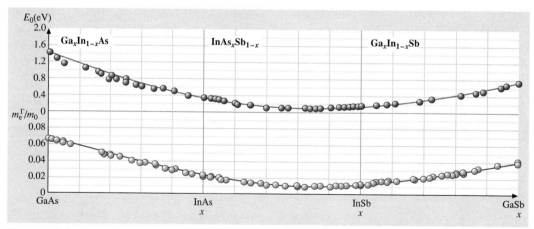

Fig. 31.5 Variation of the lowest direct gap energy E_0 ($T = 300\,K$) and electron effective mass m_e^Γ at the Γ-conduction bands of $Ga_xIn_{1-x}As$, $InAs_xSb_{1-x}$ and $Ga_xIn_{1-x}Sb$ ternaries. The experimental data are gathered from various sources. The *solid lines* are calculated from (31.2) using the binary endpoint values and bowing parameters in Tables 31.12 and 31.13 (E_0) and those in Tables 31.17 and 31.18 (m_e^Γ)

Table 31.14 Bandgap energies E_0 for some III–V quaternaries at 300 K

Quaternary	E_0 (eV)
$Ga_xIn_{1-x}P_yAs_{1-y}$/InP	$0.75 + 0.48y + 0.12y^2$
$Ga_xIn_{1-x}As_ySb_{1-y}$/GaSb	$0.290 - 0.165x + 0.60x^2$
$Ga_xIn_{1-x}As_ySb_{1-y}$/InAs	$0.36 - 0.23x + 0.54x^2$
$Al_xGa_yIn_{1-x-y}P$/GaAs*	$1.899 + 0.563x + 0.12x^2$
$Al_xGa_yIn_{1-x-y}As$/InP	$0.75 + 0.75x$
$InP_xAs_ySb_{1-x-y}$/InAs	$0.576 - 0.22y$

* The lowest indirect gap energy for this quaternary alloy can be obtained via $E_g^X = 2.20 - 0.09x$ eV

Table 31.15 Higher-lying bandgap energies, E_1 and E_2, for some III–V binaries at 300 K

Binary	E_1 (eV)	E_2 (eV)
AlN	7.76	8.79
AlP	4.30	4.63
AlAs	3.62–3.90	4.853, 4.89
AlSb	2.78–2.890	4.20–4.25
α-GaN	6.9	8.0
β-GaN	7.0	7.6
GaP	3.71	5.28
GaAs	2.89–2.97	4.960–5.45
GaSb	2.05	4.08–4.20
InN	5.0	7.6
InP	3.17	4.70 (E_0')
InAs	2.50	4.70
InSb	1.80	3.90

Table 31.16 Bowing parameters used in the calculation of the higher-lying bandgap energies, E_1 and E_2, for some cubic III–V ternaries

Ternary	C_{A-B} (eV)	
	E_1	E_2
(Al,Ga)P	0	0
(Al,In)P	0	0
(Ga,In)P	−0.86	0
(Al,Ga)As	−0.39	0
(Al,In)As	−0.38	
(Ga,In)As	−0.51	−0.27
(Al,Ga)Sb	−0.31	−0.34
(Al,In)Sb	−0.25	
(Ga,In)Sb	−0.33	−0.24
Ga(N,P)	0	0
Ga(N,As)	0	0
Ga(P,As)	0	0
Ga(As,Sb)	−0.59	−0.19
In(P,As)	−0.26	0
In(As,Sb)	≈ −0.55	≈ −0.6

* In those cases where no value is listed, linear variation should be assumed

ternaries from (31.2). Note that the density of states mass m_e^α for electrons in the conduction band minima $\alpha = \Gamma$, X, and L can be obtained from

$$m_e^\alpha = N^{2/3} m_{t\alpha}^{2/3} m_{l\alpha}^{1/3}, \quad (31.16)$$

where N is the number of equivalent α minima ($N = 1$ for the Γ minimum, $N = 3$ for the X minima, and $N = 4$ for the L minima). The two masses m_l and m_t in (31.16) are called the longitudinal and transverse masses, respectively. The density of states effective mass m_e^α is used to calculate the density of states. The conductivity effective mass m_c^α, which can be used for calculating the conductivity (mobility), is also given by

$$m_c^\alpha = \frac{3m_{t\alpha}m_{l\alpha}}{m_{t\alpha} + 2m_{l\alpha}}. \quad (31.17)$$

Since $m_{t\Gamma} = m_{l\Gamma}$ at the $\alpha = \Gamma$ minimum of cubic semiconductors, we have the relation $m_e^\Gamma = m_c^\Gamma$. In the case of wurtzite semiconductors, we have the relation $m_e^\Gamma \neq m_c^\Gamma$, but the difference is very small.

The composition dependence of the electron effective mass m_e^Γ at the Γ-conduction bands of $Ga_xIn_{1-x}As$, $InAs_xSb_{1-x}$ and $Ga_xIn_{1-x}Sb$ ternaries is plotted in Fig. 31.5. The solid lines are calculated from (31.2) using the binary endpoint values and bowing parameters in Tables 31.17 and 31.18. For conventional semiconductors, the values of the effective mass are known to decrease with decreaseing bandgap energy (Fig. 31.5). This is in agreement with a trend predicted by the $k \cdot p$ theory [31.2]. In III–V–N alloys, the electron effective mass has been predicted to increase with increasing nitrogen composition in the low composition range [31.19]. This behavior is rather unusual, and in fact is opposite to what is seen in conventional semiconductors. However, a more recent study suggested that the effective electron mass in GaN_xAs_{1-x} decreases from $0.084m_0$ to $0.029m_0$ as x increases from 0 to 0.004 [31.20]. We also summarize in Table 31.19 the composition dependence of m_e^Γ, determined for $Ga_xIn_{1-x}P_yAs_{1-y}$ and $Al_xGa_yIn_{1-x-y}As$ quaternaries lattice-matched to InP.

Hole Effective Mass

The effective mass can only be clearly defined for an isotropic parabolic band. In the case of III–V materials, the valence bands are warped from spherical symmetry some distance away from the Brillouin zone center

Table 31.17 Electron effective mass at the Γ-conduction band (m_e^Γ) and density of states (m_e^α) and conductivity masses (m_c^α) at the X-conduction and L-conduction bands of some III–V binaries. ZB = zinc blende

Binary	m_e^Γ/m_0	Density of states mass		Conductivity mass	
		m_e^X/m_0	m_e^L/m_0	m_c^X/m_0	m_c^L/m_0
AlN	0.29*	–	–	–	–
AlN (ZB)	0.26*	0.78*		0.37*	
AlP	0.220*	1.14*		0.31*	
AlAs	0.124	0.71	0.78	0.26*	0.21*
AlSb	0.14	0.84	1.05*	0.29	0.28*
α-GaN	0.21	–	–	–	–
β-GaN	0.15	0.78*		0.36*	
GaP	0.114	1.58	0.75*	0.37	0.21*
GaAs	0.067	0.85	0.56	0.32	0.11
GaSb	0.039	1.08*	0.54	0.44*	0.12
InN	0.07	–	–	–	–
InP	0.07927	1.09*	0.76*	0.45*	0.19*
InAs	0.024	0.98*	0.94*	0.38*	0.18*
InSb	0.013				

* Theoretical

Table 31.18 Bowing parameter used in the calculation of the electron effective mass m_e^Γ at the Γ-conduction bands of some III–V ternaries

Ternary	C_{A-B} (m_0)
(Ga,In)P	−0.019
(Al,In)As	0
(Ga,In)As	−0.0049
(Ga,In)Sb	−0.0092
Ga(P,As)	0
In(P,As)	0
In(As,Sb)	−0.030

Table 31.19 Electron effective mass m_e^Γ at the Γ-conduction bands of some III–V quaternaries

Quaternary	m_e^Γ/m_0
$Ga_xIn_{1-x}P_yAs_{1-y}$/InP	$0.043 + 0.036y$
$Al_xGa_yIn_{1-x-y}As$/InP	$0.043 + 0.046x - 0.017x^2$

(Γ). Depending on the measurement or calculation technique employed, different values of hole masses are then possible experimentally or theoretically. Thus, it is always important to choose the correct definition of the effective hole mass which appropriate to the physical phenomenon considered.

We list in Table 31.20 the density of states heavy hole (m_{HH}^*), the averaged light hole (m_{LH}^*), and spin orbit splitoff effective hole masses (m_{SO}) in some cubic III–V semiconductors. These masses are, respectively, defined using Luttinger's valence band parameters γ_i by

$$m_{HH}^* = \frac{(1 + 0.05\gamma_h + 0.0164\gamma_h^2)^{2/3}}{\gamma_1 - \overline{\gamma}}, \quad (31.18)$$

$$m_{LH}^* = \frac{1}{\gamma_1 + \overline{\gamma}}, \quad (31.19)$$

$$m_{SO} = \frac{1}{\gamma_1} \quad (31.20)$$

with

$$\overline{\gamma} = (2\gamma_2^2 + 2\gamma_3^2)^{1/2}, \quad \gamma_h = \frac{6(\gamma_3^2 - \gamma_2^2)}{\overline{\gamma}(\gamma_1 - \overline{\gamma})}. \quad (31.21)$$

Only a few experimental studies have been performed on the effective hole masses in III–V alloys, e.g., the $Ga_xIn_{1-x}P_yAs_{1-y}$ quaternary [31.2]. While some data imply a bowing parameter, the large uncertainties in existing determinations make it difficult to conclusively state that such experimental values are preferable to a linear interpolation. The binary endpoint data listed in Table 31.20 enable us to estimate alloy values using the linear interpolation scheme.

31.6.3 Deformation Potential

The deformation potentials of the electronic states at the Brillouin zone centers of semiconductors play an important role in many physical phenomena. For example,

Table 31.20 Density of states heavy hole (m^*_{HH}), averaged light hole (m^*_{LH}), and spin orbit splitoff effective hole masses (m_{SO}) in some cubic III–V semiconductors. ZB = zinc blende

Material	m^*_{HH}/m_0	m^*_{LH}/m_0	m_{SO}/m_0
AlN (ZB)	1.77*	0.35*	0.58*
AlP	0.63*	0.20*	0.29*
AlAs	0.81*	0.16*	0.30*
AlSb	0.9	0.13	0.317*
β-GaN	1.27*	0.21*	0.35*
GaP	0.52	0.17	0.34
GaAs	0.55	0.083	0.165
GaSb	0.37	0.043	0.12
InP	0.69	0.11	0.21
InAs	0.36	0.026	0.14
InSb	0.38	0.014	0.10

* Theoretical

the splitting of the heavy hole and light hole bands at the Γ point of the strained substance can be explained by the shear deformation potentials, b and d. The lattice mobilities of holes are also strongly affected by these potentials. Several experimental data have been reported on the deformation potential values for III–V alloys, e.g., $Al_xGa_{1-x}As$ [31.12], GaP_xAs_{1-x} [31.21] and $Al_xIn_{1-x}As$ [31.22]. Due to the large scatter in the experimental binary endpoint values, it is very difficult to establish any evolution of the deformation potentials with composition. We list in Table 31.21 the recommended values for the conduction band (a_c) and valence band deformation potentials (a_v, b, d) of some cubic III–V binaries. The deformation potentials for some wurtzite III–V semiconductors are also collected in Table 31.22. Until more precise data become available, we suggest employing the linear interpolation expressions in order to estimate the parameter values of these poorly explored properties.

Table 31.21 Conduction-band (a_c) and valence-band deformation potentials (a_v, b, d) for some cubic III–V binaries. ZB = zinc blende

Binary	Conduction band	Valence band		
	a_c (eV)	a_v (eV)	b (eV)	d (eV)
AlN (ZB)	−11.7*	−5.9*	−1.7*	−4.4*
AlP	−5.54*	3.15*	−1.5*	
AlAs	−5.64*	−2.6*	−2.3*	
AlSb	−6.97*	1.38*	−1.35	−4.3
β-GaN	−21.3*	−13.33*	−2.09*	−1.75*
GaP	−7.14*	1.70*	−1.7	−4.4
GaAs	−11.0	−0.85	−1.85	−5.1
GaSb	−9	0.79*	−2.4	−5.4
InP	−11.4	−0.6	−1.7	−4.3
InAs	−10.2	1.00*	−1.8	−3.6
InSb	−15	0.36*	−2.0	−5.4

* Theoretical

Table 31.22 Conduction-band (D_i) and valence-band deformation potentials (C_i) for some wurtzite III–V binaries (in eV)

Binary	Conduction band		Valence band							
	D_1	D_2	C_1	D_1–C_1	C_2	D_2–C_2	C_3	C_4	C_5	C_6
AlN	−10.23*	−9.65*	−12.9*		−8.4*		4.5*	−2.2*	−2.6*	−4.1*
α-GaN	−9.47*	−7.17*	−41.4	−3.1	−33.3	−11.2	8.2	−4.1	−4.7	
InN				−4.05*		−6.67*	4.92*	−1.79*		

* Theoretical

31.7 Optical Properties

31.7.1 The Reststrahlen Region

It should be noted that in homopolar semiconductors like Si and Ge, the fundamental vibration has no dipole moment and is infrared inactive. In heteropolar semiconductors, such as GaAs and InP, the first-order dipole moment gives rise to a very strong absorption band associated with optical modes that have a k vector of essentially zero (i.e., long-wavelength optical phonons). This band is called the reststrahlen band. Below this band, the real part of the dielectric constant asymptotically approaches the static or low-frequency dielectric constant ε_s. The optical constant connecting the reststrahlen near-infrared spectral range is called the high-frequency or optical dielectric constant ε_∞. The value of ε_∞ is, therefore, measured for frequencies well above the long-wavelength LO phonon frequency but below the fundamental absorption edge.

Table 31.23 Static (ε_s) and high-frequency dielectric constants (ε_∞) for some cubic III–V binaries. ZB = zinc blende

Binary	ε_s	ε_∞
AlN (ZB)	8.16*	4.20
AlP	9.6	7.4
AlAs	10.06	8.16
AlSb	11.21	9.88
β-GaN	9.40*	5.35*
GaP	11.0	8.8
GaAs	12.90	10.86
GaSb	15.5	14.2
InP	12.9	9.9
InAs	14.3	11.6
InSb	17.2	15.3

* Calculated or estimated

Table 31.24 Static (ε_s) and high-frequency dielectric constants (ε_∞) for some wurtzite III–V binaries

Binary	$E \perp c$		$E \parallel c$	
	ε_s	ε_∞	ε_s	ε_∞
AlN	8.3	4.4	8.9	4.8
α-GaN	9.6	5.4	10.6	5.4
InN	13.1*	8.4*	14.4*	8.4*

* Estimated

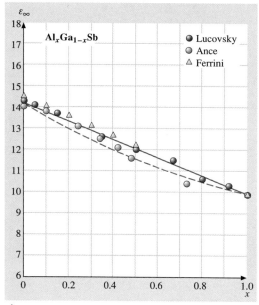

Fig. 31.6 High-frequency dielectric constant ε_∞ as a function of x for the $Al_xGa_{1-x}Sb$ ternary. The experimental data are taken from *Lucovsky* et al. [31.23] (*solid circles*), *Ance and Mau* [31.24] (*open circles*), and *Ferrini* et al. [31.25] (*open triangles*). The *solid* and *dashed* lines are, respectively, calculated from (31.1) and (31.6) (ternary) with the binary endpoint values in Table 31.23

The general properties of ε_s and ε_∞ for a specific family of compounds, namely III–V and II–VI compounds, suggest that the dielectric constants in alloy semiconductors could be deduced by using the linear interpolation method [31.26]. The simplest linear interpolation method is to use (31.1), (31.3) or (31.5). The linear interpolation scheme based on the Clausius–Mosotti relation can also be obtained from (31.6). In Fig. 31.6, we show the interpolated ε_∞ as a function of x for the $Al_xGa_{1-x}Sb$ ternary. The solid and dashed lines are, respectively, calculated from (31.1) and (31.6) (ternary). The experimental data are taken from *Lucovsky* et al. [31.23], *Ance and Mau* [31.24], and *Ferrini* et al. [31.25]. The binary endpoint values used in the calculation are listed in Table 31.23. These two methods are found to provide almost the same interpolated values. Table 31.24 also lists the ε_s and ε_∞ values for some wurtzite III–V binary semiconductors.

The optical spectra observed in the reststrahlen region of alloy semiconductors can be explained by the following multioscillator model [31.12]:

$$\varepsilon(\omega) = \varepsilon_\infty + \sum_j \frac{S_j \omega_{\mathrm{TO}j}^2}{\omega_{\mathrm{TO}j}^2 - \omega^2 - \mathrm{i}\omega\gamma_j}, \quad (31.22)$$

where $S_j = \varepsilon_\infty (\omega_{\mathrm{LO}j}^2 - \omega_{\mathrm{TO}j}^2)$ is the oscillator strength, $\omega_{\mathrm{TO}j}$ ($\omega_{\mathrm{LO}j}$) is the TO (LO) phonon frequency, and γ_j is the damping constant of the j-th lattice oscillator. We show in Fig. 31.7, as an example, the optical spectra in the reststrahlen region of the $\mathrm{Al}_x\mathrm{Ga}_{1-x}\mathrm{As}$ ternary. As expected from the two-mode behavior of the long-wavelength optical phonons, the $\varepsilon(\omega)$ spectra of $\mathrm{Al}_x\mathrm{Ga}_{1-x}\mathrm{As}$ exhibit two main optical resonances: GaAs-like and AlAs-like.

31.7.2 The Interband Transition Region

The optical constants in the interband transition regions of semiconductors depend fundamentally on the electronic energy band structure of the semiconductors. The relation between the electronic energy band structure and $\varepsilon_2(E)$ is given by

$$\varepsilon_2(E) = \frac{4e^2\hbar^2}{\pi\mu^2 E^2} \int \mathrm{d}\mathbf{k} \, |P_{\mathrm{cv}}(\mathbf{k})|^2 \delta[E_{\mathrm{c}}(\mathbf{k}) - E_{\mathrm{v}}(\mathbf{k}) - E], \quad (31.23)$$

where μ is the combined density of states mass, the Dirac δ function represents the spectral joint density of states between the valence-band [$E_{\mathrm{v}}(\mathbf{k})$] and conduction-band states [$E_{\mathrm{c}}(\mathbf{k})$], differing by the energy $E = \hbar\omega$ of the incident light, $P_{\mathrm{cv}}(\mathbf{k})$ is the momentum matrix element between the valence-band and conduction-band states, and the integration is performed over the first Brillouin zone. The Kramers–Kronig relations link $\varepsilon_2(E)$ and $\varepsilon_1(E)$ in a manner that means that $\varepsilon_1(E)$ can be calculated at each photon energy if $\varepsilon_2(E)$ is known explicitly over the entire photon energy

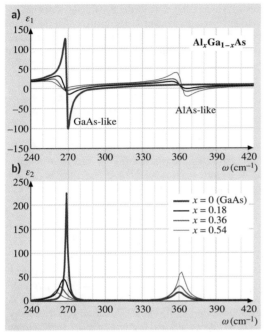

Fig. 31.7 $\varepsilon(\omega)$ spectra in the reststrahlen region of the $\mathrm{Al}_x\mathrm{Ga}_{1-x}\mathrm{As}$ ternary

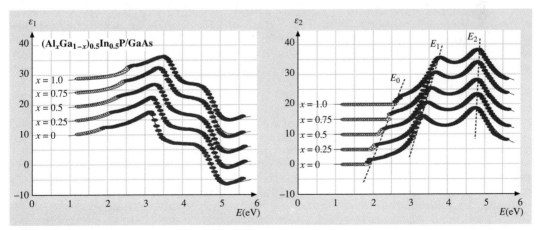

Fig. 31.8 $\varepsilon(E)$ spectra for $\mathrm{Al}_x\mathrm{Ga}_y\mathrm{In}_{1-x-y}\mathrm{P}/\mathrm{GaAs}$ at room temperature. The experimental data are taken from *Adachi* [31.27]; *open* and *solid circles*. The *solid lines* represent the theoretical fits for the MDF calculation

range, and vice versa. The Kramers–Kronig relations are of fundamental importance in the analysis of optical spectra [31.9].

The refractive indices and absorption coefficients of semiconductors are the basis of many important applications of semiconductors, such as light-emitting diodes, laser diodes and photodetectors. The optical constants of III–V binaries and their ternary and quaternary alloys have been presented in tabular and graphical forms [31.27]. We plot in Fig. 31.8 the $\varepsilon(E)$ spectra for $Al_xGa_yIn_{1-x-y}P/GaAs$ taken from tabulation by Adachi([31.27]; open and solid circles). The solid lines represent the theoretical fits of the model dielectric function (MDF) calculation [31.9]. The three major features of the spectra seen in Fig. 31.8 are the E_0, E_1 and E_2 structures at ≈ 2, ≈ 3.5 and ≈ 4.5 eV, respectively. It is found that the E_0 and E_1 structures move to higher energies with increasing x, while the E_2 structure does not do so to any perceptible degree. We can see that the MDF calculation enables us to calculate the optical spectra for optional compositions of alloy semiconductors with good accuracy.

31.8 Carrier Transport Properties

An accurate comparison between experimental mobility and theoretical calculation is of great importance for the determination of a variety of fundamental material parameters and carrier scattering mechanisms. There are various carrier scattering mechanisms in semiconductors, as schematically shown in Fig. 31.9. The effect of the individual scattering mechanisms on the total calculated carrier mobility can be visualized using Matthiessen's rule:

$$\frac{1}{\mu_{\text{tot}}} = \sum_i \frac{1}{\mu_i}. \qquad (31.24)$$

The total carrier mobility μ_{tot} can then be obtained from the scattering-limited mobilities μ_i of each scattering mechanism. We note that in alloy semiconductors the charged carriers see potential fluctuations as a result of the composition disorder. This kind of scattering mechanism, so-called alloy scattering, is important in some III–V ternaries and quaternaries. The alloy scattering limited mobility in ternary alloys can be formulated as

$$\mu_{\text{al}} = \frac{\sqrt{2\pi} e \hbar^4 N_{\text{al}} \alpha}{3(m_c^*)^{5/2}(kT)^{1/2}x(1-x)(\Delta U)^2}, \qquad (31.25)$$

where N_{al} is the density of alloy sites, m_c^* is the electron or hole conductivity mass, x and $(1-x)$ are the mole fractions of the binary endpoint materials, and ΔU is the alloy scattering potential. The factor α is caused by the band degeneracy and is given by $\alpha = 1$ for electrons and by $\alpha = [(d^{5/2}+d^3)/(1+d^{3/2})^2]$ for holes with $d = m_{\text{HH}}/m_{\text{LH}}$, where m_{HH} and m_{LH} are the heavy hole and light hole band masses, respectively [31.12].

Table 31.25 Hall mobilities for electrons (μ_e) and holes (μ_h) obtained at 300 K for relatively pure samples of III–V binaries (in cm^2/Vs)

Binary	μ_e	μ_h
AlP	80	450
AlAs	294	105
AlSb	200	420
α-GaN	1245	370
β-GaN	760	350
GaP	189	140
GaAs	9340	450
GaSb	12 040	1624
InN	3100	
InP	6460	180
InAs	30 000	450
InSb	77 000	1100

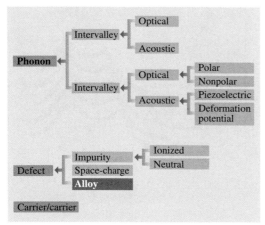

Fig. 31.9 Various possible carrier scattering mechanisms in semiconductor alloys

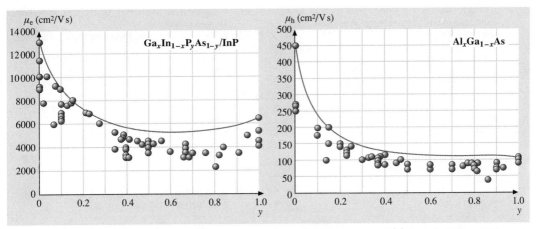

Fig. 31.10 (a) Electron Hall mobility μ_e in the $Ga_xIn_{1-x}P_yAs_{1-y}$/InP quaternary and (b) the hole Hall mobility μ_h in the $Al_xGa_{1-x}As$ ternary, respectively. The experimental data correspond to those for relatively pure samples. The *solid lines* in (a) and (b) represent the results calculated using (31.26) with $\mu_{al,0} = 3000$ and $50\,cm^2/Vs$, respectively

Let us simply express the total carrier mobility μ_{tot} in alloy $A_xB_{1-x}C$ as

$$\frac{1}{\mu_{tot}(x)} = \frac{1}{x\mu_{tot}(AC) + (1-x)\mu_{tot}(BC)} + \frac{1}{\mu_{al,0}/[x/(1-x)]} \,. \quad (31.26)$$

The first term in (31.26) comes from the linear interpolation scheme and the second term accounts for the effects of alloying.

We plot in Figs. 31.10a and 31.10b the electron Hall mobility in the $Ga_xIn_{1-x}P_yAs_{1-y}$/InP quaternary (μ_e) and the hole Hall mobility in the $Al_xGa_{1-x}As$ ternary, respectively. The experimental data correspond to those for relatively pure samples [31.28]. The solid lines in Figs. 31.10a,b represent the results calculated using (31.26) with $\mu_{al,0} = 3000$ and $50\,cm^2/Vs$, respectively. The corresponding binary endpoint values for μ_{tot} are listed in Table 31.25. For $Ga_xIn_{1-x}P_yAs_{1-y}$/InP, we have considered the quaternary to be an alloy of the constituents $Ga_{0.53}In_{0.47}As$ ($y = 0$) and InP ($y = 1.0$) and we have used the value of μ_{tot} ($Ga_{0.47}In_{0.53}As$) = $13\,000\,cm^2/Vs$. It is clear that (31.26) can successfully explain the peculiar composition dependence of the carrier mobility in the semiconductor alloys.

References

31.1 A. Mascarenhas: *Spontaneous Ordering in Semiconductor Alloys* (Kluwer Academic, New York 2002)

31.2 S. Adachi: *Physical Properties of III–V Semiconductor Compounds: InP, InAs, GaAs, GaP, InGaAs, and InGaAsP* (Wiley-Interscience, New York 1992)

31.3 S. Adachi: *Handbook on Physical Properties of Semiconductors*, III–V Compound Semiconductors, Vol. 2 (Springer, Berlin, Heidelberg 2004)

31.4 W. Martienssen (Ed): *Landolt-Börnstein, Group III/41 Semiconductors, A1α Lattice Parameters* (Springer, Berlin, Heidelberg 2001)

31.5 D. Y. Watts, A. F. W. Willoughby: J. Appl. Phys. **56**, 1869 (1984)

31.6 D. Cáceres, I. Vergara, R. González, E. Monroy, F. Calle, E. Muñoz, F. Omnès: J. Appl. Phys. **86**, 6773 (1999)

31.7 M. Krieger, H. Sigg, N. Herres, K. Bachem, K. Köhler: Appl. Phys. Lett. **66**, 682 (1995)

31.8 W. E. Hoke, T. D. Kennedy, A. Torabi: Appl. Phys. Lett. **79**, 4160 (2001)

31.9 S. Adachi: *Optical Properties of Crystalline and Amorphous Semiconductors: Materials and Fundamental Principles* (Kluwer Academic, Boston 1999)

31.10 C. Pickering: J. Electron. Mater. **15**, 51 (1986)

31.11 D. H. Jaw, Y. T. Cherng, G. B. Stringfellow: J. Appl. Phys. **66**, 1965 (1989)

31.12 S. Adachi: GaAs *and Related Materials: Bulk Semiconducting and Superlattice Properties* (World Scientific, Singapore 1994)
31.13 A. N. N. Sirota, A. M. Antyukhov, V. V. Novikov, V. A. Fedorov: Sov. Phys. Dokl. **26**, 701 (1981)
31.14 J. L. Pichardo, J. J. Alvarado-Gil, A. Cruz, J. G. Mendoza, G. Torres: J. Appl. Phys. **87**, 7740 (2000)
31.15 I. Kudman, R. J. Paff: J. Appl. Phys. **43**, 3760 (1972)
31.16 J. Bąk-Misiuk, H. G. Brühl, W. Paszkowicz, U. Pietsch: Phys. Stat. Sol. A **106**, 451 (1988)
31.17 S. Adachi: J. Appl. Phys. **54**, 1844 (1983)
31.18 I. Vurgaftman, J. R. Meyer, L. R. Ram-Mohan: J. Appl. Phys. **89**, 5815 (2001)
31.19 I. A. Buyanova, W. M. Chen, B. Monemar: MRS Internet J. Nitride Semicond. Res. **6**, 2 (2001)
31.20 D. L. Young, J. F. Geisz, T. J. Coutts: Appl. Phys. Lett. **82**, 1236 (2003)
31.21 Y. González, G. Armelles, L. González: J. Appl. Phys. **76**, 1951 (1994)
31.22 L. Pavesi, R. Houdré, P. Giannozzi: J. Appl. Phys. **78**, 470 (1995)
31.23 G. Lucovsky, K. Y. Cheng, G. L. Pearson: Phys. Rev. B **12**, 4135 (1975)
31.24 C. Ance, N. Van Mau: J. Phys. C **9**, 1565 (1976)
31.25 R. Ferrini, M. Galli, G. Guizzetti, M. Patrini, A. Bosacchi, S. Franchi, R. Magnanini: Phys. Rev. B **56**, 7549 (1997)
31.26 S. Adachi: J. Appl. Phys. **53**, 8775 (1982)
31.27 S. Adachi: *Optical Constants of Crystalline and Amorphous Semiconductors: Numerical Data and Graphical Information* (Kluwer Academic, Boston 1999)
31.28 M. Sotoodeh, A. H. Khalid, A. A. Rezazadeh: J. Appl. Phys. **87**, 2890 (2000)

32. Group III Nitrides

Optical, electrical and mechanical properties of group III nitrides, including of AlN, GaN, InN and their ternary and quaternary compounds are discussed. The driving force for semiconductor nitrides is device applications for emitters and detectors in the visible and ultraviolet (UV) portions of the optical spectrum and high-power amplifiers. Further advances in electronic and optoelectronic devices, which are imperative, require better understanding and precise measurements of the mechanical, thermal, electrical and optical properties of nitride semiconductors. Information available in the literature regarding many of the physical properties of nitrides, especially AlN and InN, is still in the process of evolution, and naturally in the subject of some controversy. This is, in part, a consequence of measurements having been performed on samples of widely varying quality. When possible, these spurious discrepancies have been disregarded. For other materials, too few measurements are available to yield a consensus, in which case the available data are simply reported. The aim of this work is to present the latest available data obtained by various experimental observations and theoretical calculations.

32.1	Crystal Structures of Nitrides	755
32.2	Lattice Parameters of Nitrides	756
32.3	Mechanical Properties of Nitrides	757
32.4	Thermal Properties of Nitrides	761
	32.4.1 Thermal Expansion Coefficients	761
	32.4.2 Thermal Conductivity	762
	32.4.3 Specific Heat	764
32.5	Electrical Properties of Nitrides	766
	32.5.1 Low-Field Transport	766
	32.5.2 High-Field Transport	775
32.6	Optical Properties of Nitrides	777
	32.6.1 Gallium Nitride	778
	32.6.2 Aluminium Nitride	786
	32.6.3 Indium Nitride	789
32.7	Properties of Nitride Alloys	791
32.8	Summary and Conclusions	794
References		795

During the last three decades, developments in the field of group III nitrides have been spectacular, with major breakthroughs taking place in the 1990s. They have been viewed as a highly promising material system for electronic and optoelectronic applications. As members of the group III nitrides family, AlN, GaN, InN and their alloys are all wide-band-gap materials and can crystallize in both wurtzite and zincblende polytypes. The band gaps of the wurtzite polytypes are direct and range from a possible value of ≈ 0.8 eV for InN, to 3.4 eV for GaN, and to 6.1 eV for AlN. GaN alloyed with AlN and InN may span a continuous range of direct-band-gap energies throughout much of the visible spectrum, well into ultraviolet (UV) wavelengths. This makes the nitride system attractive for optoelectronic applications, such as light-emitting diodes (LEDs), laser diodes (LDs), and UV detectors. Commercialization of bright blue and green LEDs and the possibility of yellow LEDs paved the way for developing full-color displays. If the three primary-color LEDs, including red, produced by the InGaAlAs system are used in place of incandescent light bulbs in some form of a color-mixing scheme, they would provide not only compactness and longer lifetime, but also lower power consumption for the same luminous flux output. Additional possible applications include use in agriculture as light sources for accelerated photosynthesis, and in health care for diagnosis and treatment. Unlike display and lighting applications, digital information storage and reading require coherent light sources because the diffraction-limited optical storage density increases approximately quadratically with decreasing wavelength. The nitride material system, when adapted to semiconductor lasers in blue and UV wavelengths, offers increased data storage density, possibly as high as 50 Gb per disc with 25 Gb promised soon in the Blu-Ray system. Other equally attractive applications envisioned include printing and surgery.

When used as UV sensors in jet engines, automobiles, and furnaces (boilers), the devices would allow optimal fuel efficiency and control of effluents for a cleaner environment. Moreover, visible-blind and solar-blind nitride-based photodetectors are also an ideal candidate for a number of applications including early missile-plume detection, UV astronomy, space-to-space communication, and biological effects.

Another area gaining a lot of attention for group III–V nitrides is high-temperature/high-power electronic applications, such as radar, missiles, and satellites as well as in low-cost compact amplifiers for wireless base stations, due to their excellent electron transport properties, including good mobility and high saturated drift velocity. The strongest feature of the group III nitrides compared to other wide-band-gap counterparts is the heterostructure technology that it can support. Quantum wells, modulation-doped heterointerfaces, and heterojunction structures can all be made in this system, giving access to new spectral regions for optical devices and new operational regimes for electronic devices. Other attractive properties of the nitrides include high mechanical and thermal stability, and large piezoelectric constants.

One of the main difficulties that have hindered group III nitride research is the lack of a lattice-matched and thermally compatible substrate material. A wide variety of materials have been studied for nitride epitaxy, including insulating metal oxides, metal nitrides, and other semiconductors. In practice, properties other than the lattice constants and thermal compatibility, including the crystal structure, surface finish, composition, reactivity, chemical, and electrical properties, are also important in determining suitability as a substrate. The substrate employed determines the crystal orientation, polarity, polytype, surface morphology, strain, and the defect concentration of the epitaxial films. The most promising results on more conventional substrates so far have been obtained on sapphire, and SiC. Also coming on the scene are thick freestanding GaN templates. Group III–V nitrides have been grown on Si, NaCl, GaP, InP, SiC, W, ZnO, $MgAl_2O_4$, TiO_2, and MgO. Other substrates have also been used for nitride growth, including Hf, $LiAlO_2$ and $LiGaO_2$. Lateral (lattice constant a) mismatched substrates lead to substantial densities of misfit and threading dislocations in broad-area epitaxially deposited GaN on foreign substrate, in the range $10^9 – 10^{10}$ cm^{-2}. An appropriate surface preparation such as nitridation, deposition of a low-temperature (LT) AlN or GaN buffer layer, selective epitaxy followed by a type of coalescence called lateral epitaxial overgrowth (LEO) or epitaxial lateral overgrowth (ELOG) can reduce dislocation densities down to 10^6 cm^{-2}. However, these numbers are still high compared to extended-defect densities of essentially zero for silicon homoepitaxy, and $10^2 – 10^4$ cm^{-2} for gallium arsenide homoepitaxy. Vertical (lattice constant c) mismatch creates additional crystalline defects besetting the layers, including inversion domain boundaries and stacking faults. In addition, mismatch of thermal expansion coefficients between the epitaxial films and the substrate induces stress, which can cause crack formation in the film and substrate for thick films during cooling from the deposition temperature. A high density of defects, which increases the laser threshold current, causes reverse leakage currents in junctions, depletes sheet charge-carrier density in heterojunction field-effect transistors, reduces the charge-carrier mobility and thermal conductivity, and is detrimental to device applications and the achievement of their optimal performance. Thus, substrates capable of supporting better-quality epitaxial layers are always needed to realize the full potential of nitride-based devices.

Nearly every major crystal-growth technique has been developed, including molecular beam epitaxy (MBE), hydride vapor-phase epitaxy (HVPE), and metalorganic chemical vapor deposition (MOCVD), in relation to nitride semiconductors. Several modifications to the conventional MBE method have been implemented for group III nitride growth: growth with ammonia or hydrazine (the latter is not attractive due to safety reasons and success of ammonia), plasma-assisted MBE (PAMBE), metalorganic MBE (MOMBE), pulsed laser deposition (PLD), etc. Among other methods, radio-frequency (RF) and electron–cyclotron resonance (ECR) plasma sources are the most commonly employed devices to activate the neutral nitrogen species in the MBE environment. Although all of these epitaxial methods contend with problems related to the lack of native GaN substrates, and difficulty with nitrogen incorporation, remarkable progress in the growth of high-quality epitaxial layers of group III nitrides by a variety of methods has been achieved.

Although many applications based on nitride semiconductors has emerged and some of them are commercially available, as discussed throughout this chapter, there are many contradictions in identification of the basic physical properties of these materials. In this respect, they are not yet mature. Additionally, knowledge of the fundamental properties is crucial not only from the physics point of view but also when understanding and optimizing the device structures for better performance. In this chapter, therefore, we present the updated fundamental properties of GaN, AlN and InN, including structural, mechanical, thermal, electrical, and optical properties. The aim is to assist readers newly entering this field and other interested researchers in accessing the most-recent available data. The reader is also urged to peruse the following publications for more detail information in several aspects of ongoing research in group III nitrides. These consist of books [32.1,2], edited books and handbooks [32.3–10], and review papers [32.11–29].

32.1 Crystal Structures of Nitrides

The crystal structures shared by the group III nitrides are wurtzite, zincblende, and rocksalt. At ambient conditions, the thermodynamically stable phase is wurtzite for bulk AlN, GaN, and InN. The cohesive energy per bond in wurtzite variety is 2.88 eV (63.5 kcal/mol), 2.20 eV (48.5 kcal/mol), and 1.93 eV (42.5 kcal/mol) for AlN, GaN, and InN, respectively [32.30]. Although the calculated energy difference $\Delta E_{\text{W–ZB}}$ between wurtzite and zincblende lattice is small (-18.41 meV/atom for AlN, -9.88 meV/atom for GaN, and -11.44 meV/atom for InN) [32.31] the wurtzite form is energetically preferable for all three nitrides compared to zincblende. The wurtzite structure has a hexagonal unit cell with two lattice parameters a and c in the ratio of $c/a = \sqrt{8/3} = 1.633$ and belongs to the space group of $P6_3mc$. The structure is composed of two interpenetrating hexagonal close-packed (hcp) sublattices, each of which consists of one type of atom displaced with respect to each other along the three-fold c-axis by an amount $u = 3/8 = 0.375$ in fractional coordinates. Each sublattice includes four atoms per unit cell and every atom of one kind (group III atom) is surrounded by four atoms of the other kind (nitrogen), or vice versa, these being coordinated at the edges of a tetrahedron. For actual nitrides, the wurtzite structure deviates from the ideal arrangement by changing the c/a ratio or the u value [32.31]. It should be pointed out that a strong correlation exists between the c/a ratio and the u parameter; when the c/a ratio decreases, the u parameter increases in such a way that those four tetrahedral distances remain nearly constant through a distortion of tetrahedral angles due to long-range polar interactions. These two slightly different bond lengths will be equal if the following relation holds;

$$u = \left(\frac{1}{3}\right)\left(\frac{a^2}{c^2}\right) + \frac{1}{4}. \tag{32.1}$$

Since the c/a ratio also correlates with the difference between the electronegativities of the two constituents, components with the greatest differences show largest departure from the ideal c/a ratio [32.32]. These two parameters were obtained experimentally by using the four-circle diffractometry technique. For GaN, the c/a ratio and the value of u are measured as 1.627 and 0.377, respectively, which are close to the ideal value [32.33]. AlN deviates significantly from the ideal parameters: $c/a = 1.601$ and $u = 0.382$. Consequently, the interatomic distance and angles differ by 0.01 Å and 3°, respectively. For InN, no reliable data are available due to the lack of single-crystal InN with a suitable size for single-crystal diffractometry measurement.

A phase transition to the rocksalt (NaCl) structure in group III nitrides takes place at very high external pressures. The reason for this is that the reduction of the lattice dimensions causes the inter-ionic Coulomb interaction to favor ionicity over the covalent nature. The structural phase transition was experimentally observed at the following pressure values: 22.9 GPa for AlN [32.34], 52.2 GPa for GaN [32.35], and 12.1 GPa for InN [32.36]. The space-group symmetry of the rocksalt type of structure is $Fm3m$, and the structure is six-fold coordinated. However, rocksalt group III nitrides cannot be stabilized by epitaxial growth.

The zincblende structure is metastable and can be stabilized only by heteroepitaxial growth on cubic substrates, such as cubic SiC [32.37], Si [32.38], MgO [32.39], and GaAs [32.40], reflecting topological compatibility to overcome the intrinsic tendency to form the wurtzite phase. In the case of highly mismatched substrates, there is usually a certain amount of zincblende phase of nitrides separated by crystallographic defects from the wurtzite phase. The symmetry of the zincblende structure is given by the space group $F\bar{4}3m$ and composed of two interpenetrating face-centered cubic (fcc) sublattices shifted by one quarter of a body diagonal. There are four atoms per unit cell and every atom of one type (group III nitrides) is tetrahedrally coordinated with four atoms of other type (nitrogen), and vice versa. The overall equivalent bond length is about 1.623 Å for zincblende structures.

Because of the tetrahedral coordination of wurtzite and zincblende structures, the four nearest neighbors and twelve next-nearest neighbors have the same bond distance in both structures. The main difference between these two structures lies in the stacking sequence of close-packed diatomic planes. The wurtzite structure consists of triangularly arranged alternating biatomic close-packed (0001) planes, for example Ga and N pairs, thus the stacking sequence of the (0001) plane is AaBbAaBb in the ⟨0001⟩ direction. In contrast, the zincblende structure consists of triangularly arranged atoms in the close-packed (111) planes along the ⟨111⟩ direction with a 60° rotation that causes a stacking order of AaBbCcAaBbCc. Small and large letters stand for the the two different kinds of constituents.

Since none of the three structures described above possess inversion symmetry, the crystal exhibits crystallographic polarity; close-packed (111) planes in the zincblende and rocksalt structures and the corresponding (0001) basal planes in the wurtzite structure differ from the ($\bar{1}\bar{1}\bar{1}$) and (000$\bar{1}$) planes, respectively. In general, group III (Al, Ga, or In)-terminated planes are denoted as (0001) A plane (referred to as Ga polarity) and group V (N)-terminated planes are designated as (000$\bar{1}$) B plane (referred to as N polarity). Many properties of the material also depend on its polarity, for example growth, etching, defect generation and plasticity, spontaneous polarization, and piezoelectricity. In wurtzite nitrides, besides the primary polar plane (0001) and associated direction ⟨0001⟩, which is the most commonly used surface and direction for growth, many other secondary planes and directions exist in the crystal structure.

32.2 Lattice Parameters of Nitrides

Like other semiconductors [32.41–43], the lattice parameters of nitride-based semiconductors depend on the following factors [32.44]:

1. free-electron concentration, acting via the deformation potential of a conduction-band minimum occupied by these electrons,
2. the concentration of foreign atoms and defects, and the difference between their ionic radii and the substituted matrix ion,
3. external strains (for example, those induced by substrate) and
4. temperature.

The lattice parameters of any crystalline materials are commonly and most accurately measured by high-resolution X-ray diffraction (HRXRD) usually at a standard temperature of 21 °C [32.45] by using the Bond method [32.46] for a set of symmetrical and asymmetrical reflections. In ternary compounds, the technique is also used for determining the composition, however, taking the strain into consideration is of crucial issues pertinent to heteroepitaxy. For nitrides, the composition can be determined with an accuracy of about 0.1% or less, down to a mole fraction of about 1%, by taking into account the elastic parameters of all nitrides and lattice parameters of AlN and InN. Since these factors may distort the lattice constants from their intrinsic values, there is wide dispersion in reported values. Table 32.1 shows a comparison of measured and calculated lattice parameters reported by several groups for AlN, GaN, and InN crystallized in the wurtzite structure.

AlN crystal has a molar mass of 20.495 g/mol when it crystallizes in the hexagonal wurtzite structure. The lattice parameters range from 3.110–3.113 Å for the a parameter and from 4.978–4.982 Å for the c parameter as reported. The c/a ratio thus varies between 1.600 and 1.602. The deviation from that of the ideal wurtzite crystal is probably due to lattice stability and ionicity. Although the cubic form of AlN is hard to obtain, several reports suggested the occurrence of a metastable zincblende polytype AlN with a lattice parameter of $a = 4.38$ Å [32.47], which is consistent with the theoretically estimated value [32.48]. The lattice parameter of a pressure-induced rocksalt phase of AlN is 4.043–4.045 Å at room temperature [32.49, 50].

GaN crystallized in the hexagonal wurtzite (WZ) structure with four atoms per cell and has a molecular weight of 83.7267 g/mol. At room temperature, the lattice parameters of WZ-GaN platelets prepared under high pressure and high temperatures with an electron concentration of 5×10^{19} cm^{-3} are $a = (3.1890 \pm 0.0003)$ Å and $c = (5.1864 \pm 0.0001)$ Å [32.44]. For GaN powder, a and c values are in the range 3.1893–3.190, and 5.1851–5.190 Å, respectively. It has been reported that free charge is the dominant factor responsible for expanding the lattice proportional to the deformation potential of the conduction-band minimum and inversely proportional to carrier density and bulk modulus. Point defects such as gallium antisites, nitrogen vacancies, and extended defects, such as threading dislocations, also increase the lattice constant of group III nitrides to a lower extent in the heteroepitax-

Table 32.1 Measured and calculated lattice constants of AlN, GaN and InN

Compound	Sample	a (Å)	c (Å)	Ref.
AlN	Bulk crystal	3.1106	4.9795	[32.53]
	Powder	3.1130	4.9816	[32.54]
	Epitaxial layer on SiC	3.110	4.980	[32.55]
	Pseudopotential LDA	3.06	4.91	[32.56]
	FP-LMTO LDA	3.084	4.948	[32.57]
GaN	Homoepitaxial layers [LFEC (low-free electron concentration)]	3.1885	5.1850	[32.58]
	Homoepitaxial layers [HFEC (high-free electron concentration)]	3.189	5.1864	[32.44]
	Relaxed layer on sapphire	3.1892	5.1850	[32.59]
	Powder	3.1893	5.1851	[32.54]
	Relaxed layer on sapphire	3.1878	5.1854	[32.60]
	GaN substrate	3.1896	5.1855	[32.61]
	Pseudotential LDA	3.162	5.142	[32.56]
	FP-LMTO LDA	3.17	5.13	[32.57]
InN	Powder	3.538	5.703	[32.62]
	Pseudopotential LDA	3.501	5.669	[32.56]
	FP-LMTO LDA	3.53	5.54	[32.57]

LDA: Local density approximation; FP-LMTO: pseudopotential linear muffin-tin orbital

ial layers [32.44]. For the zincblende polytype of GaN, the calculated lattice constant based on the measured Ga−N bond distance in WZ-GaN, is $a = 4.503$ Å while the measured values vary between 4.49 and 4.55 Å, indicating that the calculated result lies within acceptable limits [32.37], Si [32.38] MgO [32.39], and GaAs [32.40]. A high-pressure phase transition from the WZ to the rocksalt structure decreases the lattice constant down to $a_0 = 4.22$ Å in the rocksalt phase [32.51]. This is in agreement with the theoretical result of $a_0 = 4.098$ Å obtained from first-principles non-local pseudopotential calculations [32.52].

Due to the difficulties in synthesis and crystal growth, the number of experimental results concerning the physical properties of InN is quite small, and some have only been measured on non-ideal thin films, typically ordered polycrystalline with crystallites in the 50–500 nm range. Indium nitride normally crystallizes in the wurtzite (hexagonal) structure, like the other compounds of this family, and has a molecular weight of 128.827 g/mol. The measured lattice parameters using a powder technique are in the range of $a = 3.530$–3.548 Å and $c = 5.960$–5.704 Å with a consistent c/a ratio of about 1.615 ± 0.008. The ratio approaches the ideal value of 1.633 in samples having a low density of nitrogen vacancies [32.63]. Recently, Paszkowicz [32.64] reported basal and perpendicular lattice parameters of 3.5378 and 5.7033 Å, respectively, for wurtzite-type InN synthesized using a microwave plasma source of nitrogen, having a c/a ratio far from the ideal value. The single reported measurement yields a lattice constant of $a_0 = 4.98$ Å in the zincblende (cubic) form InN occurring in films containing both polytypes [32.63].

32.3 Mechanical Properties of Nitrides

The mechanical properties of materials involve various concepts such as hardness, stiffness constants, Young's and bulk modulus, yield strength, etc. However, the precise determination of the mechanical properties of the group III nitrides is hindered due to the lack of high-quality large single crystals. However, attempts to estimate and measure the mechanical properties of thin and thick (separated from substrate) epitaxial layers and bulk crystal of nitrides have been made repeatedly. It has been claimed that the most precise technique used to determine the elastic moduli of compound materials is ultrasonic measurement. Unfortunately, this ultrasonic pulse-echo method requires thick single crystalline samples, about 1 cm thick, to enable measurement of the timing of plane-wave acoustic pulses with sufficient resolution, which makes it almost inapplicable to the group III nitrides. As an optical technique, Brillouin scattering allows the determination of the elastic constants and hence of the bulk moduli through the interaction of light with thermal excitation in a material, in particular acoustic phonons in a crystal. Various forms of X-ray diffraction, such as energy dispersive X-ray diffraction (EDX), angular dispersive X-ray diffraction (ADX) and X-ray absorption spectroscopy (XAS) can also be employed to determine the pressure dependence of the lattice parameters. From these, the experimental equation of state (EOS), (a widely used one is Murnaghan's equation of state) and hence directly the bulk modulus, assuming that it has a linear dependence with the pressure P, can be deduced as [32.65]:

$$V = V_0 \left(1 + \frac{B'P}{B}\right)^{-\frac{1}{B'}}, \qquad (32.2)$$

where B and V_0 represent the bulk modulus and unit volume at ambient pressure, respectively, and B' is the derivative of B with respect to pressure. X-ray diffraction leads to the determination of the isothermal bulk modulus, whereas Brillouin scattering leads to the adiabatic bulk modulus. Nevertheless in solids other than molecular solids there is no measurable difference between these two thermodynamic quantities. Besides the experimental investigation many theoretical calculations have been performed on structural and mechanical properties of group III nitrides. Most of the calculations are based on density-functional theory within the local density approximation (LDA) using various types of exchange correlation functionals, and either plane-wave expansion for the pseudopotentials or the linear muffin-tin orbital (LMTO) method.

In hexagonal crystals, there exist five independent elastic constants, C_{11}, C_{33}, C_{12}, C_{13} and C_{44}. C_{11} and C_{33} correspond to longitudinal modes along the [1000] and [0001] directions, respectively. C_{44} and $C_{66} = (C_{11} - C_{12})/2$ can be determined from the speed of sound of transverse modes propagating along the [0001] and [1000] directions, respectively. The remaining constant, C_{13}, is present in combination with four other moduli in the velocity of modes propagating in less-symmetrical directions, such as [0011]. The bulk modulus is related to the elastic constants by [32.66]

$$B = \frac{(C_{11} + C_{12})C_{33} - 2C_{13}^2}{C_{11} + C_{12} + 2C_{33} - 4C_{13}}. \qquad (32.3)$$

In the isotropic approximation, the Young's modulus E and shear modulus G can also be evaluated using the relations $E = 3B(1 - 2\nu)$ and $G = E/2(1 + \nu)$, respectively. The term ν is the Poisson's ratio and is given by $\nu = C_{13}/(C_{11} + C_{12})$ [32.67].

The micro- and nanoindentation methods are widely used in the determination of the hardness of group III nitrides over a wide range of size scales and temperature. Hardness measurements are usually carried out on the (0001) surface of the crystal using a conventional pyramidal or spherical diamond tip, or alternatively, with a sharp triangular indenter (Berhovich). Depth-sensing indentation measurements provide complete information on the hardness and pressure-induced phase transformation of semiconductor materials. Table 32.2 shows the measured and calculated mechanical parameters reported by several groups for AlN, GaN and InN crystallized in wurtzite structure.

From the widely scattered experimental results presented in Table 32.2, the quality of the crystals is clearly one of the main problems for the precise determination of the physical properties of the group III nitrides. This is true especially for InN, where no elastic moduli could be measured, due to difficulties in synthesis and crystal growth. The difference between elastic moduli measured with the same technique (Brillouin scattering) in GaN is further proof that the quality and nature (bulk single crystal or epitaxial layer) of the samples is of primary importance. Nevertheless, with the notable exception of InN, group III nitrides can be considered as hard and incompressible material family members. Their elastic and bulk moduli are of the same order of magnitude as those of diamond. The hardness of semiconductors is often suggested to be dependent on the bonding distance or shear modulus. Indeed, the softest material InN has a smaller shear modulus and larger bonding distance (0.214 nm) compared to GaN (0.196 nm) and AlN (0.192 nm). The temperature dependence of the hardness shows that macroscopic dislocation motion and plastic deformation of GaN and AlN may start at around 1100 °C. The yield strength of bulk single-crystal GaN is found to be 100–300 MPa at 900 °C. The yield strength of AlN was deduced to be ≈ 300 MPa at 1000 °C [32.68].

Most applications of group III nitrides depend on the high thermal conductivity of the material, and a funda-

Table 32.2 Some mechanical properties of wurtzite AlN, GaN, and InN obtained by several experimental techniques and theoretical calculations. The units are in GPa

Parameters	AlN (GPa)	GaN (GPa)	InN (GPa)
C_{11}	345[a], 411[b], 396[c], 398[d]	296[o], 390[s], 377[t], 370[v], 373[w], 367[c], 396[d]	190[g1], 223[c], 271[d]
C_{12}	125[a], 149[b], 137[c], 140[d]	120[o], 145[s], 160[t], 145[v], 141[w], 135[c], 144[d]	104[g1], 115[c], 124[d]
C_{13}	120[a], 90[b], 108[c], 127[d]	158[o], 106[s], 114[t], 106[v], 80[w], 103[c], 100[d]	121[g1], 92[c], 94[d]
C_{33}	395[a], 389[b], 373[c], 382[d]	267[o], 398[s], 209[t], 398[v], 387[w], 405[c], 392[d]	182[g1], 224[c], 200[d]
C_{44}	118[a], 125[b], 116[c], 96[d]	24[o], 105[s], 81[t], 105[v], 94[w], 95[c], 91[d]	10[g1], 48[c], 46[d]
Poisson's ratio ν	0.287[f], 0.216[g]	0.38[f1], 0.372[o1]	
Bulk modulus B	201[e], 210[b], 208[h], 160[i], 207[c], 218[d]	195[o], 210[s], 245[y], 237[z], 188[a1], 202[c], 207[d]	139[g1], 125[z], 141[c], 147[d], 146[b1]
dB/dP	5.2[j], 6.3[k], 5.7[l], 3.74[m], 3.77[n]	4[y], 4.3[z], 3.2[a1], 4.5[b1], 2.9[c1]	12.7[z], 3.4[b1]
Young's modulus E	308[i], 295[e], 374[p]	150[o], 295[d1]	
Shear modulus	154[p], 131[i], 117[e]	121[r]	43[g1]
Yield strength σ_Y	0.3 at 1000 °C[r]	15[d1] 0.1–0.2 at 900 °C[e1]	
Hardness	Micro-hardness: 17.7[r] Nano-hardness: 18.0[r]	Micro-hardness: 10.2[r] Nano-hardness: 18–20[d1]	Nano-hardness: 11.2[h1]

[a] Ultrasonic measurement on thin film [32.69]; [b] Brillouin scattering on single crystal [32.70]; [c] Calculated using pseudopotential LDA [32.71]; [d] Calculated using FP-LMTO LDA [32.56]; [e] Ultrasonic measurement on thin-film AlN [32.72]; [f] {0001}, c-plane calculated [32.73]; [g] {112$\bar{}$0}, r-plane calculated [32.73]; [h] ADX on single-crystal AlN [32.74]; [i] Ultrasonic measurement on sintered, isotropic, polycrystalline AlN ceramic [32.75]; [j] Ultrasonic measurement on sintered, isotropic, polycrystalline AlN ceramic [32.75]; [k] ADX on single-crystal AlN [32.74]; [l] EDXD on polycrystalline AlN [32.49]; [m] Calculated using plane-wave pseudopotential [32.76]; [n] Calculated using Keating–Harrison model [32.77]; [o] Temperature-dependent X-ray diffraction on polycrystalline GaN [32.78]; [p] Hardness measurement on single-crystal AlN [32.67]; [r] Hardness measurement on bulk single-crystal AlN [32.67]; [s] Brillouin spectroscopy on bulk GaN [32.66]; [t] Resonance ultrasound method on GaN plate [32.79]; [v] Surface-acoustic-wave measurement on GaN grown on sapphire [32.80]; [w] Brillouin spectroscopy on GaN substrate grown by LEO [32.61]; [y] X-ray absorption spectroscopy on GaN [32.35]; [z] X-ray diffraction on bulk GaN [32.36]; [a1] EDX on bulk GaN [32.34]; [b1] Calculated using FP-LMTO [32.56]; [c1] Calculated using plane-wave pseudopotential [32.52]; [d1] Nanoindentation on bulk GaN [32.81]; [e1] Hardness on single-crystal GaN [32.82]; [f1] {0001}, c-plane using Bond's X-ray method on heteroepitaxially grown GaN [32.59]; [g1] Temperature-dependent X-ray measurements on powder InN [32.78]; [h1] Hardness measurement for InN grown on sapphire [32.83]; [o1] {0001}, c-plane estimated from elastic constants [32.78].

mental understanding of the thermal properties requires precise knowledge of the vibrational modes on the single crystal. Infrared reflection and Raman spectroscopies have been employed to derive zone-center and some zone-boundary phonon modes in nitrides. The A_1 and E_1 branches are both Raman- and infrared-active, the E_2 branches are Raman-active only, and the B_1 branches are inactive. The A_1 and E_1 modes are each split into longitudinal optic (LO) and transverse optic (TO) components, giving a total of six Raman peaks. Table 32.3 gives a list of observed zone-center optical-phonon wave numbers along with those calculated from several techniques employed for AlN, GaN, and InN.

The phonon dispersion spectrum of AlN has a total of twelve branches: three acoustic and nine optical. Perlin et al. measured the effect of pressure on the Raman shift of a single-crystal AlN sample synthesized at high pressure and high temperature and showed that the pressure dependence of the three observed peaks could be fitted to a quadratic law up to 14 GPa [32.84]. McNeil et al. reported the complete set of Raman-active phonon modes of AlN on single crystals grown by the sublimation recondensation method and noted that Raman peaks and widths are influenced by oxygen-related defects [32.70]. Recently, vibrational properties of epitaxial AlN [deposited on silicon and sapphire substrates at ≈ 325 K by ion-beam-assisted deposition (IBAD)]

Table 32.3 Optical phonon frequencies of wurtzite AlN, GaN, and InN at the center of the Brillouin zone in units of cm^{-1}

Symmetry	AlN (cm^{-1})	GaN (cm^{-1})	InN (cm^{-1})
A_1-TO	614[a], 667[b], 607[c], 612[d], 601[e]	533[f], 531[g], 544[d], 534[e], 533[h]	480[h], 445[i], 440[i]
E_1-TO	673[a], 667[b], 679[d], 650[e]	561[f], 558[g], 566[d], 556[e], 559[h]	476[h], 472[i], 472[i]
A_1-LO	893[a], 910[b]	735[f], 733[g], 737[j]	580[h], 588[i]
E_1-LO	916[a], 910[b], 924[c],	743[f], 740[g], 745[j]	570[h]
E_2-(low)	252[a], 241[c], 247[d], 228[e]	144[f], 144[g], 185[d], 146[e]	87[h], 104[i]
E_2-(high)	660[a], 660[b], 665[c], 672[d], 638[e]	569[f], 567[g], 557[d], 560[e]	488[h], 488[i], 483[i]
B_1-(low)	636[d], 534[e]	526[d], 335[e]	200[h], 270[i]
B_1-(high)	645[d], 703[e]	584[d], 697[e]	540[h], 530[i]

[a] Raman scattering on sublimation recondensation AlN [32.70]
[b] Raman scattering on whisker AlN [32.87]
[c] Raman scattering on synthesized AlN by [32.84]
[d] Calculated using first-principle total energy [32.88]
[e] Calculated using pseudopotential LDA [32.89]
[f] Raman scattering on bulk GaN [32.90]
[g] Raman scattering on GaN substrate grown by LEO [32.61]
[h] Raman study on InN grown on sapphire and calculation based on the pairwise interatomic potentials and rigid-ion Coulomb interaction [32.91]
[i] Raman study on polycrystalline and faceted platelets of InN and calculation using FP-LMTO LDA by [32.92]
[j] Raman study on high-quality freestanding GaN templates grown by HVPE [32.93]

have been investigated by *Ribeiro* et al. [32.85]. Raman scattering measurements revealed interesting features and they argued that, due to the extremely weak Raman signal usually exhibited by AlN films, misidentification of some vibration modes can lead to incorrect interpretation of the crystalline quality of AlN films, in which some of the previous mentioned features have been erroneously ascribed [32.86].

The wide spread of studied material has led to some uncertainty in phonon frequencies in GaN, especially of the LO modes. Coupling to plasmons in highly doped material and stress-induced effects due to lattice mismatch with the substrate might play a role in interpretation of the observed phonon frequencies. Moreover, the strong bond in GaN and the light N atoms result in high phonon frequencies that limit the range of observable impurity-related local vibrational modes to even lighter elements at higher frequencies. So far, few reports have appeared for the infrared and Raman modes, which have been associated with local vibrational modes of impurities, dopants, and hydrogen complexes [32.94, 95]. The hydrostatic pressure dependence of the zone-center phonon modes has also been determined in n-type bulk GaN [32.35] and low-residual-doping GaN grown on sapphire [32.96]. The first- and second-order pressure coefficients of the phonon modes have been derived using a polynomial fit.

Raman and infrared spectroscopy studies in InN samples, some grown on (0001) and some on ($1\bar{1}02$) sapphire substrates, have been undertaken. The infrared (IR) data for A_1(TO) at 448 cm^{-1} and E_1(TO) at 476 cm^{-1} correlate well with the Raman measurements. All six Raman-active modes in the spectra of InN have been observed, with five of them appearing in one InN sample grown on ($1\bar{1}02$) sapphire. *Kaczmarczyk* et al. [32.97] also studied the first- and second-order Raman-scattering of both hexagonal and cubic InN grown on GaN and GaAs, respectively, covering the acoustic and optical phonon and overtone region. They obtained good agreement with a theoretical model developed by using a modified valence force. The high quality of the samples gives credence to the Raman frequency values as also evidenced by the narrow line widths of all the Raman lines [6.2 cm^{-1} for E_2 (high) and 11.6 cm^{-1} for A_1(LO)]. Details of all the modes are shown in Table 32.3.

32.4 Thermal Properties of Nitrides

32.4.1 Thermal Expansion Coefficients

The lattice parameters of semiconductors are temperature dependent and quantified by thermal expansion coefficients (TEC), which are defined as $\Delta a/a$ or α_a and $\Delta c/c$ or α_c, in and out of plane, respectively. They are dependent on stoichiometry, the presence of extended defects, as well as the free-carrier concentration. As in the case of the lattice parameter, a large scatter in the published data exists for the TEC, particularly for nitrides as they are grown on foreign substrates with different thermal and mechanical properties.

The temperature dependence of the lattice constants a and c, and the thermal expansion coefficients of hexagonal AlN parallel (α_c) and perpendicular (α_a) to the c-axis are shown Fig. 32.1, which can be fitted by the following polynomials within the temperature range $293 < T < 1700$ K;

$$\Delta a/a_0 = -8.679 \times 10^{-2} + 1.929 \times 10^{-4} T \\ + 3.400 \times 10^{-7} T^2 - 7.969 \times 10^{-11} T^3$$

and

$$\Delta c/c_0 = -7.006 \times 10^{-2} + 1.583 \times 10^{-4} T \\ + 2.719 \times 10^{-7} T^2 - 5.834 \times 10^{-11} T^3 .$$
(32.4)

Using X-ray techniques across a broad temperature range (77–1269 K), it has been noted that the thermal expansion of AlN is isotropic with a room-temperature value of 2.56×10^{-6} K^{-1} [32.100]. The thermal expansion coefficients of AlN measured by Yim and Paff have mean values of $\Delta a/a = 4.2 \times 10^{-6}$ K^{-1} and $\Delta c/c = 5.3 \times 10^{-6}$ K^{-1} [32.101]. For AlN powder, the expansion coefficients of 2.9×10^{-6} K^{-1} and 3.4×10^{-6} K^{-1} has been reported for the a and c parameters, respectively [32.10].

Thermal expansion of single-crystal wurtzite GaN has been studied in the temperature range 300–900 K [32.102] and 80–820 K [32.103]. Maruska and Tietjen reported that the lattice constant a changes linearly with temperature with a mean coefficient of thermal expansion of $\Delta a/a = \alpha_a = 5.59 \times 10^{-6}$ K^{-1}. Meanwhile, the expansion of the lattice constant

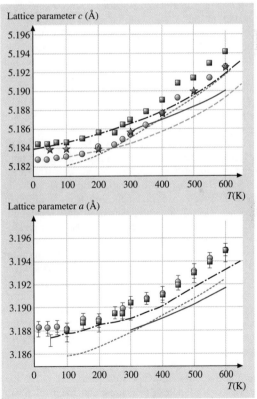

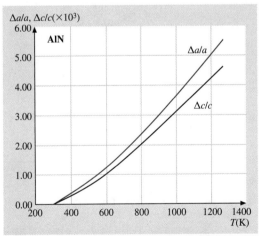

Fig. 32.1 Variation of the thermal expansion coefficient of AlN on temperature, in and out of the c-plane. After [32.98]

Fig. 32.2 Lattice parameters c and a of a homoepitaxially grown GaN layer, the corresponding substrate and a Mg-doped bulk crystal as a function of temperature in comparison with literature data. After [32.99]

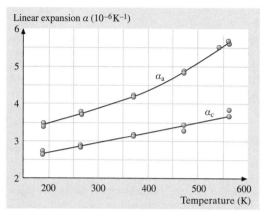

Fig. 32.3 The temperature dependence of linear expansion coefficients in InN. After [32.103]

c shows a superlinear dependence on temperature parallel to the c-axis with a mean coefficients of thermal expansion of $\Delta c/c = \alpha_c = 3.17 \times 10^{-6}\,\mathrm{K}^{-1}$ and $7.75 \times 10^{-6}\,\mathrm{K}^{-1}$, over the temperature ranges 300–700 K and 700–900 K, respectively. Sheleg and Savastenko reported a thermal expansion coefficient near 600 K of $(4.52 \pm 0.5) \times 10^{-6}\,\mathrm{K}^{-1}$ and $(5.25 \pm 0.05) \times 10^{-6}\,\mathrm{K}^{-1}$ for the perpendicular and parallel directions, respectively. Leszczynski and Walker reported α_a values of 3.1×10^{-6} and $6.2 \times 10^{-6}\,\mathrm{K}^{-1}$, for the temperature ranges of 300–350 K and 700–750 K, respectively [32.104]. The α_c values in the same temperature ranges, were 2.8×10^{-6} and $6.1 \times 10^{-6}\,\mathrm{K}^{-1}$, respectively. Being grown on various substrates with different thermal expansion coefficients leads to different dependencies of the lattice parameter on temperature. The temperature dependence of the GaN lattice parameter has been measured for a bulk crystal (grown at high pressure) with a high free-electron concentration ($5 \times 10^{19}\,\mathrm{cm}^{-3}$), a slightly strained homoepitaxial layer with a low free-electron concentration (about $10^{17}\,\mathrm{cm}^{3}$), and a heteroepitaxial layer (also with a small electron concentration) on sapphire. Figure 32.2 shows the lattice parameters of various GaN samples along with the theoretical calculation as a function of temperature for comparison.

The linear thermal expansion coefficients measured at five different temperatures between 190 K and 560 K [32.103] indicate that, along both the parallel and perpendicular directions to the c-axis of InN, these coefficients increase with increasing temperature, as shown in Fig. 32.3. The values range from $3.40 \times 10^{-6}\,\mathrm{K}^{-1}$ at 190 K to $5.70 \times 10^{-6}\,\mathrm{K}^{-1}$ at 560 K and $2.70 \times 10^{-6}\,\mathrm{K}^{-1}$ at 190 K to $3.70 \times 10^{-6}\,\mathrm{K}^{-1}$ at 560 K for the perpendicular and parallel directions, respectively.

32.4.2 Thermal Conductivity

GaN and other group III nitride semiconductors are considered for high-power/high-temperature electronic and optoelectronic devices where a key issue is thermal dissipation. Consequently, the thermal conductivity (κ), which is the kinetic property determined by the contributions from the vibrational, rotational, and electronic degrees of freedom, is an extremely important material property. The heat transport is predominantly determined by phonon–phonon Umklapp scattering, and phonon scattering by point and extended defects, such as vacancies (including lattice distortions caused by them), impurities such as oxygen, and isotope fluctuations (mass fluctuation). For pure crystals, phonon–phonon scattering is the limiting process. For most groups III nitrides, due to their imperfection, point defects play a significant role for single crystals, .

The thermal conductivity κ of AlN at room temperature has been theoretically estimated to be 3.19 W/cmK for pure AlN single crystals [32.106]. Values of κ meas-

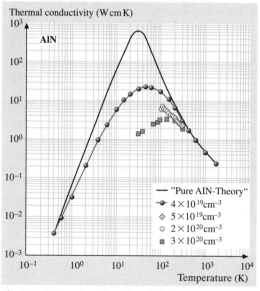

Fig. 32.4 Thermal conductivity of single-crystal AlN. The *solid line* alone indicates the theory whereas the others represent measurements of AlN with various concentrations of O. The lower the O concentration, the higher the thermal conductivity. After [32.105]

ured at 300 K are 2.5 [32.107] and 2.85 W/cmK [32.100] for single crystals obtained by the sublimation technique, while measurements made on polycrystalline specimens generally yield lower values (reported maximum value of $\kappa = 2.2$ W/cmK) [32.108], due to enhanced phonon O-related point defects and grain-boundary scattering. A more recent prediction of 5.4 W/cmK for the thermal conductivity of AlN is much larger than the value measured by *Witek* [32.109]. The measured thermal conductivity as a function of temperature in bulk AlN containing some amount of O is plotted in Fig. 32.4. Also shown is a series of samples with estimated concentrations of O, indicating an overall reduction in the thermal conductivity of AlN with O contamination. In the temperature range of interest, where many of the devices would operate, the thermal conductivity in the sample containing the least amount of O assumes a $T^{-1.25}$ dependence. Recently, even higher thermal conductivity values in the range 3.0–3.3 W/cmK were also reported by [32.110] and [32.111] for freestanding and 300–800 μm-thick AlN samples grown by HVPE originally on Si (111) substrates with a 10^8 cm^{-2} dislocation density.

The thermal conductivity of GaN layers grown on sapphire substrates by HVPE were measured by [32.112] as a function of temperature (25–360 K) using a heat-flow method. The room-temperature thermal conductivity was measured as $\kappa \cong 1.3$ W/cmK along the c-axis. This value is slightly smaller than the intrinsic value of 1.7 W/cmK for pure GaN predicted by [32.106] and much smaller than the $\kappa \approx 4.10$ W/cmK calculated by [32.109]. More recently, the room-temperature thermal conductivity of GaN was computed as $\kappa = 2.27$ W/cmK, assuming that there is no isotope scattering [32.105]. This prediction is very close to the measured value of 2.3 W/cmK at room temperature, which increases to over 10 W/cmK at 77 K, in high-quality freestanding GaN samples using a steady-state four-probe method. As can be seen in Fig. 32.5, the measured thermal conductivity of GaN in the temperature range 80–300 K has a $T^{-1.22}$ temperature dependence. This slope is typical of pure adamantine crystals below the Debye temperature, indicating acoustic phonon transport where the phonon–phonon scattering is a combination of acoustic–acoustic and acoustic–optic interactions. This temperature dependence strongly suggests that the thermal conductivity depends mainly on intrinsic phonon–phonon scattering and not on phonon–impurity scattering.

A newer method, scanning thermal microscopy (SThM), has been applied to the measurement of room-temperature thermal conductivity on both fully and partially coalesced lateral epitaxial overgrown (LEO) GaN/sapphire (0001) samples. A correlation between low threading-dislocation density and high thermal conductivity values was established. The reduction in the thermal conductivity with increased dislocation density is expected as threading dislocations degrade the velocity of sound and increase phonon scattering in the material. The highest GaN κ values using this method, in the 2.0–2.1 W/cmK range, were found in the regions of the samples that were laterally grown and thus contained a low density of threading dislocations. This compares with a value of 2.3 W/cmK in a freestanding sample measured by the steady-state four-probe method discussed earlier. An explanation for the dramatic increase from to $\kappa \approx 1.3$ W/cmK for the early samples to 2.3 W/cmK for the freestanding sample may be related to the extended-defect concentration (D_d) and the differences in background doping. The effect of dislocation density on the thermal conductivity has been calculated [32.113, 114] by showing that κ remains fairly independent of D_d up to some characteristic value D_d^{char}, after which it decreases by about a factor of two for ev-

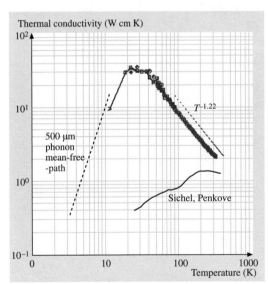

Fig. 32.5 The thermal conductivity of a 200-μm-thick freestanding GaN sample as a function of temperature. The *dashed line* indicates a calculation using the boundary scattering limit for a phonon mean free path of 500 μm. Also shown is the $T^{-1.22}$ dependence in the range 80–300 K, and earlier results from [32.112] measured using a 400-μm HVPE sample. After [32.105]

ery decade increase in D_d. The thermal conductivity has also been correlated to doping levels in hydride vapor-phase epitaxy (HVPE) n-GaN/sapphire (0001) by SThM on two sets of samples [32.115, 116]. In both sets of data the thermal conductivity decreased linearly with log n, where n is the electron concentration, the variation being about a factor of two decrease in κ for every decade increase in n.

InN single crystals of a size suitable for thermal conductivity measurements have not been obtained. The only measurement of thermal conductivity, which has a room temperature value of $\kappa = 0.45$ W/cmK, was made on InN ceramics by using the laser-flash method [32.118]. This value is much lower than the thermal conductivity data obtained from the Leibfried–Schloman scaling parameter [32.117]. A value of about (0.8 ± 0.2) W/cmK is predicted assuming that the thermal conductivity is limited by intrinsic phonon–phonon scattering, but this value may be reduced by oxygen contamination and phonon scattering at defects or increased by very high electronic concentrations.

32.4.3 Specific Heat

The specific heat of a semiconductor has contributions from lattice vibrations, free carriers, and point and extended defects. For good-quality semi-insulating crystal, it is only determined by the lattice vibrations. Due to the lack of defect-free crystals of group III nitrides the specific heat measurements are affected by contributions from free carriers and defects, especially at low temperatures. The specific heat C_p of AlN in the temperature interval 298–1800 K has been approximated [32.119] using the expression

$$C_p = 45.94 + 3.347 \times 10^{-3} T \\ - 14.98 \times 10^{-5} T^2 \text{ J/(molK)}. \quad (32.5)$$

For the higher temperature range 1800–2700 K, an approximation [32.120] using the specific heat of $C_p = 51.5$ J/(molK) at $T = 1800$ K, and the estimated value of $C_p = 58.6$ J/(molK) at $T = 2700$ K, yielded

$$C_p = 37.34 + 7.86 \times 10^{-3} T \text{ J/(molK)}. \quad (32.6)$$

However, since the free electrons (very effective at low temperatures), impurities, defects (inclusive of point defects), and lattice vibrations contribute to the specific heat, these expressions are very simplistic. The Debye expression for the temperature dependence of the specific heat in a solid at constant pressure (C_p) can be expressed as

$$C_p = 18R \left(\frac{T}{\theta_D}\right)^3 \cdot \int_0^{x_D} \frac{x^4 e^x}{(e^x - 1)^2} dx, \quad (32.7)$$

where $x_D \equiv \theta_D/T$, and $R = 8.3144$ J/(mol K) is the molar gas constant. The coefficient in front of the term R has been multiplied by 2 to take into account the two constituents making up the group III nitrides. By fitting the measured temperature-dependent heat capacity to the Debye expression, one can obtain the Debye temperature θ_D related to the heat capacity.

The specific heat obtained from the above approximations coupled with the measured values for constant pressure from the literature are shown in Fig. 32.6 along with the calculated specific heat using the Debye equation for Debye temperature values of 800–1100 K with 50 K increments. The best fit between the data and Debye specific-heat expression for insulators indicates a Debye temperature of 1000 K, which is in good agreement with the value of 950 K reported in [32.117].

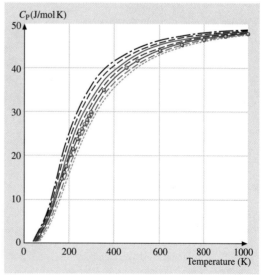

Fig. 32.6 Molar specific heat at constant pressure C_p of AlN versus temperature. *Open circles* represent the experimental data. The *solid lines* are calculation based on the Debye model for Debye temperatures θ_D in the range 800–1100 K with 50 K increments. The data can be fitted with the Debye expression for $\theta_D = 1000$ K, which compares with the value 950 K reported by Slack et al. The data are taken from [32.117]

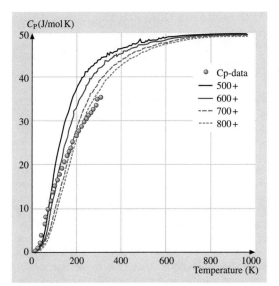

Fig. 32.7 Molar specific heat at constant pressure C_p, of GaN versus temperature. *Open circles* represent the experimental data. The *solid lines* are calculation based on the Debye model for Debye temperatures of θ_D of 500, 600, 700, and 800 K. Unfortunately it is difficult to discern a Debye temperature that is effective over a large temperature range because of the high concentration of defects and impurities in GaN used. The data are taken from [32.122]

To a first approximation the temperature dependence of the specific heat of WZ-GaN at constant pressure (C_p) can be expressed by the following phenomenological expression [32.121]

$$C_p = 9.1 + 2.15 \times 10^{-3} T \text{ cal/(molK)}. \quad (32.8)$$

The specific heat of WZ-GaN has been studied in the temperature range 5–60 K [32.122] and also in the temperature range 55–300 K [32.124], and is discussed in [32.123].

The experimental data from [32.122] and [32.124] are plotted in Fig. 32.7. Also shown in the figure is the calculated specific heat using the Debye expression for Debye temperatures of 500, 600, 700 and 800 K. It is clear that the quality of the data and or sample prevent attainment of a good fit between the experimental data and the Debye curve. Consequently, a Debye temperature with sufficient accuracy cannot be determined. It is easier to extract a Debye temperature using data either near very low temperatures or well below the Debye temperature where the specific heat has a simple cubic

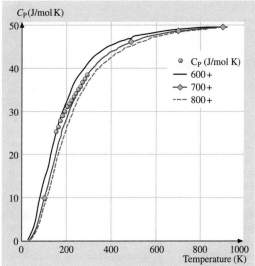

Fig. 32.8 The specific heat of InN with experimental data points, albeit over a small range, and the 600, 700, and 800 K Debye temperature fits. After [32.123]

dependence on temperature [32.125]

$$C_p = 234R \left(\frac{T}{\theta_D}\right)^3. \quad (32.9)$$

Unfortunately, for GaN the samples contain large densities of free carriers and defects, which compromise the application of the Debye specific-heat expression. Consequently, a good fit to the data is not obtained and the Debye temperature extracted in this way is not dependable. Compared to GaN, the Debye temperature obtained in this way for AlN appears more dependable due to a much better fit.

Using the Debye expression for the specific heat and these data, the Debye temperature of InN was obtained as 660 K. The resulting specific-heat curve and the experimental data are plotted in Fig. 32.8. Since the temperature range of these measurements is rather narrow, it is difficult to compare these results and the Debye curve. Good-quality pure InN crystals are extremely difficult to grow and deviations from the Debye curve indicate that the InN samples have significant contributions from non vibrational modes. Using the calculated phonon density of states, the lattice specific heat of InN at constant volume C_v was calculated [32.126] and compared to the experimental values of *Krukowski* et al. [32.123]. This exercise led to the determination of the Debye temperature

θ_D as a function of temperature following the approach of [32.125]. The estimates for θ_D are about 670 K, 580 K, and 370 K at 300 K, 150 K and 0 K, respectively.

32.5 Electrical Properties of Nitrides

GaN and related nitrides being direct and large-band-gap materials lend themselves to a variety of electronic and optoelectronic applications. Advantages associated with a large band gap include higher breakdown voltages, ability to sustain large electric fields, lower noise generation, and high-temperature and high-power operation. Small effective masses in the conduction band minimum lead to reasonably low field mobility, higher satellite-energy separation, and high phonon frequency. Their excellent thermal conductivity, large electrical breakdown fields, and resistance to hostile environments also support the group III nitrides as a material of choice for such applications. The electron transport in semiconductors, including nitrides, can be considered at low and high electric field conditions.

1. At sufficiently low electric fields, the energy gained by the electrons from the applied electric field is small compared to their thermal energy, and therefore the energy distribution of the electrons is unaffected by such a low electric field. Since the scattering rates determining the electron mobility depend on the electron distribution function, electron mobility remains independent of the applied electric field, and Ohm's law is obeyed.
2. When the electric field is increased to a point where the energy gained by electrons from the external field is no longer negligible compared to the thermal energy of the electron, the electron distribution function changes significantly from its equilibrium value. These electrons become hot electrons characterized by an electron temperature larger than the lattice temperature. Furthermore, as the dimensions of the device are shrunk to submicron range, transient transport occurs when there is minimal or no energy loss to the lattice. The transient transport is characterized by the onset of ballistic or velocity-overshoot phenomenon. Since the electron drift velocity is higher than its steady-state value one can design a device operating at higher frequency.

32.5.1 Low-Field Transport

The Hall effect is the most widely used technique to measure the transport properties and assess the quality of epitaxial layers. For semiconductor materials, it yields the carrier concentration, its type, and carrier mobility. More specifically, experimental data on Hall measurements over a wide temperature range (4.2–300 K) give information on impurities, imperfections, uniformity, scattering mechanism, etc. The Hall coefficient and resistivity are experimentally determined and then related to the electrical parameters through $R_H = r_H/ne$ and $\mu_H = R_H/\rho$, where n is the free-carrier concentration, e is the unit of electrical charge, μ_H is the Hall mobility, and r_H is the Hall scattering factor. The drift mobility is the average velocity per unit electric field in the limit of zero electric field and is related to the Hall mobility through the Hall scattering factor by $\mu_H = r_H \mu$. The Hall scattering factor depends on the details of the scattering mechanism, which limits the drift velocity. As the carriers travel through a semiconductor, they encounter various scattering mechanisms that govern the carrier mobility in the electronic system. The parameter for characterizing the various scattering mechanisms is the relaxation time τ, which determines the rate of change in electron momentum as it moves about in the semiconductor crystal. Mobility is related to the scattering time by

$$\mu = \frac{q\langle\tau\rangle}{m^*}, \qquad (32.10)$$

where μ^* is the electron effective mass, q is the electronic charge, and $\langle\tau\rangle$ is the relaxation time averaged over the energy distribution of electrons. The total relaxation time, τ_T when various scattering mechanisms are operative is given by Matthiessen's rule

$$\frac{1}{\tau} = \sum_i \frac{1}{\tau_i}, \qquad (32.11)$$

where i represents each scattering process. The major scattering mechanisms that generally governs the electron transport in group III–V semiconductors is also valid for group III nitrides. They are briefly listed as follows:

1. Ionized-impurity scattering is due to the deflection of free carriers by the long-range Coulomb potential of the charged centers caused by defects or intentionally doped impurities. This can be considered as

a local perturbation of the band edge, which affects the electron motion.
2. Polar longitudinal-optical (LO) phonon scattering is caused by the interaction of a moving charge with the electric field induced by electric polarization associated with lattice vibration due to the ionic nature of the bonds in polar semiconductors such as nitrides.
3. Acoustic phonon scattering through the deformation potential arises from the energy change of the band edges induced by strain associated with acoustic phonons, where the scattering rate increases with the wave vectors of the phonons.
4. Piezoelectric scattering arises from the electric fields that are produced by the strain associated with phonons in a crystal without inversion symmetry, particularly in wide-band-gap nitrides.
5. Because of the high density of dislocations and native defects induced by nitrogen vacancies in GaN, dislocation scattering and scattering through nitrogen vacancies has also been considered as a possible scattering mechanism. Dislocation scattering is due to the fact that acceptor centers are introduced along the dislocation line, which capture electrons from the conduction band in an n-type semiconductor. The dislocation lines become negatively charged and a space-charge region is formed around it, which scatters electrons traveling across the dislocations, thus reducing the mobility.

Gallium Nitride

Electron mobility in GaN is one of the most important parameters associated with the material, with a great impact on devices. It has been the subject of intensive studies in recent years from both the experimental and theoretical points of view. Experimental investigation of the temperature-dependent carrier mobility and concentration can be used to determine the fundamental material parameters and understand the carrier scattering mechanism along with an accurate comparison with theory [32.127, 128]. Compared to other group III–V semiconductors, such as GaAs, GaN possesses many unique material and physical properties, as discussed in the previous section. However, the lack of high-quality material, until very recently, prevented detailed investigations of carrier transport. The earlier transport investigations had to cope with poor crystal quality and low carrier mobility, well below predictions [32.129, 130]. Early MBE layers exhibited mobilities as high as $580\,\text{cm}^2/\text{Vs}$ on SiC substrates, which at that time were not as commonly used as in recent times [32.131]. Typically, however, MBE-grown films produce much lower mobility values of $100–300\,\text{cm}^2/\text{Vs}$ [32.132]. Different models were used to explain the observed low electron mobilities in GaN, especially at low temperatures. Scattering of electrons at charged dislocation lines [32.132–136] and scattering through elevated levels of point defects [32.137, 138], such as nitrogen vacancies [32.139, 140] were considered as a possible mechanisms responsible for these observations. These scattering mechanisms were investigated by studying the temperature dependence of the carrier concentration and electron mobility. It has been argued that mobility is related to the dislocation density (N_{dis}) and free-carrier concentration (n) via a $\mu_{\text{dis}} \propto \sqrt{n}/N_{\text{dis}}$ relationship [32.135]. At low carrier concentrations ($< 5 \times 10^{17}\,\text{cm}^{-3}$), the mobility decreases due to charged dislocation scattering, while at higher carrier concentrations ionized impurities are the dominant mechanism determining the mobility. The temperature dependence of the mobility for samples where dislocations play a dominant role shows that the mobility increases monotonically with temperature, following a $T^{2/3}$ dependence. Electron mobility limited by nitrogen-vacancy scattering was taken into account in n-type GaN grown by MOVPE by *Zhu* and *Sawaki* [32.139] and *Chen* et al. [32.140]. A good fit was obtained between the calculated and experimental results. The estimated mobility shows a $T^{-2/2}$ temperature dependence and it was argued that the measured mobility is dominated by ionized-impurity and dislocation scattering at low temperatures, but polar optical phonon and nitrogen-vacancy scattering at high temperatures.

Hall mobility and electron concentration in undoped GaN were investigated as a function of the thickness of buffer layers and epilayers. *Nakamura* [32.141] made Hall-effect measurements on undoped GaN layers grown by MOVPE on GaN buffer/sapphire substrates. As the thickness of the buffer layer increased from 100 Å, the mobility also increased up to a thickness of 200 Å. At larger thicknesses the mobility began to decrease. The value of the mobility was $600\,\text{cm}^2/\text{Vs}$ at room temperature for a 200 Å thick layer. The electron concentration was a minimum at 520 nm and increased monotonically as the buffer thickness increased. The mechanism responsible for this observation was not clearly established. *Götz* et al. [32.142] have studied the effect of the layer thickness on the Hall mobility and electron concentration in unintentionally doped n-type GaN films grown by HVPE on sapphire substrates, which is pretreated with either ZnO or GaCl. They found that the mobility increased and the carrier concentration decreased as the thickness of the epilay-

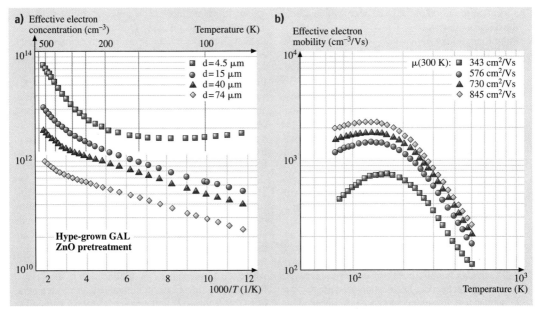

Fig. 32.9a,b Effective electron concentration versus reciprocal temperature (**a**) and effective electron mobility versus temperature (**b**), as determined from Hall-effect measurements under the assumption of uniform film properties. After [32.142]

ers increased, as shown in Fig. 32.9. They related this variation to a nonuniform distribution of electrically active defects through the film thickness for both types of films. For GaCl-pretreated sapphire the presence of a highly conductive, 200-nm-thick near-interface layer was assumed to account for the observed phenomena. For ZnO-pretreated sapphire, the Hall-effect data indicated a continuous reduction of the defect density with increasing film thickness.

Since GaN layers are often grown on foreign substrates with very different properties, a degenerate layer forms at the interface caused by extended defects and impurities. Experiments show that, even for thick GaN grown by HVPE, the degenerate interfacial layer makes an important contribution to the Hall conductivity. *Look* and *Molnar* [32.143] investigated the Hall effect in the temperature range 10–400 K in HVPE-grown layers on sapphire substrate by assuming a thin, degenerate n-type region at the GaN/sapphire interface. This degenerate interfacial region dominates the electrical properties below 30 K, but also significantly affects those properties even at 400 K, and can cause a second, deeper donor to falsely appear in the analysis. The curve of mobility versus temperature is also affected in that the whole curve is shifted downward from the true, bulk curve. A model consisting of two layers was constructed to interpret these observations and the result is shown in Fig. 32.10.

Hall mobilities for electron concentrations in the range $\approx 10^{16}$–10^{19} cm^{-3} for undoped and intentionally doped (the commonly used donors Si and Ge, which substitute for Ga, are shallow donors with almost identical activation energies for ionization) GaN layers grown by different growth techniques, Fig. 32.11. The most clearly observed trend is that the mobility shows no signs of leveling down to the lowest carrier concentration reported, and the mobility values are practically the same, irrespective of the growth techniques and dopants used, which reflects transport properties inherent to GaN, not to the extrinsic effects. The room-temperature mobility measured by *Nakamura* et al. [32.144] is 600 cm^2/Vs for an electron concentration of $\approx 3 \times 10^{16}$ cm^{-3}, and decreases slowly with increasing carrier concentration, reaching a value of about 100 cm^2/Vs at a carrier concentration of 3×10^{18} cm^{-3}. For an electron concentration of 1×10^{18} cm^{-3}, the mobility is 250 cm^2/Vs. A higher room-temperature mobility of 845 cm^2/Vs at an electron concentration of $\approx 6 \times 10^{16}$ cm^{-3} was achieved by *Götz* et al. [32.142]. In later publications, due to the advent of high-quality samples grown by several

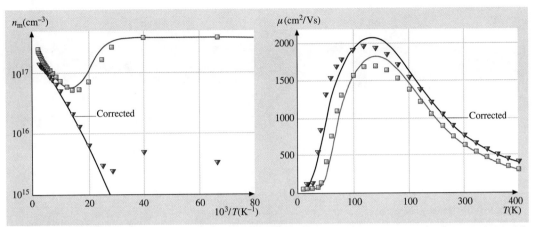

Fig. 32.10 (a) Uncorrected Hall-concentration data (*squares*) and fit (*solid line*), and corrected data (*triangles*) and fit (*dashed line*) versus inverse temperature. (b) Uncorrected Hall-mobility data (*squares*) and fit (*solid line*), and corrected data (*triangles*) and fit (*dashed line*) versus temperature. After [32.134]

growth techniques, there is a significant improvement in reported room-temperature mobility values. *Nakamura* et al. [32.145] and *Binari* and *Dietrich* [32.146] were able to obtain 900 cm^2/Vs room-temperature mobility at an electron concentration of 3×10^{16} cm^{-3} and 5×10^{16} cm^{-3}, respectively. Recently, even higher electron mobilities of 1100 cm^2/Vs at room temperature and 1425 cm^2/Vs at 273 K were reported in [32.93, 147] for a 200-μm-thick freestanding n-type GaN template grown by HVPE. This achievement was attributed to the excellent crystalline structure of the GaN sample with low levels of compensation and the defect-related scattering. A quantitative comparison with theoretical calculations demonstrates that the one-layer and one-donor conductance model is sufficient to account for the measured data in the entire temperature range without considering any dislocation scattering and any adjustable parameter other than the acceptor concentration. The measured temperature-dependent Hall mobility, carrier concentration and Hall scattering factor is shown in Fig. 32.12 along with the best-fit theoretical calculation based on an iterative solution of the Boltzmann equation. As shown, quantitative agreement with the measured mobility in the entire temperature range was obtained to within about 30%. *Heying* et al. [32.148] investigated both the morphology and electrical properties of homoepitaxial GaN layers grown by MBE as a function of Ga/N ratio. GaN films grown with higher Ga/N ratios (intermediate regime) showed fewer pits with areas of atomically flat surface, which gives the highest mobility ≈ 1191 cm^2/Vs reported so far at room tem-

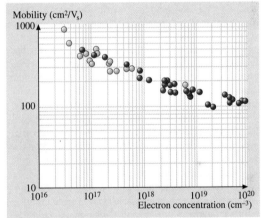

Fig. 32.11 The 300 K Hall mobility versus free electron concentration for GaN from various groups using both MOVPE and MVE. The *open circles* are from unintentionally doped samples and the *solid circles* are from samples doped with either Si or Ge. After [32.150]

perature. *Koleske* et al. [32.149] investigated the effect of the AlN nucleation layer on the transport properties of GaN films grown on 6H− and 4H−SiC substrate. Room-temperature electron mobilities of 876, 884, and 932 cm^2/Vs were obtained on 6H−SiC, 4H−SiC, and 3.5° off-axis 6H−SiC substrates. They attributed the observed high electron mobilities to the improved AlN morphology and reduction in screw-dislocation density near the AlN/GaN interface.

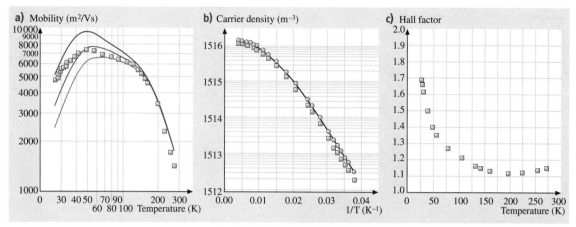

Fig. 32.12 (a) The measured Hall-mobility data (*solid squares*) from the GaN template grown by HVPE as a function of temperature. The *solid line* is the calculated result using $N_a = 2.4 \times 10^{15}$ cm^{-3}, representing the best fit to the measured results. The *upper* and *lower* dotted lines are the calculated results using $N_a = 1.4 \times 10^{15}$ and 3.4×10^{15} cm^{-3}. (b) The measured Hall densities n_H (*solid squares*) as a function of reciprocal temperature from the GaN template grown by HVPE. The *open circles* represent the carrier density corrected by the Hall factor, $n = n_H r_H$. The *solid line* is the fit to the theoretical expression of charge balance with hole and neutral acceptor densities neglected. (c) Temperature dependence of the calculated Hall factor, r_H. After [32.93]

The thermal activation energy of free carriers has also been extracted from the temperature-dependent Hall concentration and mobility measurements by fitting the simple exponential dependence of the carrier temperature on inverse temperature, by the two-band model and by other theoretical fitting techniques. It could be concluded that the activation energy of n-type GaN free carriers lies in the range 14–36 meV, depending on the extent of screening [32.93, 144, 151, 152]. In the dilute limit, the values are close to 30 meV.

In order to analyze the mobility data, one must understand the scattering processes that dominate mobility at different temperatures. Monte Carlo simulation of the electron velocity in GaN as a function of electric field at different doping concentration and temperature predicted a peak drift velocity of 2×10^7 cm/s at an electric field of $\approx 1.4 \times 10^5$ V/cm for an electron concentration of 10^{17} cm^{-3} [32.153, 154]. These values show that an electron mobility as high as 900 cm^2/Vs could be achieved in the case of uncompensated GaN at room temperature with $\approx 10^{17}$ cm^{-3} doping concentration. As discussed above, in the case of high-quality samples with very low compensation, a mobility of even more than 900 cm^2/Vs at room temperature with a similar doping concentration has been reported. *Albrecht* et al. [32.155] calculated the electron mobility for different concentrations of ionized impurities and at different temperatures by using a Monte Carlo simulation technique based on empirical pseudopotential band-structure calculations. For practical use they have also derived an analytical expression describing the dependence of the mobility on temperature and ionized-impurity concentration as

$$\frac{1}{\mu_e} = a \left(\frac{N_I}{10^{17}\, \text{cm}^{-3}} \right) \ln\left(1 + \beta_{CW}^2\right) \left(\frac{T}{300\, \text{K}} \right)^{-1.5} + b \left(\frac{T}{300\, \text{K}} \right)^{1.5} + c \frac{1}{\exp(\Theta/T) - 1}, \quad (32.12)$$

where

$$\Theta = \frac{\hbar \omega_{LO}}{k_B} = 1065\, \text{K},$$

$$\beta_{WC}^2 = 3.00 \left(\frac{T}{300\, \text{K}} \right)^2 \left(\frac{N_I}{10^{17}\, \text{cm}^{-3}} \right),$$

$$N_I = (1 + k_c)/N_D$$

and

$a = 2.61 \times 10^{-4}$ Vs/cm^2,
$b = 2.90 \times 10^{-4}$ Vs/cm^2 and
$c = 1.70 \times 10^{-2}$ Vs/cm^2.

Here N_D is the ionized donor concentration in cm^{-3} and $k_c = N_A/N_D$ is the compensation ratio. By the mobility calculated using this analytical expression and that calculated by Monte Carlo simulation, a reasonable agreement with a maximum error of 6% is realized

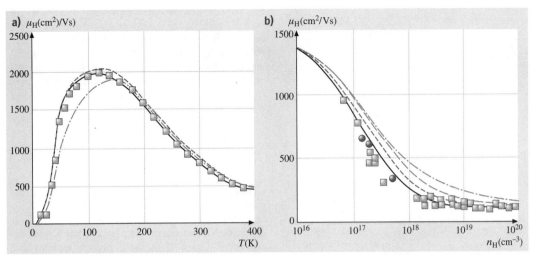

Fig. 32.13 (a) Temperature dependence of the calculated electron Hall mobility. The *dashed curve* shows the calculated mobility using the BH (Brook–Herring) technique. The *dotted curve* shows the calculated mobility using phase-shift analysis with multi-ion screening correction. The *solid curves* show the calculated mobility using phase-shift analysis with multi-ion screening correction and electron–plasmon scattering. (b) Electron concentration dependence of the calculated electron Hall mobility for compensation ratio $r = 0.2$. The *dashed curve* shows the calculated mobility using BH technique. The *dashed–dotted curve* shows the calculated mobility using phase-shift analysis without multi-ion screening correction. The *dotted curve* shows the calculated mobility using phase-shift analysis with multi-ion screening correction. The *solid curve* shows the calculated mobility using phase-shift analysis with multi-ion screening correction and electron–plasmon scattering. Experimental data are taken from different references. After [32.156]

between the temperature range 300–600 K, and ionized donor concentration of 10^{16}–10^{18}.

The calculation of the low-field electron mobility in GaN has also been carried out by using different calculation techniques. *Chin* et al. [32.129] have used the variational principle to calculate low-field electron mobilities as a function of temperature for carrier concentrations of 10^{16}, 10^{17}, and 10^{18} cm^{-3} with the compensation ratio as a parameter. GaN exhibits maximum mobilities in the range 100–200 K, depending on the electron density and compensation ratio, with lower electron densities peaking at lower temperature. This behavior is related to the interplay of piezoelectric acoustic-phonon scattering at low carrier concentration and ionized-impurity scattering at higher carrier concentrations. The maximum polar-mode optical-phonon scattering-limited room-temperature mobility in GaN is found to be about 1000 cm^2/Vs. Although a degree of correlation is achieved with the experimental data, there is a disparity, which is attributed to the structural imperfection and overestimated compensation ratio. Typical compensation ratios observed for MOCVD- and MBE-grown films are about 0.3, although a lower ratio of ≈ 0.24 was reported for HVPE-grown crystals. Compensation reduces the electron mobility in GaN for a given electron concentration. *Rode* and *Gaskill* [32.130] have used an iterative technique, which takes into account all major scattering mechanisms, for low-field electron mobility in GaN for the dependence of the mobility on the electron concentration, but not on temperature. The result was applied to the Hall-mobility data published by *Kim* et al. [32.157] and a good fit between theory and experiment is demonstrated within 2.5% for the lowest-doped samples with free-electron concentrations of 7.24×10^{17} and 1.74×10^{17} cm^{-3}. However, there is a significant disagreement for the more-heavily-doped samples having much higher free-electron concentrations. The Born approximation applied to the ionized-impurity scattering might be the reason for poor fitting at high electron concentrations. By assuming uncompensated material and carrier freeze out onto donors, the ionization energy was theoretically determined to be about 45 and 57 meV for the best-fitted samples with free-electron concentrations of 7.24×10^{17} and 1.74×10^{17} cm^{-3}, respectively. Recent developments in

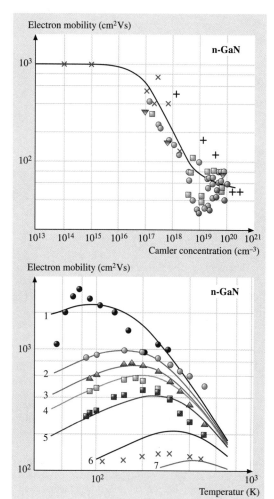

Fig. 32.14 (a) Low-field electron mobility as a function of doping concentration in GaN at room temperature. The curve represents the best approximate equation. The symbols (*crosses, open circles,* and *pluses*) are experimental data taken from different references data for wurtzite GaN and zincblende. (b) Temperature dependencies of low-field electron mobility in wurtzite GaN at different values of doping concentration. Curves represent the best approximate equations. 1: $N = 3 \times 10^{16}$ cm^{-3}, 2: $N = 10^{17}$ cm^{-3}, 3: $N = 1.5 \times 10^{17}$ cm^{-3}, 4: $N = 2 \times 10^{17}$ cm^{-3}, 5: $N = 3.5 \times 10^{17}$ cm^{-3}, 6: $N = 10^{18}$ cm^{-3}, 7: $N = 3 \times 10^{18}$ cm^{-3}. After [32.169]

calculation over a wide range of temperatures and electron concentrations, as shown in Fig. 32.13. They also investigated the effect of the degenerate layer at the GaN/substrate interface to extract reliable experimental values of the bulk electron mobility and concentration.

Besides the numerical simulation techniques mentioned above, recently *Mnatsakanov* et al. [32.169] derived a simple analytical approximation to describe the temperature and carrier concentration dependencies of the low-field mobility in wide temperature ($50 \leq T \leq 1000$ K) and carrier concentration ($10^{14} \leq N \leq 10^{19}$ cm^{-3}) ranges. At the first step of this model, an adequate approximation of the doping-level dependence of the mobility at room temperature is used on the base of the Goughey–Thomas approximation [32.170]

$$\mu_i(N) = \mu_{\min,i} + \frac{\mu_{\max,i} - \mu_{\min,i}}{1 + \left(\frac{N}{N_{g,i}}\right)^{\gamma_i}}, \quad (32.13)$$

where $i = $ n, p for electrons and holes, respectively, $\mu_{\min,i}$, $\mu_{\max,i}$, $N_{g,i}$ and γ_i are the model parameters dependent on the type of semiconductor materials, and N is the doping concentration. Figure 32.14a shows the comparison between the calculated low-field electron mobility as a function of doping level and some experimental data on the room-temperature electron mobility in GaN. The proposed approximation provides rather good agreement with the experimental data. For the temperature-dependent mobility calculation, the authors derived the following equation by taking into account the main scattering mechanisms:

$$\mu_i(N,T) = \mu_{\max,i}(T_0) \frac{B_i(N)\left(\frac{T}{T_0}\right)^{\beta_i}}{1 + B_i(N)\left(\frac{T}{T_0}\right)^{\alpha_i+\beta_i}}, \quad (32.14)$$

the major growth techniques like HVPE, MOCVD, and MBE for GaN and other group III nitride semiconductors have led to the growth of high-quality epitaxial layers [32.158–160], which allowed the comparison of the Hall data with theory to commence. *Dhar* and *Gosh* [32.156] have calculated the temperature and doping dependencies of the electron mobility using an iterative technique, in which the scattering mechanisms have been treated beyond the Born approximation. The compensation ratio was used as a parameter with a realistic charge-neutrality condition. They tested their model with the experimental Hall data taken from the literature [32.94, 143–145, 161–168]. Reasonable agreement was achieved between these data points and the

where

$$B_i(N) = \left(\frac{\mu_{\min,i} + \mu_{\max,i} \left(\frac{N_{g,i}}{N} \right)^{\gamma_i}}{\mu_{\max,i} - \mu_{\min,i}} \right) \Bigg|_{T=T_0}.$$

Figure 32.14b presents the calculated temperature-dependent electron mobility and experimental data reported in the literature and a good agreement is realized. Ionized-impurity scattering is the dominant mechanism at low temperatures, the mobility increases with increasing temperature, at high temperatures mobility is limited by polar optical-phonon scattering, and the mobility decrease with increasing temperature. Finally, many material and physical parameters of GaN were not available for some of the previous simulations where those parameters were treated as adjustable parameters. Needless to say, reliable parameters are required in the calculation of the electron mobility and in the interpretation of experimental results to gain more accuracy.

Aluminium Nitride

Due to the low intrinsic carrier concentration, and the deep native-defect and impurity energy levels (owing to the wide band gap of AlN ≈ 6.2 eV at 300 K), the electrical transport properties of AlN have not been studied extensively and have usually been limited to resistivity measurements. Resistivities in the range $\rho = 10^7 - 10^{13}$ Ωcm have been reported for unintentionally doped AlN single crystals [32.171, 172], a value consistent with other reports [32.173–175]. The conductivity exhibited an Arrhenius behavior for all crystals and the activation energies were reported to be 1.4 eV for temperatures of 330 and 400 K and 0.5 eV between 300 and 330 K. Intentional doping of AlN has resulted in both n- and p-type AlN by introducing Hg and Se, respectively. *Gorbatov* and *Kamyshon* [32.176] obtained the n-type conductivity of polycrystalline AlN with the incorporation Si. Unintentionally doped n-AlN films grown by a modified physical-transport technique by *Rutz* [32.177] had a resistivity as low as 400 Ω cm. Although the source of the electrons has not been determined, *Rutz* et al. [32.178] observed an interesting transition in their AlN films in which the resistivity abruptly decreased by two orders of magnitude with an increase in the applied bias. This observation found applications in switchable resistive memory elements that are operated at 20 MHz. It has been concluded by *Fara* et al. [32.179], who reported the theoretical evidence, based on ab initio density-functional calculations, for acceptors, donors, and native defects in AlN,

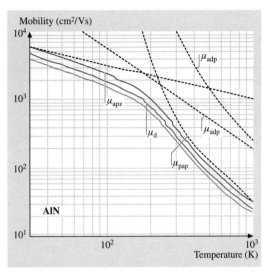

Fig. 32.15 The phonon-limited electron drift mobility in AlN plotted as a function of temperature for $m^* = 0.48$. The *upper* and *lower solid curves* are the phonon-limited electron drift mobility calculated using $m^* = 0.42$ and 0.52, respectively. After [32.129]

for the semi-insulating character of AlN that acceptors are deeper (Be ≈ 0.25, Mg ≈ 0.45) and less soluble in AlN than in GaN, and both the extrinsic donors Si$_{Al}$ and C$_{Al}$, and the native donor V$_N$ are found to be deep (about 1–3 eV below the conduction). Consequently, doped AlN may in general turn out to be semi-insulating, when attained using Al-rich conditions, in agreement with the known doping difficulties of high-Al-content AlGaN alloys.

As far as the mobility of AlN is concerned, in early studies, *Edwards* et al. [32.171] and *Kawabe* et al. [32.172] carried out some Hall measurements in p-type AlN, which produced a very rough estimate of the hole mobility of $\mu_p = 14$ cm^2/Vs at 290 K. Later work on carbon-doped material has resulted in highly conductive p-type AlN with hole mobilities on the order of 60 cm^2/Vs for carbon doping of 10^{18} cm^{-3} [32.180]. Theoretical calculations to estimate the mobility in AlN have only been made by *Chin* et al. [32.129]. Since AlN is an insulator with extremely low carrier concentration, only phonon-limited scattering processes were considered in the calculation of the temperature-dependent mobility, as shown in Fig. 32.15. The mobility was found to decline rapidly at high temperatures, with a value of about 2000 cm^2/Vs at 77 K, dropping to 300 cm^2/Vs at 300 K for optical-phonon-limited mobility.

Indium Nitride

InN suffers from the lack of a suitable substrate material, causing a high concentration of extended defects in the films. Sapphire substrate is usually used for InN growth, but the difference in lattice constants between InN and sapphire is as large as 25% and even more than 19% with the other nitrides. A large disparity of the atomic radii of In and N is an additional contributing factor to the difficulty of obtaining InN of good quality. Because of all these factors, the electron mobilities obtained from various films have varied very widely. Electrical properties vary also substantially with the choice of growth techniques. A range of film-deposition techniques, such as reactive-ion sputtering, reactive radio-frequency (RF) magneton sputtering, metalorganic vapor-phase epitaxy (MOVPE) and MBE, has been used. Table 32.4 gives updated values of electron mobility and concentration in InN films grown by various growth techniques.

The transport properties reported in the literature are mostly based on Hall-effect measurements, assuming the Hall scattering factor to be unity. In majority, electron mobilities often remain relatively poor, despite predicted values as high as $3000\,\text{cm}^2/\text{Vs}$ at room temperature for InN [32.198]. It is widely believed that nitrogen vacancies lead to large background electron concentrations, which is responsible for the observed low electron mobility. An empirical linear relationship between the electron mobility and electron concentration can be deduced from the table for a series of InN films, although not all films exhibit this type of behavior. A systematic study carried out by *Tansley* et al. [32.199] indicates that the electron concentration decreases as the nitrogen density is increased in the growth plasma in reactive-ion sputtering. Although the reported time is relatively old, the maximum mobility of $2700\,\text{cm}^2/\text{Vs}$ at an electron concentration of $5 \times 10^{16}\,\text{cm}^{-3}$ was reported for RF reactive-ion sputtered growth of InN. Early study of the electron mobility of InN as a function of the growth temperature indicates that the mobility of InN grown by ultrahigh-vacuum electron–cyclotron resonance-radio-frequency magnetron sputtering (UHVECR-RMS) can be as much as four times the mobility of conventionally grown (vacuum-deposited) InN [32.187]. However, more recent work indicates a progressive improvement in electrical properties of InN films grown by vacuum-deposition techniques, including MBE and MOVPE. Values of the electron mobility as high as $1420\,\text{cm}^2/\text{Vs}$ at an electron concentration of $1.4 \times 10^{18}\,\text{cm}^{-3}$ were re-

Table 32.4 A compilation of electron mobilities obtained in wide-band-gap InN on different substrates and for various deposition conditions

Growth method	Carrier concentration (cm^{-3})	Electron mobility (cm/Vs)	Ref.
Reactive ion sputtering	7.0×10^{18}	250	[32.181]
Reactive-ion sputtering	2.1×10^{17}	470	[32.182]
Reactive-ion sputtering	8.0×10^{16}	1300	[32.182]
Reactive-ion sputtering	5.5×10^{16}	2700	[32.182]
Reactive-ion sputtering	2.0×10^{20}	9	[32.183]
RF magnetron sputtering	–	44	[32.184]
Reactive-ion sputter	$\approx 10^{20}$	60	[32.185]
Plasma-assisted MBE	$\approx 10^{20}$	229	[32.63]
ECR-assisted MOMBE	2.0×10^{20}	100	[32.186]
ECR-assisted reactive-ion sputtering	–	80	[32.187]
Reactive sputter	6.0×10^{18}	363	[32.188]
MOVPE	5.0×10^{19}	700	[32.159]
Migration-enhanced epitaxy	3.0×10^{18}	542	[32.189]
RF MBE	3.0×10^{19}	760	[32.190]
MOMBE	8.8×10^{18}	500	[32.191]
MBE	$2-3 \times 10^{18}$	800	[32.192]
Reactive-ion sputtering	$\approx 10^{19}$	306	[32.193]
RF MBE	1.0×10^{19}	830	[32.194]
Plasma-assisted MBE	1.6×10^{18}	1180	[32.195]
MBE	4×10^{17}	2100	[32.196]
Plasma-assisted MBE	1.4×10^{18}	1420	[32.197]

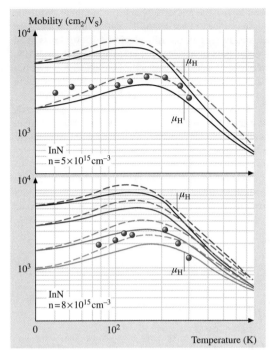

Fig. 32.16a,b The electron drift and Hall mobilities of InN as a function of temperature for (**a**) $n = 5 \times 10^{16}$ cm^{-3} for compensation ratios of 0.00 and 0.60; and (**b**) 8×10^{16} cm^{-3} for compensation ratios of 0.00, 0.30, 0.60, and 0.75; all experimental data are from [32.129]

ported for InN layers grown by plasma-assisted MBE using a low-temperature-grown GaN intermediate layer and a low-temperature-grown InN buffer layer. Hall measurements in InN films grown on AlN buffer layers, which are in turn grown on sapphire, indicated an electron mobility of 2100 cm^2/Vs with a relatively low electron concentration (4×10^{17} cm^{-3}) at room temperature in material grown by MBE. Very high inadvertent-donor concentrations ($> 10^{18}$ cm^{-3}) seem to be one of the major problems for further progress of device applications of InN. O_N and Si_{In} and possible interstitial H have been proposed to be the likely dominant defects responsible for high electron concentration for state-of-the-art MBE-grown InN, based on their low formation energies [32.200, 201].

The electron mobility in InN has been calculated using the variational principle for a range of temperatures, carrier concentrations, and compensation ratios [32.129]. Figure 32.16 shows the theoretical results with experimental mobility values taken from *Tansley* and *Foley* [32.182]. The calculated peak mobilities are found to be 25 000, 12 000, and 8000 cm^2/Vs for 10^{16}, 10^{17}, and 10^{18} cm^{-3}, respectively, at different temperatures (100–200 K), depending on the electron density and the compensation ratio. This is due to the interplay of piezoelectric acoustic-phonon scattering at low concentrations and ionized-impurity scattering at high temperatures. These two mechanisms are the dominant scattering mechanisms below 200 K, while polar-mode optical-phonon scattering is the most significant process above this temperature. The low-concentration limit for room-temperature mobility in uncompensated InN is estimated to be 4400 cm^2/Vs.

32.5.2 High-Field Transport

Ensemble Monte Carlo simulations have been the popular tools for the theoretical investigation of steady-state electron transport in nitrides. In particular, the steady-state velocity–field characteristics have been determined for AlN [32.203, 204], GaN [32.154, 155, 205–208], and InN [32.209, 210]. These reports show that the

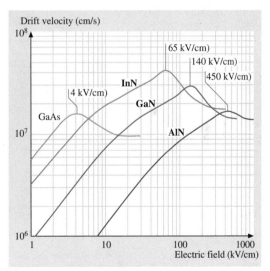

Fig. 32.17 The velocity–field characteristics associated with wurtzite GaN, InN, AlN, and zincblende GaAs. In all cases, we set the temperature to 300 K and the doping concentration to 10^{17} cm^{-3}. The critical fields at which the peak drift velocity is achieved for each velocity–field characteristic are clearly marked: 140 kV/cm for GaN, 65 kV/cm for InN, 450 kV/cm for AlN, and 4 kV/cm for GaAs. After [32.202]

drift velocity initially increases with the applied electric field, reaches a maximum and then decreases with further increases in the field strength. Inter-valley electron transfer plays a dominant role at high electric fields, leading to a strongly inverted electron distribution and to a large negative differential resistance (NDR). The reduction in the drift velocities was attributed to the transfer of electrons from the high-mobility Γ-valley to the low-mobility satellite X-valley. The onset electric field and peak drift velocities, however, show some disparity among the reported calculations due to the variety of degree of approximation and material physical constants used. A typical velocity–field characteristic for bulk group III nitrides at room temperature is shown in Fig. 32.17, along with the well-studied GaAs data used to test the author's Monte Carlo model. With the doping concentration set to 10^{17} cm^{-3}, InN has the highest steady-state peak drift velocity: 4.2×10^7 cm/s at an electric field of 65 kV/cm. In the case of GaN and AlN, steady-state peak drift velocities are rather low and occur at larger electric fields: 2.9×10^7 cm/s at 140 kV/cm for GaN, and 1.7×10^7 cm/s at 450 kV/cm for AlN.

Another interesting aspect of electron transport is its transient behavior, which is relevant to short-channel devices with dimensions smaller than 0.2 μm, where a significant overshoot is expected to occur in the electron velocity over the steady-state drift velocity.

Transient electron transport and velocity overshoot in both wurtzite and zincblende GaN, InN, and AlN were studied theoretically by a number of groups. *Foutz*

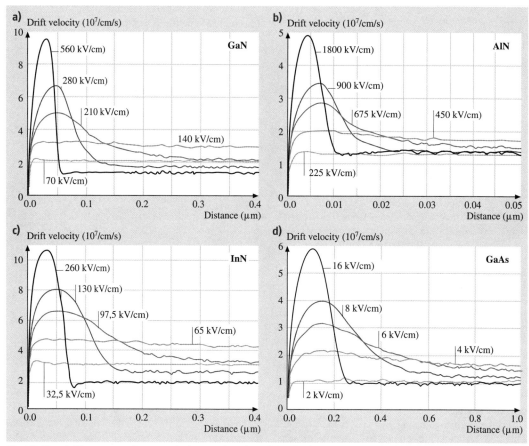

Fig. 32.18a–d The average electron velocity as a function of the displacement for various applied fields for the cases of (**a**) GaN, (**b**) InN, (**c**) AlN, and (**d**) GaAs. In all cases, we have assumed an initial zero-field electron distribution, a crystal temperature of 300 K, and a doping concentration of 10^{17} cm^{-3}. After [32.202]

et al. [32.202] employed both Monte Carlo simulations and one-dimensional energy–momentum balance techniques. They used a three-valley model for the conduction band by taking the main scattering mechanisms, such as ionized impurity, polar optical-phonon, acoustic-phonon through deformation potential and piezoelectric, and inter-valley scatterings into account. In particular, they examined how electrons, initially in equilibrium, respond to the instant application of a constant electric field. Figure 32.18 shows the average velocity of the electrons in AlN, GaN, and InN as a function of distance. According to their calculation, electron velocity overshoot only occurs when the electric field exceeds a certain critical value unique to each material and it lasts over a certain distance dependent on applied field. These critical fields are points where the highest steady-state peak drift velocities are achieved and being reported as $65\,\text{kV/cm}$, $140\,\text{kV/cm}$, and $450\,\text{kV/cm}$ with corresponding peak velocities of $2.9\times 10^7\,\text{cm/s}$, $1.7\times 10^7\,\text{cm/s}$, and $1.6\times 10^7\,\text{cm/s}$ for InN, GaN, and AlN, respectively. Among them InN exhibits the highest peak overshoot velocity, on the order of $10^8\,\text{cm/s}$ at $260\,\text{kV/cm}$, and the longest overshoot relaxation distance, on the order of $0.8\,\mu\text{m}$ at $65\,\text{kV/cm}$. To optimize device performance by only minimizing the transit time over a given distance is prevented by a trade-off between the peak overshoot velocity and distance taken to achieve steady state. The upper bound for the cutoff frequency of InN- and GaN-based HFETs (heterojunction field effect transistor) benefits from larger applied fields and accompanying large velocity overshoot when the gate length is less than $0.3\,\mu\text{m}$ in GaN and $0.6\,\mu\text{m}$ in InN based devices. However, all measured cutoff frequencies are gate-length-dependent and well below these expectations, indicating that devices operate in the steady-state regime and other effects, such as real-space transfer, should also be considered. On the other hand, there is some controversy in the reports related to the onset of velocity overshoot in nitride semiconductors. For example, *Rodrigues* et al. [32.211] reported that overshoot onsets at $10\,\text{kV/cm}$ in InN, $20\,\text{kV/cm}$ in GaN, and $60\,\text{kV/cm}$ in AlN by using a theoretical model based on a nonlinear quantum kinetic theory, which compares the relation between the carriers' relaxation rate of momentum and energy.

Experimental investigations of transient transport in group III nitrides are very limited and few results are reported by using different techniques. *Wraback* et al. [32.212] employed a femtosecond time-resolved electroabsorption technique to study transient electron velocity overshoot for transport in the AlGaN/GaN heterojunction p-i-n photodiode structures. It has been reported that electron velocity overshoot can be observable at electric fields as low as $105\,\text{kV/cm}$. Velocity overshoot increases with electric fields up to $\approx 320\,\text{kV/cm}$ with a peak velocity of $7.25\times 10^7\,\text{cm/s}$ relaxing within the first $0.2\,\text{ps}$ after photoexcitation. The increase in electron transit time across the device and the decrease in peak velocity overshoot with increasing field beyond $320\,\text{kV/cm}$ is attributed to a negative differential resistivity region of the steady-state velocity–field characteristic in this high-field range. *Collazo* et al. [32.213] used another experimental technique based on the measurement of the energy distribution of electrons which were extracted into vacuum through a semitransparent Au electrode, after their transportation through intrinsic AlN heteroepitaxial films using an electron spectrometer. They observed electron velocity overshoot as high as five times the saturation velocity and a transient length of less than $80\,\text{nm}$ at a field of $510\,\text{kV/cm}$. In order to design an electronic device that is expected to operate at high power and high frequency, one should consider obtaining benefit from velocity overshoot effect in group III nitrides semiconductor heterojunctions. A systematic investigation of InN, GaN, AlN and their alloys as a function of various parameters in dynamic mode would be very beneficial for the development of higher-performance, next-generation electronic and optoelectronic devices.

32.6 Optical Properties of Nitrides

The optical properties of a semiconductor are connected with both intrinsic and extrinsic effects. Intrinsic optical transitions take place between the electrons in the conduction band and holes in the valance band, including excitonic effects due to the Coulomb interaction. Excitons are classified into free and bound excitons. In high-quality samples with low impurity concentrations, the free exciton can also exhibit excited states, in addition to their ground-state transitions. Extrinsic properties are related to dopants or defects, which usually create discrete electronic states in the band gap, and therefore influence both optical absorption and emission processes. The electronic states of the bound excitons (BEs) depend strongly on the semiconductor material, in particular its band structure. In theory, excitons could be bound to neutral or charged donors and acceptors. A basic assumption in the description of the principal bound exciton states for neutral donors and acceptors is a dominant coupling of the like particles in the BE states [32.214]. For a shallow neutral donor-bound exciton (DBE), for example, the two electrons in the BE state are assumed to pair off into a two-electron state with zero spin. The additional hole is then assumed to be weakly bound in the net hole-attractive Coulomb potential set up by this bound two-electron aggregate. Similarly, neutral shallow acceptor-bound excitons (ABE) are expected to have a two-hole state derived from the topmost valence band and one electron interaction. These two classes of bound excitons are by far the most important cases for direct-band-gap materials like group III nitrides. Other defect-related transitions could be seen in optical spectra such as free-to-bound (electron–acceptor), bound-to-bound (donor–acceptor) and so-called yellow luminescence. Several experimental techniques are used for the investigation of the optical properties of group III nitrides, including optical absorption, transmission, photoreflection, spectroscopic ellipsometry, photoluminescence, time-resolved photoluminescence, cathodoluminescence, calorimetric spectroscopy, pump-probe spectroscopy, etc. In this section, we will only present some important optical properties of GaN and available data related to AlN and InN.

32.6.1 Gallium Nitride

Free Exciton in GaN

The wurtzite GaN conduction band (Γ_7^c) is mainly constructed from the s state of gallium, whereas the valance band is mainly constructed from the p state of nitrogen. Under the influence of the crystal-field and spin-orbit interactions, the six-fold degenerate Γ_{15} level splits into a highest Γ_9^v, upper Γ_7^v, and lower Γ_7^v levels. The near-band-gap intrinsic absorption and emission spectrum is therefore expected to be dominated by transitions from these three valance bands. The related free-exciton transitions from the conduction band to these three valance bands or vice versa are usually denoted by A $\equiv \Gamma_7^c \leftrightarrow \Gamma_9^v$ (also referred to as the heavy hole), B $\equiv \Gamma_7^c \leftrightarrow \Gamma_7^v$ the upper one (also referred to as the light hole), and C $\equiv \Gamma_7^c \leftrightarrow \Gamma_7^v$ the lower one (also referred to as the crystal-field split band), respectively. The optical properties of these excitonic transitions will be discussed in some detail here. In ideal wurtzite crystals, i.e. strain-free, these three exciton states obey the following selection rules in optical one-photon processes:

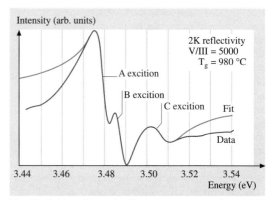

Fig. 32.19 Experimental reflectance spectrum taken at 2 K (*solid line*) for a 2-μm GaN film grown on sapphire along with a theoretical fit using a damped-oscillator model. After [32.215]

all excitons are allowed in the σ polarization ($E \perp c$ and $k \perp c$-axis), but the C exciton is quite weak. The C exciton is strongly allowed in the π polarization ($E \parallel c$ and $k \perp c$), however, where the B exciton is weakly observable, the A exciton is forbidden in this geometry. In the α polarization ($E \perp c$ and $k \parallel c$) all three transitions are clearly observable [32.216]. Each of these fundamental excitons states are expected to have a fine structure due both to exciton polariton longitudinal–transverse splitting and the splitting caused by the electron–hole exchange interaction, which are on the order of 1–2 meV [32.217]. Until very recently, this splitting has nearly been impossible to observe due to limitations in the spectroscopic line width of the free exciton on the order of 1 meV in the best sample so far available [32.218]. The optical spectroscopy of the intrinsic excitons can be measured by employing the low-temperature photoluminescence (PL), absorption, and/or derivative technique like photoreflectance (PR) and calorimetric absorption and reflection techniques [32.219, 220]. These measurements pave the way for the determination of exciton binding energies, exciton Bohr radii, the dielectric constant, and with the aid of the quasi-cubic model, spin–orbit and crystal-field parameters. There are several reports in the literature on reflectance studies for thick GaN epilayers [32.216, 221–224] as well as on homoepitaxial layers [32.225, 226] and bulk GaN [32.227]. *Monemar et al.* [32.220], who have examined numbers of thin and thick layers on various substrates, including homoepitaxial layers on GaN substrates, concluded that the A, B, and C exciton lines in GaN relaxed to an accuracy of ±2 meV are 3.478, 3.484, and 3.502 eV, respectively, at

2 K. An example of a reflectance spectrum obtained in the α polarization is given in Fig. 32.19 for a 2-μm-thick GaN epilayer on sapphire. The corresponding excitonic transition energies were evaluated with a classical model involving a damped-oscillator transition [32.215].

Calorimetric absorption or reflection is a different experimental technique that has recently been employed to measure the fundamental exciton resonance spectra for a thick GaN epilayer grown on sapphire at 43 mK (Fig. 32.20). The values for the A, B, and C excitons agree with those quoted above for reflection measurement, apart from a small increase of about 2 meV in the case of calorimetric measurement [32.228].

A very powerful technique for studying exciton structure is photoluminescence. Figure 32.21 displays an example of a PL spectrum in the range of fundamental excitons recently taken at 2 K for an approximately strain-free thick GaN epilayer grown on sapphire by HVPE. The PL spectrum is typically dominated by strong emission related to donor-bound excitons (bound excitons will be further discussed later). At higher energies, the $n = 1$ ground states of A and B excitons

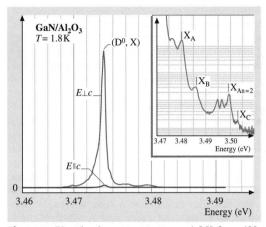

Fig. 32.21 Photoluminescence spectra at 1.8 K for a 400-μm-thick GaN grown on a sapphire substrate. The spectrum is dominated by the donor-bound exciton, but the intrinsic exciton states are also resolved at higher energies (*inset*). After [32.228]

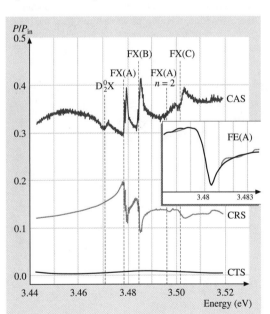

Fig. 32.20 Calorimetric absorption (CAS), reflection (CRS), and transmission (CTS) spectra of the 400-μm-thick HVPE GaN/sapphire layer ($T = 45$ mK). This sample is assumed to be nearly strain-free. In the *inset*, a fit to the reflection spectra of the FE(A) (free exciton) exciton is shown. After [32.228]

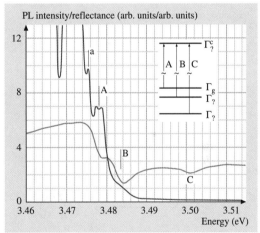

Fig. 32.22 Photoluminescence spectrum (*solid line*) taken at 4.2 K for a homoepitaxial GaN layer, showing a doublet structure of the A exciton. Corresponding reflectance spectrum for the sample are also shown (*dashed line*). After [32.229]

are clearly observed with a less-intense C exciton, as well as the $n = 2$ excited state of the A exciton [32.230]. The splitting between the A, B, and C exciton ground states, Δ_{AB} and Δ_{AC}, are (5.5 ± 0.1) and (22.0 ± 0.1) meV. With the aid of these exciton energies, the corresponding values for the splittings of the three top valance bands have been estimated using a quasi-cubic approximation model. Values of $\Delta_{cr} \approx 20 \pm 2$ meV for the crystal-field splitting and $\Delta_{so} \approx 10 \pm 2$ meV for the spin–orbit splitting were reported [32.220]. *Edwards* et al. [32.171] and *Rodina* et al. [32.230] have reported different values for the energy splitting parameters: $\Delta_{cr} \approx \Delta_{so} \approx 16 \pm 2$ meV, and $\Delta_{cr} \approx 12.3 \pm 0.1$ meV, and $\Delta_{so} \approx 18.5 \pm 0.1$ meV, respectively.

Additionally, fine structure of exciton lines was also reported in the energy region near the band edge of GaN using various optical measurements by several groups [32.233–236]. The excitonic spectra taken at 4.2 K for homoepitaxial GaN layers, shown in Fig. 32.22, indicate that the A exciton line splits into two components about 2 meV apart, with the lowest component at about 3.477 eV. A possible interpretation is that these two components actually correspond to recombination from the lower and upper polariton branches. Similarly, *Gil* et al. [32.237] have also deduced the longitudinal–transverse splitting by reflectance line-shape fitting to the data. For each exciton, the longitudinal transverse splitting is calculated as 2.9 meV and 1.8 meV for the A and B excitons, respectively. This compares, within the experimental accuracy, with the splitting of 2.4 and 1.8 meV between the energies of the dips in the PL bands at 3.4894 and 3.4978 eV, and transverse excitons as shown in Fig. 32.22.

The values of the near-band-gap exciton energies are strongly sensitive to built-in strain, which commonly occurs when the GaN is grown on a foreign substrate with heteroepitaxy. GaN grown on sapphire substrates usually experiences a compressive strain, which in turn, increases the band gap, and hence increases in A, B, and C exciton energies and splittings, compared to the case of unstrained bulk GaN, are consistently observed. Upshifts of the A and B excitons by as much as 20 meV have been observed at 2 K, and the C exciton has been found to shift as much as 50 meV [32.222]. An example of the systematic shifts observed in reflectance data for A, B, and C excitons is shown in Fig. 32.23. On the other hand, a different situation is expected to occur for GaN growth on another widely used substrate, SiC, due to the tensile strain, which in turn is expected to lead to a decrease in the overall exciton energies (and the band

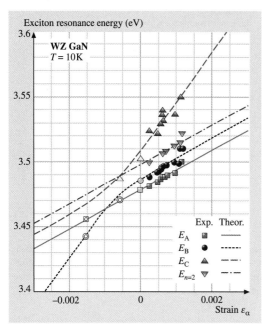

Fig. 32.23 The strain dependence of the free-exciton resonance energies in wurtzite GaN grown on sapphire substrate. The strain was obtained from the measured lattice parameter values for each sample. Theoretical modeling of the strain dependence of exciton energies are also shown. [32.231]

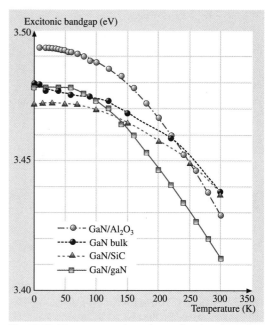

Fig. 32.24 Temperature dependence for the A exciton energy in samples grown on different substrates. After [32.225, 232]

Table 32.5 Reported exciton energies in GaN using different experimental techniques

FX(A)	FX(B)	FX(C)	Substrate	Strain	Growth	Technique	Ref.
3.474	3.481	3.501	Sapphire	Compressive	VPE	Reflectance	[32.214]
3.475	3.481	3.493	Sapphire	Compressive	VPE	PLE	[32.221]
3.480	3.486	3.503	Sapphire	Compressive	HVPE	PL	[32.230]
3.4799	3.4860	3.5025	Sapphire	Compressive	HVPE	Calorimetric reflection	[32.228]
3.4770	3.4865	3.5062	Sapphire	Compressive	MOVPE	Reflectance	[32.215]
3.4775	3.4845	3.5062	Sapphire	Compressive	MOVPE	Reflectance	[32.222]
3.485	3.493	3.518	Sapphire	Compressive	MOCVD	Reflectance	[32.238]
3.491	3.499	3.528	Sapphire	Compressive	MOCVD	Reflectance	[32.223]
3.4903	3.4996	3.525	Sapphire	Compressive	MBE	Reflectance	[32.224]
3.484	3.490	3.512	Sapphire	Compressive	MOCVD	Contactless electro reflectance	[32.239]
3.479	3.486		Sapphire	Compressive	MOCVD	PL	[32.240]
3.488	3.496		Sapphire	Compressive	MOCVD	PL	[32.241]
3.4857	3.4921		Sapphire	Compressive	MOCVD	PL	[32.242]
3.483	3.489		Sapphire	Compressive	MBE	PL	[32.243]
3.476	3.489	3.511	ZnO	Compressive	RAMBE	Reflectance	[32.244]
3.480	3.493		ZnO	Compressive	RMBE	PL	[32.245]
3.470	3.474	3.491	6H−SiC	Tensile	MOCVD	Reflectance	[32.246]
3.470		3.486	6H−SiC	Tensile	MOVPE	PL	[32.247]
3.472			6H−SiC	Tensile	HVPE	PL	[32.248]
3.4771	3.4818	3.4988	GaN	Unstrained	MOVPE	Reflectance	[32.226]
3.4776	3.4827	3.5015	GaN	Unstrained	MOCVD	Reflectance	[32.225]
3.478	3.484	3.502	GaN	Unstrained	MOCVD	Reflectance	[32.229]
3.4772	3.4830	3.4998	Bulk GaN	Unstrained	Na−Ga melt	Reflectance	[32.249]
3.490	3.500	3.520	Freestanding	Unstrained	HVPE	Contactless electro reflectance	[32.250]

gap), and also a decrease in the A, B, and C splittings. As a summary, a comparison of free-exciton energies obtained from reflectance PL and other optical techniques for heteroepitaxial GaN grown by several techniques on various substrates, producing different degrees of strain, and unstrained homoepitaxy as well as bulk GaN is given in Table 32.5.

It should be pointed out that considerable variation exists between different reports in the literature [32.232, 238, 251]. The temperature dependence of the intrinsic-exciton energies is also dependent on the particular sample and local strain. Figure 32.24 presents a comparison between the temperature dependence of the exciton energies for three samples, presumably relaxed (bulk GaN), under compression (grown on sapphire) and under tension (grown on SiC). The temperature dependence of excitonic resonance (in the absence of localization) can be described by the Varshni empirical relation

$$E_g(T) = E_0(0) - \alpha T^2/(\beta + T), \quad (32.15)$$

where $E_0(0)$ is the transition energy at 0 K, and α and β are the temperature coefficients. The free-exciton transitions are the dominating PL process at room temperature in GaN. This has been well established by PL spectral data over a wide range of temperatures for nominally undoped samples. However, the PL intensity at the position of the band gap at 300 K is considerably lower than the A exciton intensity [32.252].

Free-exciton transitions in wide-band-gap materials have a characteristic coupling to LO phonons [32.253]. As expected from the theory of LO-phonon coupling for exciton polaritons the first two replicas are strongest. The 3^-LO and 4^-LO replicas are also clearly observable. As predicted by theory, the characteristic temperature dependence of relative ratios of the intensities of the LO-phonon replicas is linearly proportional to the temperature and this behavior is confirmed for the A exciton for temperatures $T < 100$ K [32.248].

The binding energy of excitons determines the energy of the band gap of the material, but the strength of the binding is also an important factor for the thermal stability of the excitons. *Chen* et al. [32.254] have calculated the binding energies as $\Delta E_A^b = \Delta E_B^b = 20$ meV for the A and B excitons and $\Delta E_C^b = 18$ meV for the C free exciton. By using the values of the effective electron and hole masses, *Chichibu* et al. [32.241] reported the calculated binding energy of the A exciton as 27 meV, which is close to the early estimation of 27.7 meV reported by *Mahler* and *Schroder* [32.255]. There have been very scattered experimental values reported for the free-exciton binding energies in GaN, as recently reviewed [32.256]. The scatter in the data is probably due to measurements on highly strained layers, and in some case misinterpretations of some features in the optical spectra. There is, in general, no feature at the band-edge position in the optical spectra, therefore, the exciton binding energy has to be obtained indirectly using the temperature dependence of the free-exciton transition in photoluminescence [32.240] or the position of the excited states if the positions of the excited states of the excitons are known [32.242, 243, 257]. The more recent data on both reflectance and photoluminescence (PL) reported for homoepitaxial samples appear convincing.

A value of 25 meV is reported for the A exciton, and very similar values are reported for the B and C excitons [32.226, 258]. The $n = 2$ transitions are clearly resolved in these spectra, giving confidence that these results will be accurate for pure unstrained material, with a precision of about 1 meV.

The temporal behavior of excitons is of importance for emitters in that it provides a window on the dynamics of recombination processes. They should ideally be fast to be able to compete efficiently with nonradiative processes, such as multi-phonon emission and defect-related nonradiative centers. PL transient data represent an excellent tool to study both recombination rates of radiative and nonradiative processes in GaN. The temporal behavior at low excitation levels indicates processes, depending on the sample, that are as fast as about 35 ps for free excitons. Typical transients are shown in Fig. 32.25, which were obtained at different excitation intensities for a GaN epilayer grown on sapphire. Decay times at 2 K were found to vary between 60 ps and 115 ps, depending on the excitation intensity, indicating strong defect participation under these excitation intensities. The decay in luminescence intensity can more reasonably be described with a combination of a fast decay followed by a slower process with a decay time of about 300 ps at longer times. This slow process may be associated with weak localization of the free excitons, perhaps due to potential fluctuation induced by the inhomogeneous strain field.

Bound Excitons in GaN

The A, B, and C excitons discussed above represent intrinsic processes, as they do not involve pathways requiring extrinsic centers. In GaN, the neutral shallow donor-bound exciton (DBE), or I_2, is often dominant because of the presence of donors, due to doped impurities and/or shallow donor-like defects. In samples containing acceptors, the acceptor-bound exciton (ABE), or I_1, is observed. The recombination of bound excitons typically gives rise to sharp lines, with a photon energy characteristic for each defect. As in the free-exciton case, some fine structure is also expected in bound exciton lines, which are usually on the order of or below 1 meV for shallow donor- and acceptor-bound excitons. A characteristic phonon coupling, which can involve both lattice modes and defect-related vibrational modes, is also seen for each particular bound exciton spectrum. The photon energy region for DBE spectra is about 3.470–3.4733 eV at 2 K for strain-free GaN [32.218, 226]. Thick heteroepitaxial layers grown by HVPE give the best spectroscopic characteristic,

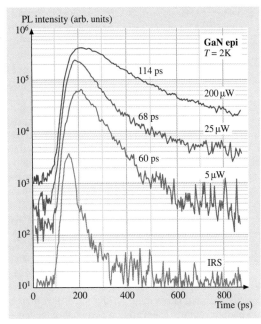

Fig. 32.25 Photoluminescence transients at 2 K for a GaN epilayer grown on sapphire measured at different excitation power intensity. IRS denotes the instrumental response in the experiment. [32.220]

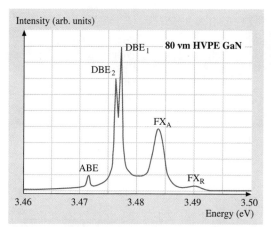

Fig. 32.26 Photoluminescence spectrum of a thick HVPE GaN layer showing two donor BE lines and an acceptor BE. The spectrum is upshifted by the compressive strain by about 6 meV compared with strain-free GaN. After [32.266]

with an optical line width of about 1 meV or less, as seen in Fig. 32.26. It is commonly argued that oxygen and silicon are the two most dominant residual shallow donors in GaN. In homoepitaxial GaN samples grown by MOVPE on a pretreated GaN single crystal, a narrower spectral line width down to 0.1 meV for the DBE peaked at 3.4709 eV has been reported [32.226]. In this case, in addition to unidentified sharp satellite lines on both sides of the main DBE peak, shallow DBE is the dominant line and has recently in the literature been assigned to substitutional Si donors [32.259]. Therefore, the high-energy line would then be O-related; O is known to be a typical contaminant in GaN grown by all techniques. Another high-quality MBE-grown GaN epitaxy on a GaN substrate shows at least three well-resolved peaks in the DBE region, at about 3.4709, 3.4718, and 3.4755 eV [32.260, 261]. In addition to the mostly accepted Si- and O-related shallow donor states, another donor is needed to identify the third line. It has been argued that these two levels (3.4709 and 3.4718 eV) could be attributed to exchange-split components of the same donor. Strain-splitting has also been considered in some cases. The most energetic one at 3.4755 eV is related to the B valance band involved in DBE. The localization energy for donor-bound excitons is about 6–7 meV. The corresponding binding energy of the dominating shallow donor electron has been estimated as about 35.5 meV from IR absorption data. An activation energy for electron conduction of

30–40 meV has been reported for O implanted in GaN by *Pearton* et al. [32.262]. Recent data from a combination of electrical and optical measurements on Si-doped GaN samples indicate that the Si donor has a binding energy of about 20 meV, while a deeper donor (35 meV) was also present, suggested to be O-related [32.151]. *Monemar* [32.263] estimated the binding energy of the donor electron from the two-electron satellites, which involves the radiative recombination of one electron with a hole, leaving the remaining neutral donor with a second electron in an excited $n = 2$ state. Assuming ideal effective-mass behavior, a binding energy of about

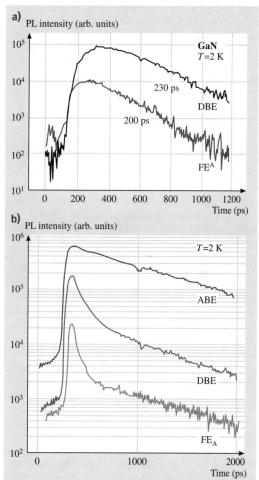

Fig. 32.27a,b Photoluminescence transients for a 400-μm-thick GaN layer (**a**) and for a thin strained layer (**b**) both grown on sapphire. After [32.266]

29 meV for the dominant donor in a 400-μm-thick HVPE GaN layer grown on sapphire has been estimated. By using the same technique for the MBE-grown GaN on a freestanding GaN template, *Reshchikov* et al. [32.264] determined the binding energies of two distinct shallow donors as 28.8 and 32.6 meV, which were attributed to Si and O, respectively. These combined data also provide strong evidence for a tentative identification of the 3.4709 eV DBE as being due to the neutral silicon donor. It is likely that the often-observed second DBE peak at 3.4718 eV in unstrained GaN is then related to the O donor, but this remains to be proven. In his review, *Viswanath* [32.265] listed commonly observed and reported peak positions and localization energies of DBE in GaN grown by various growth techniques on different substrates under compressive or tensile strain.

Another important property of DBEs is the recombination dynamics. The low-temperature decay rate for the DBE PL line gives the radiative lifetime of the DBE. Figure 32.27 shows photoluminescence transients measurements for a thick HVPE-grown GaN layer ($\approx 100\,\mu$m), with low dislocation density (in the 10^7 cm^{-2} range), as well as for a thin strained layer grown on sapphire. In HVPE-grown thick layers, a low intensity decay time of about 200 ps is observed at 2 K, which is a typical radiative lifetime for DBEs related to shallow neutral donors. In thin strained heteroepitaxial layers with a higher defect density, a shorter lifetime is observed [32.267–269], indicating the effect of excitation transfer from the DBE to lower-energy states before the radiative recombination takes place. Similarly, homoepitaxial GaN layers also often show a fast DBE decay at low excitation density, due to excitation transfer to point defects [32.220]. A slow component, which is indicative of a slower radiative process overlapping the DBE transition, is also observed in the DBE decay dynamics [32.220, 257]. A recent observation on the DBE decay time for n-type GaN layers grown on SiC substrates indicated a clear trend versus the energy position of the DBE line, which in turn correlated with the strain in the layer. The biaxial strain in the GaN layer strongly affects the top valence band states, which in turn is reflected in the DBE wavefunction [32.258], affecting the oscillator strength and hence the observed radiative lifetime [32.227].

In the energy region below the principle DBE line there is usually a rich spectrum of bound excitons assumed to be acceptor-related. However, the situation for acceptor BEs (ABE or I_1) in GaN is somewhat less clear than for the donors. The most prominent neutral ABE is found at about 3.466 eV in strain-free GaN and

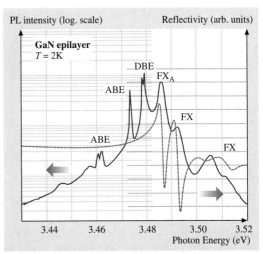

Fig. 32.28 Low-temperature PL spectrum (*solid line*) of a HVPE GaN layer shown on a semi-logarithmic scale. The reflectivity spectrum (*dashed line*) of the same sample taken at normal incidence geometry is plotted on a linear scale. After [32.266]

attributed to the exciton bound to a shallow acceptor, probably Mg_{Ga} (Fig. 32.28). Furthermore, this peak has been found to be dominant in slightly Mg-doped GaN samples [32.270–272]. An alternative interpretation of this BE line, which is related to charged donor-bound excitons, has recently been proposed by several groups [32.273–276]. The other main peak at about 3.455 eV has been attributed to Zn, but this is not yet justified. The broader peak at 3.44 eV is probably due to a low-energy acoustic-phonon wing connected to the main peak at 3.455 eV, which is very characteristic for ABEs [32.277, 278]. Also, phonon replicas are seen in the spectrum, indicating a rather strong LO-phonon coupling, much stronger than for the DBE. Since most experiments with doped samples up to now have been performed on heteroepitaxial material, there is a difficulty in determining the peaks, where the ABE peak position strongly depends on the strain. A binding energy of 19 meV was reported for Mg-doped GaN with an A exciton position at 3.499 eV and the Mg-related ABE at 3.480 eV at 2 K [32.279] indicating that ABEs, being deeper, may not follow a constant distance to the A free-exciton position (as in the case for the most shallow DBEs) in strained layers. In unstrained samples, like Mg-doped homoepitaxial GaN epilayers, the binding energy is found to be 11–12 meV [32.226, 280, 281], under the assumption that the 3.466 eV BE is Mg-related.

An additional deeper, probably acceptor-related, BE PL line in GaN is also observed at 3.461 eV (a doublet structure) [32.269]. However, due to the strain-induced shift in an upwards direction in the spectrum, this might be the same ABE as observed in some homoepitaxial layers at 3.455 eV with a line width of about 95 μeV [32.226], indicating a BE binding energy of about 21 meV. Although the identity of this acceptor has not been established, it is close to the position observed from the dominant ABE in Zn-doped samples, if strain shifts are considered [32.279]. Therefore, it might be due to residual Zn acceptors, present as contaminants in many samples, but this is not certain.

In contrast to the case of DBEs, the decay curves are usually clean exponential for the ABEs, as seen in Fig. 32.27b, and presumably reflect the radiative lifetimes of the BEs at the lowest temperatures, before thermalization occurs. The observed value of radiative lifetimes is about 0.7 ns for the shallowest ABE peaked at 3.466 eV, as compared to the much longer time, 3.6 ns, for the deeper acceptors with an ABE peak in the range 3.455–3.46 eV [32.269]. This corresponds to an oscillator strength of the order of 1, very similar to the shallow acceptor BEs in CdS [32.282]. To distinguish band-edge exciton features, the temperature-dependent luminescence experiment allows one to discriminate free versus donor-bound or acceptor-bound excitons. With increasing temperature, all peaks related to the bound excitons quench due to thermal delocalization, while the quenching of the free excitons is negligible up to 50 K. Excitons bound to the donor quench faster than excitons bound to the acceptor, which is consistent with their binding energies [32.264].

Donor–Acceptor Transitions in GaN

Due to compensation in semiconductors both ionized donors and acceptors could be present in the material. Nonequilibrium carriers generated by optical excitation can be trapped at the donor and acceptor sites, causing them to be neutral. While reaching equilibrium, they can relax their excess energies through radiative recombination of some electrons on the neutral donor sites with holes on the neutral acceptor sites, a process termed donor–acceptor pair (DAP) transition. DAP spectra are very common examples of radiative recombination in GaN. A no-phonon replica is observed at about 3.26 eV at low temperature, with well-resolved LO-phonon replicas towards the lower energy [32.283–285]. The temperature dependence of the spectrum reveals the evolution from a DAP pair spectrum at low temperature to a free-to-bound conduction-band-to-acceptor transition

peaked 3.273 eV at higher temperature (120 K) due to thermal ionization of the shallow donors ($E_D \approx 29$ meV) into the conduction band. At intermediate temperatures, both processes may be resolved. The acceptor binding energy has also been estimated as about 230 meV from this measurement. The identity of this acceptor is not clear, but it has been suggested that it is due to carbon acceptor substitution on nitrogen sides [32.285], while others claimed that it is simply related to Mg acceptors on Ga sites. PL of Mg-doped wurtzite GaN has been studied and showed the DAP transition at 3.26 eV at 4.2 K along with its LO-phonon replicas. At higher Mg concentrations, the PL is dominated by a deep-level broad band with its peak at 2.95 eV. The blue emission is clearly observed for Mg concentrations of 5×10^{19}–2×10^{20} cm^{-3}. In addition to the peak position for the identification of this transition, further confirmation is provided by a blue shift at a rate of about 2–3 meV per decade of intensity as the excitation intensity increased [32.286].

The behavior of radiative lifetimes and time-resolved spectra has been used as a further test for the DAP transition. It has been observed that carrier dynamics of 3.21 and 2.95 eV emissions in relatively heavily doped Mg-doped p-type GaN epilayers [32.287] exhibited nonexponential lifetimes on the sub-nanoseconds scale, comparable to that of a band-to-impurity transition in a highly n-type GaN epilayer involving a donor and the valence band [32.242, 288]. *Smith* et al. [32.287] observed the temporal behavior of the 3.21-eV emission band, which follows a power law with an exponent greater than 1.0 at longer delay times. All these indications lead one to conclude that the 3.21-eV line corresponds to the conduction-band-to-impurity transition involving shallow Mg impurities, while that at about 2.95 eV is attributed to the conduction-band-to-impurity transitions involving deep-level centers (or complexes).

Defect-Related Transitions in GaN: Yellow Luminescence (YL)

A broad PL band peaking at 2.2 ± 0.1 eV, so-called yellow emission, is almost systematically observed in undoped or n-GaN. As studied extensively by *Ogino* and *Aoki* [32.289] and more recently by others [32.290–292] there seems to be agreement that transitions from the conduction band or a shallow donor to a deep acceptor are responsible for this band. Another interpretation of the yellow band as being due to transitions between a deep donor and a shallow acceptor was also proposed to explain the results of magnetic-resonance experiments [32.293]. The exact position of this band

and its line width differ slightly in numerous publications. The nature of the deep donor has not been established, but several candidates such as the nitrogen vacancy [32.294], gallium vacancy or its complexes with a shallow donor [32.291, 292] or carbon [32.289, 295] have been suggested. The issue of whether the YL is related to a point defect or to a distribution of states in the gap is still an open question. *Shalish* et al. [32.296] invoked a broad distribution of acceptor-like surface states to account for the YL band. A redshift of the YL band with decreasing energy of the below-gap excitation may also indicate that the broadening of the YL band is due to emission from several closely spaced traps [32.297]. On the other hand, the temperature dependence of the bandwidth and the photoluminescence excitation spectrum have been quantitatively explained by using a configuration-coordinate model that attributes the YL band to a point defect with strong electron–phonon interaction [32.289, 298]. Theoretical calculations predict a low formation energy and the deep acceptor levels for the isolated Ga vacancy, possibly as a complex with a shallow donor impurity such as Si and O ($V_{Ga}Si_{Ga}$ and $V_{Ga}O_N$) [32.291, 292]. It has also been demonstrated that the formation of these defects is much more favorable at a threading-edge dislocation [32.299]. The intensity of the YL band has been found to correlate with dislocation-related defects [32.300].

The transient PL decay times of the YL has been investigated by different research groups. *Hofmann* et al. [32.301] and *Korotkov* et al. [32.302] reported rather long and nonexponential decay of the YL in the range 0.1–100 μs at low temperatures, which was quantitatively described in the Thomas–Hopfield model [32.303] of the donor–acceptor pair (DAP) recombination. A very long-lived emission decay time of about 300 ms at 10 K has been observed by *Seitz* et al. [32.304]. In contrast, very fast decay times of about 1 ns at 2 K and 20 ps at room temperature have been reported and related to a strong contribution from free-to-bound transitions and DAP recombination by *Godlewski* et al. [32.305] and *Haag* et al. [32.306], respectively. *Reshchikov* et al. [32.307] performed time-dependent PL of YL in freestanding high-quality GaN templates grown by HVPE, which led them to conclude that, at temperatures below 40 K, the time decay of the 2.4 eV yellow peak is nonexponential and can be explained in the framework of the Thomas–Hopfield model for DAP-type recombination involving shallow donors. At elevated temperatures, the decay becomes exponential with two components, leading to the suggestion that the transitions from the conduction band to two deep acceptors are involved.

32.6.2 Aluminium Nitride

The optical properties of AlN have been investigated in many forms, including powders, sintered ceramics, polycrystals and single-crystal samples. Since an AlN lattice has a very large affinity to oxygen dissolution, oxygen contamination is hard to eliminate in AlN, in which optical properties are influenced by oxygen-related defects. Some oxygen is dissolved in the AlN lattice while the remainder forms an oxide coating on the surface of each powder grain.

AlN doped with oxygen was found to emit a series of broad luminescence bands at near-ultraviolet frequencies at room temperature, no matter whether the sample was powdered, single crystal, or sintered ceramic. In an early study of the luminescence properties, *Pacesova* and *Jastrabik* [32.308] observed two broad emission lines centered near 3.0 and 4.2 eV, more than 0.5 eV wide, for samples contaminated with 1–6% oxygen under steady-state excitation. *Youngman* and *Harris* [32.309] and *Harris* et al. [32.310] found broad peaks centered at 2.7 and 3.8 eV in large single-crystal AlN with an oxygen content of 380 ppm. Oxygen-related luminescence spectra in AlN are very sensitive to sample preparation, particularly oxygen-impurity content. They observed an emission-peak shift in the PL spectra and a drastic increase in luminescence intensity below a critical oxygen content of 0.75%. Others reported different emission peaks for different forms of AlN samples with various oxygen contents. A microscopic model explaining the results was proposed by *Harris* et al. [32.310] and supported by the study of *Katsikini* et al. [32.311] and *Pastrnak* et al. [32.312]. Below the critical concentration, oxygen substitutes into nitrogen sites (O_N) with subsequent formation of Al atom vacancies (V_{Al}), while at higher oxygen content a new defect based on octahedrally coordinated Al forms.

Besides native oxygen defects or intentionally oxygen-doped AlN, other impurities that have been widely investigated are manganese and, more recently, rare-earth metals, such as erbium. In an early study, *Karel* and coworkers [32.313] reported a number of sharp emission peaks in the visible region from Mn-doped AlN. These peaks are interpreted as arising from phonon emission, in all likelihood due to localized Mn-ion vibration, associated with electronic transitions experienced by Mn^{4+} ions located at Al sites. This work was further extended by *Karel* and

Mares [32.314] and *Archangelskii* et al. [32.315], who observed a red band emission (600 nm) with Mn^{4+} and a green band (515 nm) with Mn^{2+}. Recently, particular attention has been paid to Er-doped AlN due to the observed strong photoluminescence at $1.54\,\mu m$, which is important for optical-fiber communication systems. In rare-earth atoms, the luminescence transition comes from the core electrons falling energetically from excited states to the ground states. For example, Er-doped AlN gives a luminescence due to transitions between the weak crystal-field-split levels of $Er^{3+}\,{}^4I_{13/2}$ and ${}^4I_{15/2}$ multiplets [32.316–318]. *Pearton* et al. [32.319] reported luminescence enhancement in AlN(Er) samples treated in a hydrogen plasma by passivating the defects in AlN. This increase shows no quenching on heating the sample up to 300 °C. Due to its wide band gap the $1.54\text{-}\mu m$ luminescence is more stable in AlN than in GaN.

In recent work, band-edge emission of high-quality AlN grown by MOCVD on sapphire substrate has been investigated by *Li* et al. [32.320]. Band-edge emission lines at 5.960 and 6.033 eV have been observed at room temperature and 10 K, respectively, as shown in Fig. 32.29. The peak integrated emission intensity of the deep-impurity-related emission centered at around 3.2 and 4.2 eV is only about 1% and 3%, respectively, of that of the band-edge transition at room temperature. The PL emission properties of AlN have been compared with those of GaN and it was shown that the optical quality as well as the quantum efficiency of AlN epilayers is as good as that of GaN. The same group [32.322] has also studied deep UV picosecond time-resolved photoluminescence spectroscopy to investigate the observed optical transition in steady-state PL measurements. Two PL emission lines at 6.015 and 6.033 eV were attributed to donor-bound exciton and

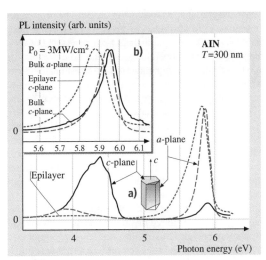

Fig. 32.30 (a) PL spectra of single-crystal AlN for *A*- and *C*-planes and AlN epilayers at room temperature under an excitation power density of $1\,MW/cm^2$. Inset (b) shows wurtzite-type crystal *c*-axes, *C*-plane, and *A*-plane. *Inset* shows room-temperature near-band emission spectra of various AlN samples under very high excitation. After [32.321]

free-exciton transitions, respectively, from which the binding energy of the donor-bound excitons in AlN epilayers was determined to be around 16 meV.

Kuokstis et al. [32.321] reported similar results by employing temperature-dependent PL measurements for *A*-plane and *C*-plane bulk single crystals and MOCVD-grown epitaxial layers on sapphire. The spectra consisted of a long wave band probably from an oxygen-related deep-level transition as discussed above and an intense

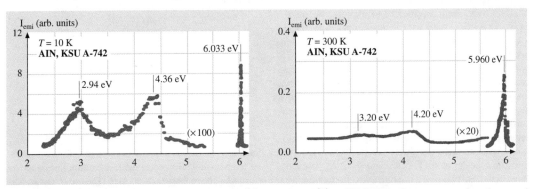

Fig. 32.29a,b PL spectra of AlN epilayers measured (**a**) at 10 K and (**b**) at 300 K. The spectra cover a broad spectral range from 2.2 to 6.2 eV. After [32.320]

near-band emission, as seen in Fig. 32.30. The near-band emissions were attributed to processes involving free excitons at 6.065 eV and their LO-phonon (≈ 5.066 eV) replicas at high temperatures and the bound excitons at 6.034 eV at low temperatures. These results, together with other well-known physical properties of AlN, may considerably expand future prospects for the application of group III nitride materials.

Time-resolved PL measurements revealed that the recombination lifetimes were around 80 ps for bound excitons and 50 ps for free excitons [32.322]. A free-exciton binding energy of 80 meV was also deduced from the temperature dependencies of both the free-exciton radiative decay time and emission intensity.

On the other hand, optical reflectance (OR) measurements [32.323] exhibited distinct reflectance anomalies at photon energies just above the multiple-internal-reflection fringes, and the spectral line shape was fitted considering A, B, and C exciton transitions. The fit gave values for the A and B excitons at 0 K of 6.211 and 6.266 eV, giving a crystal-field splitting (Δ_{cr}) of approximately 55 meV. The AlN film exhibited an excitonic emission even at 300 K, which is due to the small Bohr radius of the excitons and the large longitudinal optical-phonon energies.

Cathodoluminescence and electroluminescence emission properties of AlN have also been investigated by several groups. *Rutz* [32.177] reported a broad near-UV band in the electroluminescence spectra, extending from 215 nm into the blue end of the visible region. Several peaks or humps between 2.71 and 3.53 eV, which were attributed to nitrogen vacancies, interstitial Al impurities, and oxygen impurities or defects, were observed in cathodoluminescence measurements for epitaxial AlN films grown on C-plane sapphire [32.324]. In the investigation of Youngman and Harris, 4.64-eV photons were used specifically to explore the below-band transitions such as the peak near 4 eV. Their investigation showed a blue shift with increased O concentration in the peak in question. *Hossian* et al. [32.325] and *Tang* et al. [32.326] measured the cathodoluminescence at 300, 77, and 4.2 K for undoped AlN thin films grown on sapphire and SiC by LP-MOCVD (low pressure). From these samples, a strong luminescence peak surrounded by two weaker peaks in the near band-edge region, near 6 eV, was observed. For AlN on sapphire, this near-band-edge transition can be further resolved into three peaks at 6.11, 5.92, and 5.82 eV. They believe that these two peaks are due to excitonic transition because of their exciton-like temperature behavior, i. e. the energy position of these peaks increases and the

line width becomes narrower as the temperature is decreased. Recently, *Shishkin* et al. [32.327] has also observed a sharp, strong near-band-edge peak at 5.99 eV, which was tentatively assigned to the optical recombination of a donor-bound exciton, accompanied by weak one- and two-longitudinal optical-phonon replicas in the cathodoluminescence spectrum of a 1.25-μm-thick RMBE-grown (reactive molecular beam epitaxy) AlN film on C-plane sapphire.

Not much data are available on the temperature dependence of the optical band gap of AlN. *Guo* and *Yoshida* [32.328] investigated the variation of the band gap with temperature in AlN crystalline films. The band-gap energy increases linearly from room temperature down to about 150 K with a temperature coefficient of 0.55 meV/K. Another work was carried out by *Kuokstis* et al. [32.321] and similar temperature behavior was observed for single-crystal C-plane AlN with a temperature coefficient of 0.53 meV/K between 150 and 300 K (Fig. 32.31). These two observations are close to each other and also close to the value reported by *Tang* et al. [32.326] (0.51 meV/K). However, below 150 K, in all measurements, a new structure starts to from, where the band gap does not change linearly with temperature and has a very small temperature coefficient, which is unusual for semiconductors.

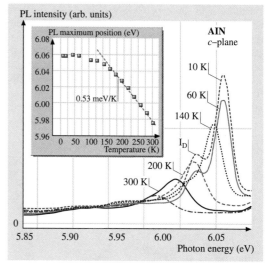

Fig. 32.31 Single-crystal AlN C-plane PL spectra at different temperatures. B denotes the free-exciton line, whereas I_D marks the bound-exciton line. The *inset* shows the dependence of the exciton line position on temperature. After [32.321]

In the case of absorption, the absorption coefficient in the near band edge was measured by using straightforward optical transmission and reflection experiments. In early studies, *Yim* et al. [32.174] characterized AlN by optical absorption and determined the room-temperature band gap to be direct with a value of 6.2 eV. *Perry* and *Rutz* [32.329] performed temperature-dependent optical absorption on a $1 \times 1 \mathrm{cm}^2$ epitaxial single-crystal AlN sample, measuring a band gap of 6.28 eV at 5 K compared to their room-temperature value of 6.2 eV, resulting from a straight-line fit. Several groups have reported comparable values whereas others have produced questionable values considerably below 6.2 eV, probably due to oxygen contamination or nonstoichiometry. In addition to the band-edge absorption, a much lower energy absorption peak at 2.86 eV (although some variation in the peak position has been recorded from 2.8 to 2.9 eV) is likely to be due to nitrogen vacancies or nonstoichiometry, as proposed by *Cox* et al. [32.173]. *Yim* and *Paff* [32.101] also observed a broad emission-spectrum range of 2–3 eV with a peak at about 2.8 eV. This peak does not correlate with the presence of oxygen. The oxygen absorption region lies between 3.5 eV and 5.2 eV, as originally found by *Pastrnak* and *Roskovcova* [32.330]. The exact position of this particular peak appears to change with the oxygen content from 4.3 eV at low oxygen levels to 4.8 eV at high oxygen levels.

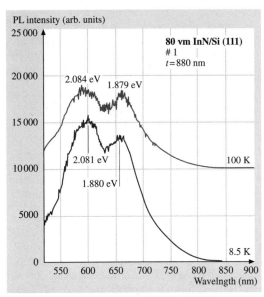

Fig. 32.32 PL spectra at 8.5 and 100 K of InN films grown on Si substrate. After [32.331]

Measurements of the refractive index of AlN have been carried out in amorphous, polycrystalline, and single-crystal epitaxial thin films [32.332–335]. The values of the refractive index are in the range $n = 1.9$–2.34 with a generally accepted value of $n = 2.15 \pm 0.05$ at visible wavelengths. These values are found to increase with increasing structural order, varying between 1.8 to 1.9 for amorphous films, 1.9 to 2.1 for polycrystalline films, and 2.1 to 2.2 for single-crystal epitaxial films. The spectral dependence and the polarization dependence of the index of refraction have been measured and showed a near-constant refractive index in the wavelength range 400–600 nm. The reported refractive index generally agrees with low- and high-frequency dielectric constants determined from infrared measurements. In the long-wavelength range, the low-frequency dielectric constant of AlN (ε_0) lies in the range 8.5–9.1 [32.336, 337], and most of the values fall within the range 8.5 ± 0.2, while the high-frequency dielectric constant (ε_∞) is in the range 4.6–4.8 [32.98, 338].

32.6.3 Indium Nitride

Among the group III nitrides, InN is not very well investigated. The band gap has been frequently measured by transmission experiments. In early studies, due to the large background electron concentrations of the samples and their low crystalline quality, the excitonic features, neither in the absorption spectra nor in the band-edge photoluminescence spectra, had been observed. However, a number of groups have performed optical measurements on InN [32.339–342]. Early experimental studies suggested that the band gap of wurtzite InN ranges from 1.7–2.07 eV at room temperature. Table 32.1 lists some experimental determinations of the InN band gap. They were estimated from the absorption spectra, which can lead to an overestimation of the band gap if the sample quality is poor or if the sample is highly doped. Despite this, the value of 1.89 eV for the InN band gap obtained by *Tansley* and *Foley* [32.343] is widely cited in the literature. They explained this experimental scatter by considering residual electron concentration (which changes the band gap of the semiconductor, as discussed before) and attributed this variation to the combined effect of band-tailing and band-filling. The value of 1.89 eV is often used as an end-point value to interpret experimentally measured composition dependence of the band gap of InN alloys [32.344]. A few studies of the inter-band optical absorption performed on InN thin films deposited by sputtering techniques [32.343] and

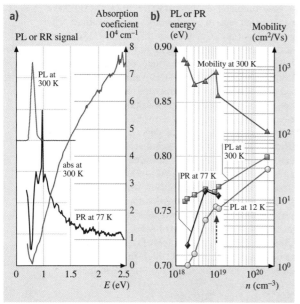

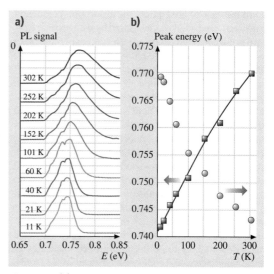

Fig. 32.33 (a) Optical absorption (300 K), PL (300 K), and PR (77 K) spectra of a typical InN sample. This sample is undoped with a room-temperature electron concentration of 5.48×10^{18} cm^{-3}. The spike on the PR spectrum at 0.97 eV is an artifact due to the light source used in the PR measurement. (b) Room-temperature mobility, PL peak energy (300 and 12 K), and the critical energy determined by PR (77 K) as a function of free-electron concentration. After [32.349]

MOCVD [32.328] were found consistent with a fundamental energy gap of about 2 eV. Recent progress of growth techniques using molecular-beam epitaxy has led to improved InN samples, which show photoluminescence as well as a clear absorption edge. Yodo et al. [32.331] observed weak photoluminescence peaks with energies in the range 1.81–2.16 eV on InN grown on Si substrates (Fig. 32.32). However, in one such case, an emission centered at 1.86 eV at a temperature below 20 K was seen, while the reflectance measurement shows a strong plasma reflection at 0.7 eV. Another branch of studies show that in improved InN films strong photoluminescence transitions at energies around 1 eV appear [32.345–348]. These new measurements have challenged the previous widely accepted band-gap value and suggest that the actual fundamental band gap of InN is much smaller, between 0.7 and 1.1 eV. The real band gap of InN is still under investigation and undetermined.

Figure 32.33 shows the room-temperature electron mobility, the peak energy of PL, and the transition energy

Fig. 32.34 (a) PL spectra as a function of temperature for the sample shown in Fig. 32.20a. PL spectra are normalized to a constant peak height. (b) PL peak energy and PL integrated intensity (log scale) as a function of temperature. The *line* through the peak energy data is a guide to the eye. After [32.349]

determined by PR as functions of electron concentration. The transition energies increase with increasing free-electron concentration, indicating that the transitions from higher-energy occupied states in the conduction band contribute significantly to the PL spectrum. The absorption coefficient increases gradually with increasing photon energy, and at the photon energy of 1 eV it reaches a value of more than 10^4 cm^{-1}. This high value is consistent with inter-band absorption in semiconductors. Moreover, the integrated PL intensity increased linearly with excitation intensity over three orders of magnitude, lending more credence to the concept that the observed nonsaturable peak is related to the fundamental inter-band transitions. The free-electron concentration in this sample was measured by the Hall effect to be 5.48×10^{18} cm^{-3}, which also shows intense room-temperature luminescence at energies close to the optical absorption edge. Additionally, the 77-K photoreflectance (PR) spectrum exhibits a transition feature at 0.8 eV with a shape characteristic for direct-gap interband transitions. Consistent with the absorption data, no discernible change in the PR signal near 1.9 eV is seen. The simultaneous observations of the absorption edge, PL, and PR features at nearly the same energy indicate that this energy position of ≈ 0.78 eV led *Wu* et al. [32.349] to argue that this is the fundamental band gap of InN. This value is very close to the fundamental gap for InN reported by *Davydov* et al. [32.345].

A few groups [32.350, 351] have reported the temperature dependence of the band gap of InN, and indicated a band-gap temperature coefficient of $1.3–1.8 \times 10^{-4}$ eV/K at room temperature by using $E_g(300\,\text{K}) = 1.89$ eV. The temperature dependence of PL spectra of recently investigated InN samples indicated a lower band-gap energy of 0.7 eV, as shown in Fig. 32.34. There is a small blue shift (nearly linear at 0.1 meV/K) observed in the peak energy position of the temperature-dependent PL. In addition to that, the integrated intensity of the PL decreases by ≈ 20 times as the temperature is increased from 11 K to room temperature. The data in Fig. 32.34a also show a considerable increase in the line width of the PL spectra. The full width at half maximum increases from 35 to 70 meV when the temperature increases from 11 K to room temperature.

Therefore, they concluded that there is no significant shift of the PL spectra, as the temperature-induced line broadening can easily account for the observed small upward shift of the PL line maximum.

In the transparency region, the refractive index was measured using the interference spectrum observed in reflectivity or absorption [32.352, 353]. In the visible and UV region ($E > 2$ eV) optical constants were determined from reflectivity or spectroscopic ellipsometry [32.354, 355]. Although the most frequently quoted value for the static dielectric constant is $\varepsilon_0 = 15.3$, the values reported for the high-frequency dielectric constant are scattered between 6.6 and 8.4, as well as the refractive index measured in the vicinity of the band gap. *Tyagai* et al. [32.356] were able to estimate an effective mass of $m*_e = 0.11\,m_0$ and an index of refraction $n = 3.05 \pm 0.05$ in the vicinity of the band gap. The long-wavelength limit of the refractive index was reported to be 2.88 ± 0.15.

32.7 Properties of Nitride Alloys

Many important GaN-based devices involve heterostructures to achieve improved device performance. Ternary alloys of wurtzite polytypes of GaN AlN, and InN have been obtained in continuous alloy systems whose direct band gap ranges from ≈ 0.8 eV for InN to 6.1 eV for AlN. Many of these properties, such as the energy band gap, effective masses of the electrons and holes, and the dielectric constant, are dependent on the alloy composition. AlN and GaN are reasonably well lattice-matched (3.9%) and, owing to the large band-gap difference between GaN and AlN, for many devices only small amounts of AlN are needed in the GaN lattice to provide sufficient carrier and optical field confinement. The compositional dependence of the lattice constant, the direct energy gap, electrical and cathodoluminescence (CL) properties of the AlGaN alloys were measured [32.358]. In general, the compositional dependence of the optical band gap of ternary alloys can be predicted by the following empirical expression;

$$E_g(x) = xE_g(\text{AlN}) + (1-x)E_g(\text{GaN}) - x(1-x)b\,, \tag{32.16}$$

where $E_g(\text{AlN})$, $E_g(\text{GaN})$ are the optical band gap of AlN and GaN, respectively, and x and b are the AlN molar fraction and bowing parameter, respectively. In order to determine this relation precise characterization of both the band gap and alloy composition is important. Wide dispersion in the bowing parameters ranging from -0.8 eV (upward bowing) to $+2.6$ eV (downward bowing) has been reported [32.359–364]. *Yun* et al. [32.357]

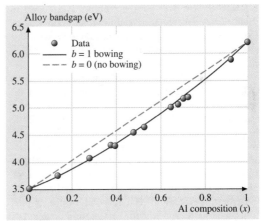

Fig. 32.35 Experimental data for the energy band gap of AlGaN ($0 \leq x \leq 1$) plotted as a function of Al composition (*solid circle*), and the least-squares fit (*solid line*), giving a bowing parameter of $b = 1.0$ eV. The *dashed line* shows the case of zero bowing. As the quality of the near-50:50 alloy layers get better, giving rise to sharper X-ray and PR data, smaller bowing parameters may ensue. Already, bowing parameters as low as 0.7 have been reported. After [32.357]

revisited the bowing parameter using X-ray and analytical techniques, such as secondary-ion mass spectroscopy (SIMS) and Rutherford back-scattering (RBS), for composition determination, reflectance and absorption for band-gap determination. The results of this study are depicted in Fig. 32.35 in the form of AlGaN band gap versus composition along with a least-square fit to the data, solid circles, yields a bowing parameter of $b = 1.0$ eV for the entire range of alloy compositions. Widening the X-ray diffraction peaks for alloy composition around the midway point has been attributed to be a most likely source of error in determining the bowing parameter. On the other hand, the validity of the characterization techniques used in determination the optical properties of AlGaN alloys is deeply affected by the material crystalline quality and purity.

As far as the electrical and doping issues are concerned, Hall measurement for n-$Al_{0.09}Ga_{0.91}N$ demonstrated a carrier concentration of 5×10^{18} cm^{-3} and a mobility of 35 cm^2/Vs at room temperature [32.365]. Other Hall measurements [32.366] on Mg-doped p-$Al_{0.08} \cdot Ga_{0.92}N$ grown by OMVPE (organometallic vapor phase epitaxy) addressed the temperature dependence of the mobility. They indicate that the hole mobility decreases with increasing temperature, reaching a value of about 9 cm^2/Vs for a doping density of 1.48×10^{19} cm^{-3}. This low mobility is ascribed to a high carrier concentration and the inter-grain scattering present in the samples. While the lattice constant was studied, it was observed to be almost linearly dependent on the AlN mole fraction in AlGaN.

Until recently the resistivity of unintentionally doped AlGaN was believed to increase so rapidly with increasing AlN mole fraction that AlGaN became almost insulating for AlN mole fractions exceeding 20%. As the AlN mole fraction increased from 0 to 30%, the n-type carrier concentration dropped from 10^{20} to 10^{17} cm^{-3}, and the mobility increased from 10 to 30 cm^2/Vs. An increase in the native-defect ionization energies with increasing AlN may possibly be responsible for this variation. The respond of the dopant atoms such as Si and Mg to the variation of the AlN mole fraction in AlGaN has not been well understood yet. It was suggested that dopant atom moves deeper into the forbidden energy band gap as the AlN mole fraction increases. For example, Hall-effect measurements show that the activation energy of Si donor increases linearly from 0.02 eV in GaN to 0.32 eV in AlN [32.367]. However, devices such as lasers, which depend critically on the overall device series resistance and require low-resistivity p-type material, will probably be restricted by the ability to dope high-mole-fraction AlGaN. Fortunately, the emergence of InGaN coupled with the fact that good optical-field confinement can be obtained with low-AlN-mole-fraction AlGaN mitigate this problem enormously, and the potential is very bright for laser development in this material system.

The ternary InGaN is used mostly for quantum wells, strained to some extent depending on the level of phase segregation, etc., in the active regions of LEDs and lasers, which can emit in the violet or blue wavelength range. Needless to say, high-efficiency blue and green LEDs utilizing InGaN active layers are commercially available. However, added complexities such as phase separation and other inhomogeneities due to the large disparity between Ga and In make the determination of the band gap of InGaN versus composition a very difficult task, not to mention the controversy regarding the band gap of InN. The compositional dependence of the InGaN band gap is a crucial parameter in the design of any heterostructure utilizing this material. Similarly to the case of AlGaN, the energy band gap of $In_xGa_{1-x}N$ over $0 < x < 1$ can be expressed by (32.16) using the band gap of InN instead of AlN band gap. When a bandgap of ≈ 1.9 eV for InN is assumed as the end-point value for InN in regard to the InGaN ternary, large and/or more than one bowing parameters are required to fit the compositional dependence of the band-gap energy. An earlier investigation of InGaN bowing parameter for alloys with small concentrations of InN [32.368] led to a bowing parameter of 1.0, which is in disagreement with 3.2 reported by *Amano* et al. [32.369], who also took into consideration strain and piezoelectric fields and arrived at a value of 3.2. A bowing parameter of 2.5 eV was obtained from optical absorption measurements and a value of 4.4 eV was obtained from the position of the emission peak [32.370]. *Nagatomo* et al. [32.371] noted that the $In_xGa_{1-x}N$ lattice constant varies linearly with the In mole fraction up to at least $x = 0.42$, but it violates Vegard's law for $x > 0.42$, which may be caused by erroneous determination of the composition and is very relevant the problem at hand. Recent observations indicated that these alloys show strong infrared PL signal as expected from an InN band gap of ≈ 0.8 eV, extending the emission spectrum of the $In_{1-x}Ga_xN$ system to near infrared. [32.347] revisited the dependence of the InGaN band gap on composition by considering ≈ 0.8 eV for the band gap of the end binary InN. Figure 32.36 shows the compositional dependence of the band gap of InGaN, determined by photomodu-

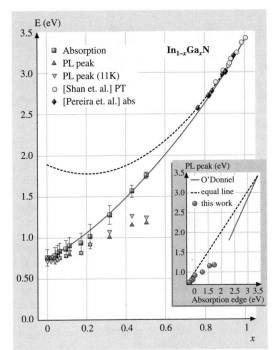

Fig. 32.36 PL peak energy and band gap determined by optical absorption as a function of composition, including previously reported data for the Ga-rich end of the ternary which is not as affected by the large InN band gap previously used. The *solid curve* shows the fit to the band gap energies (abs and PT) using a bowing parameter $b = 1.43$ eV. The *dashed curve* is the fit to the band-gap energies on the Ga-rich side assuming a band gap of 1.9 eV for InN. *Inset*: PL peak energy plotted against absorption edge energy. The *solid line* is a least-squares fit to experimental data on the Ga-rich side. The *dashed straight line* shows the relation when the Stokes shift is zero. After [32.245] and courtesy of *Wladek Walukiewicz*

lated transmission [32.372] optical absorption [32.373] measurements, as a function of GaN fraction. The compositional dependence of the band gap in the entire compositional range can also be fit by a bowing parameter of $b = 1.43$ eV. Also shown in the figure with a dashed line is the fit to the empirical expression using an energy of 1.9 eV for InN and bowing parameter of 2.63 eV to demonstrate that it does represent the Ga-rich side of the compositions well. However, the bowing parameter that is good for the entire compositional range is 1.43 eV, utilizing 0.77 eV for the band gap of InN.

There are isolated reports on electrical properties of the InGaN alloy. *Nagatoma* et al. [32.371] reported high-resistivity InGaN films with In compositions as high as 0.42. Background electron concentration as low as 10^{16} cm^{-3} has been reported for MOVPE-grown InGaN films. Conductivity control of both n-type and p-type InGaN was reported by *Akasaki* and *Amano* [32.374]. The growth and mobility of p-InGaN was also discussed by *Yamasaki* et al. [32.375]. *Yoshimoto* et al. [32.376] studied the effect of growth conditions on the carrier concentration and transport properties of In$_x$Ga$_{1-x}$N. They observed that increasing the deposition temperature of In$_x$Ga$_{1-x}$N grown both on sapphire and ZnO with $x \approx 0.2$ results in a decrease in carrier concentration and increase in electron mobility. *Nakamura* and *Mukai* [32.150] discovered that the film quality of In$_x$Ga$_{1-x}$N grown on high-quality GaN films could be significantly improved. Thus, it may be concluded that the major challenge for obtaining high-mobility InGaN is to find a compromise value for the growth temperature, since InN is unstable at typical GaN deposition temperatures. This growth temperature would undoubtedly be a function of the dopant atoms, as well as the method (MBE, OMVPE, etc.) used for the growth. This is evident from a study by Nakamura and coworkers, who have since expanded the study of InGaN employing Si [32.377] and Cd [32.378].

In$_{1-x}$Al$_x$N is an important compound that can provide a lattice-matched barrier to GaN, low-fraction AlGaN and InGaN, and consequently, lattice-matched AlInN/AlGaN or AlInN/InGaN heterostructures would result. The growth and electrical properties of this semiconductor have not yet been extensively studied, as the growth of this ternary is also challenging because of the different thermal stability, lattice constant, and cohesive energy of AlN and InN. *Kim* et al. [32.379] deposited thin AlInN films and observed an increase of In content in AlInN of up to 8% by lowering the substrate temperature to 600 °C. Radio-frequency (RF) sputtering was employed to grow InAlN alloy by *Starosta* [32.380], and later *Kubota* et al. [32.342]. *Kistenmacher* et al. [32.381], on the other hand, used RF-magnetron sputtering (RF-MS) from a composite metal target to grow InAlN at 300 °C. It was observed that the energy band gap of this alloy varies between 2.0 eV and 6.20 eV for x between 0 and 1 [32.342]. Optical properties of 1-μm-thick Al$_{1-x}$In$_x$N layers for x values up to the range of 0.19–0.44 have been investigated by absorption and photoluminescence [32.382]. Figure 32.37 shows results of the photoluminescence spectra from Al$_{1-x}$In$_x$N layers. From the absorption spectra the band

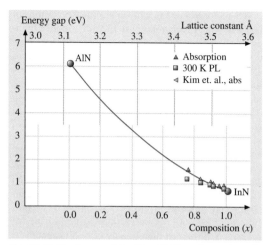

Fig. 32.37 Band gap of $In_xAl_{1-x}N$ films for various In concentrations. Courtesy of *Wladek Walukiewicz*

gaps are found to shift gradually with varying In content. Tailing of the band gap edge is also found to occur, probably due to fluctuation of the In content and grain boundaries. The PL peak-energy position shifts towards the lower-energy region with increasing In content as expected [32.383]. Considering that PL spectral line widths are much larger than expected from a completely random alloy, mechanisms other than alloy broadening such as microscopic phase separation might be invoked to account for such broadening. The small amount which the PL peak shifts (the same is true for the absorption edge) with mole fraction, x, might reflect the immiscibility of AlN in InN (or vice versa).

Transport data for InAlN are extremely scarce. The carrier concentration and the mobility of $In_{1-x}Al_xN$ for $x = 0.04$ were reported as 2×10^{20} cm^{-3} and 35 cm^2/Vs, respectively, and 8×10^{19} cm^{-3} and 2 cm^2/Vs for $x = 0.25$, respectively [32.384]. The mobility was found to decrease substantially with an increase in the Al mole fraction and increase with an increase in the In mole fraction, in close analogy with the parent compounds AlN and InN.

By alloying InN with GaN and AlN, the band gap of the resulting alloy(s) can be increased from 1.9 eV (≈ 0.8 eV if we use the recently determined InN band gap) to a value of 6.1 eV, which is critical for making high-efficiency visible-light sources and detectors. In addition, the band gap of this quaternary can be changed while keeping the lattice constant matched to GaN [32.385, 386]. In quaternary alloys of nitrides the N atoms constitute an anion sublattice while the three group III elements (In, Ga, Al) constitute the cation sublattice. Use of this quaternary material allows almost independent control of the band offset in AlInGaN-based heterostructures. However, among other difficulties brought about by the four-component system, the optimal growth temperature is important to optimize and control, as aluminium-based compounds generally require higher growth temperatures, while In-based compounds require lower temperatures. Higher temperatures are also desirable to reduce the incorporation of O into the growing film as oxides of Ga and In desorb from the surface. The growth temperature will therefore govern the limits of In and Al incorporation into the AlGaInN quaternary alloy [32.385]. Quaternary alloys, $(Ga_{1-x}Al_x)In_{1-y}N$, are expected to exist in the entire composition range $0 < x < 1$ and $0 < y < 1$. Unfortunately, as is the case for the InGaN alloy, the incorporation of indium into these quaternary alloys is not easy. *Ryu* et al. [32.387] reported on optical emission in this quaternary system and AlInGaN/AlInGaN multiple quantum wells grown by pulsed metalorganic chemical-vapor deposition. A strong blue shift with excitation intensity was observed in both the quaternary layers and quantum wells, which was attributed to localization. This would imply that the structures are of inhomogeneous nature and or presence of band-tail states indicative or early stages of material development and or serious technological problems involved.

32.8 Summary and Conclusions

AlN exhibits many useful mechanical and electronic properties. For example, hardness, high thermal conductivity, resistance to high temperature and caustic chemicals combined with, in noncrystalline form, a reasonable thermal match to Si and GaAs, make AlN an attractive material for electronic packaging applications.

The wide band gap is also the reason for AlN to be touted as an insulating material in semiconductor device applications. Piezoelectric properties make AlN suitable for surface-acoustic-wave devices and sensor applications. However, the majority of interest in this semiconductor in the context of electronic and optoelectronic devices stems from its ability to form alloys with GaN, producing AlGaN and allowing the fabrication of electronic and optical devices based on AlGaN/GaN and AlGaN/InGaN, the latter of which can be active from green wavelengths well into the ultraviolet. AlN also forms a crucial component of the nitride-based AlInGaN quaternary which allows tuning of the band gap independently of the composition over a reasonably wide range of band gaps. This way, lattice-matched conditions to the underlying epitaxial structure can be maintained while being able to adjust the band gap.

AlN is not a particularly easy material to investigate because of the high reactivity of aluminium with oxygen in the growth vessel. Early measurements indicated that oxygen-contaminated material can lead to errors in the energy band gap and, depending on the extent of contamination, in the lattice constant. Only recently has the attainment of contamination-free deposition environments coupled with advanced procedures allowed researchers to grow higher-quality AlN consistently. Consequently, many of the physical properties of AlN have been reliably measured and bulk AlN synthesized.

Although GaN has been studied far more extensively than the other group III nitrides, there is still a great need for further investigations even to approach the level of understanding of technologically important materials such as Si and GaAs. GaN growth often suffers from relatively high background n-type carrier concentrations due to native defects and possibly impurities. The lack of commercially available native substrates exacerbates this situation. These, together with the difficulties in obtaining p-type doping, and the arcane fabrication processes, catalyzed the early bottlenecks, stymieing progress. Recent burgeoning interest has led to improvements in crystal-growth and processing technologies, and allowed many difficulties encountered earlier to be overcome. Consequently, a number of laboratories have begun to obtain high-quality GaN consistently with room-temperature background electron concentrations as low as 5×10^{16} cm^{-3}. The successful development of schemes leading to p-type GaN has led to the demonstration of excellent p-n-junction LEDs in the UV, violet, blue, green and even yellow bands of the visible spectrum with brightness suitable for outdoor displays, CW (continuous wave) lasers, power modulation doped field-effect transistors (MODFETs), and UV detectors, including those for the solar-blind region. Still, much work remains to be done in the determination of the fundamental physical properties of GaN.

InN has not received the experimental attention given to GaN and AlN. This is probably due to difficulties in growing high-quality crystalline InN samples, the poor luminescence properties of InN, and the existence of alternative, well-characterized/developed semiconductors such as AlGaAs and In (Ga, Al)AsP, which have energy band gaps close to what was thought to be the band gap of InN. More recent data and calculations appear to indicate that the band gap of InN is close to 0.7–0.8 eV. Even with this new band gap, InN does not escape competition in the sense that another well-developed semiconductor, InGaAs, covers that region. Consequently, practical applications of InN are restricted to its alloys with GaN and AlN, in addition to tandem solar cells. The growth of high-quality InN and the enumeration of its fundamental physical properties remain, for the present, a purely scientific enterprize. InN is not different from GaN and AlN in the sense that it suffers from the same lack of a suitable substrate material and, in particular, a high native-defect concentration. Moreover, early InN layers may have been polycrystalline and contained large concentrations of O and impurities and/or point defects resulting in large electron concentrations. The large band gap measured in early InN layers may have been caused by O, leading to oxides of In with large band gaps as well as a Burstein–Moss blue shift due to high electron concentrations.

References

32.1 H. Morkoç: *Nitride Semiconductors and Devices*, its 2nd edition will appear within 2004 (Springer, Heidelberg 1999)

32.2 S. Nakamura, S. Pearton, G. Fasol: *The Blue Laser Diodes* (Springer, Berlin Heidelberg New York 2001)

32.3 P. Ruterana, M. Albrecht, J. Neugebauer: *Nitride Semiconductors – Handbook on Materials and Devices* (Wiley, Weinheim 2003)

32.4 S.J. Pearton: *GaN and Related Materials* (Gordon and Breach, New York 1997)

32.5 B. Gil: *Group III Nitride Semiconductor Compounds* (Clarendon, Oxford 1998)

32.6 J.I. Pankove, T.D. Moustakas: *GaN*, Vol.1 (Academic, New York 1998)

32.7 T.D. Moustakas, I. Akasaki, B.A. Monemar: III–V nitrides, Mater. Res. Soc. Symp. Proc. **449**, 482 (1997, 1998)

32.8   M. E. Levinshtein, S. L. Rumyantsev, M. S. Shur: *Properties of Advanced Semiconductor Materials* (Wiley, New York 2001)
32.9   J. H. Edgar: *Properties of Group III Nitrides* (INSPEC, London 1994)
32.10   J. H. Edgar, S. T. Strite, I. Akasaki, H. Amano, C. Wetzel: *Properties, Processing and Applications of Gallium Nitride and Related Semiconductors* (INCPEC, London 1999)
32.11   S. N. Mohammad, W. Kim, A. Salvador, H. Morkoc: MRS Bull. **22**, 22 (1997)
32.12   S. Strite, H. Morkoç: J. Vac. Sci. Technol. B **10**, 1237 (1992)
32.13   H. Morkoc, S. Strite, G. B. Gao, M. E. Lin, B. Sverdlov, M. Burns: J. Appl. Phys. **76**, 1363 (1994)
32.14   S. N. Mohammad, H. Morkoç: Proc. Quant. Electr. **20**, 361 (1996)
32.15   H. Morkoç, S. N. Mohammad: Science **267**, 51 (1995)
32.16   I. Akasaki, H. Amano: Tech. Dig. Int. Electron. Devices Meet. **96**, 231 (1996)
32.17   S. Nakamura: MRS Bull. **22**, 29 (1997)
32.18   S. Nakamura: Sel. Top. Quant. Electron. **3**, 712 (1997)
32.19   S. Nakamura: Science **281**, 956 (1998)
32.20   S. J. Pearton, J. C. Zolper, R. J. Shul, F. Ren: J. Appl. Phys. **86**, 1 (1999)
32.21   S. J. Pearton, F. Ren, J. C. Zolper, R. J. Shul: Mater. Sci. Eng. **R30**, 55 (2000)
32.22   M. A. Khan, Q. Chen, J. W. Yang, C. J. Sun: Inst. Phys. Conf. Ser. **142**, 985 (1995)
32.23   M. A. Khan, Q. Chen, J. Yang, M. Z. Anwar, M. Blasingame, M. S. Shur: Tech. Dig. Int. Electron. Devices Meet. **96**, 27 (1996)
32.24   M. S. Shur, M. A. Khan: MRS Bull. **22**, 44 (1997)
32.25   L. Liu, J. H. Edgar: Mater. Sci. Eng. **R37**, 61 (2002)
32.26   S. C. Jain, M. Willander, J. Narayan, R. Van Overstraeten: J. Appl. Phys. **87**, 965 (2000)
32.27   B. Monemar: Semicond. Semimetals **50**, 305 (1998)
32.28   O. Ambacher: J. Phys. D: Appl. Phys. **31**, 2653 (1998)
32.29   I. Akasaki, H. Amano, I. Suemune: Inst. Phys. Conf. Ser. **142**, 7 (1996)
32.30   W. A. Harris: *Electronic Structure and Properties of Solids* (Dover, New York 1980) pp. 174–179
32.31   C. Y. Yeh, Z. W. Lu, S. Froyen, A. Zunger: Phys. Rev. B **46**, 10086 (1992)
32.32   G. A. Jeffery, G. S. Parry, R. L. Mozzi: J. Chem. Phys. **25**, 1024 (1956)
32.33   H. Schulz, K. H. Theimann: Solid State Commun. **23**, 815 (1977)
32.34   Q. Xia, H. Xia, A. L. Ruoff: J. Appl. Phys. **73**, 8198 (1993)
32.35   P. Perlin, C. Jauberthie-Carillon, J. P. Itie, A. San Miguel, I. Grzegory, A. Polian: Phys. Rev. B **45**, 83 (1992)
32.36   M. Ueno, M. Yoshida, A. Onodera, O. Shimommura, K. Takemura: Phys. Rev. B **49**, 14 (1994)
32.37   M. J. Paisley, Z. Sitar, J. B. Posthill, R. F. Davis: J. Vac. Sci. Technol. **7**, 701 (1989)
32.38   T. Lei, M. Fanciulli, R. J. Molnar, T. D. Moustakas, R. J. Graham, J. Scanlon: Appl. Phys. Lett. **59**, 944 (1991)
32.39   R. C. Powell, N. E. Lee, Y. W. Kim, J. E. Greene: J. Appl. Phys. **73**, 189 (1993)
32.40   M. Mizita, S. Fujieda, Y. Matsumoto, T. Kawamura: Jpn. J. Appl. Phys. **25**, L945 (1986)
32.41   M. Leszczynski, J. Bak-Misiuk, J. Domagala, J. Muszalski, M. Kaniewska, J. Marczewski: Appl. Phys. Lett. **67**, 539 (1995)
32.42   M. Leszczynski, E. Litwin-Staszewska, J. Bak-Misiuk, J. Domagala: Acta Phys. Pol. **88**, 837 (1995)
32.43   G. S. Cargill, A. Segmuller, T. F. Kuech, T. N. Theis: Phys. Rev. B **46**, 10078 (1992)
32.44   M. Leszczynski, T. Suski, P. Perlin, H. Teisseyre, I. Grzegory, M. Bockowski, J. Jun, S. Polowski, K. Pakula, J. M. Baranowski, C. T. Foxon, T. S. Cheng: Appl. Phys. Lett. **69**, 73 (1996)
32.45   J. Härtwing, S. Groswing: Phys. Stat. Solidi **115**, 369 (1989)
32.46   W. L. Bond: Acta Crystallogr. **13**, 814 (1960)
32.47   I. Petrov, E. Mojab, R. Powell, J. Greene, L. Hultman, J.-E. Sundgren: Appl. Phys. Lett. **60**, 2491 (1992)
32.48   M. E. Sherwin, T. J. Drummond: J. Appl. Phys. **69**, 8423 (1991)
32.49   Q. Xia, H. Xia, A. L. Ruoff: Phys. Rev. B **74**, 12925 (1993)
32.50   H. Vollstadt, E. Ito, M. Akaishi, S. Akimoto, O. Fukunaga: Proc. Jpn. Acad. B **66**, 7 (1990)
32.51   A. Munoz, K. Kunc: Phys. Rev. B **44**, 10372 (1991)
32.52   P. E. Van Camp, V. E. Van Doren, J. T. Devreese: Solid State Commun. **81**, 23 (1992)
32.53   M. Tanaka, S. Nakahata, K. Sogabe, H. Nakata, M. Tabioka: Jpn. J. Appl. Phys. **36**, L1062 (1997)
32.54   H. Angerer, D. Brunner, F. Freudenberg, O. Ambacher, M. Stutzmann, R. Höpler, T. Metzger, E. Born, G. Dollinger, A. Bergmaier, S. Karsch, H.-J. Körner: Appl. Phys. Lett. **71**, 1504 (1997)
32.55   J. Domagala, M. Leszczynski, P. Prystawko, T. Suski, R. Langer, A. Barski, M. Bremser: J. Alloy Compd. **286**, 284 (1999)
32.56   K. Kim, W. R. L. Lambrecht, B. Segall: Phys. Rev. B **53**, 16310 (1996). erratum: [32.388]
32.57   A. F. Wright, J. S. Nelson: Phys. Rev. **51**, 7866 (1995)
32.58   M. Leszczynski, P. Prystawko, T. Suski, B. Lucznik, J. Domagala, J. Bak-Misiuk, A. Stonert, A. Turos, R. Langer, A. Barski: J. Alloy Compd. **286**, 271 (1999)
32.59   T. Detchprohm, K. Hiramatsu, K. Itoh, I. Akasaki: Jpn. J. Appl. Phys. **31**, L1454 (1992)
32.60   M. Leszczynski, H. Teisseyre, T. Suski, I. Grzegory, M. Bockowski, J. Jun, S. Polowski, J. Major: J. Phys. D **69**, A149 (1995)
32.61   T. Deguchi, D. Ichiryu, K. Toshikawa, K. Sekiguchi, T. Sota, R. Matsuo, T. Azuhata, M. Yamaguchi, T. Yagi, S. Chichibu, S. Nakamura: J. Appl. Phys. **86**, 1860 (1999)

32.62 W. Paszkowicz, J. Adamczyk, S. Krukowski, M. Leszczynski, S. Porowski, J. A. Sokolowski, M. Michalec, W. Lasocha: Philos. Mag. A **79**, 1145 (1999)

32.63 S. Strite, D. Chandrasekhar, D. J. Smith, J. Sariel, H. Chen, N. Teraguchi: J. Cryst. Growth **127**, 204 (1993)

32.64 W. Paszkowicz: Powder Diffr. **14**, 258 (1999)

32.65 F. D. Murnaghan: Proc. Natl. Acad. Sci. **30**, 244 (1944)

32.66 A. Polian, M. Grimsditch, I. Grzegory: J. Appl. Phys. **79**, 3343 (1996)

32.67 I. Yonenaga, T. Shima, M. H. F. Sluiter: Jpn. J. Appl. Phys. **41**, 4620 (2002)

32.68 I. Yonenaga: MRS Internet J. Nitride Semicond. Res. **7**, 6 (2002)

32.69 T. Tsubouchi, N. Mikoshiba: IEEE Trans. Sonics Ultroson. **SU 32**, 634 (1985)

32.70 L. E. McNeil, M. Grimsditch, R. H. French: J. Am. Ceram. Soc. **76**, 1132 (1993)

32.71 A. F. Wright: J. Appl. Phys. **82**, 2833 (1997)

32.72 T. Tsubouchi, K. Sugai, N. Mikoshiba: *Ultrosonic Symposium Preceedings* (IEEE, New York 1981) p. 375

32.73 R. Thokala, J. Chaudhuri: Thin Solid Films **266**, 189 (1995)

32.74 M. Ueno, A. Onodera, O. Shimomura, K. Takemura: Phys. Rev. B **45**, 10123 (1992)

32.75 D. Gerlich, S. L. Dole, G. A. Slack: J. Phys. Chem. Solid **47**, 437 (1986)

32.76 P. E. Van Camp, V. E. Van Doren, J. T. Devreese: Phys. Rev. B **44**, 9056 (1991)

32.77 E. Ruiz, S. Alvarez, P. Alemany: Phys. Rev. B **49**, 7617 (1994)

32.78 V. A. Savastenko, A. V. Sheleg: Phys. Stat. Sol. **A 48**, K135 (1978)

32.79 R. B. Schwarz, K. Khachataryan, E. R. Weber: Appl. Phys. Lett. **70**, 1122 (1997)

32.80 C. Deger, E. Born, H. Angerer, O. Ambacher, M. Stutzmann, J. Hornstein, E. Riha, G. Fischeruer: Appl. Phys. Lett. **72**, 2400 (1998)

32.81 R. Nowak, M. Pessa, M. Suganuma, M. Leszczynski, I. Grzegory, S. Porowski, F. Yoshida: Appl. Phys. Lett. **75**, 2070 (1999)

32.82 I. Yonenaga, K. Motoki: J. Appl. Phys. **90**, 6539 (2001)

32.83 J. H. Edgar, C. H. Wei, D. T. Smith, T. J. Kistenmacher, W. A. Bryden: J. Mater. Sci. **8**, 307 (1997)

32.84 P. Perlin, A. Polian, T. Suski: Phys. Rev. B **47**, 2874 (1993)

32.85 C. T. M. Ribeiro, F. Alvarez, A. R. Zanatta: Appl. Phys. Lett. **81**, 1005 (2002)

32.86 Z. M. Ren, Y. F. Lu, H. Q. Ni, T. Y. F. Liew, B. A. Cheong, S. K. Chow, M. L. Ng, J. P. Wang: J. Appl. Phys. **88**, 7346 (2000)

32.87 O. Brafman, G. Lengyel, S. S. Mitra, P. J. Gielisse, J. N. Plendl, L. C. Mansur: Solid State Commun. **6**, 523 (1968)

32.88 K. Shimada, T. Sota, K. Suzuki: J. Appl. Phys. **84**, 4951 (1993)

32.89 K. Miwa, A. Fukumoto: Phys. Rev. B **48**, 7897 (1993)

32.90 T. Azuhata, T. Sota, K. Suzuki, S. Nakamura: J. Phys.: Condens. Matter **7**, L129 (1995)

32.91 V. Yu. Davydov, V. V Emtsev, I. N. Goncharuk, A. N. Smirnov, V. D. Petrikov, V. V. Mamutin, V. A. Vekshin, S. V. Ivanov, M. B. Smirnov, T. Inushima: Appl. Phys. Lett. **75**, 3297 (1999)

32.92 J. S. Dyck, K. Kim, S. Limpijumnong, W. R. L. Lambrecht, J. Kash, J. C. Angus: Solid State Commun. **114**, 355 (2000)

32.93 D. Huang, F. Yun, M. A. Reshchikov, D. Wang, H. Morkoç, D. L. Rode, L. A. Farina, Ç. Kurdak, K. T. Tsen, S. S. Park, K. Y. Lee: Solid State Electron. **45**, 711 (2001)

32.94 W. Götz, N. M. Johnson, D. P. Bour, M. G. McCluskey, E. E. Haller: Appl. Phys. Lett. **69**, 3725 (1996)

32.95 C. G. Van de Walle: Phys. Rev. B **56**, R10020 (1997)

32.96 C. Wetzel, A. L. Chen, J. W. Suski, J. W. Ager III, W. Walukiewicz: Phys. Stat. Sol. B **198**, 243 (1996)

32.97 G. Kaczmarczyk, A. Kaschner, S. Reich, A. Hoffmann, C. Thomsen, D. J. As, A. P. Lima, D. Schikora, K. Lischka, R. Averbeck, H. Riechert: Appl. Phys. Lett. **76**, 2122 (2000)

32.98 K. M. Taylor, C. Lenie: J. Electrochem. Soc. **107**, 308 (1960)

32.99 V. Kirchner, H. Heinke, D. Hommel, J. Z. Domagala, M. Leszczynski: Appl. Phys. Lett. **77**, 1434 (2000)

32.100 G. A. Slack, S. F. Bartram: J. Appl. Phys. **46**, 89 (1975)

32.101 W. M. Yim, R. J. Paff: J. Appl. Phys. **45**, 1456 (1974)

32.102 H. P. Maruska, J. J. Tietjen: Appl. Phys. Lett. **15**, 327 (1969)

32.103 A. U. Sheleg, V. A. Savastenko: Vestsi Akad. Nauk, Set. Fiz.-Mat. Nauk USSR **3**, 126 (1976)

32.104 M. Leszczynski, J. F. Walker: Appl. Phys. Lett. **62**, 1484 (1993)

32.105 G. L. Slack, L. J. Schowalter, D. Morelli, J. A. Freitas Jr.: J. Cryst. Growth **246**, 287 (2002)

32.106 G. A. Slack: J. Phys. Chem. Solids **34**, 321 (1973)

32.107 G. A. Slack, T. F. McNelly: J. Cryst. Growth **42**, 560 (1977)

32.108 K. Watari, K. Ishizaki, F. Tsuchiya: J. Mater. Sci. **28**, 3709 (1993)

32.109 A. Witek: Diamond Relat. Mater. **7**, 962 (1998)

32.110 A. Nikolaev, I. Nikitina, A. Zubrilov, M. Mynbaeva, Y. Melnik, V. Dmitriev: Mater. Res. Soc. Symp. Proc. **595**, 6.5.1 (2000)

32.111 D. I. Florescu, V. M. Asnin, F. H. Pollak: Compound Semiconductor **7**, 62 (2001)

32.112 E. K. Sichel, J. I. Pankove: J. Phys. Chem. Solids **38**, 330 (1977)

32.113 D. Kotchetkov, J. Zou, A. A. Balandin, D. I. Florescu, F. H. Pollak: Appl. Phys. Lett. **79**, 4316 (2001)

32.114 J. Zou, D. Kotchetkov, A. A. Balandin, D. I. Florescu, F. H. Pollak: J. Appl. Phys. **92**, 2534 (2002)

32.115 D. I. Florescu, V. M. Asnin, F. H. Pollak, R. J. Molnar: Mater. Res. Soc. Symp. Proc. **595**, 3.89.1 (2000)

32.116 D. I. Florescu, V. M. Asnin, F. H. Pollak, R. J. Molnar, C. E. C. Wood: J. Appl. Phys. **88**, 3295 (2000)

32.117 G. A. Slack, R. A. Tanzilli, R. O. Pohl, J. W. Vandersande: J. Phys. Chem. Solids **48**, 641 (1987)

32.118 S. Krukowski, A. Witek, J. Adamczyk, J. Jun, M. Bockowski, I. Grzegory, B. Lucznik, G. Nowak, M. Wroblewski, A. Presz, S. Gierlotka, S. Stelmach, B. Palosz, S. Porowski, P. Zinn: J. Phys. Chem. Solids **59**, 289 (1998)

32.119 A. D. Mah, E. G. King, W. W. Weller, A. U. Christensen: Bur. Mines, Rept. Invest. **RI-5716**, 18 (1961)

32.120 V. P. Glushko, L. V. Gurevich, G. A. Bergman, I. V. Weitz, V. A. Medvedev, G. A. Chachkurov, V. S. and Yungman: *Thermodinamicheskiie Swoistwa Indiwidualnych Weshchestw (the old USSR)*, Vol. 1 (Nauka, Moscow 1979) p. 164

32.121 I. Basin, O. Knacke, O. Kubaschewski: *Thermochemical Properties of Inorganic Substances* (Springer, Berlin, Heidelberg 1977)

32.122 V. I. Koshchenko, A. F. Demidienko, L. D. Sabanova, V. E. Yachmenev, V. E. Gran, A. E. Radchenko: Inorg. Mater. **15**, 1329 (1979)

32.123 S. Krukowski, M. Leszczynski, S. Porowski: Thermal properties of the Group III nitrides. In: *Properties, Processing and Applications of Gallium Nitride and Related Semiconductors*, EMIS Datareviews Series, No. 23, ed. by J. H. Edgar, S. Strite, I. Akasaki, H. Amano, C. Wetzel (INSPEC, The Institution of Electrical Engineers, Stevenage, UK 1999) p. 23

32.124 A. F. Demidienko, V. I. Koshchenko, L. D. Sabanova, V. E. Gran: Russ. J. Phys. Chem. **49**, 1585 (1975)

32.125 J. C. Nipko, C.-K. Loong, C. M. Balkas, R. F. Davis: Appl. Phys. Lett. **73**, 34 (1998)

32.126 V. Davydov, A. Klochikhin, S. Ivanov, J. Aderhold, A. Yamamoto: Growth and properties of InN. In: *Nitride Semiconductors – Handbook on Materials and Devices*, ed. by P. Ruterana, M. Albrecht, J. Neugebauer (Wiley, New York 2003)

32.127 D. L. Rode, R. K. Willardson, A. C. Beer (Eds.): *Semiconductors and Semimetals*, Vol. 10 (Academic, New York 1975) pp. 1–90

32.128 K. Seeger: *Semiconductor Physics*, 2 edn. (Springer, Berlin Heidelberg New York 1982)

32.129 V. W. L. Chin, T. L. Tansley, T. Osotchan: J. Appl. Phys. **75**, 7365 (1994)

32.130 D. L. Rode, D. K. Gaskill: Appl. Phys. Lett. **66**, 1972 (1995)

32.131 M. E. Lin, B. Sverdlov, G. L. Zhou, H. Morkoç: Appl. Phys. Lett. **62**, 3479 (1993)

32.132 H. M. Ng, D. Doppalapudi, T. D. Moustakas, N. G. Weimann, L. F. Eastman: Appl. Phys. Lett. **73**, 821 (1998)

32.133 N. G. Weimann, L. F. Eastman, D. Doppalapudi, H. M. Ng, T. D. Moustakas: J. Appl. Phys. **83**, 3656 (1998)

32.134 D. C. Look, J. R. Sizelove: Phys. Rev. Lett. **82**, 1237 (1999)

32.135 H. W. Choi, J. Zhang, S. J. Chua: Mater. Sci. Semicond. Process. **4**, 567 (2001)

32.136 J. Y. Shi, L. P. Yu, Y. Z. Wang, G. Y. Zhang, H. Zhang: Appl. Phys. Lett. **80**, 2293 (2002)

32.137 Z. Q. Fang, D. C. Look, W. Kim, Z. Fan, A. Botchkarev, H. Morkoc: Appl. Phys. Lett. **72**, 2277 (1998)

32.138 K. Wook, A. E. Botohkarev, H. Morkoc, Z. Q. Fang, D. C. Look, D. J. Smith: J. Appl. Phys. **84**, 6680 (1998)

32.139 Q. S. Zhu, N. Sawaki: Appl. Phys. Lett. **76**, 1594 (2000)

32.140 Z. Chen, Y. Yuan, Da-C. Lu, X. Sun, S. Wan, X. Liu, P. Han, X. Wang, Q. Zhu, Z. Wang: Solid State Electron. **46**, 2069 (2002)

32.141 S. Nakamura: Jpn. J. Appl. Phys. **30**, L1705 (1991)

32.142 W. Götz, L. T. Romano, J. Walker, N. M. Johnson, R. J. Molnar: Appl. Phys. Lett. **72**, 1214 (1998)

32.143 D. C. Look, R. J. Molnar: Appl. Phys. Lett. **70**, 3377 (1997)

32.144 S. Nakamura, T. Mukai, M. Senoh: Jpn. J. Appl. Phys. **31**, 2883 (1992)

32.145 S. Nakamura, T. Mukai, M. Senoh: J. Appl. Phys. **71**, 5543 (1992)

32.146 S. C. Binari, H. C. Dietrich: In: *GaN and Related Materials*, ed. by S. J. Pearton (Gordon and Breach, New York 1997) pp. 509–534

32.147 F. Yun, M. A. Reshchikov, K. Jones, P. Visconti, S. S. Park, K. Y. Lee: Solid State Electron. **44**, 2225 (2000)

32.148 B. Heying, I. Smorchkova, C. Poblenz, C. Elsass, P. Fini, S. Den Baars, U. Mishra, J. S. Speck: Appl. Phys. Lett. **77**, 2885 (2000)

32.149 D. D. Koleske, R. L. Henry, M. E. Twigg, J. C. Culbertson, S. C. Binari, A. E. Wickenden, M. Fatemi: Appl. Phys. Lett. **80**, 4372 (2000)

32.150 S. Nakamura, T. Mukai: Jpn. J. Appl. Phys. **31**, L1457 (1992)

32.151 W. Götz, N. M. Johnson, C. Chen, H. Liu, C. Kuo, W. Imler: Appl. Phys. Lett. **68**, 3144 (1996)

32.152 M. Ilegams, H. C. Montgomery: J. Phys. Chem. Solids **34**, 885 (1973)

32.153 M. A. Littlejohn, J. R. Hauser, M. Glisson: Appl. Phys. Lett. **26**, 625 (1975)

32.154 U. V. Bhapkar, M. S. Shur: J. Appl. Phys. **82**, 1649 (1997)

32.155 J. D. Albrecht, R. P. Wang, P. P. Ruden, M. Farahmand, K. F. Brennan: J. Appl. Phys. **83**, 1446 (1998)

32.156 S. Dhar, S. Ghosh: J. Appl. Phys. **86**, 2668 (1999)

32.157 J. G. Kim, A. C. Frenkel, H. Liu, R. M. Park: Appl. Phys. Lett. **65**, 91 (1994)

32.158 R. J. Molnar, W. Götz, L. T. Romano, N. M. Johnson: J. Cryst. Growth **178**, 147 (1997)

32.159 S. Yamaguchi, M. Kariya, S. Nitta, T. Takeuchi, C. Wetzel, H. Amano, I. Akasaki: J. Appl. Phys. **85**, 7682 (1999)

32.160 H. Morkoç: IEEE J. Select. Top. Quant. Electron. **4**, 537 (1998)

32.161 D. C. Look, D. C. Reynolds, J. W. Hemsky, J. R. Sizelove, R. L. Jones, R. J. Molnar: Phys. Rev. Lett. **79**, 2273 (1997)

32.162 S. Keller, B. P. Keller, Y. F. Wu, B. Heying, D. Kapolnek, J. S. Speck, U. K. Mishra, S. P. Den Baars: Appl. Phys. Lett. **68**, 1525 (1996)

32.163 S. Nakamura, T. Mukhai, M. Senoh: J. Appl. Phys. **75**, 7365 (1997)

32.164 M. A. Khan, R. A. Skogman, R. G. Schulze, M. Gershenzon: Appl. Phys. Lett. **42**, 430 (1983)

32.165 I. Akasaki, H. Amano: Mater. Res. Soc. Symp. Proc. **242**, 383 (1992)

32.166 T. Matsuoka: Mater. Res. Soc. Symp. Proc. **395**, 39 (1995)

32.167 S. Sinharoy, A. K. Aggarwal, G. Augustine, L. B. Rawland, R. L. Messham, M. C. Driver, R. H. Hopkins: Mater. Res. Soc. Symp. Proc. **395**, 157 (1995)

32.168 R. J. Molnar, R. Aggarwal, Z. L. Lian, E. R. Brown, I. Melngailis, W. Götz, L. T. Romano, N. M. Johnson: Mater. Res. Soc. Symp. Proc. **395**, 157 (1995)

32.169 T. T. Mnatsakanov, M. E. Levinshtein, L. I. Pomortseva, S. N. Yurkov, G. S. Simin, M. A. Khan: Solid State Electron. **47**, 111 (2003)

32.170 D. M. Caughey, R. E. Thomas: Proc. IEEE **55**, 2192 (1967)

32.171 J. Edwards, K. Kawabe, G. Stevens, R. H. Tredgold: Solid State Commun. **3**, 99 (1965)

32.172 K. Kawabe, R. H. Tredgold, Y. Inyishi: Electr. Eng. Jpn. **87**, 62 (1967)

32.173 G. A. Cox, D. O. Cummins, K. Kawabe, R. H. Tredgold: J. Phys. Chem. Solilds **28**, 543 (1967)

32.174 W. M. Yim, E. J. Stotko, P. J. Zanzucchi, J. Pankove, M. Ettenberg, S. L. Gilbert: J. Appl. Phys. **44**, 292 (1973)

32.175 S. Yoshida, S. Misawa, Y. Fujii, S. Takada, H. Hayakawa, S. Gonda, A. Itoh: J. Vac. Sci. Technol. **16**, 990 (1979)

32.176 A. G. Gorbatov, V. M. Kamyshoc: Sov. Powder, Metall. Met. Ceram. **9**, 917 (1970)

32.177 R. F. Rutz: Appl. Phys. Lett. **28**, 379 (1976)

32.178 R. F. Rutz, E. P. Harrison, J. J. Cuome: IBMJ. Res. Sev. **17**, 61 (1973)

32.179 A. Fara, F. Bernadini, V. Fiorentini: J. Appl. Phys. **85**, 2001 (1999)

32.180 K. Wongchotigul, N. Chen, D. P. Zhang, X. Tang, M. G. Spencer: Mater. Lett. **26**, 223 (1996)

32.181 H. J. Hovel, J. J. Cuomo: Appl. Phys. Lett. **20**, 71 (1972)

32.182 T. L. Tansley, C. P. Foley: Electron. Lett. **20**, 1066 (1984)

32.183 M. J. Brett, K. L. Westra: Thin Solid Films **192**, 227 (1990)

32.184 J. S. Morgan, T. J. Kistenmacher, W. A. Bryden, S. A. Ecelberger: Proc. Mater. Res. Soc. **202**, 383 (1991)

32.185 T. J. Kistenmacher, W. A. Bryden: Appl. Phys. Lett. **59**, 1844 (1991)

32.186 C. R. Abernathy, S. J. Pearton, F. Ren, P. W. Wisk: J. Vac. Sci. Technol. B **11**, 179 (1993)

32.187 W. R. Bryden, S. A. Ecelberger, M. E. Hawley, T. J. Kistenmacher: MRS Proc. **339**, 497 (1994)

32.188 T. Maruyama, T. Morishita: J. Appl. Phys. **76**, 5809 (1994)

32.189 H. Lu, W. J. Schaff, J. Hwang, H. Wu, W. Yeo, A. Pharkya, L. F. Eastman: Appl. Phys. Lett. **77**, 2548 (2000)

32.190 Y. Saito, N. Teraguchi, A. Suzuki, T. Araki, Y. Nanishi: Jpn. J. Appl. Phys. **40**, L91 (2001)

32.191 J. Aderhold, V. Yu. Davydov, F. Fedler, H. Klausing, D. Mistele, T. Rotter, O. Semchinova, J. Stemmer, J. Graul: J. Cryst. Growth **221**, 701 (2001)

32.192 H. Lu, W. J. Schaff, J. Hwang, H. Wu, G. Koley, L. F. Eastman: Appl. Phys. Lett. **79**, 1489 (2001)

32.193 Motlan, E. M. Goldys, T. L. Tansley: J. Cryst. Growth **241**, 165 (2002)

32.194 Y. Saito, T. Yamaguchi, H. Kanazawa, K. Kano, T. Araki, Y. Nanishi, N. Teraguchi, A. Suzuki: J. Cryst. Growth **237–239**, 1017 (2002)

32.195 M. Higashiwaki, T. Matsui: Jpn. J. Appl. Phys. **41**, L540 (2002)

32.196 H. W. Lu, J. Schaff, L. F. Eastman, J. Wu, W. Walukiewicz, K. M. Yu, J. W. Auger III, E. E. Haller, O. Ambacher: *Conference Digest of the 44th Electronic Materials Conference*, Santa Barbara, p. 2 (2002)

32.197 M. Higashiwaki, T. Matsui: J. Cryst. Growth **252**, 128 (2003)

32.198 T. L. Tansley, C. P. Foley, J. S. Blakemore (Ed.): *Proc. 3rd Int. Conf. on Semiinsulating III–V Materials*, Warm Springs, OR 1984 (Shiva, London 1985)

32.199 T. L. Tansley, R. J. Egan, E. C. Horrigan: Thin Solid Films **164**, 441 (1988)

32.200 C. Stampfl, C. G. Van de Walle, D. Vogel, P. Kruger, J. Pollmann: Phys. Rev. BR **61**, 7846 (2000)

32.201 D. C. Look, H. Lu, W. J. Schaff, J. Jasinski, Z. Liliental-Weber: Appl. Phys. Lett. **80**, 258 (2002)

32.202 B. E. Foutz, S. K. O'Leary, M. S. Shur, L. F. Eastman: J. Appl. Phys. **85**, 7727 (1999)

32.203 J. D. Albrecht, R. P. Wang, P. P. Ruden, M. Farahmand, K. F. Brennan: J. Appl. Phys. **83**, 4777 (1998)

32.204 S. K. O'Leary, B. E. Foutz, M. S. Shur, U. V. Bhapkar, L. F. Eastman: Solid State Commun. **105**, 621 (1998)

32.205 B. Gelmont, K. Kim, M. Shur: J. Appl. Phys. **74**, 1818 (1993)

32.206 N. S. Mansour, K. W. Kim, M. A. Littlejohn: J. Appl. Phys. **77**, 2834 (1995)

32.207 J. Kolnik, I. H. Oguzman, K. F. Brennan, R. Wang, P. P. Ruden, Y. Wang: J. Appl. Phys. **78**, 1033 (1995)

32.208 M. Shur, B. Gelmont, M. A. Khan: J. Electron. Mater. **25**, 777 (1996)

32.209 S. K. O'Leary, B. E. Foutz, M. S. Shur, U. V. Bhapkar, L. F. Eastman: J. Appl. Phys. **83**, 826 (1998)

32.210 E. Bellotti, B. K. Doshi, K. F. Brennan, J. D. Albrecht, P. P. Ruden: J. Appl. Phys. **85**, 916 (1999)

32.211 C. G. Rodrigues, V. N. Freire, A. R. Vasconcellos, R. Luzzi: Appl. Phys. Lett. **76**, 1893 (2000)

32.212 M. Wraback, H. Shen, S. Rudin, E. Bellotti: Phys. Stat. Sol. (b) **234**, 810 (2002)

32.213 R. Collazo, R. Schesser, Z. Sitar: Appl. Phys. Lett. **81**, 5189 (2002)

32.214 B. Monemar, U. Lindefelt, W. M. Chen: Physica B **146**, 256 (1987)

32.215 M. Tchounkeu, O. Briot, B. Gil, J. P. Alexis, R. L. Aulombard: J. Appl. Phys. **80**, 5352 (1996)

32.216 R. Dingle, D. D. Sell, S. E. Stokowski, M. Ilegems: Phys. Rev. **B 4**, 1211 (1971)

32.217 E. L. Ivchenko: *Excitons* (North Holland, Amsterdam 1982) p. 141

32.218 B. Monemar, J. P. Bergman, I. A. Buyanova: Optical characterization of GaN and related materials. In: *GaN and Related Material*, ed. by S. J. Pearton (Golden and Breach, Amsterdam 1997) p. 85

32.219 J. J. Song, W. Shan: In: *Group III Nitride Semiconductor Compounds*, ed. by B. Gil (Clarendon, Oxford 1998) pp. 182–241

32.220 B. Monemar, J. P. Bergman, I. A. Buyanova: In: *GaN and Related Materials Semiconductor Compounds*, ed. by S. J. Pearton (Gordon and Breach, New York 1998) pp. 85–139

32.221 B. Monemar: Phys. Rev. B **10**, 676 (1974)

32.222 B. Gil, O. Briot, R. L. Aulombard: Phys. Rev. B **52**, R17028 (1995)

32.223 W. Shan, B. D. Little, A. J. Fischer, J. J. Song, B. Goldenberg, W. G. Perry, M. D. Bremser, R. F. Davis: Phys. Rev. B **54**, 16369 (1996)

32.224 M. Smith, G. D. Chen, J. Y. Lin, H. X. Jiang, A. Salvador, W. K. Kim, O. Aktas, A. Botchkarev, H. Morkoç: Appl. Phys. Lett. **67**, 3387 (1995)

32.225 K. P. Korona, A. Wysmolek, K. Pakula, R. Stepniewski, J. M. Baranowski, I. Grzegory, B. Lucznik, M. Wroblewski, S. Porowski: Appl. Phys. Lett. **69**, 788 (1996)

32.226 K. Kornitzer, T. Ebner, M. Grehl, K. Thonke, R. Sauer, C. Kirchner, V. Schwegler, M. Kamp, M. Leszczynski, I. Grzegory, S. Porowski: Phys. Stat. Sol. (b) **216**, 5 (1999)

32.227 B. J. Skromme, K. Palle, C. D. Poweleit, H. Yamane, M. Aoki, F. J. Disalvo: J. Cryst. Growth **246**, 299 (2002)

32.228 L. Eckey, L. Podloswski, A. Goldner, A. Hoffmann, I. Broser, B. K. Meyer, D. Volm, T. Streibl, K. Hiramatsu, T. Detcprohm, H. Amano, I. Akasaki: Ins. Phys. Conf. Ser. **142**, 943 (1996)

32.229 K. Pakula, A. Wysmolek, K. P. Korona, J. M. Baranowski, R. Stepniewski, I. Grzegory, M. Bockowski, J. Jun, S. Krukowski, M. Wroblewski, S. Porowski: Solid State Commun. **97**, 919 (1996)

32.230 A. V. Rodina, M. Dietrich, A. Goldner, L. Eckey, A. L. L. Efros, M. Rosen, A. Hoffmann, B. K. Meyer: Phys. Stat. Sol. (b) **216**, 216 (1999)

32.231 A. Shikanai, T. Azuhata, T. Sota, S. Chichibu, A. Kuramata, K. Horino, S. Nakamura: J. Appl. Phys. **81**, 417 (1997)

32.232 B. Monemar, J. P. Bergman, I. A. Buyanova, W. Li, H. Amano, I. Akasaki: MRS Int. J. Nitride Semicond. Res. **1**, 2 (1996)

32.233 S. F. Chichibu, K. Torii, T. Deguchi, T. Sota, A. Setoguchi, H. Nakanishi, T. Azuhata, S. Nakamura: Appl. Phys. Lett. **76**, 1576 (2000)

32.234 J. F. Muth, J. H. Lee, I. K. Shmagin, R. M. Kolbas, H. C. Casey Jr., B. P. Keller, U. K. Mishra, S. P. DenBaars: Appl. Phys. Lett. **71**, 2572 (1997)

32.235 R. Stepniewski, K. P. Korona, A. Wysmolek, J. M. Baranowski, K. Pakula, M. Potemski, G. Martinez, I. Grzegory, S. Porowski: Phys. Rev. B **56**, 15151 (1997)

32.236 W. Shan, A. J. Fischer, S. J. Hwang, B. D. Little, R. J. Hauenstein, X. C. Xie, J. J. Song, D. S. Kim, B. Goldenberg, R. Horning, S. Krishnankutty, W. G. Perry, M. D. Bremser, R. F. Davis: J. Appl. Phys. **83**, 455 (1998)

32.237 B. Gil, S. Clur, O. Briot: Solid State Commun. **104**, 267 (1997)

32.238 W. Shan, R. J. Hauenstein, A. J. Fischer, J. J. Song, W. G. Perry, M. D. Bremser, R. F. Davis, B. Goldenberg: Appl. Phys. Lett. **66**, 985 (1995)

32.239 C. F. Li, Y. S. Huang, L. Malikova, F. H. Pollak: Phys. Rev. B **55**, 9251 (1997)

32.240 A. K. Viswanath, J. I. Lee, D. Kim, C. R. Lee, J. Y. Leam: Phys. Rev. B **58**, 16333 (1998)

32.241 S. Chichibu, T. Azuhata, T. Sota, S. Nakamura: J. Appl. Phys. **79**, 2784 (1996)

32.242 M. Smith, G. D. Chen, J. Y. Lin, H. X. Jiang, M. A. Khan, C. J. Sun, Q. Chen, J. W. Yang: J. Appl. Phys. **79**, 7001 (1996)

32.243 M. Smith, G. D. Chen, J. Y. Lin, H. X. Jiang, A. Salvador, B. N. Sverdlov, A. Botchkarev, H. Morkoç: Appl. Phys. Lett. **66**, 3474 (1995)

32.244 F. Hamdani, A. Botchkarev, H. Tang, W. K. Kim, H. Morkoç: Appl. Phys. Lett. **71**, 3111 (1997)

32.245 F. Hamdani, A. Botchkarev, W. Kim, H. Morkoç, M. Yeadon, J. M. Gibson, S. C. Y. Tsen, D. J. Smith, D. C. Reynolds, D. C. Look, K. Evans, C. W. Litton, W. C. Mitchel, P. Hemenger: Appl. Phys. Lett. **70**, 467 (1997)

32.246 W. Shan, A. J. Fischer, J. J. Song, G. E. Bulman, H. S. Kong, M. T. Leonard, W. G. Perry, M. D. Bremser, B. Goldenberg, R. F. Davis: Appl. Phys. Lett. **69**, 740 (1996)

32.247 S. Chichibu, T. Azuhata, T. Sota, H. Amano, I. Akasaki: Appl. Phys. Lett. **70**, 2085 (1997)

32.248 I. A. Buyanova, J. P. Bergman, B. Monemar, H. Amano, I. Akasaki: Appl. Phys. Lett. **69**, 1255 (1996)

32.249 B. J. Skromme, K. C. Palle, C. D. Poweleit, H. Yamane, M. Aoki, F. J. Disalvo: Appl. Phys. Lett. **81**, 3765 (2002)

32.250 Y.S. Huang, Fred H. Pollak, S.S. Park, K.Y. Lee, H. Morkoç: J. Appl. Phys. **94**, 899 (2003)
32.251 B.K. Meyer: In: *Free and Bound Excitons in GaN Epitaxial Films*, MRS Proc., Vol. 449, ed. by F.A. Ponce, T.D. Moustakas, I. Akasaki, B.A. Monemar (Materials Research Society, Pittsburgh, Pennsylvania 1997) p. 497
32.252 B. Monemar, J.P. Bergman, I.A. Buyanova, H. Amano, I. Akasaki, T. Detchprohm, K. Hiramatsu, N. Sawaki: Solid State Electron. **41**, 239 (1995)
32.253 D. Kovalev, B. Averboukh, D. Volm, B.K. Meyer: Phys. Rev. B **54**, 2518 (1996)
32.254 G.D. Chen, M. Smith, J.Y. Lin, H.X. Jiang, S.-H. Wei, M.A. Khan, C.J. Sun: Appl. Phys. Lett. **68**, 2784 (1996)
32.255 G. Mahler, U. Schroder: Phys. Stat. Sol. (b) **61**, 629 (1974)
32.256 B. Monemar: In: *Gallium Nitride I*, ed. by J.I. Pankove, T.D. Moustakas (Academic, San Diego 1998) p. 305
32.257 D.C. Reynolds, D.C. Look, W. Kim, O. Aktas, A. Botchkarev, A. Salvador, H. Morkoç, D.N. Talwar: J. Appl. Phys. **80**, 594 (1996)
32.258 R. Stepniewski, M. Potemski, A. Wysmolek, K. Pakula, J.M. Baranowski, J. Lusakowski, I. Grzegory, S. Porowski, G. Martinez, P. Wyder: Phys. Rev. B **60**, 4438 (1999)
32.259 G. Neu, M. Teisseire, E. Frayssinet, W. Knap, M.L. Sadowski, A.M. Witowski, K. Pakula, M. Leszczynski, P. Prystawsko: Appl. Phys. Lett. **77**, 1348 (2000)
32.260 J.M. Baranowski, Z. Liliental-Weber, K. Korona, K. Pakula, R. Stepniewski, A. Wysmolek, I. Grzegory, G. Nowak, S. Porowski, B. Monemar, P. Bergman: *III–V Nitrides*, Vol. 449 (MRS Proc., Pittsburg, PA 1997) p. 393
32.261 J.M. Baranowski and S. Porowski, Proc. 23rd Int. Conf. on Physics of Semiconductors, Berlin, p.497 (1996)
32.262 S.J. Pearton, C.R. Abernathy, J.W. Lee, C.B. Vartuli, C.B. Mackenzi, J.D. Ren, R.G. Wilson, J.M. Zavada, R.J. Shul, J.C. Zolper: Mater. Res. Soc. Symp. Proc. **423**, 124 (1996)
32.263 B. Monemar: J. Mater. Sci.: Mater. Electron. **10**, 227 (1999)
32.264 M. Reshchikov, D. Huang, F. Yun, L. He, H. Morkoç, D.C. Reynolds, S.S. Park, K.Y. Lee: Appl. Phys. Lett. **79**, 3779 (2001)
32.265 A.K. Viswanath: Semicond. Semimetals **73**, 63 (2002)
32.266 B. Monemar, P.P. Paskov, T. Paskova, J.P. Bergman, G. Pozina, W.M. Chan, P.N. Hai, I.A. Buyanova, H. Amano, I. Akasaki: Mater. Sci. Eng. B **93**, 112 (2002)
32.267 G. Pozina, N.V. Edwards, J.P. Bergman, T. Paskova, B. Monemar, M.D. Bremser, R.F. Davis: Appl. Phys. Lett. **78**, 1062 (2001)
32.268 S. Pau, J. Kuhl, M.A. Khan, C.J. Sun: Phys. Rev. B **58**, 12916 (1998)
32.269 G. Pozina, J.P. Bergman, T. Paskova, B. Monemar: Appl. Phys. Lett. **75**, 412 (1999)
32.270 M. Leroux, B. Beaumont, N. Grandjean, P. Lorenzini, S. Haffouz, P. Vennegues, J. Massies, P. Gibart: Mater. Sci. Eng. B **50**, 97 (1997)
32.271 M. Leroux, N. Grandjean, B. Beaumont, G. Nataf, F. Semond, J. Massies, P. Gibart: J. Appl. Phys. **86**, 3721 (1999)
32.272 B.J. Skromme, G.L. Martinez: Mater. Res. Soc. Symp. **595**, W9.8. (1999)
32.273 D.C. Reynolds, D.C. Look, B. Jogai, V.M. Phanse, R.P. Vaudo: Solid State Commun. **103**, 533 (1997)
32.274 B. Santic, C. Merz, U. Kaufmann, R. Niebuhr, H. Obloh, K. Bachem: Appl. Phys. Lett. **71**, 1837 (1997)
32.275 A.K. Viswanath, J.I. Lee, S. Yu, D. Kim, Y. Choi, C.H. Hong: J. Appl. Phys. **84**, 3848 (1998)
32.276 R.A. Mair, J. Li, S.K. Duan, J.Y. Lin, H.X. Jiang: Appl. Phys. Lett. **74**, 513 (1999)
32.277 D.G. Thomas, J.J. Hopfield: Phys. Rev. **128**, 2135 (1962)
32.278 H. Saito, S. Shionoya, E. Hanamura: Solid State Commun. **12**, 227 (1973)
32.279 U. Kaufmann, M. Kunzer, C. Merz, I. Akasaki, H. Amano: Mater. Res. Soc. Symp. Proc. **395**, 633 (1996)
32.280 K.P. Korona, J.P. Bergman, B. Monemar, J.M. Baranowski, K. Pakula, L. Gregory, S. Porowski: Mater. Sci. Forum **258–263**, 1125 (1997)
32.281 G. Neu, M. Teisseire, N. Grandjean, H. Lahreche, B. Beaumont, I. Grzegory, S. and Porowski: Proc. Phys. **87**, 1577 (2001)
32.282 C.H. Henry, K. Nassau: Phys. Rev. B **1**, 1628 (1970)
32.283 O. Lagerstedt, B. Monemar: J. Appl. Phys. **45**, 2266 (1974)
32.284 R. Dingle, M. Ilegems: Solid State Commun. **9**, 175 (1971)
32.285 S. Fischer, C. Wetzel, E.E. Haller, B.K. Meyer: Appl. Phys. Lett. **67**, 1298 (1995)
32.286 M.A.L. Johnson, Z. Yu, C. Boney, W.C. Hughes, J.W. Cook Jr, J.F. Schetzina, H. Zao, B.J. Skromme, J.A. Edmond: MRS Proc. **449**, 271 (1997)
32.287 M. Smith, G.D. Chen, J.Y. Lin, H.X. Jiang, A. Salvador, B.N. Sverdlov, A. Botchkarev, H. Morkoc, B. Goldenberg: Appl. Phys. Lett. **68**, 1883 (1996)
32.288 G.D. Chen, M. Smith, J.Y. Lin, H.X. Jiang, A. Salvador, B.N. Sverdlov, A. Botchkarev, H. Morkoc: J. Appl. Phys. **79**, 2675 (1995)
32.289 T. Ogino, M. Aoki: Jpn. J. Appl. Phys. **19**, 2395 (1980)
32.290 K. Saarinen, T. Laine, S. Kuisma, J. Nissilä, P. Hautojärvi, L. Dobrzynski, J.M. Baranowski, K. Pakula, R. Stepniewski, A. Wojdak, A. Wysmolek, T. Suski, M. Leszczynski, I. Grzegory, S. Porowski: Phys. Rev. Lett. **79**, 3030 (1997)

32.291 J. Neugebauer, C. G. Van de Walle: Appl. Phys. Lett. **69**, 503 (1996)

32.292 T. Mattila, R. M. Nieminen: Phys. Rev. **55**, 9571 (1997)

32.293 E. R. Glaser, T. A. Kennedy, K. Doverspike, L. B. Rowland, D. K. Gaskill, J. A. Freitas Jr, M. Asif Khan, D. T. Olson, J. N. Kuznia, D. K. Wickenden: Phys. Rev. B **51**, 13326 (1995)

32.294 P. Perlin, T. Suski, H. Teisseyre, M. Leszczynski, I. Grzegory, J. Jun, S. Porowski, P. Boguslawski, J. Berholc, J. C. Chervin, A. Polian, T. D. Moustakas: Phys. Rev. Lett. **75**, 296 (1995)

32.295 R. Zhang, T. F. Kuech: Appl. Phys. Lett. **72**, 1611 (1998)

32.296 I. Shalish, L. Kronik, G. Segal, Y. Rosenwaks, Y. Shapira, U. Tisch, J. Salzman: Phys. Rev. B **59**, 9748 (1999)

32.297 E. Calleja, F. J. Sanchez, D. Basak, M. A. Sanchez-Garsia, E. Munoz, I. Izpura, F. Calle, J. M. G. Tijero, J. L. Sanchez-Rojas, B. Beaumont, P. Lorenzini, P. Gibart: Phys. Rev. B **55**, 4689 (1997)

32.298 M. A. Reshchikov, F. Shahedipour, R. Y. Korotkov, M. P. Ulmer, B. W. Wessels: Physica B **273-274**, 103 (1999)

32.299 J. Elsner, R. Jones, M. I. Heggie, P. K. Sitch, M. Haugk, Th. Frauenheim, S. Öberg, P. R. Briddon: Phys. Rev. B **58**, 12571 (1998)

32.300 F. A. Ponce, D. P. Bour, W. Gotz, P. J. Wright: Appl. Phys. Lett. **68**, 57 (1996)

32.301 D. M. Hofmann, D. Kovalev, G. Steude, B. K. Meyer, A. Hoffmann, L. Eckey, R. Heitz, T. Detchprom, H. Amano, I. Akasaki: Phys. Rev. B **52**, 16702 (1995)

32.302 R. Y. Korotkov, M. A. Reshchikov, B. W. Wessels: Physica B **273-274**, 80 (1999)

32.303 D. G. Thomas, J. J. Hopfield, W. M. Augustyniak: Phys. Rev. A **140**, 202 (1965)

32.304 R. Seitz, C. Gaspar, T. Monteiro, E. Pereira, M. Leroux, B. Beaumont, P. Gibart: MRS Internet J. Nitride Semicond. Res. **2**, article 36 (1997)

32.305 M. Godlewski, V. Yu. Ivanov, A. Kaminska, H. Y. Zuo, E. M. Goldys, T. L. Tansley, A. Barski, U. Rossner, J. L. Rouvicre, M. Arlery, I. Grzegory, T. Suski, S. Porowski, J. P. Bergman, B. Monemar: Mat. Sci. Forum **258-263**, 1149 (1997)

32.306 H. Haag, B. Hönerlage, O. Briot, R. L. Aulombard: Phys. Rev. B **60**, 11624 (1999)

32.307 M. A. Reshchikov, F. Yun, H. Morkoç, S. S. Park, K. Y. Lee: Appl. Phys. Lett. **78**, 2882 (2001)

32.308 S. Pacesova, L. Jastrabik: Czech. J. Phys. B **29**, 913 (1979)

32.309 R. A. Youngman, J. H. Harris: J. Am. Ceram. Soc. **73**, 3238 (1990)

32.310 J. H. Harris, R. A. Youngman, R. G. Teller: J. Mater. Res. **5**, 1763 (1990)

32.311 M. Katsikini, E. C. Paloura, T. S. Cheng, C. T. Foxon: J. Appl. Phys. **82**, 1166 (1997)

32.312 J. Pastrnak, S. Pacesova, L. Roskovcova: Czech. J. Phys. **B24**, 1149 (1974)

32.313 F. Karel, J. Pastrnak, J. Hejduk, V. Losik: Phys. Stat. Sol. **15**, 693 (1966)

32.314 F. Karel, J. Mares: Czech. J. Phys. B **22**, 847 (1972)

32.315 G. E. Archangelskii, F. Karel, J. Mares, S. Pacesova, J. Pastrnak: Phys. Stat. Sol. **69**, 173 (1982)

32.316 R. G. Wilson, R. N. Schwartz, C. R. Abernathy, S. J. Pearton, N. Newman, M. Rubin, T. Fu, J. M. Zavada: Appl. Phys. Lett. **65**, 992 (1994)

32.317 J. D. MacKenzie, C. R. Abernathy, S. J. Pearton, U. Hömmerich, X. Wu, R. N. Schwartz, R. G. Wilson, J. M. Zavada: Appl. Phys. Lett. **69**, 2083 (1996)

32.318 X. Wu, U. Hömmerich, J. D. MacKenzie, C. R. Abernathy, S. J. Pearton, R. G. Wilson, R. N. Schwartz, J. M. Zavada: J. Lumin. **72-74**, 284 (1997)

32.319 S. J. Pearton, J. D. MacKenzie, C. R. Abernathy, U. Hömmerich, X. Wu, R. G. Wilson, R. N. Schwartz, J. M. Zavada, F. Ren: Appl. Phys. Lett. **71**, 1807 (1997)

32.320 J. Li, K. B. Nam, M. L. Nakarmi, J. Y. Lin, H. X. Jiang: Appl. Phys. Lett. **81**, 3365 (2002)

32.321 E. Kuokstis, J. Zhang, Q. Fareed, J. W. Yang, G. Simin, M. A. Khan, R. Gaska, M. Shur, C. Rojo, L. Schowalter: Appl. Phys. Lett. **81**, 2755 (2002)

32.322 K. B. Nam, J. Li, M. L. Nakarmi, J. Y. Lin, H. X. Jiang: Appl. Phys. Lett. **82**, 1694 (2003)

32.323 T. Onuma, S. F. Chichibu, T. Sota, K. Asai, S. Sumiya, T. Shibata, M. Tanaka: Appl. Phys. Lett. **81**, 652 (2002)

32.324 M. Morita, K. Tsubouchi, N. Mikoshiba: Jpn. J. Appl. Phys. **21**, 1102 (1982)

32.325 F. R. B. Hossain, X. Tang, K. Wongchotigul, M. G. Spencer: Proc. SPIE **42**, 2877 (1996)

32.326 X. Tang, F. R. B. Hossian, K. Wongchotigul, M. G. Spencer: Appl. Phys. Lett. **72**, 1501 (1998)

32.327 Y. Shishkin, R. P. Devaty, W. J. Choyke, F. Yun, T. King, H. Morkoç: Phys. Stat. Sol. (a) **188**, 591 (2001)

32.328 Q. Guo, A. Yoshida: Jpn. J. Appl. Phys. **33**, 2453 (1994)

32.329 P. B. Perry, R. F. Rutz: Appl. Phys. Lett. **33**, 319 (1978)

32.330 J. Pasternak, L. Roskovcova: Phys. Stat. Sol. **26**, 591 (1968)

32.331 T. Yodo, H. Yona, H. Ando, D. Nosei, Y. Harada: Appl. Phys. Lett. **80**, 968 (2002)

32.332 J. Bauer, L. Biste, D. Bolze: Phys. Stat. Sol. **39**, 173 (1977)

32.333 R. G. Gordon, D. M. Hoffmann, U. Riaz: J. Mater. Res. **6**, 5 (1991)

32.334 H. Demiryont, L. R. Thompson, G. J. Collins: Appl. Opt. **25**, 1311 (1986)

32.335 W. J. Meng, J. A. Sell, G. L. Eesley: J. Appl. Phys. **74**, 2411 (1993)

32.336 I. Akasaki, M. Hashimoto: Solid State Commun. **5**, 851 (1967)

32.337 A. T. Collins, E. C. Lightowlers, P. J. Dean: Phys. Rev. **158**, 833 (1967)

32.338 A. J. Noreika, M. H. Francombe, S. A. Zeitman: J. Vac. Sci. Technol. **6**, 194 (1969)

32.339 K. Osamura, N. Nakajima, Y. Murakami, P. H. Shingu, A. Ohtsuki: Solid State Commun. **46**, 3432 (1975)

32.340 T. Inushima, T. Yaguchi, A. Nagase, T. Shiraishi: Ins. Phys. Conf. Ser. **142**, 971 (1996)

32.341 A. Wakahara, T. Tsuchiya, A. Yoshida: J. Cryst. Growth **99**, 385 (1990)

32.342 K. Kubota, Y. Kobayashi, K. Fujimoto: J. Appl. Phys. **66**, 2984 (1989)

32.343 T. L. Tansley, C. P. Foley: J. Appl. Phys. **59**, 3241 (1986)

32.344 S. Yamaguchi, M. Kariya, S. Nitta, T. Takeuchi, C. Wetzel, H. Amano, I. Akasaki: Appl. Phys. Lett. **76**, 876 (2000)

32.345 V. Yu. Davydov, A. A. Klochikhin, R. P. Seisyan, V. V. Emtsev, S. V. Ivanov, F. Bechstedt, J. Furthmuller, H. Harima, A. V. Mudryi, J. Aderhold, O. Semchinova, J. Graul: Phys. Stat. Solidi (b)R **229**, 1 (2002)

32.346 T. Inushima, V. V. Mamutin, V. A. Vekshin, S. V. Ivanov, T. Sakon, M. Motokawa, S. Ohoya: J. Crystal Growth **227-228**, 481 (2001)

32.347 J. Wu, W. Walukiewicz, K. M. Yu, J. W. Ager III, E. E. Haller, H. Lu, W. J. Schaff: Appl. Phys. Lett. **80**, 4741 (2002)

32.348 J. Wu, W. Walukiewicz, W. Shan, K. M. Yu, J. W. Ager III, E. E. Haller, H. Lu, W. J. Schaff: Phys. Rev. B **60**, 201403 (2002)

32.349 J. Wu, W. Walukiewicz, K. M. Yu, J. W. Ager III, E. E. Haller, H. Lu, W. J. Schaff, Y. Saito, Y. Nanishi: Appl. Phys. Lett. **80**, 3967 (2002)

32.350 K. Osamura, S. Naka, Y. Murakami: J. Appl. Phys. **46**, 3432 (1975)

32.351 A. Wakahara, T. Tsuchida, A. Yoshida: Vacuum **41**, 1071 (1990)

32.352 K. L. Westra, M. J. Brett: Thin Solid Films **192**, 234 (1990)

32.353 J. W. Trainor, K. Rose: J. Electron. Meter. **3**, 821 (1974)

32.354 Q. Guo, O. Kato, M. Fujisawa, A. Yoshida: Solid State Commun. **83**, 721 (1992)

32.355 Q. Guo, H. Ogawa, A. Yoshida: J. Electron. Spectrosc. Relat. Phenom. **79**, 9 (1996)

32.356 V. A. Tyagai, O. V. Snitko, A. M. Evstigneev, A. N. Krasiko: Phys. Stat. Sol. (b) **103**, 589 (1981)

32.357 F. Yun, M. A. Reshchikov, L. He, T. King, H. Morkoç, S. W. Novak, L. Wei: J. Appl. Phys. Rapid Commun. **92**, 4837 (2002)

32.358 S. Yoshida, S. Misawa, S. Gonda: J. Appl. Phys. **53**, 6844 (1982)

32.359 S. A. Nikishin, N. N. Faleev, A. S. Zubrilov, V. G. Antipov, H. Temkin: Appl. Phys. Lett. **76**, 3028 (2000)

32.360 W. Shan, J. W. Ager III, K. M. Yu, W. Walukiewicz, E. E. Haller, M. C. Martin, W. R. McKinney, W. Yang: J. Appl. Phys. **85**, 8505 (1999)

32.361 Ü. Özgür, G. Webb-Wood, H. O. Everitt, F. Yun, H. Morkoç: Appl. Phys. Lett. **79**, 4103 (2001)

32.362 J. Wagner, H. Obloh, M. Kunzer, M. Maier, K. Kohler, B. Johs: J. Appl. Phys. **89**, 2779 (2000)

32.363 H. Jiang, G. Y. Zhao, H. Ishikawa, T. Egawa, T. Jimbo, M. J. Umeno: Appl. Phys. **89**, 1046 (2001)

32.364 T. J. Ochalski, B. Gil, P. Lefebvre, M. Grandjean, M. Leroux, J. Massies, S. Nakamura, H. Morkoç: Appl. Phys. Lett. **74**, 3353 (1999)

32.365 M. A. Khan, J. M. Van Hove, J. N. Kuznia, D. T. Olson: Appl. Phys. Lett. **58**, 2408 (1991)

32.366 T. Tanaka, A. Watanabe, H. Amano, Y. Kobayashi, I. Akasaki, S. Yamazaki, M. Koike: Appl. Phys. Lett. **65**, 593 (1994)

32.367 M. Stutzmann, O. Ambacher, A. Cros, M. S. Brandt, H. Angerer, R. Dimitrov, N. Reinacher, T. Metzger, R. Hopler, D. Brunner, F. Freudenberg, R. Handschuh, Ch. Deger: presented at the E-MRS Straßburg, Symposium L (1997)

32.368 S. Nakamura, T. Mukai: J. Vac. Sci. Technol. A **13**, 6844 (1995)

32.369 H. Amano, T. Takeuchi, S. Sota, H. Sakai, I. Akasaki: In: *III-V nitrides*, Vol. 449, ed. by F. A. Ponce, T. D. Moustakas, I. Akasaki, B. Menemar (MRS Proc., Pittsburgh, Pennsylvania 1997) p. 1143

32.370 K. P. O'Donnell, R. W. Martin, C. Trager-Cowan, M. E. White, K. Esona, C. Deatcher, P. G. Middleton, K. Jacobs, W. van der Stricht, C. Merlet, B. Gil, A. Vantomme, J. F. W. Mosselmans: Mater. Sci. Eng. B **82**, 194 (2001)

32.371 T. Nagatomo, T. Kuboyama, H. Minamino, O. Omoto: Jpn. J. Appl. Phys. **28**, L1334 (1989)

32.372 W. Shan, W. Walukiewicz, E. E. Haller, B. D. Little, J. J. Song, M. D. McCluskey, N. M. Johnson, Z. C. Feng, M. Schurman, R. A. Stall: J. App. Phys. **84**, 4452 (1998)

32.373 S. Pereira, M. R. Correia, T. Monteiro, E. Pereira, E. Alves, A. D. Sequeira, N. Franco: Appl. Phys. Lett. **78**, 2137 (2001)

32.374 I. Akasaki, H. Amano: Jpn. J. Appl., Phys. **36**, 5393 (1997)

32.375 S. Yamasaki, S. Asami, N. Shibata, M. Koike, K. Manabe, T. Tanaka, H. Amano, I. Akasaki: Appl. Phys. Lett. **66**, 1112 (1995)

32.376 N. Yoshimoto, T. Matsuoka, A. Katsui: Appl. Phys. Lett. **59**, 2251 (1991)

32.377 S. Nakamura, T. Mukai, M. Seno: Jpn. J. Appl. Phys. **31**, L16 (1993)

32.378 S. Nakamura, N. Iwasa, S. Nagahama: Jpn. J. Appl. Phys. **32**, L338 (1993)

32.379 K. S. Kim, A. Saxler, P. Kung, R. Razeghi, K. Y. Lim: Appl. Phys. Lett. **71**, 800 (1997)

32.380 K. Starosta: Phys. Status Solidi A **68**, K55 (1981)

32.381 T. J. Kistenmacher, S. A. Ecelberger, W. A. Bryden: J. Appl. Phys. **74**, 1684 (1993)

32.382 S. Yamaguchi, M. Kariya, S. Nitta, T. Takeuchi, C. Wetzel, H. Amano, I. Akasaki: Appl. Phys. Lett. **73**, 830 (1998)

32.383 G. Davies: Phys. Rep. **176**, 83 (1989)

32.384 W. R. Bryden, T. J. Kistenmacher: Electrical transport properties of InN, GaInN and AlInN. In: *Properties of Group III Nitrides*, ed. by J. H. Edgar (INSPEC, London 1994)

32.385 S. M. Bedair, F. G. McIntosh, J. C. Roberts, E. L. Piner, K. S. Boutros, N. A. El-Masry: J. Crystal Growth **178**, 32 (1997)

32.386 S. N. Mohammad, A. Salvador, H. Morkoç: Proc. IEEE **83**, 1306 (1995)

32.387 M.-Y. Ryu, C. Q. Chen, E. Kuokstis, J. W. Yang, G. Simin, M. A. Khan: Appl. Phys. Lett. **80**, 3730 (2002)

32.388 K. Kim, W. R. L. Lambrecht, B. Segall: Phys. Rev. B **53**, 7018 (1997)

33. Electron Transport Within the III–V Nitride Semiconductors, GaN, AlN, and InN: A Monte Carlo Analysis

The III–V nitride semiconductors, gallium nitride, aluminium nitride, and indium nitride, have been recognized as promising materials for novel electronic and optoelectronic device applications for some time now. Since informed device design requires a firm grasp of the material properties of the underlying electronic materials, the electron transport that occurs within these III–V nitride semiconductors has been the focus of considerable study over the years. In an effort to provide some perspective on this rapidly evolving field, in this paper we review analyses of the electron transport within these III–V nitride semiconductors. In particular, we discuss the evolution of the field, compare and contrast results obtained by different researchers, and survey the current literature. In order to narrow the scope of this chapter, we will primarily focus on electron transport within bulk wurtzite gallium nitride, aluminium nitride, and indium nitride for this analysis. Most of our discussion will focus on results obtained from our ensemble semi-classical three-valley Monte Carlo simulations of the electron transport within these materials, our results conforming with state-of-the-art III–V nitride semiconductor orthodoxy. Steady-state and transient electron transport results are presented. We conclude our discussion by presenting some recent developments on the electron transport within these materials.

33.1 Electron Transport Within Semiconductors and the Monte Carlo Simulation Approach 806
    33.1.1 The Boltzmann Transport Equation 807
    33.1.2 Our Ensemble Semi-Classical Monte Carlo Simulation Approach 808
    33.1.3 Parameter Selections for Bulk Wurtzite GaN, AlN, and InN 808
33.2 Steady-State and Transient Electron Transport Within Bulk Wurtzite GaN, AlN, and InN 810
    33.2.1 Steady-State Electron Transport Within Bulk Wurtzite GaN 811
    33.2.2 Steady-State Electron Transport: A Comparison of the III–V Nitride Semiconductors with GaAs 812
    33.2.3 Influence of Temperature on the Electron Drift Velocities Within GaN and GaAs 812
    33.2.4 Influence of Doping on the Electron Drift Velocities Within GaN and GaAs 815
    33.2.5 Electron Transport in AlN 816
    33.2.6 Electron Transport in InN 818
    33.2.7 Transient Electron Transport 820
    33.2.8 Electron Transport: Conclusions... 822
33.3 Electron Transport Within III–V Nitride Semiconductors: A Review 822
    33.3.1 Evolution of the Field 822
    33.3.2 Recent Developments 824
    33.3.3 Future Perspectives 825
33.4 Conclusions 826
References 826

The III–V nitride semiconductors, gallium nitride (GaN), aluminium nitride (AlN), and indium nitride (InN), have been known as promising materials for novel electronic and optoelectronic device applications for some time now [33.1–4]. In terms of electronics, their wide energy gaps, large breakdown fields, high thermal conductivities, and favorable electron transport characteristics, make GaN, AlN, and InN, and alloys of these materials, ideally suited for novel high-power and high-frequency electron device applications. On the optoelectronics front, the direct nature of the energy gaps associated with GaN, AlN, and InN, make this family of materials, and its alloys, well suited for novel optoelectronic device applications in the visible and ultraviolet frequency range. While initial efforts to study these materials were hindered by growth difficulties, recent improvements in material quality have made the realization of a number of

III–V nitride semiconductor-based electronic [33.5–9] and optoelectronic [33.9–12] devices possible. These developments have fueled considerable interest in these III–V nitride semiconductors.

In order to analyze and improve the design of III–V nitride semiconductor-based devices, an understanding of the electron transport that occurs within these materials is necessary. Electron transport within bulk GaN, AlN, and InN has been examined extensively over the years [33.13–32]. Unfortunately, uncertainty in the material parameters associated with GaN, AlN, and InN remains a key source of ambiguity in the analysis of the electron transport within these materials [33.32]. In addition, some recent experimental [33.33] and theoretical [33.34] developments have cast doubt upon the validity of widely accepted notions upon which our understanding of the electron transport mechanisms within the III–V nitride semiconductors, GaN, AlN, and InN, has evolved. Another confounding matter is the sheer volume of research activity being performed on the electron transport within these materials, presenting the researcher with a dizzying array of seemingly disparate approaches and results. Clearly, at this critical juncture at least, our understanding of the electron transport within the III–V nitride semiconductors, GaN, AlN, and InN, remains in a state of flux.

In order to provide some perspective on this rapidly evolving field, we aim to review analyses of the electron transport within the III–V nitride semiconductors, GaN, AlN, and InN, within this paper. In particular, we will discuss the evolution of the field and survey the current literature. In order to narrow the scope of this review, we will primarily focus on the electron transport within bulk wurtzite GaN, AlN, and InN for the purposes of this paper. Most of our discussion will focus upon results obtained from our ensemble semi-classical three-valley Monte Carlo simulations of the electron transport within these materials, our results conforming with state-of-the-art III–V nitride semiconductor orthodoxy. We hope that researchers in the field will find this review useful and informative.

We begin our review with the Boltzmann transport equation, which underlies most analyses of the electron transport within semiconductors. The ensemble semi-classical three-valley Monte Carlo simulation approach that we employ in order to solve this Boltzmann transport equation is then discussed. The material parameters corresponding to bulk wurtzite GaN, AlN, and InN are then presented. We then use these material parameter selections and our ensemble semi-classical three-valley Monte Carlo simulation approach to determine the nature of the steady-state and transient electron transport within the III–V nitride semiconductors. Finally, we present some recent developments on the electron transport within these materials.

This paper is organized in the following manner. In Sect. 33.1, we present the Boltzmann transport equation and our ensemble semi-classical three-valley Monte Carlo simulation approach that we employ in order to solve this equation for the III–V nitride semiconductors, GaN, AlN, and InN. The material parameters, corresponding to bulk wurtzite GaN, AlN, and InN, are also presented in Sect. 33.1. Then, in Sect. 33.2, using results obtained from our ensemble semi-classical three-valley Monte Carlo simulations of the electron transport within these III–V nitride semiconductors, we study the nature of the steady-state electron transport that occurs within these materials. Transient electron transport within the III–V nitride semiconductors is also discussed in Sect. 33.2. A review of the III–V nitride semiconductor electron transport literature, in which the evolution of the field is discussed and a survey of the current literature is presented, is then featured in Sect. 33.3. Finally, conclusions are provided in Sect. 33.4.

33.1 Electron Transport Within Semiconductors and the Monte Carlo Simulation Approach

The electrons within a semiconductor are in a perpetual state of motion. In the absence of an applied electric field, this motion arises as a result of the thermal energy that is present, and is referred to as thermal motion. From the perspective of an individual electron, thermal motion may be viewed as a series of trajectories, interrupted by a series of random scattering events. Scattering may arise as a result of interactions with the lattice atoms, impurities, other electrons, and defects. As these interactions lead to electron trajectories in all possible directions, i.e., there is no preferred direction, while individual electrons will move from one location to another, when taken as an ensemble, and assuming that the electrons are in thermal equilibrium, the overall electron distribution will remain static. Accordingly, no net current flow occurs.

With the application of an applied electric field, E, each electron in the ensemble will experience a force, $-qE$. While this force may have a negligible impact upon the motion of any given individual electron, taken as an ensemble, the application of such a force will lead to a net aggregate motion of the electron distribution. Accordingly, a net current flow will occur, and the overall electron ensemble will no longer be in thermal equilibrium. This movement of the electron ensemble in response to an applied electric field, in essence, represents the fundamental issue at stake when we study the electron transport within a semiconductor.

In this section, we provide a brief tutorial on the issues at stake in our analysis of the electron transport within the III–V nitride semiconductors, GaN, AlN, and InN. We begin our analysis with an introduction to the Boltzmann transport equation. This equation describes how the electron distribution function evolves under the action of an applied electric field, and underlies the electron transport within bulk semiconductors. We then introduce the Monte Carlo simulation approach to solving this Boltzmann transport equation, focusing on the ensemble semi-classical three-valley Monte Carlo simulation approach used in our simulations of the electron transport within the III–V nitride semiconductors. Finally, we present the material parameters corresponding to bulk wurtzite GaN, AlN, and InN.

This section is organized in the following manner. In Sect. 33.1.1, the Boltzmann transport equation is introduced. Then, in Sect. 33.1.2, our ensemble semi-classical three-valley Monte Carlo simulation approach to solving this Boltzmann transport equation is presented. Finally, in Sect. 33.1.3, our material parameter selections, corresponding to bulk wurtzite GaN, AlN, and InN, are presented.

33.1.1 The Boltzmann Transport Equation

An electron ensemble may be characterized by its distribution function, $f(r, p, t)$, where r denotes the position, p represents the momentum, and t indicates time. The response of this distribution function to an applied electric field, E, is the issue at stake when one investigates the electron transport within a semiconductor. When the dimensions of the semiconductor are large, and quantum effects are negligible, the ensemble of electrons may be treated as a continuum, so the corpuscular nature of the individual electrons within the ensemble, and the attendant complications which arise, may be neglected. In such a circumstance, the evolution of the distribution function, $f(r, p, t)$, may be determined using the Boltzmann transport equation. In contrast, when the dimensions of the semiconductor are small, and quantum effects are significant, then the Boltzmann transport equation, and its continuum description of the electron ensemble, is no longer valid. In such a case, it is necessary to adopt quantum transport methods in order to study the electron transport within the semiconductor [33.35].

For the purposes of this analysis, we will focus on the electron transport within bulk semiconductors, i.e., semiconductors of sufficient dimensions so that the Boltzmann transport equation is valid. *Ashcroft* and *Mermin* [33.36] demonstrated that this equation may be expressed as

$$\frac{\partial f}{\partial t} = -\dot{p} \cdot \nabla_p f - \dot{r} \cdot \nabla_r f + \left.\frac{\partial f}{\partial t}\right|_{\text{scat}}. \quad (33.1)$$

The first term on the right-hand side of (33.1) represents the change in the distribution function due to external forces applied to the system. The second term on the right-hand side of (33.1) accounts for the electron diffusion which occurs. The final term on the right-hand side of (33.1) describes the effects of scattering.

Owing to its fundamental importance in the analysis of the electron transport within semiconductors, a number of techniques have been developed over the years in order to solve the Boltzmann transport equation. Approximate solutions to the Boltzmann transport equation, such as the displaced Maxwellian distribution function approach of *Ferry* [33.14] and *Das* and *Ferry* [33.15] and the nonstationary charge transport analysis of *Sandborn* et al. [33.37], have proven useful. Low-field approximate solutions have also proven elementary and insightful [33.17, 20, 38]. A number of these techniques have been applied to the analysis of the electron transport within the III–V nitride semiconductors, GaN, AlN, and InN [33.14, 15, 17, 20, 38, 39]. Alternatively, more sophisticated techniques have been developed which solve the Boltzmann transport equation directly. These techniques, while allowing for a rigorous solution of the Boltzmann transport equation, are rather involved, and require intense numerical analysis. They are further discussed by *Nag* [33.40].

For studies of the electron transport within the III–V nitride semiconductors, GaN, AlN, and InN, by far the most common approach to solving the Boltzmann transport equation has been the ensemble semi-classical Monte Carlo simulation approach. Of the III–V nitride semiconductors, the electron transport within GaN has been studied the most extensively using this ensemble Monte Carlo simulation approach [33.13, 16, 18, 19, 21, 22, 27, 29, 32], with AlN [33.24, 25, 29] and InN [33.23, 28, 29, 31] less so. The Monte Carlo simulation approach has also been used to study the electron transport within the two-dimensional electron gas of the AlGaN/GaN interface which occurs in high electron mobility AlGaN/GaN field-effect transistors [33.41, 42].

At this point, it should be noted that the complete solution of the Boltzmann transport equation requires the resolution of both steady-state and transient responses. Steady-state electron transport refers to the electron transport that occurs long after the application of an applied electric field, i.e., once the electron ensemble has settled to a new equilibrium state (we are not necessarily referring to thermal equilibrium here, since thermal equilibrium is only achieved in the absence of an applied electric field). As the distribution function is difficult to visualize quantitatively, researchers typically study the dependence of the electron drift velocity (the average electron velocity determined by statistically averaging over the entire electron ensemble) on the applied electric field in the analysis of steady-state electron transport; in other words, they determine the velocity–field characteristic. Transient electron transport, by way of contrast, refers to the transport that occurs while the electron ensemble is evolving into its new equilibrium state. Typically, it is characterized by studying the dependence of the electron drift velocity on the time elapsed, or the distance displaced, since the electric field was initially applied. Both steady-state and transient electron transport within the III–V nitride semiconductors, GaN, AlN, and InN, are reviewed within this paper.

33.1.2 Our Ensemble Semi-Classical Monte Carlo Simulation Approach

For the purposes of our analysis of the electron transport within the III–V nitride semiconductors, GaN, AlN, and InN, we employ ensemble semi-classical Monte Carlo simulations. A three-valley model for the conduction band is employed. Nonparabolicity is considered in the lowest conduction band valley, this nonparabolicity being treated through the application of the Kane model [33.43].

In the Kane model, the energy band of the Γ valley is assumed to be nonparabolic, spherical, and of the form

$$\frac{\hbar^2 k^2}{2m^*} = E(1 + \alpha E), \quad (33.2)$$

where $\hbar k$ denotes the crystal momentum, E represents the energy above the minimum, m^* is the effective mass, and the nonparabolicity coefficient, α, is given by

$$\alpha = \frac{1}{E_g}\left(1 - \frac{m^*}{m_e}\right)^2, \quad (33.3)$$

where m_e and E_g denote the free electron mass and the energy gap, respectively [33.43].

The scattering mechanisms considered in our analysis are (1) ionized impurity, (2) polar optical phonon, (3) piezoelectric [33.44, 45], and (4) acoustic deformation potential. Intervalley scattering is also considered. Piezoelectric scattering is treated using the well established zinc blende scattering rates, and so a suitably transformed piezoelectric constant, e_{14}, must be selected. This may be achieved through the transformation suggested by *Bykhovski* et al. [33.44, 45]. We also assume that all donors are ionized and that the free electron concentration is equal to the dopant concentration. The motion of three thousand electrons is examined in our steady-state electron transport simulations, while the motion of ten thousand electrons is considered in our transient electron transport simulations. The crystal temperature is set to 300 K and the doping concentration is set to 10^{17} cm^{-3} in all cases, unless otherwise specified. Electron degeneracy effects are accounted for by means of the rejection technique of *Lugli* and *Ferry* [33.46]. Electron screening is also accounted for following the Brooks–Herring method [33.47]. Further details of our approach are discussed in the literature [33.16, 21–24, 29, 32, 48].

33.1.3 Parameter Selections for Bulk Wurtzite GaN, AlN, and InN

The material parameter selections, used for our simulations of the electron transport within the III–V nitride semiconductors, GaN, AlN, and InN, are tabulated in Table 33.1. These parameter selections are the same as those employed by *Foutz* et al. [33.29]. While the band structures corresponding to bulk wurtzite GaN, AlN, and InN are still not agreed upon, the band structures of *Lambrecht* and *Segall* [33.49] are adopted for the purposes of this analysis. For the case of bulk wurtzite GaN, the analysis of *Lambrecht* and *Segall* [33.49] suggests that the lowest point in the conduction band is located at the center of the Brillouin zone, at the Γ point, the first upper conduction band valley minimum also occurring at the Γ point, 1.9 eV above the lowest point in the conduction band, the second upper conduction

Table 33.1 The material parameter selections corresponding to bulk wurtzite GaN, AlN, and InN. These parameter selections are from *Foutz* et al. [33.29]

Parameter	GaN	AlN	InN
Mass density (g/cm^3)	6.15	3.23	6.81
Longitudinal sound velocity (cm/s)	6.56×10^5	9.06×10^5	6.24×10^5
Transverse sound velocity (cm/s)	2.68×10^5	3.70×10^5	2.55×10^5
Acoustic deformation potential (eV)	8.3	9.5	7.1
Static dielectric constant	8.9	8.5	15.3
High-frequency dielectric constant	5.35	4.77	8.4
Effective mass (Γ_1 valley)	$0.20\, m_e$	$0.48\, m_e$	$0.11\, m_e$
Piezoelectric constant, e_{14} (C/cm^2)	3.75×10^{-5}	9.2×10^{-5}	3.75×10^{-5}
Direct energy gap (eV)	3.39	6.2	1.89
Optical phonon energy (meV)	91.2	99.2	89.0
Intervalley deformation potentials (eV/cm)	10^9	10^9	10^9
Intervalley phonon energies (meV)	91.2	99.2	89.0

Table 33.2 The valley parameter selections corresponding to bulk wurtzite GaN, AlN, and InN. These parameter selections are from *Foutz* et al. [33.29]. These parameters were originally determined from the band structural calculations of *Lambrecht* and *Segall* [33.49].

	Valley number	1	2	3
GaN	Valley location	Γ_1	Γ_2	L–M
	Valley degeneracy	1	1	6
	Effective mass	$0.2\, m_e$	m_e	m_e
	Intervalley energy separation (eV)	–	1.9	2.1
	Energy gap (eV)	3.39	5.29	5.49
	Nonparabolicity (eV^{-1})	0.189	0.0	0.0
AlN	Valley location	Γ_1	L–M	K
	Valley degeneracy	1	6	2
	Effective mass	$0.48\, m_e$	m_e	m_e
	Intervalley energy separation (eV)	–	0.7	1.0
	Energy gap (eV)	6.2	6.9	7.2
	Nonparabolicity (eV^{-1})	0.044	0.0	0.0
InN	Valley location	Γ_1	A	Γ_2
	Valley degeneracy	1	1	1
	Effective mass	$0.11\, m_e$	m_e	m_e
	Intervalley energy separation (eV)	–	2.2	2.6
	Energy gap (eV)	1.89	4.09	4.49
	Nonparabolicity (eV^{-1})	0.419	0.0	0.0

band valley minima occurring along the symmetry lines between the L and M points, 2.1 eV above the lowest point in the conduction band; see Table 33.2. For the case of bulk wurtzite AlN, the analysis of *Lambrecht* and *Segall* [33.49] suggests that the lowest point in the conduction band is located at the center of the Brillouin zone, at the Γ point, the first upper conduction band valley minima occurring along the symmetry lines between the L and M points, 0.7 eV above the lowest point in the conduction band, the second upper conduction band valley minima occurring at the K points, 1 eV above the lowest point in the conduction band; see Table 33.2. For the case of bulk wurtzite InN, the analysis of *Lambrecht* and *Segall* [33.49] suggests that the lowest point in the conduction band is located at the center of the Brillouin zone, at the Γ point, the first upper conduction band valley minimum occurring at the A point, 2.2 eV above the lowest point in the conduction band, the second upper conduction band valley minimum occurring at the Γ point, 2.6 eV above the lowest point in the conduction band; see Table 33.2. We ascribe an effective mass equal to the free electron mass, m_e, to all of the upper conduction band valleys. The nonparabolicity coefficient, α, corresponding to each upper conduction band valley is set to zero, so the upper conduction band valleys are assumed to be completely parabolic. For our simulations of the electron transport within gallium arsenide (GaAs), the material parameters employed are mostly from *Littlejohn* et al. [33.50], although it should be noted that the mass density, the energy gap, and the sound velocities are from *Blakemore* [33.51].

It should be noted that the energy gap associated with InN has been the subject of some controversy since 2002. The pioneering experimental results of *Tansley* and *Foley* [33.52], reported in 1986, suggested that InN has an energy gap of 1.89 eV. This value has been used extensively in Monte Carlo simulations of the electron transport within this material since that time [33.23, 28, 29, 31]; typically, the influence of the energy gap on the electron transport occurs through its impact on the nonparabolicity coefficient, α. In 2002, *Davydov* et al. [33.53], *Wu* et al. [33.54], and *Matsuoka* et al. [33.55], presented experimental evidence which instead suggests a considerably smaller energy gap for InN, around 0.7 eV. As this new result is still the subject of some controversy, we adopt the traditional *Tansley* and *Foley* [33.52] energy gap value for the purposes of our present analysis, noting that even if the newer value for the energy gap was adopted, it would only change our electron transport results marginally; the sensitivity of the velocity–field characteristic associated with bulk wurtzite GaN to variations in the nonparabolicity coefficient, α, has been explored, in detail, by *O'Leary* et al. [33.32].

The band structure associated with bulk wurtzite GaN has also been the focus of some controversy. In particular, *Brazel* et al. [33.56] employed ballistic electron emission microscopy measurements in order to demonstrate that the first upper conduction band valley occurs only 340 meV above the lowest point in the conduction band for this material. This contrasts rather dramatically with more traditional results, such as the calculation of *Lambrecht* and *Segall* [33.49], which instead suggest that the first upper conduction band valley minimum within wurtzite GaN occurs about 2 eV above the lowest point in the conduction band. Clearly, this will have a significant impact upon the results. While the results of *Brazel* et al. [33.56] were reported in 1997, electron transport simulations adopted the more traditional intervalley energy separation of about 2 eV until relatively recently. Accordingly, we have adopted the more traditional intervalley energy separation for the purposes of our present analysis. The sensitivity of the velocity–field characteristic associated with bulk wurtzite GaN to variations in the intervalley energy separation has been explored, in detail, by *O'Leary* et al. [33.32].

33.2 Steady-State and Transient Electron Transport Within Bulk Wurtzite GaN, AlN, and InN

The current interest in the III–V nitride semiconductors, GaN, AlN, and InN, is primarily being fueled by the tremendous potential of these materials for novel electronic and optoelectronic device applications. With the recognition that informed electronic and optoelectronic device design requires a firm understanding of the nature of the electron transport within these materials, electron transport within the III–V nitride semiconductors has been the focus of intensive investigation over the years. The literature abounds with studies on steady-state and transient electron transport within these materials [33.13–34, 38, 39, 41, 42, 48]. As a result of this intense flurry of research activity, novel III–V nitride semiconductor-based devices are starting to be deployed in today's commercial products. Future developments in the III–V nitride semiconductor field will undoubtedly require an even deeper understanding of the electron transport mechanisms within these materials.

In the previous section, we presented details of the Monte Carlo simulation approach that we employ for the analysis of the electron transport within the III–V nitride semiconductors, GaN, AlN, and InN. In this section, an overview of the steady-state and transient electron transport results we obtained from these Monte Carlo simulations is provided. In the first part of this section, we focus upon bulk wurtzite GaN. In particular, the velocity–field characteristic associated with this material will be examined in detail. Then, an overview of our steady-state electron transport results, corresponding to the three III–V nitride semiconductors under consideration in this analysis, will be given, and a comparison with the more conventional III–V compound semiconductor, GaAs, will be presented. A comparison between the temperature dependence of the velocity–field characteristics associated with GaN and GaAs will then be presented, and our Monte Carlo results will be used to account for the differences in behavior. A similar analysis will be presented for the doping dependence. Next, detailed simulation results for AlN and InN will be presented. Finally, the transient electron transport that occurs within the III–V nitride semiconductors, GaN, AlN, and InN, is determined and compared with that in GaAs.

This section is organized in the following manner. In Sect. 33.2.1, the velocity–field characteristic associated with bulk wurtzite GaN is presented and analyzed. Then, in Sect. 33.2.2, the velocity-field characteristics associated with the III–V nitride semiconductors under consideration in this analysis will be compared and contrasted with that of GaAs. The sensitivity of the velocity–field characteristic associated with bulk wurtzite GaN to variations in the crystal temperature will then be examined in Sect. 33.2.3, and a comparison with that corresponding to GaAs presented. In Sect. 33.2.4, the sensitivity of the velocity–field characteristic associated with bulk wurtzite GaN to variations in the doping concentration level will be explored, and a comparison with that corresponding to GaAs presented. The velocity–field characteristics associated with AlN and InN will then be examined in Sect. 33.2.5 and Sect. 33.2.6, respectively. Our transient electron transport analysis results are then presented in Sect. 33.2.7. Finally, the conclusions of this electron transport analysis are summarized in Sect. 33.2.8.

33.2.1 Steady-State Electron Transport Within Bulk Wurtzite GaN

Our examination of results begins with GaN, the most commonly studied III–V nitride semiconductor. The velocity–field characteristic associated with this material is presented in Fig. 33.1. This result was obtained through our Monte Carlo simulations of the electron transport within this material for the bulk wurtzite GaN parameter selections specified in Table 33.1 and Table 33.2; the crystal temperature was set to 300 K and the doping concentration to 10^{17} cm^{-3}. We see that for applied electric fields in excess of 140 kV/cm, the electron drift velocity decreases, eventually saturating at 1.4×10^7 cm/s for high applied electric fields. By examining the results of our Monte Carlo simulation further, an understanding of this result becomes clear.

First, we discuss the results at low applied electric fields, i.e., applied electric fields of less than 30 kV/cm. This is referred to as the linear regime of electron trans-

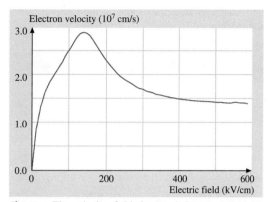

Fig. 33.1 The velocity–field characteristic associated with bulk wurtzite GaN. Like many other compound semiconductors, the electron drift velocity reaches a peak, and at higher applied electric fields it decreases until it saturates

port as the electron drift velocity is well characterized by the low-field electron drift mobility, μ, in this regime, i.e., a linear low-field electron drift velocity dependence on the applied electric field $v_d = \mu E$, applies in this regime. Examining the distribution function for this regime, we find that it is very similar to the zero-field distribution function with a slight shift in the direction opposite to the applied electric field. In this regime, the average electron energy remains relatively low, with most of the energy gained from the applied electric field being transferred into the lattice through polar optical phonon scattering.

If we examine the average electron energy as a function of the applied electric field, shown in Fig. 33.2, we see that there is a sudden increase at around 100 kV/cm. In order to understand why this increase occurs, we note that the dominant energy loss mechanism for many of the III–V compound semiconductors, including GaN, is polar optical phonon scattering. When the applied electric field is less than 100 kV/cm, all of the energy that the electrons gain from the applied electric field is lost through polar optical phonon scattering. The other scattering mechanisms, i.e., ionized impurity scattering, piezoelectric scattering and acoustic deformation potential scattering, do not remove energy from the electron ensemble: they are elastic scattering mechanisms. However, beyond a certain critical applied electric field strength, the polar optical phonon scattering mechanism can no longer remove all of the energy gained from the applied electric field. Other scattering mechanisms must start to play a role if the electron ensemble is to remain in equilibrium. The average electron energy increases until intervalley scattering begins and an energy balance is re-established.

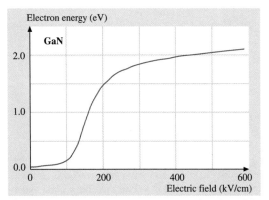

Fig. 33.2 The average electron energy as a function of the applied electric field for bulk wurtzite GaN. Initially, the average electron energy remains low, only slightly higher than the thermal energy, $\frac{3}{2}k_B T$. At $100\,\text{kV/cm}$, however, the average electron energy increases dramatically. This increase is due to the fact that the polar optical phonon scattering mechanism can no longer absorb all of the energy gained from the applied electric field

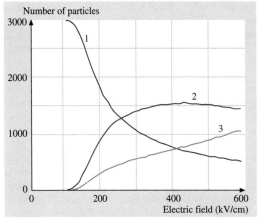

Fig. 33.3 The valley occupancy as a function of the applied electric field for the case of bulk wurtzite GaN. Soon after the average electron energy increases, electrons begin to transfer to the upper valleys of the conduction band. Three thousand electrons were employed for this simulation. The valleys are labeled 1, 2, and 3, in accordance with their energy minima; the lowest energy valley is valley 1, the next higher energy valley is valley 2, and the highest energy valley is valley 3

As the applied electric field is increased beyond $100\,\text{kV/cm}$, the average electron energy increases until a substantial fraction of the electrons have acquired enough energy in order to transfer into the upper valleys. As the effective mass of the electrons in the upper valleys is greater than that in the lowest valley, the electrons in the upper valleys will be slower. As more electrons transfer to the upper valleys (Fig. 33.3), the electron drift velocity decreases. This accounts for the negative differential mobility observed in the velocity–field characteristic depicted in Fig. 33.1.

Finally, at high applied electric fields, the number of electrons in each valley saturates. It can be shown that in the high-field limit the number of electrons in each valley is proportional to the product of the density of states of that particular valley and the corresponding valley degeneracy. At this point, the electron drift velocity stops decreasing and achieves saturation.

Thus far, electron transport results corresponding to bulk wurtzite GaN have been presented and discussed qualitatively. It should be noted, however, that the same phenomenon that occurs in the velocity–field characteristic associated with GaN also occurs for the other III–V nitride semiconductors, AlN and InN. The importance of polar optical phonon scattering when determining the nature of the electron transport within the III–V nitride semiconductors, GaN, AlN, and InN, will become even more apparent later, as it will be used to account for much of the electron transport behavior within these materials.

33.2.2 Steady-State Electron Transport: A Comparison of the III–V Nitride Semiconductors with GaAs

Setting the crystal temperature to $300\,\text{K}$ and the level of doping to $10^{17}\,\text{cm}^{-3}$, the velocity–field characteristics associated with the III–V nitride semiconductors under consideration in this analysis – GaN, AlN, and InN – are contrasted with that of GaAs in Fig. 33.4. We see that each of these III–V compound semiconductors achieves a peak in its velocity–field characteristic. InN achieves the highest steady-state peak electron drift velocity, $4.1 \times 10^7\,\text{cm/s}$ at an applied electric field of $65\,\text{kV/cm}$. This contrasts with the case of GaN, $2.9 \times 10^7\,\text{cm/s}$ at $140\,\text{kV/cm}$, and that of AlN, $1.7 \times 10^7\,\text{cm/s}$ at $450\,\text{kV/cm}$. For GaAs, the peak electron drift velocity of $1.6 \times 10^7\,\text{cm/s}$ occurs at a much lower applied electric field than that for the III–V nitride semiconductors (only $4\,\text{kV/cm}$).

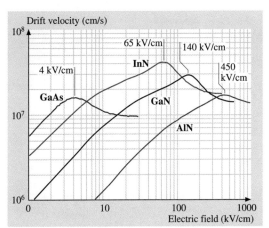

Fig. 33.4 A comparison of the velocity–field characteristics associated with the III–V nitride semiconductors, GaN, AlN, and InN, with that associated with GaAs. After [33.29] with permission, copyright AIP

33.2.3 Influence of Temperature on the Electron Drift Velocities Within GaN and GaAs

The temperature dependence of the velocity–field characteristic associated with bulk wurtzite GaN is now examined. Figure 33.5a shows how the velocity–field characteristic associated with bulk wurtzite GaN varies as the crystal temperature is increased from 100 to 700 K, in increments of 200 K. The upper limit, 700 K, is chosen as it is the highest operating temperature that may be expected for AlGaN/GaN power devices. To highlight the difference between the III–V nitride semiconductors with more conventional III–V compound semiconductors, such as GaAs, Monte Carlo simulations of the electron transport within GaAs have also been performed under the same conditions as GaN. Figure 33.5b shows the results of these simulations. Note that the electron drift velocity for the case of GaN is much less sensitive to changes in temperature than that associated with GaAs.

To quantify this dependence further, the low-field electron drift mobility, the peak electron drift velocity, and the saturation electron drift velocity are plotted as a function of the crystal temperature in Fig. 33.6, these results being determined from our Monte Carlo simulations of the electron transport within these materials. For both GaN and GaAs, it is found that all of these electron transport metrics diminish as the crystal temperature is increased. As may be seen through an inspection of Fig. 33.5, the peak and saturation electron drift velocities do not drop as much in GaN as they do in GaAs in response to increases in the crystal temperature. The low-field electron drift mobility in GaN, however, is seen to fall quite rapidly with temperature, this drop being particularly severe for temperatures at and below room temperature. This property will likely have an impact on high-power device performance.

Delving deeper into our Monte Carlo results yields clues to the reason for this variation in temperature dependence. First, we examine the polar optical phonon scattering rate as a function of the applied electric field strength. Figure 33.7 shows that the scattering rate only

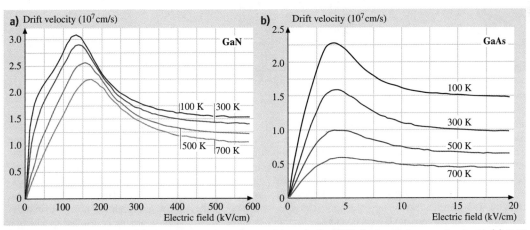

Fig. 33.5a,b A comparison of the temperature dependence of the velocity–field characteristics associated with (a) GaN and (b) GaAs. GaN maintains a higher electron drift velocity with increased temperatures than GaAs does

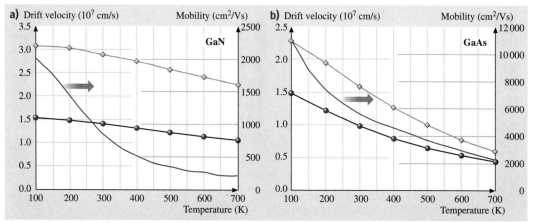

Fig. 33.6a,b A comparison of the temperature dependence of the low-field electron drift mobility (*solid lines*), the peak electron drift velocity (*diamonds*), and the saturation electron drift velocity (*solid points*) for (**a**) GaN and (**b**) GaAs. The low-field electron drift mobility of GaN drops quickly with increasing temperature, but its peak and saturation electron drift velocities are less sensitive to increases in temperature than GaAs

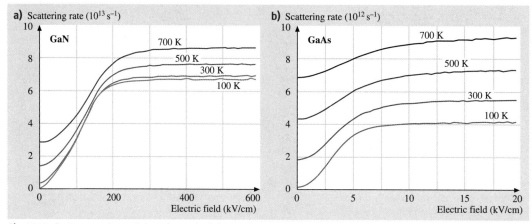

Fig. 33.7a,b A comparison of the polar optical phonon scattering rates as a function of the applied electric field strength for various crystal temperatures for (**a**) GaN and (**b**) GaAs. Polar optical phonon scattering is seen to increase much more quickly with temperature in GaAs

increases slightly with temperature for the case of GaN, from 6.7×10^{13} s^{-1} at 100 K to 8.6×10^{13} s^{-1} at 700 K, for high applied electric field strengths. Contrast this with the case of GaAs, where the rate increases from 4.0×10^{12} s^{-1} at 100 K to more than twice that amount at 700 K, 9.2×10^{12} s^{-1}, at high applied electric field strengths. This large increase in the polar optical phonon scattering rate for the case of GaAs is one reason for the large drop in the electron drift velocity with increasing temperature for the case of GaAs.

A second reason for the variation in temperature dependence of the two materials is the occupancy of the upper valleys, shown in Fig. 33.8. In the case of GaN, the upper valleys begin to become occupied at roughly the same applied electric field strength, 100 kV/cm, independent of temperature. For the case of GaAs, however, the upper valleys are at a much lower energy than those in GaN. In particular, while the first upper conduction band valley minimum is 1.9 eV above the lowest point in the conduction band in GaN, the first upper conduc-

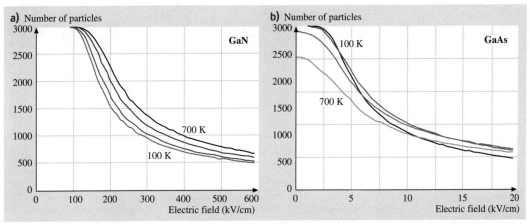

Fig. 33.8a,b A comparison of the number of particles in the lowest energy valley of the conduction band, the Γ valley, as a function of the applied electric field for various crystal temperatures, for the cases of (**a**) GaN and (**b**) GaAs. In GaAs, the electrons begin to occupy the upper valleys much more quickly, causing the electron drift velocity to drop as the crystal temperature is increased. Three thousand electrons were employed for these steady-state electron transport simulations

tion band valley is only 290 meV above the bottom of the conduction band in GaAs [33.51]. As the upper conduction band valleys are so close to the bottom of the conduction band for the case of GaAs, the thermal energy (at 700 K, $k_B T \simeq 60$ meV) is enough in order to allow for a small fraction of the electrons to transfer into the upper valleys even before an electric field is applied. When electrons occupy the upper valleys, intervalley scattering and the upper valleys' larger effective masses reduce the overall electron drift velocity. This is another reason why the velocity–field characteristic associated with GaAs is more sensitive to variations in crystal temperature than that associated with GaN.

33.2.4 Influence of Doping on the Electron Drift Velocities Within GaN and GaAs

One parameter that can be readily controlled during the fabrication of semiconductor devices is the doping concentration. Understanding the effect of doping

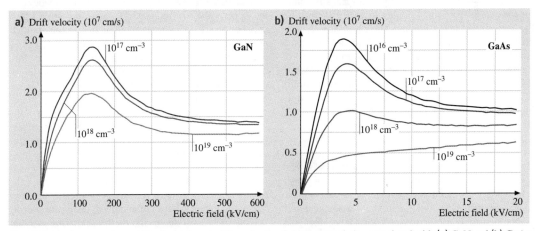

Fig. 33.9a,b A comparison of the dependence of the velocity–field characteristics associated with (**a**) GaN and (**b**) GaAs on the doping concentration. GaN maintains a higher electron drift velocity with increased doping levels than GaAs does

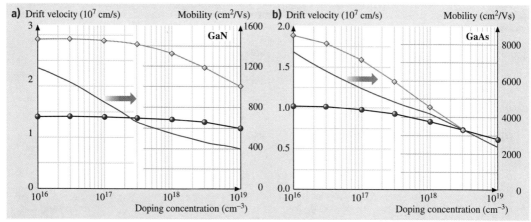

Fig. 33.10a,b A comparison of the low-field electron drift mobility (*solid lines*), the peak electron drift velocity (*diamonds*), and the saturation electron drift velocity (*solid points*) for (**a**) GaN and (**b**) GaAs as a function of the doping concentration. These parameters are more insensitive to increases in doping in GaN than in GaAs

on the resultant electron transport is also important. In Fig. 33.9, the velocity–field characteristic associated with GaN is presented for a number of different doping concentration levels. Once again, three important electron transport metrics are influenced by the doping concentration level: the low-field electron drift mobility, the peak electron drift velocity, and the saturation electron drift velocity; see Fig. 33.10. Our simulation results suggest that for doping concentrations of less than 10^{17} cm^{-3}, there is very little effect on the velocity–field characteristic for the case of GaN. However, for doping concentrations above 10^{17} cm^{-3}, the peak electron drift velocity diminishes considerably, from 2.9×10^7 cm/s for the case of 10^{17} cm^{-3} doping to 2.0×10^7 cm/s for the case of 10^{19} cm^{-3} doping. The saturation electron drift velocity within GaN is found to only decrease slightly in response to increases in the doping concentration. The effect of doping on the low-field electron drift mobility is also shown. It is seen that this mobility drops significantly in response to increases in the doping concentration level, from 1200 cm^2/Vs at 10^{16} cm^{-3} doping to 400 cm^2/Vs at 10^{19} cm^{-3} doping.

As we did for temperature, we compare the sensitivity of the velocity-field characteristic associated with GaN to doping with that associated with GaAs. Figure 33.9 shows this comparison. For the case of GaAs, it is seen that the electron drift velocities decrease much more with increased doping than those associated with GaN. In particular, for the case of GaAs, the peak electron drift velocity decreases from 1.8×10^7 cm/s at 10^{16} cm^{-3} doping to 0.6×10^7 cm/s at 10^{19} cm^{-3} doping. For GaAs, at the higher doping levels, the peak in the velocity-field characteristic disappears completely for sufficiently high doping concentrations. The saturation electron drift velocity decreases from 1.0×10^7 cm/s at 10^{16} cm^{-3} doping to 0.6×10^7 cm/s at 10^{19} cm^{-3} doping. The low-field electron drift mobility also diminishes dramatically with increased doping, dropping from 7800 cm^2/Vs at 10^{16} cm^{-3} doping to 2200 cm^2/Vs at 10^{19} cm^{-3} doping.

Once again, it is interesting to determine why the doping dependence in GaAs is so much more pronounced than it is in GaN. Again, we examine the polar optical phonon scattering rate and the occupancy of the upper valleys. Figure 33.11 shows the polar optical phonon scattering rates as a function of the applied electric field, for both GaN and GaAs. In this case, however, due to screening effects, the rate drops when the doping concentration is increased. The decrease, however, is much more pronounced for the case of GaAs than for GaN. It is believed that this drop in the polar optical phonon scattering rate allows for upper valley occupancy to occur more quickly in GaAs rather than in GaN (Fig. 33.12). For GaN, electrons begin to occupy the upper valleys at roughly the same applied electric field strength, independent of the doping level. However, for the case of GaAs, the upper valleys are occupied more quickly with greater doping. When the upper valleys are occupied, the electron drift velocity decreases due to intervalley scattering and the larger effective mass of the electrons within the upper valleys.

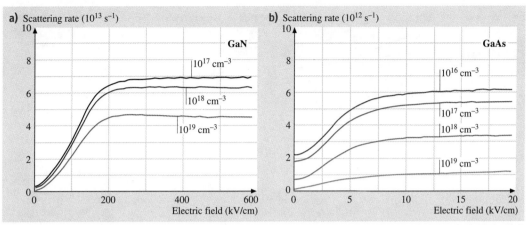

Fig. 33.11a,b A comparison of the polar optical phonon scattering rates as a function of the applied electric field, for both (a) GaN and (b) GaAs, for various doping concentrations

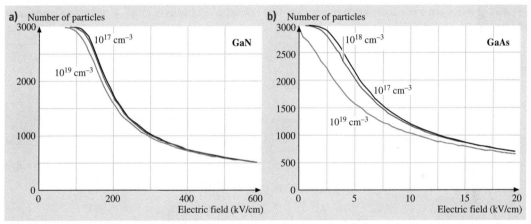

Fig. 33.12a,b A comparison of the number of particles in the lowest valley of the conduction band, the Γ valley, as a function of the applied electric field, for both (a) GaN and (b) GaAs, for various doping concentration levels. Three thousand electrons were employed for these steady-state electron transport simulations

33.2.5 Electron Transport in AlN

AlN has the largest effective mass of the III–V nitride semiconductors considered in this analysis. Accordingly, it is not surprising that this material exhibits the lowest electron drift velocity and the lowest low-field electron drift mobility. The sensitivity of the velocity–field characteristic associated with AlN to variations in the crystal temperature may be examined by considering Fig. 33.13. As with the case of GaN, the velocity–field characteristic associated with AlN is extremely robust to variations in the crystal temperature. In particular, its peak electron drift velocity, which is 1.8×10^7 cm/s at 100 K, only decreases to 1.2×10^7 cm/s at 700 K. Similarly, its saturation electron drift velocity, which is 1.5×10^7 cm/s at 100 K, only decreases to 1.0×10^7 cm/s at 700 K. The low-field electron drift mobility associated with AlN also diminishes in response to increases in the crystal temperature, from $375 \text{ cm}^2/\text{Vs}$ at 100 K to $40 \text{ cm}^2/\text{Vs}$ at 700 K.

The sensitivity of the velocity–field characteristic associated with AlN to variations in the doping concentration may be examined by considering Fig. 33.14. It

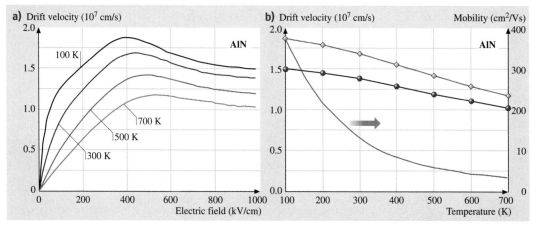

Fig. 33.13a,b The velocity–field characteristic associated with AlN (**a**) for various crystal temperatures. The trends in the low-field mobility (*solid line*), the peak electron drift velocity (*diamonds*), and the saturation electron drift velocity (*solid points*), are also shown. AlN exhibits its peak electron drift velocity at very high applied electric fields. AlN has the lowest peak electron drift velocity and the lowest low-field electron drift mobility of the III–V nitride semiconductors considered in this analysis (**b**)

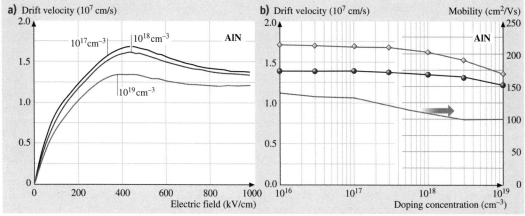

Fig. 33.14a,b The velocity–field characteristic associated with AlN for various doping concentrations (**a**). The trends in the low-field electron drift mobility (*solid line*), the peak electron drift velocity (*diamonds*), and the saturation electron drift velocity (*solid points*), are also shown (**b**)

is noted that the variations in the velocity–field characteristic associated with AlN in response to variations in the doping concentration are not as pronounced as those which occur in response to variations in the crystal temperature. Quantitatively, the peak electron drift velocity drops from 1.7×10^7 cm/s at 10^{17} cm^{-3} doping to 1.3×10^7 cm/s at 10^{19} cm^{-3} doping. Similarly, its saturation electron drift velocity drops from 1.4×10^7 cm/s at 10^{17} cm^{-3} doping to 1.2×10^7 cm/s at 10^{19} cm^{-3} doping. The influence of doping on the low-field electron drift mobility associated with AlN is also observed to be not as pronounced as for the case of crystal temperature. Figure 33.14b shows that the low-field electron drift mobility associated with AlN decreases from 140 cm^2/Vs at 10^{16} cm^{-3} doping to 100 cm^2/Vs at 10^{19} cm^{-3} doping.

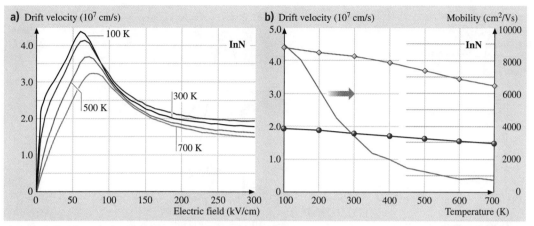

Fig. 33.15a,b The velocity–field characteristic associated with InN for various crystal temperatures (**a**). The trends in the low-field electron drift mobility (*solid line*), the peak electron drift velocity (*diamonds*), and the saturation electron drift velocity (*solid points*), are also shown. (**b**) InN has the highest peak electron drift velocity and the highest low-field electron drift mobility of the III–V nitride semiconductors considered in this analysis

33.2.6 Electron Transport in InN

InN has the smallest effective mass of the three III–V nitride semiconductors considered in this analysis. Accordingly, it is not surprising that it exhibits the highest electron drift velocity and the highest low-field electron drift mobility. The sensitivity of the velocity-field characteristic associated with InN to variations in the crystal temperature may be examined by considering Fig. 33.15. As with the cases of GaN and AlN, the velocity–field characteristic associated with InN is extremely robust to increases in the crystal temperature. In particular, its peak electron drift velocity, which is 4.4×10^7 cm/s at 100 K, only decreases to 3.2×10^7 cm/s at 700 K. Similarly, its saturation electron drift velocity, which is 2.0×10^7 cm/s at 100 K, only decreases to 1.5×10^7 cm/s at 700 K. The low-field electron drift mobility associated with InN also

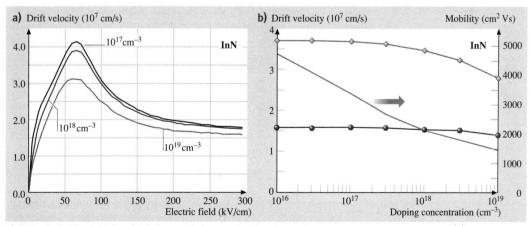

Fig. 33.16a,b The velocity–field characteristic associated with InN for various doping concentrations (**a**). The trends in the low-field electron drift mobility (*solid line*), the peak electron drift velocity (*diamonds*), and the saturation electron drift velocity (*solid points*), are also shown (**b**)

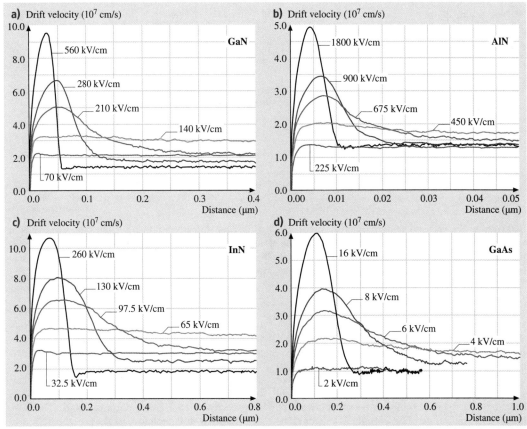

Fig. 33.17a–d The electron drift velocity as a function of the distance displaced since the application of the electric field for various applied electric field strengths, for the cases of (**a**) GaN, (**b**) AlN, (**c**) InN, and (**d**) GaAs. In all cases, we have assumed an initial zero field electron distribution, a crystal temperature of 300 K, and a doping concentration of 10^{17} cm^{-3}. After [33.29] with permission, copyright AIP

diminishes in response to increases in the crystal temperature, from about 9000 cm^2/Vs at 100 K to below 1000 cm^2/Vs at 700 K.

The sensitivity of the velocity–field characteristic associated with InN to variations in the doping concentration may be examined by considering Fig. 33.16. These results suggest a similar robustness to the doping concentration for the case of InN. In particular, it is noted that for doping concentrations below 10^{17} cm^{-3}, the velocity–field characteristic associated with InN exhibits very little dependence on the doping concentration. When the doping concentration is increased above 10^{17} cm^{-3}, however, the peak electron drift velocity diminishes. Quantitatively, the peak electron drift velocity decreases from 4.1×10^7 cm/s at 10^{17} cm^{-3} doping to 3.2×10^7 cm/s at 10^{19} cm^{-3} doping. The saturation electron drift velocity only drops slightly, however, from 1.8×10^7 cm/s at 10^{17} cm^{-3} doping to 1.5×10^7 cm/s at 10^{19} cm^{-3} doping. The low-field electron drift mobility, however, drops significantly with doping, from 4700 cm^2/Vs at 10^{16} cm^{-3} doping to 1500 cm^2/Vs at 10^{19} cm^{-3} doping.

33.2.7 Transient Electron Transport

Steady-state electron transport is the dominant electron transport mechanism in devices with larger dimensions. For devices with smaller dimensions, however, transient electron transport must also be considered when evaluating device performance. *Ruch* [33.57] demonstrated, for both silicon and GaAs, that the transient electron drift velocity may exceed the corresponding steady-state electron drift velocity by a considerable margin for appropriate selections of the applied electric field. *Shur* and *Eastman* [33.58] explored the device implications of transient electron transport, and demonstrated that substantial improvements in the device performance can be achieved as a consequence. *Heiblum* et al. [33.59] made the first direct experimental observation of transient electron transport within GaAs. Since then there have been a number of experimental investigations into the transient electron transport within the III–V compound semiconductors; see, for example, [33.60–62].

Thus far, very little research has been invested into the study of transient electron transport within the III–V nitride semiconductors, GaN, AlN, and InN. *Foutz* et al. [33.21] examined transient electron transport within both the wurtzite and zinc blende phases of GaN. In particular, they examined how electrons, initially in thermal equilibrium, respond to the sudden application of a constant electric field. In devices with dimensions greater than 0.2 μm, they found that steady-state electron transport is expected to dominate device performance. For devices with smaller dimensions, however, upon the application of a sufficiently high electric field, they found that the transient electron drift velocity can considerably overshoot the corresponding steady-state electron drift velocity. This velocity overshoot was found to be comparable with that which occurs within GaAs.

Foutz et al. [33.29] performed a subsequent analysis in which the transient electron transport within all of the III–V nitride semiconductors under consideration in this analysis were compared with that which occurs within GaAs. In particular, following the approach of *Foutz* et al. [33.21], they examined how electrons, initially in thermal equilibrium, respond to the sudden application of a constant electric field. A key result of this study, presented in Fig. 33.17, plots the transient electron drift velocity as a function of the distance displaced since the electric field was initially applied for a number of applied electric field strengths and for each of the materials considered in this analysis.

Focusing initially on the case of GaN (Fig. 33.17a), we note that the electron drift velocity for the applied electric field strengths 70 kV/cm and 140 kV/cm reaches steady-state very quickly, with little or no velocity overshoot. In contrast, for applied electric field strengths above 140 kV/cm, significant velocity overshoot occurs. This result suggests that in GaN, 140 kV/cm is a critical field for the onset of velocity overshoot effects. As mentioned in Sect. 33.2.2, 140 kV/cm also corresponds to the peak in the velocity-field characteristic associated with GaN; recall Fig. 33.4. Steady-state Monte Carlo simulations suggest that this is the point at which significant upper valley occupation begins to occur; recall Fig. 33.3. This suggests that velocity overshoot is related to the transfer of electrons to the upper valleys. Similar results are found for the other III–V nitride semiconductors, AlN and InN, and GaAs; see Figs. 33.17b–d.

We now compare the transient electron transport characteristics for the materials. From Fig. 33.17, it is clear that certain materials exhibit higher peak overshoot velocities and longer overshoot relaxation times. It is not possible to fairly compare these different semiconductors by applying the same applied electric field strength to each of the materials, as the transient effects occur over such a disparate range of applied electric field strengths in each material. In order to facilitate such a comparison, we choose a field strength equal to twice the critical applied electric field strength for each material. Figure 33.18 shows a comparison of the velocity overshoot effects amongst the four materials considered in this analysis, i.e., GaN, AlN, InN, and

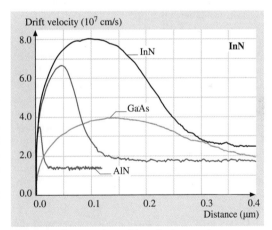

Fig. 33.18 A comparison of the velocity overshoot amongst the III–V nitride semiconductors and GaAs. The applied electric field strength chosen corresponds to twice the critical applied electric field strength at which the peak in the steady-state velocity–field characteristic occurs (Fig. 33.4), i.e., 280 kV/cm for the case of GaN, 900 kV/cm for the case of AlN, 130 kV/cm for the case of InN, and 8 kV/cm for the case of GaAs. After [33.29] with permission, copyright AIP

GaAs. It is clear that among the three III–V nitride semiconductors considered, InN exhibits superior transient electron transport characteristics. In particular, InN has the largest overshoot velocity and the distance over which this overshoot occurs, 0.3 μm, is longer than in either GaN and AlN. GaAs exhibits a longer overshoot relaxation distance, approximately 0.7 μm, but the electron drift velocity exhibited by InN is greater than that of GaAs for all distances.

33.2.8 Electron Transport: Conclusions

In this section, steady-state and transient electron transport results, corresponding to the III–V nitride semiconductors, GaN, AlN, and InN, were presented, these results being obtained from our Monte Carlo simulations of the electron transport within these materials. Steady-state electron transport was the dominant theme of our analysis. In order to aid in the understanding of these electron transport characteristics, a comparison was made between GaN and GaAs. Our simulations showed that GaN is more robust to variations in crystal temperature and doping concentration than GaAs, and an analysis of our Monte Carlo simulation results showed that polar optical phonon scattering plays the dominant role in accounting for these differences in behavior. This analysis was also performed for the other III–V nitride semiconductors considered in this analysis – AlN and InN – and similar results were obtained. Finally, we presented some key transient electron transport results. These results indicated that the transient electron transport that occurs within InN is the most pronounced of all of the materials under consideration in this review (GaN, AlN, InN, and GaAs).

33.3 Electron Transport Within III–V Nitride Semiconductors: A Review

Pioneering investigations into the material properties of the III–V nitride semiconductors, GaN, AlN, and InN, were performed during the earlier half of the twentieth century [33.63–65]. The III–V nitride semiconductor materials available at the time, small crystals and powders, were of poor quality, and completely unsuitable for device applications. Thus, it was not until the late 1960s, when *Maruska* and *Tietjen* [33.66] employed chemical vapor deposition to fabricate GaN, that interest in the III–V nitride semiconductors experienced a renaissance. Since that time, interest in the III–V nitride semiconductors has been growing, the material properties of these semiconductors improving considerably over the years. As a result of this research effort, there are currently a number of commercial devices available that employ the III–V nitride semiconductors. More III–V nitride semiconductor-based device applications are currently under development, and these should become available in the near future.

In this section, we present a brief overview of the III–V nitride semiconductor electron transport field. We start with a survey describing the evolution of the field. In particular, the sequence of critical developments that have occurred that contribute to our current understanding of the electron transport mechanisms within the III–V nitride semiconductors, GaN, AlN, and InN, is chronicled. Then, some of the current literature is presented, with particular emphasis being placed on the most recent developments in the field and how such developments are modifying our understanding of the electron transport mechanisms within the III–V nitride semiconductors, GaN, AlN, and InN. Finally, frontiers for further research and investigation are presented.

This section is organized in the following manner. In Sect. 33.3.1, we present a brief survey describing the evolution of the field. Then, in Sect. 33.3.2, the current literature is discussed. Finally, frontiers for further research and investigation are presented in Sect. 33.3.3.

33.3.1 Evolution of the Field

The favorable electron transport characteristics of the III–V nitride semiconductors, GaN, AlN, and InN, have long been recognized. As early as the 1970s, *Littlejohn* et al. [33.13] pointed out that the large polar optical phonon energy characteristic of GaN, in conjunction with its large intervalley energy separation, suggests a high saturation electron drift velocity for this material. As the high-frequency electron device performance is, to a large degree, determined by this saturation electron drift velocity [33.14], the recognition of this fact ignited enhanced interest in this material and its III–V nitride semiconductor compatriots, AlN and InN. This enhanced interest, and the developments which have transpired as a result of it, are responsible for the III–V nitride semiconductor industry of today.

In 1975, *Littlejohn* et al. [33.13] were the first to report results obtained from semi-classical Monte Carlo simulations of the steady-state electron transport within bulk wurtzite GaN. A one-valley model for the conduction band was adopted in their analysis. Steady-state electron transport, for both parabolic and nonparabolic band structures, was considered in their analysis, nonparabolicity being treated through the application of the Kane model [33.43]. The primary focus of their investigation was the determination of the velocity-field characteristic associated with GaN. All donors were assumed to be ionized, and the free electron concentration was taken to be equal to the dopant concentration. The scattering mechanisms considered were (1) ionized impurity, (2) polar optical phonon, (3) piezoelectric, and (4) acoustic deformation potential. For the case of the parabolic band, in the absence of ionized impurities, they found that the electron drift velocity monotonically increases with the applied electric field strength, saturating at a value of about 2.5×10^7 cm/s for the case of high applied electric fields. In contrast, for the case of the nonparabolic band, and in the absence of ionized impurities, a region of negative differential mobility was found, the electron drift velocity achieving a maximum of about 2×10^7 cm/s at an applied electric field strength of about $100\,\text{kV/cm}$, with further increases in the applied electric field strength resulting in a slight decrease in the corresponding electron drift velocity. The role of ionized impurity scattering was also investigated by *Littlejohn* et al. [33.13].

In 1993, *Gelmont* et al. [33.16] reported on ensemble semi-classical two-valley Monte Carlo simulations of the electron transport within bulk wurtzite GaN, this analysis improving upon the analysis of *Littlejohn* et al. [33.13] by incorporating intervalley scattering into the simulations. They found that the negative differential mobility found in bulk wurtzite GaN is much more pronounced than that found by *Littlejohn* et al. [33.13], and that intervalley transitions are responsible for this. For a doping concentration of $10^{17}\,\text{cm}^{-1}$, *Gelmont* et al. [33.16] demonstrated that the electron drift velocity achieves a peak value of about 2.8×10^7 cm/s at an applied electric field of about $140\,\text{kV/cm}$. The impact of intervalley transitions on the electron distribution function was also determined and shown to be significant. The impact of doping and compensation on the velocity-field characteristic associated with bulk wurtzite GaN was also examined.

Since these pioneering investigations, ensemble Monte Carlo simulations of the electron transport within GaN have been performed numerous times. In particular, in 1995 *Mansour* et al. [33.18] reported the use of such an approach in order to determine how the crystal temperature influences the velocity-field characteristic associated with bulk wurtzite GaN. Later that year, *Kolník* et al. [33.19] reported on employing full-band Monte Carlo simulations of the electron transport within bulk wurtzite GaN and bulk zinc blende GaN, finding that bulk zinc blende GaN exhibits a much higher low-field electron drift mobility than bulk wurtzite GaN. The peak electron drift velocity corresponding to bulk zinc blende GaN was found to be only marginally greater than that exhibited by bulk wurtzite GaN. In 1997, *Bhapkar* and *Shur* [33.22] reported on employing ensemble semi-classical three-valley Monte Carlo simulations of the electron transport within bulk and confined wurtzite GaN. Their simulations demonstrated that the two-dimensional electron gas within a confined wurtzite GaN structure will exhibit a higher low-field electron drift mobility than bulk wurtzite GaN, by almost an order of magnitude, this being in agreement with experiment. In 1998, *Albrecht* et al. [33.27] reported on employing ensemble semi-classical five-valley Monte Carlo simulations of the electron transport within bulk wurtzite GaN, with the aim of determining elementary analytical expressions for a number of electron transport metrics corresponding to bulk wurtzite GaN, for the purposes of device modeling.

Electron transport within the other III–V nitride semiconductors, AlN and InN, has also been studied using ensemble semi-classical Monte Carlo simulations of the electron transport. In particular, by employing ensemble semi-classical three-valley Monte Carlo simulations, the velocity–field characteristic associated with bulk wurtzite AlN was studied and reported by *O'Leary* et al. [33.24] in 1998. They found that AlN exhibits the lowest peak and saturation electron drift velocities of the III–V nitride semiconductors considered in this analysis. Similar simulations of the electron transport within bulk wurtzite AlN were also reported by *Albrecht* et al. [33.25] in 1998. The results of *O'Leary* et al. [33.24] and *Albrecht* et al. [33.25] were found to be quite similar. The first known simulation of the electron transport within bulk wurtzite InN was the semi-classical three-valley Monte Carlo simulation of *O'Leary* et al. [33.23], reported in 1998. InN was demonstrated to have the highest peak and saturation electron drift velocities of the III–V nitride semiconductors. The subsequent ensemble full-band Monte Carlo simulations of *Bellotti* et al. [33.28], reported in 1999, produced results similar to those of *O'Leary* et al. [33.23].

The first known study of transient electron transport within the III–V nitride semiconductors was that performed by *Foutz* et al. [33.21], reported in 1997. In this study, ensemble semi-classical three-valley Monte Carlo simulations were employed in order to determine how the electrons within wurtzite and zinc blende GaN, initially in thermal equilibrium, respond to the sudden application of a constant electric field. The velocity overshoot that occurs within these materials was examined. It was found that the electron drift velocities that occur within the zinc blende phase of GaN are slightly greater than those exhibited by the wurtzite phase owing to the slightly higher steady-state electron drift velocity exhibited by the zinc blende phase of GaN. A comparison with the transient electron transport that occurs within GaAs was made. Using the results from this analysis, a determination of the minimum transit time as a function of the distance displaced since the application of the applied electric field was performed for all three materials considered in this study: wurtzite GaN, zinc blende GaN, and GaAs. For distances in excess of $0.1\,\mu$m, both phases of GaN were shown to exhibit superior performance (reduced transit time) when contrasted with that associated with GaAs.

A more general analysis, in which transient electron transport within GaN, AlN, and InN was studied, was performed by *Foutz* et al. [33.29], and reported in 1999. As with their previous study, *Foutz* et al. [33.29] determined how electrons, initially in thermal equilibrium, respond to the sudden application of a constant electric field. For GaN, AlN, InN, and GaAs, it was found that the electron drift velocity overshoot only occurs when the applied electric field exceeds a certain critical applied electric field strength unique to each material. The critical applied electric field strengths, $140\,\mathrm{kV/cm}$ for the case of wurtzite GaN, $450\,\mathrm{kV/cm}$ for the case of AlN, $65\,\mathrm{kV/cm}$ for the case of InN, and $4\,\mathrm{kV/cm}$ for the case of GaAs, were shown to correspond to the peak electron drift velocity in the velocity-field characteristic associated with each of these materials; recall Fig. 33.4. It was found that InN exhibits the highest peak overshoot velocity, and that this overshoot lasts over prolonged distances compared with AlN, InN, and GaAs. A comparison with the results of experiment was performed.

In addition to Monte Carlo simulations of the electron transport within these materials, a number of other types of electron transport studies have been performed. In 1975, for example, *Ferry* [33.14] reported on the determination of the velocity-field characteristic associated with wurtzite GaN using a displaced Maxwellian distribution function approach. For high applied electric fields, *Ferry* [33.14] found that the electron drift velocity associated with GaN monotonically increases with the applied electric field strength (it does not saturate), reaching a value of about 2.5×10^7 cm/s at an applied electric field strength of $300\,\mathrm{kV/cm}$. The device implications of this result were further explored by *Das* and *Ferry* [33.15]. In 1994, *Chin* et al. [33.17] reported on a detailed study of the dependence of the low-field electron drift mobilities associated with the III–V nitride semiconductors, GaN, AlN, and InN, on crystal temperature and doping concentration. An analytical expression for the low-field electron drift mobility, μ, determined using a variational principle, was used for the purposes of this analysis. The results obtained were contrasted with those from experiment. A subsequent mobility study was reported in 1997 by *Look* et al. [33.38]. Then, in 1998, *Weimann* et al. [33.26] reported on a model for determining how the scattering of electrons by the threading dislocations within bulk wurtzite GaN influence the low-field electron drift mobility. They demonstrated why the experimentally measured low-field electron drift mobility associated with this material is much lower than that predicted from Monte Carlo analyses: threading dislocations were not taken into account in the Monte Carlo simulations of the electron transport within the III–V nitride semiconductors, GaN, AlN, and InN.

While the negative differential mobility exhibited by the velocity-field characteristics associated with the III–V nitride semiconductors, GaN, AlN, and InN, is widely attributed to intervalley transitions, and while direct experimental evidence confirming this has been presented [33.67], *Krishnamurthy* et al. [33.34] suggest that the inflection points in the bands located in the vicinity of the Γ valley are primarily responsible for the negative differential mobility exhibited by wurtzite GaN instead. The relative importance of these two mechanisms (intervalley transitions and inflection point considerations) were evaluated by *Krishnamurthy* et al. [33.34], for both bulk wurtzite GaN and an AlGaN alloy.

33.3.2 Recent Developments

There have been a number of interesting recent developments in the study of the electron transport within the III–V nitride semiconductors which have influenced the direction of thought in this field. On the experimental front, in 2000 *Wraback* et al. [33.33] reported on the use of a femtosecond optically detected time-of-flight experimental technique in order to experimentally determine the velocity–field characteristic associated with bulk wurtzite GaN. They found that the peak electron drift velocity, 1.9×10^7 cm/s, is achieved at an applied electric field strength of 225 kV/cm. No discernible negative differential mobility was observed. *Wraback* et al. [33.33] suggested that the large defect density characteristic of the GaN samples they employed, which were not taken into account in Monte Carlo simulations of the electron transport within this material, accounts for the difference between this experimental result and that obtained using simulation. They also suggested that decreasing the intervalley energy separation from about 2 eV to 340 meV, as suggested by the experimental results of *Brazel* et al. [33.56], may also account for these observations.

The determination of the electron drift velocity from experimental measurements of the unity gain cut-off frequency, f_t, has been pursued by a number of researchers. The key challenge in these analyses is the de-embedding of the parasitics from the experimental measurements so that the true intrinsic saturation electron drift velocity may be obtained. *Eastman* et al. [33.68] present experimental evidence which suggests that the saturation electron drift velocity within bulk wurtzite GaN is about $1.2 \times 10^7 - 1.3 \times 10^7$ cm/s. A more recent report, by *Oxley* and *Uren* [33.69], suggests a value of 1.1×10^7 cm/s. The role of self-heating was also probed by *Oxley* and *Uren* [33.69] and shown to be relatively insignificant. A completely satisfactory explanation for the discrepancy between these results and those from the Monte Carlo simulations has yet to be provided.

Wraback et al. [33.70] performed a subsequent study on the transient electron transport within wurtzite GaN. In particular, using their femtosecond optically detected time-of-flight experimental technique in order to experimentally determine the velocity overshoot that occurs within bulk wurtzite GaN, they observed substantial velocity overshoot within this material. In particular, a peak transient electron drift velocity of 7.25×10^7 cm/s was observed within the first 200 fs after photoexcitation for an applied electric field strength of 320 kV/cm. These experimental results were shown to be consistent with the theoretical predictions of *Foutz* et al. [33.29].

On the theoretical front, there have been a number of recent developments. In 2001, *O'Leary* et al. [33.30] presented an elementary, one-dimensional analytical model for the electron transport within the III–V compound semiconductors, and applied it to the cases of wurtzite GaN and GaAs. The predictions of this analytical model were compared with those of Monte Carlo simulations and were found to be in satisfactory agreement. Hot-electron energy relaxation times within the III–V nitride semiconductors were recently studied by *Matulionis* et al. [33.71] and reported in 2002. *Bulutay* et al. [33.72] studied the electron momentum and energy relaxation times within the III–V nitride semiconductors and reported the results of this study in 2003. It is particularly interesting to note that their arguments add considerable credence to the earlier inflection point argument of *Krishnamurthy* et al. [33.34]. In 2004, *Brazis* and *Raguotis* [33.73] reported on the results of a Monte Carlo study involving additional phonon modes and a smaller intervalley energy separation for bulk wurtzite GaN. Their results were found to be much closer to the experimental results of *Wraback* et al. [33.33] than those found previously.

The influence of hot-phonons on the electron transport mechanisms within the III–V nitride semiconductors, GaN, AlN, and InN, has been the focus of considerable recent investigation. In particular, in 2004 *Silva* and *Nascimento* [33.74], *Gökden* [33.75], and *Ridley* et al. [33.76], to name just three, presented results related to this research focus. These results suggest that hot-phonon effects play a significant role in influencing the nature of the electron transport within the III–V nitride semiconductors, GaN, AlN, and InN. In particular, *Ridley* et al. [33.76] point out that the saturation electron drift velocity and the applied electric field strength at which the peak in the velocity–field characteristic occurs are both influenced by hot-phonon effects. The role that hot-phonons play in influencing device performance was studied by *Matulionis* and *Liberis* [33.77]. Research into the role that hot-phonons play in influencing the electron transport mechanisms within the III–V nitride semiconductors, GaN, AlN, and InN, seems likely to continue into the foreseeable future.

33.3.3 Future Perspectives

It is clear that our understanding of the electron transport within the III–V nitride semiconductors, GaN, AlN, and InN, is, at present at least, in a state of flux. A complete understanding of the electron transport mechanisms within these materials has yet to be achieved, and is the subject of intense current research. Most troubling is the discrepancy between the results of experiment and those of simulation. There are a two principal sources of uncertainty in our analysis of the electron transport mechanisms within these materials: (1) uncertainty in the material properties, and (2) uncertainty in the underlying physics. We discuss each of these subsequently.

Uncertainty in the material parameters associated with the III–V nitride semiconductors, GaN, AlN, and

InN, remains a key source of ambiguity in the analysis of the electron transport with these materials [33.32]. Even for bulk wurtzite GaN, the most studied of the III–V nitride semiconductors considered in this analysis, uncertainty in the band structure remains an issue [33.56]. The energy gap associated with InN and the effective mass associated with this material continue to fuel debate; see, for example, *Davydov* et al. [33.53], *Wu* et al. [33.54], and *Matsuoka* et al. [33.55]. Variations in the experimentally determined energy gap associated with InN, observed from sample to sample, further confound matters. Most recently, *Shubina* et al. [33.78] suggested that nonstoichiometry within InN may be responsible for these variations in the energy gap. Further research will have to be performed in order to confirm this. Given this uncertainty in the band structures associated with the III–V nitride semiconductors, it is clear that new simulations of the electron transport will have to be performed once researchers have settled on the appropriate band structures. We thus view our present results as a baseline, the sensitivity analysis of *O'Leary* et al. [33.32] providing some insights into how variations in the band structures will impact upon the results.

Uncertainty in the underlying physics is considerable. The source of the negative differential mobility remains a matter to be resolved. The presence of hot-phonons within these materials, and how such phonons impact upon the electron transport mechanisms within these materials, remains another point of contention. It is clear that a deeper understanding of these electron transport mechanisms will have to be achieved in order for the next generation of III–V nitride semiconductor-based devices to be properly designed.

33.4 Conclusions

In this paper, we reviewed analyses of the electron transport within the III–V nitride semiconductors GaN, AlN, and InN. In particular, we have discussed the evolution of the field, surveyed the current literature, and presented frontiers for further investigation and analysis. In order to narrow the scope of this review, we focused on the electron transport within bulk wurtzite GaN, AlN, and InN for the purposes of this paper. Most of our discussion focused upon results obtained from our ensemble semi-classical three-valley Monte Carlo simulations of the electron transport within these materials, our results conforming with state-of-the-art III–V nitride semiconductor orthodoxy.

We began our review with the Boltzmann transport equation, since this equation underlies most analyses of the electron transport within semiconductors. A brief description of our ensemble semi-classical three-valley Monte Carlo simulation approach to solving the Boltzmann transport equation was then provided. The material parameters, corresponding to bulk wurtzite GaN, AlN, and InN, were then presented. We then used these material parameter selections, and our ensemble semi-classical three-valley Monte Carlo simulation approach, to determine the nature of the steady-state and transient electron transport within the III–V nitride semiconductors. Finally, we presented some recent developments on the electron transport within these materials, and pointed to fertile frontiers for further research and investigation.

References

33.1 S. Strite, H. Morkoç: J. Vac. Sci. Technol. B **10**, 1237 (1992)
33.2 H. Morkoç, S. Strite, G. B. Gao, M. E. Lin, B. Sverdlov, M. Burns: J. Appl. Phys. **76**, 1363 (1994)
33.3 S. N. Mohammad, H. Morkoç: Prog. Quantum Electron. **20**, 361 (1996)
33.4 S. J. Pearton, J. C. Zolper, R. J. Shul, F. Ren: J. Appl. Phys. **86**, 1 (1999)
33.5 M. A. Khan, J. W. Yang, W. Knap, E. Frayssinet, X. Hu, G. Simin, P. Prystawko, M. Leszczynski, I. Grzegory, S. Porowski, R. Gaska, M. S. Shur, B. Beaumont, M. Teisseire, G. Neu: Appl. Phys. Lett. **76**, 3807 (2000)
33.6 X. Hu, J. Deng, N. Pala, R. Gaska, M. S. Shur, C. Q. Chen, J. Yang, G. Simin, M. A. Khan, J. C. Rojo, L. J. Schowalter: Appl. Phys. Lett. **82**, 1299 (2003)
33.7 W. Lu, V. Kumar, E. L. Piner, I. Adesida: IEEE Trans. Electron Dev. **50**, 1069 (2003)
33.8 A. Jiménez, Z. Bougrioua, J. M. Tirado, A. F. Braña, E. Calleja, E. Muñoz, I. Moerman: Appl. Phys. Lett. **82**, 4827 (2003)

33.9 A.A. Burk Jr., M.J. O'Loughlin, R.R. Siergiej, A.K. Agarwal, S. Sriram, R.C. Clarke, M.F. MacMillan, V. Balakrishna, C.D. Brandt: Solid-State Electron. **43**, 1459 (1999)

33.10 M. Umeno, T. Egawa, H. Ishikawa: Mater. Sci. Semicond. Process. **4**, 459 (2001)

33.11 A. Krost, A. Dadgar: Phys. Status Solidi A **194**, 361 (2002)

33.12 C.L. Tseng, M.J. Youh, G.P. Moore, M.A. Hopkins, R. Stevens, W.N. Wang: Appl. Phys. Lett. **83**, 3677 (2003)

33.13 M.A. Littlejohn, J.R. Hauser, T.H. Glisson: Appl. Phys. Lett. **26**, 625 (1975)

33.14 D.K. Ferry: Phys. Rev. B **12**, 2361 (1975)

33.15 P. Das, D.K. Ferry: Solid-State Electron. **19**, 851 (1976)

33.16 B. Gelmont, K. Kim, M. Shur: J. Appl. Phys. **74**, 1818 (1993)

33.17 V.W.L. Chin, T.L. Tansley, T. Osotchan: J. Appl. Phys. **75**, 7365 (1994)

33.18 N.S. Mansour, K.W. Kim, M.A. Littlejohn: J. Appl. Phys. **77**, 2834 (1995)

33.19 J. Kolník, İ.H. Oğuzman, K.F. Brennan, R. Wang, P.P. Ruden, Y. Wang: J. Appl. Phys. **78**, 1033 (1995)

33.20 M. Shur, B. Gelmont, M.A. Khan: J. Electron. Mater. **25**, 777 (1996)

33.21 B.E. Foutz, L.F. Eastman, U.V. Bhapkar, M.S. Shur: Appl. Phys. Lett. **70**, 2849 (1997)

33.22 U.V. Bhapkar, M.S. Shur: J. Appl. Phys. **82**, 1649 (1997)

33.23 S.K. O'Leary, B.E. Foutz, M.S. Shur, U.V. Bhapkar, L.F. Eastman: J. Appl. Phys. **83**, 826 (1998)

33.24 S.K. O'Leary, B.E. Foutz, M.S. Shur, U.V. Bhapkar, L.F. Eastman: Solid State Commun. **105**, 621 (1998)

33.25 J.D. Albrecht, R.P. Wang, P.P. Ruden, M. Farahmand, K.F. Brennan: J. Appl. Phys. **83**, 1446 (1998)

33.26 N.G. Weimann, L.F. Eastman, D. Doppalapudi, H.M. Ng, T.D. Moustakas: J. Appl. Phys. **83**, 3656 (1998)

33.27 J.D. Albrecht, R.P. Wang, P.P. Ruden, M. Farahmand, K.F. Brennan: J. Appl. Phys. **83**, 4777 (1998)

33.28 E. Bellotti, B.K. Doshi, K.F. Brennan, J.D. Albrecht, P.P. Ruden: J. Appl. Phys. **85**, 916 (1999)

33.29 B.E. Foutz, S.K. O'Leary, M.S. Shur, L.F. Eastman: J. Appl. Phys. **85**, 7727 (1999)

33.30 S.K. O'Leary, B.E. Foutz, M.S. Shur, L.F. Eastman: Solid State Commun. **118**, 79 (2001)

33.31 T.F. de Vasconcelos, F.F. Maia Jr., E.W.S. Caetano, V.N. Freire, J.A.P. da Costa, E.F. da Silva Jr.: J. Cryst. Growth **246**, 320 (2002)

33.32 S.K. O'Leary, B.E. Foutz, M.S. Shur, L.F. Eastman: J. Electron. Mater. **32**, 327 (2003)

33.33 M. Wraback, H. Shen, J.C. Carrano, T. Li, J.C. Campbell, M.J. Schurman, I.T. Ferguson: Appl. Phys. Lett. **76**, 1155 (2000)

33.34 S. Krishnamurthy, M. van Schilfgaarde, A. Sher, A.-B. Chen: Appl. Phys. Lett. **71**, 1999 (1997)

33.35 D.K. Ferry, C. Jacoboni (Eds.): *Quantum Transport in Semiconductors* (Plenum, New York 1992)

33.36 N.W. Ashcroft, N.D. Mermin: *Solid State Physics* (Saunders College, Philadelphia 1976)

33.37 P.A. Sandborn, A. Rao, P.A. Blakey: IEEE Trans. Electron Dev. **36**, 1244 (1989)

33.38 D.C. Look, J.R. Sizelove, S. Keller, Y.F. Wu, U.K. Mishra, S.P. DenBaars: Solid State Commun. **102**, 297 (1997)

33.39 N.A. Zakhleniuk, C.R. Bennett, B.K. Ridley, M. Babiker: Appl. Phys. Lett. **73**, 2485 (1998)

33.40 B.R. Nag: *Electron Transport in Compound Semiconductors* (Springer, Berlin, Heidelberg 1980)

33.41 M.S. Krishnan, N. Goldsman, A. Christou: J. Appl. Phys. **83**, 5896 (1998)

33.42 R. Oberhuber, G. Zandler, P. Vogl: Appl. Phys. Lett. **73**, 818 (1998)

33.43 W. Fawcett, A.D. Boardman, S. Swain: J. Phys. Chem. Solids **31**, 1963 (1970)

33.44 A. Bykhovski, B. Gelmont, M. Shur, A. Khan: J. Appl. Phys. **77**, 1616 (1995)

33.45 A.D. Bykhovski, V.V. Kaminski, M.S. Shur, Q.C. Chen, M.A. Khan: Appl. Phys. Lett. **68**, 818 (1996)

33.46 P. Lugli, D.K. Ferry: IEEE Trans. Electron Dev. **32**, 2431 (1985)

33.47 K. Seeger: *Semiconductor Physics: An Introduction*, 9th edn. (Springer, Berlin, Heidelberg 2004)

33.48 S.K. O'Leary, B.E. Foutz, M.S. Shur, L.F. Eastman: J. Mater. Sci.: Mater. Electron. **17**, 87 (2006)

33.49 W.R.L. Lambrecht, B. Segall: In: *Properties of Group III Nitrides*, EMIS Datareviews Series, ed. by J.H. Edgar (Inspec, London 1994) Chap. 4

33.50 M.A. Littlejohn, J.R. Hauser, T.H. Glisson: J. Appl. Phys. **48**, 4587 (1977)

33.51 J.S. Blakemore: J. Appl. Phys. **53**, 123 (1982)

33.52 T.L. Tansley, C.P. Foley: J. Appl. Phys. **59**, 3241 (1986)

33.53 V.Y. Davydov, A.A. Klochikhin, V.V. Emtsev, S.V. Ivanov, V.V. Vekshin, F. Bechstedt, J. Furthmüller, H. Harima, A.V. Mudryi, A. Hashimoto, A. Yamamoto, J. Aderhold, J. Graul, E.E. Haller: Phys. Status Solidi B **230**, R4 (2002)

33.54 J. Wu, W. Walukiewicz, K.M. Yu, J.W. Ager III., E.E. Haller, H. Lu, W.J. Schaff, Y. Saito, Y. Nanishi: Appl. Phys. Lett. **80**, 3967 (2002)

33.55 T. Matsuoka, H. Okamoto, M. Nakao, H. Harima, E. Kurimoto: Appl. Phys. Lett. **81**, 1246 (2002)

33.56 E.G. Brazel, M.A. Chin, V. Narayanamurti, D. Kapolnek, E.J. Tarsa, S.P. DenBaars: Appl. Phys. Lett. **70**, 330 (1997)

33.57 J.G. Ruch: IEEE Trans. Electron Dev. **19**, 652 (1972)

33.58 M.S. Shur, L.F. Eastman: IEEE Trans. Electron Dev. **26**, 1677 (1979)

33.59 M. Heiblum, M.I. Nathan, D.C. Thomas, C.M. Knoedler: Phys. Rev. Lett. **55**, 2200 (1985)

33.60 A. Palevski, M. Heiblum, C. P. Umbach, C. M. Knoedler, A. N. Broers, R. H. Koch: Phys. Rev. Lett. **62**, 1776 (1989)
33.61 A. Palevski, C. P. Umbach, M. Heiblum: Appl. Phys. Lett. **55**, 1421 (1989)
33.62 A. Yacoby, U. Sivan, C. P. Umbach, J. M. Hong: Phys. Rev. Lett. **66**, 1938 (1991)
33.63 E. Tiede, M. Thimann, K. Sensse: Chem. Berichte **61**, 1568 (1928)
33.64 W. C. Johnson, J. B. Parsons, M. C. Crew: J. Phys. Chem. **36**, 2561 (1932)
33.65 R. Juza, H. Hahn: Z. Anorg. Allg. Chem. **239**, 282 (1938)
33.66 H. P. Maruska, J. J. Tietjen: Appl. Phys. Lett. **15**, 327 (1969)
33.67 Z. C. Huang, R. Goldberg, J. C. Chen, Y. Zheng, D. B. Mott, P. Shu: Appl. Phys. Lett. **67**, 2825 (1995)
33.68 L. F. Eastman, V. Tilak, J. Smart, B. M. Green, E. M. Chumbes, R. Dimitrov, H. Kim, O. S. Ambacher, N. Weimann, T. Prunty, M. Murphy, W. J. Schaff, J. R. Shealy: IEEE Trans. Electron Dev. **48**, 479 (2001)
33.69 C. H. Oxley, M. J. Uren: IEEE Trans. Electron Dev. **52**, 165 (2005)
33.70 M. Wraback, H. Shen, J. C. Carrano, C. J. Collins, J. C. Campbell, R. D. Dupuis, M. J. Schurman, I. T. Ferguson: Appl. Phys. Lett. **79**, 1303 (2001)
33.71 A. Matulionis, J. Liberis, L. Ardaravičius, M. Ramonas, I. Matulionienė, J. Smart: Semicond. Sci. Technol. **17**, 9 (2002)
33.72 C. Bulutay, B. K. Ridley, N. A. Zakhleniuk: Phys. Rev. B **68**, 115205 (2003)
33.73 R. Brazis, R. Raguotis: Appl. Phys. Lett. **85**, 609 (2004)
33.74 A. A. P. Silva, V. A. Nascimento: J. Lumin. **106**, 253 (2004)
33.75 S. Gökden: Physica E **23**, 198 (2004)
33.76 B. K. Ridley, W. J. Schaff, L. F. Eastman: J. Appl. Phys. **96**, 1499 (2004)
33.77 A. Matulionis, J. Liberis: IEE Proc. Circ. Dev. Syst. **151**, 148 (2004)
33.78 T. V. Shubina, S. V. Ivanov, V. N. Jmerik, M. M. Glazov, A. P. Kalavarskii, M. G. Tkachman, A. Vasson, J. Leymarie, A. Kavokin, H. Amano, I. Akasaki, K. S. A. Butcher, Q. Guo, B. Monemar, P. S. Kop'ev: Phys. Status Solidi A **202**, 377 (2005)

34. II–IV Semiconductors for Optoelectronics: CdS, CdSe, CdTe

Owing to their suitable band gaps and high absorption coefficients, Cd-based compounds such as CdTe and CdS are the most promising photovoltaic materials available for low-cost high-efficiency solar cells. Additionally, because of their large atomic number, Cd-based compounds such as CdTe and CdZnTe, have been applied to radiation detectors. For these reasons, preparation techniques for these materials in the polycrystalline films and bulk single crystals demanded by these devices have advanced significantly in recent decades, and practical applications have been realized in optoelectronic devices. This chapter mainly describes the application of these materials in solar cells and radiation detectors and introduces recent progress.

34.1	Background	829
34.2	Solar Cells	829
	34.2.1 Basic Description of Solar Cells	829
	34.2.2 Design of Cd-Based Solar Cells	830
	34.2.3 Development of CdS/CdTe Solar Cells	831
	34.2.4 CdZnTe Solar Cells	834
	34.2.5 The Future of Cd-Based Solar Cells	834
34.3	Radiation Detectors	834
	34.3.1 Basic Description of Semiconductor Radiation Detectors	835
	34.3.2 CdTe and CdZnTe Radiation Detectors	835
	34.3.3 Performance of CdTe and CdZnTe Detectors	836
	34.3.4 Applications of CdTe and CdZnTe Detectors	839
34.4	Conclusions	840
	References	840

34.1 Background

Cd-based compounds are very important semiconductor materials in the II–VI family. The attraction of Cd-based binary and ternary compounds arises from their promising applications as solar cells, γ- and X-ray detectors etc. These devices made from Cd-based materials are being widely applied in many fields.

34.2 Solar Cells

With the development of human society, energy sources in the earth are being slowly exhausted and we are faced with a serious problem. The solar cell is one substitute for fossil fuels and is being realized throughout the world. For this reason, solar-cell technologies have been developed since work was started by *Becquerel* in 1839 [34.1]. The solar cell has now been applied to daily life, industry, agriculture, space exploration, military affairs etc.

Solar cells have many advantages. Firstly, sunlight as an energy source for power generation is not only limitless but can be used freely. Secondly, since light is directly converted to electricity, the conversion process is clean, noise-free and not harmful to the environment, unlike a mechanical power generator. Thirdly, solar cells need little maintenance.

34.2.1 Basic Description of Solar Cells

A solar cell is a semiconductor device that directly converts light energy into electrical energy through the photovoltaic process. The basic structure of a solar cell is shown in Fig. 34.1. A typical solar cell consists of a junction formed between an n-type and a p-type

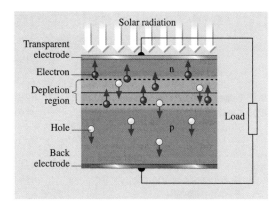

Fig. 34.1 The principle of photovoltaic devices

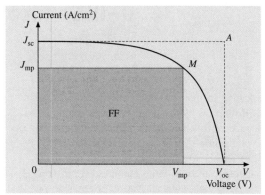

Fig. 34.2 Photovoltaic output characteristics of a solar cell

semiconductor, of the same (homojunction) or different materials (heterojunction), an antireflection coating and Ohmic collecting electrodes. When the light irradiates the surface of a solar cell, photons with an energy greater than the band gap of the semiconductor are absorbed by the semiconductor material. This absorption activates electron transitions from the valence to the conduction band, so that electron–hole pairs are generated. If these carriers can diffuse into the depletion region before they recombine, they can be separated by the applied electric field. At the p–n junction, the negative electrons diffuse into the n-type region and the positive holes diffuse into the p-type region. They are then collected by electrodes, resulting in a voltage difference between the two electrodes. When an external load is connected, electric current flows through the load. This is the origin of the solar cell's photocurrent.

Three parameters can be used to describe the performance of a solar cell: the short-circuit current density (J_{SC}), the open-circuit voltage (V_{OC}) and the fill factor (FF).

The short-circuit current density J_{SC} is the photocurrent output from a solar cell when the output terminals are short-circuited. In the ideal case, this is equal to the current density generated by the light J_L and is proportional to the incident photon flux. J_{SC} is determined by the spectral response of the device, the junction depth, and the series (internal) resistance, R_S. The open-circuit voltage V_{OC} is the voltage appearing across the output terminals of the cell when there is no load present, i. e. when $J = 0$. The relationship between V_{OC} and J_{SC} can be described by the diode equation, i. e.:

$$V_{OC} = \frac{nkT}{e} \ln\left(\frac{J_{SC}}{J_0} + 1\right), \qquad (34.1)$$

where e, n, J_0, k and T are the charge on an electron, the diode ideality factor, the reverse saturation-current density, the Boltzmann constant and the absolute temperature, respectively. The fill factor (FF) is the ratio of the maximum electrical power available from the cell [i. e. at the operating point J_m, V_m] in Fig. 34.2) to the product of V_{OC} and J_{SC}. It describes the rectangularity of the photovoltaic output characteristic.

$$FF = \frac{V_m \cdot J_m}{V_{oc} \cdot J_{sc}} = \frac{P_m}{V_{oc} \cdot J_{sc}}. \qquad (34.2)$$

The conversion efficiency (η) of a solar cell is determined as

$$\eta = \frac{P_m}{P_s} = \frac{V_{oc} \cdot J_{sc} \cdot FF}{P_s}. \qquad (34.3)$$

Where P_s is the incident illumination power and P_m is the output electricity power, both being per unit area. The relationship between the short-circuit current density J_{SC} and the open-circuit voltage V_{OC} is shown in Fig. 34.2.

34.2.2 Design of Cd-Based Solar Cells

Solar cells can be made from silicon (Si), III–V compounds such as GaAs, or II–VI compounds such as Zn-based and Cd-based compound semiconductors. Of these materials, CdTe is the most attractive because of a number of advantages. CdTe is a direct-band-gap material with a band gap of 1.54 eV at room temperature. This is very close to the theoretically calculated optimum value for solar cells. CdTe has a high absorption coefficient (above 10^5 cm^{-1} at a wavelength of 700 nm), so that approximately 90% of the incident light is absorbed by a layer thickness of only 2 μm (compared

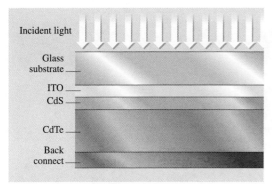

Fig. 34.3 A typical CdS/CdTe solar cell

with around 10 μm for Si), cutting down the quantity of semiconductor required. Therefore, CdTe is one of the most promising photovoltaic materials available for use in low-cost high-efficiency solar cells.

CdTe is the only material in which both n- and p-type conductivity can be easily controlled by doping acceptor or donor impurities in II–VI compound semiconductors. Furthermore, due to its high absorption coefficient and small carrier diffusion length, the junction must be formed closed to the surface, which reduces the carrier lifetime through surface recombination. In spite of this, the conversion efficiency has been gradually enhanced from 4% [34.2] to 16% [34.3] since the first CdTe-based solar cell was fabricated. CdTe solar cells can be made from polycrystalline or single-crystal material. However, it was found that this p–n homojuction structure is unstable due to the aging behavior of dopants in CdTe films and bulk single crystals [34.4]. Furthermore, it is also difficult to fabricate shallow p–n junctions with highly conducting surface layers, and high surface recombination velocities led to substantial losses that could not easily be avoided. For this reason, the CdS/CdTe heterojunction solar cell was proposed.

It has been proved that CdS/CdTe is a high-efficiency heterojunction solar cell. The theoretical calculation shows that this cell has a conversion efficiency of as high as 29% [34.5], and it is considered a promising alternative to the more widely used silicon devices. The typical structure of a CdS/CdTe solar cell is glass/indium tin oxide (ITO)/CdS/CdTe/back electrode, as seen in Fig. 34.3. The CdS/CdTe solar cell is based on the heterojunction formed between n-type CdS (only n-CdS is available) and p-type CdTe and is fabricated in a superstrate configuration where the incident light passes through the glass substrate, which has a thickness of 2–4 mm. This window glass is transparent, strong and cheap. This glass substrate also protects the active layers from the environment, and it provides the mechanical strength of the device. The outer face of the panel often has an antireflection coating to enhance absorption efficiency. A transparent conducting oxide (TCO) layer, usually tin oxide or indium tin oxide (ITO), acts as the front contact to the device. This is needed to reduce the series resistance of the device, which would otherwise arise from the thinness of the CdS layer. Usually, the polycrystalline n-type CdS layer is deposited onto the TCO. Due to its wide band gap ($E_g \approx 2.4$ eV at 300 K) it is transparent down to wavelengths of around 516 nm. The thickness of this layer is typically 100–300 nm. The CdTe layer is p-type-doped and its thickness is typically around 10 μm. Generally, the carrier concentration in CdS layer is 2–3 orders greater than that of the CdTe layer. As a result, the potential is applied mostly to the CdTe absorber layer, i.e. the depletion region is mostly within the CdTe layer. Therefore, the photogenerated carriers can be effectively separated in this active region.

The electrode preparation is very important in obtaining Ohmic contact or minimal series resistivity. Since there are no metals with a work function higher than the hole affinity of p-type CdTe (−5.78 eV) [34.6], it is difficult to produce a complete Ohmic contact on p-type CdTe surface. Many methods have been tried to fabricate Ohmic contact to p-type CdTe. Different metals or alloy and compounds such as Au, Al, and ZnTe/Cu etc. have also been used to serve as electrode materials. Although significant effort has been applied to this issue, the junction still inevitably displays some Schottky-diode characteristics. This is a problem that will be continuously studied.

34.2.3 Development of CdS/CdTe Solar Cells

The layers of CdS and CdTe for solar cells can be prepared using various techniques, which are summarized in Table 34.1. In these techniques, evaporation, close-space sublimation (CSS), screen printing and electrodeposition (ED) have been demonstrated to be very effective for fabricating high-efficiency large-area CdS/CdTe solar cells.

Study of solar cells based on CdTe single crystals and polycrystalline films were started as early as 1963 [34.7]. The cells were prepared with a p-type copper-telluride surface film as an integral part of the photovoltaic junction. The junction was thought to be a heterojunction between the p-type copper telluride and

Table 34.1 Some fabrication techniques and conversion efficiency of CdS/CdTe solar cells

Fabrication techniques of CdS and CdTe	Conversion efficiency	area	Reference
Evaporation: physical vapor deposition (PVD), and chemical vapor deposition (CVD)	11.8%	0.3 cm^2	[34.8]
Close-spaced sublimation (CSS)	16%	1.0 cm^2	[34.3]
	8.4%	7200 cm^2	[34.9]
Electrodeposition (ED)	14.2%	0.02 cm^2	[34.10]
Screen printing (SP) and sintering	12.8%	0.78 cm^2	[34.11]
Metalorganic CVD (MOCVD)	11.9%	0.08 cm^2	[34.12]
PVD vacuum evaporation	11.8%	0.3 cm^2	[34.8]
MBE	10.5%	0.08 cm^2	[34.13]
Sputtering deposition (SD)	14%	0.09 cm^2	[34.14]

the n-type CdTe, with the transition region extending well into the CdTe side. A similar structure was reported for single-crystal cells. A conversion efficiency of up to 6% was obtained in the film cells, and 7.5% for single-crystal cells.

Dutton and *Muller* [34.15] fabricated CdS/CdTe solar cells by standard vapor-deposition procedures with a very thin CdTe layer. The existence of the interfacial layer was confirmed using X-ray diffraction studies, and substantiated by diode photocurrent measurements. The results showed a threshold photon energy that corresponds to the band gap of CdTe. The thin CdTe layer resulted in excellent rectification ratios and high reverse breakdown voltages. The observed photoresponse, $I-V$ (current versus voltage), and $C-V$ (capacitance versus voltage) characteristics were consistence with those predicted by a semiconductor heterojunction model of the CdS–CdTe interface. *Mitchell* et al. [34.16] prepared a variety of CdS/CdTe heterojunction solar cells with an ITO coating and a glycerol antireflection coating by vacuum evaporation of n-CdS films onto single-crystal p-CdTe substrates. Comparisons were made between cells prepared using different substrate resistivity, substrate surface preparations, and CdS film resistivity. The mechanisms of controlling the dark junction current, photocarrier collection, and photovoltaic properties were modeled, taking account of the interface states of the junction. A conversion efficiency of 7.9% was obtained under 85 mW/cm^2 of solar simulator illumination.

At the end of the 1970s, a new process, screen printing and sintering, was developed for fabricating CdS/CdTe polycrystal solar cells by *Matsushita* and coworkers [34.11, 17–20]. This technique proved to be a very effective method for fabricating large-area CdS/CdTe film solar cells. There are several important parameters in the screen-printing technique: the viscosity of the paste, the mesh number of the screen, the snap-off distance between the screen and the substrate and the pressure and speed of the squeegee. As cadmium chloride (CdCl$_2$) has a low melting point ($T_m = 568\,°$C), and forms a eutectic with both CdS and CdTe at low temperature, and the vapor pressure at 600 °C is high enough to allow complete volatilization of CdCl$_2$ after the sintering process, CdCl$_2$ is used as an ideal flux for sintering both CdS and CdTe.

Nakayama et al. [34.18] prepared the first thin-film CdS/CdTe solar cells using the screen-printing technique (Fig. 34.4). A glass plate, which was coated successively with In$_2$O$_3$ film and CdS ceramic thin film, was used as a transparent Ohmic contact substrate. The n-type CdS film with a thickness of about 20 μm and a resistivity of 0.2 Ω cm was prepared on this substrate by the screen-printing method. The CdTe paste was printed onto the CdS layer using the same technique and then heat-treated in an N$_2$ atmosphere at temperatures of 500–800 °C. As a result, the p-type CdTe layer with a thickness of 10 μm showed a resistivity of 0.1–1 Ωcm. A silver (Ag) paste electrode was applied to the p-type

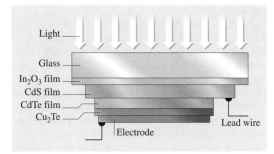

Fig. 34.4 Cross section of a ceramic thin-film CdTe solar cell

layer and an In–Ga alloy was applied to the CdS layer. The cell, with an active area of $0.36\,\text{cm}^2$, showed an intrinsic solar conversion efficiency of 8.1% under an illumination of $140\,\text{mW/cm}^2$ solar simulator (AM0). Subsequently, this group continued to work on improving CdS/CdTe solar cells [34.11, 19, 20]. From the practical point of view, a modular solar cell is necessary. For this purpose, thin-film CdS/CdTe solar cells with an efficiency of 6.3% were prepared on a borosilicate-glass substrate of $4\times4\,\text{cm}^2$ by successively screen printing and heating (sintering) of each paste of CdS, CdTe and carbon [34.19]. Although they fabricated some structures, including dividing the CdTe film on one substrate into five small cells, to prevent increases in the series resistivity, a 1-watt module only showed an efficiency of 2.9% from 25 elemental cells with a $4\times4\,\text{cm}^2$ substrate. A main reason for this low efficiency was thought to be the high series resistivity.

In order to decrease the resistivity of the carbon electrode further, this group changed the heating conditions of the carbon electrodes for CdTe and examined the effect of impurities in the carbon paste on the characteristics of solar cells [34.11]. They found that the series resistance (R_s) and the conversion efficiency (η) of solar cells exhibited strong dependence on the oxygen (O_2) partial pressure during the heating process. The results showed that the addition of Cu to the carbon paste causes R_s and the diode factor (n) to decrease, resulting in a remarkable improvement in η. Using a low-resistance contact electrode, which was made from 50 ppm added Cu carbon paste, a cell with an active area of $0.78\,\text{cm}^2$ displayed $V_{oc}= 0.754\,\text{V}$, $I_{sc}= 0.022\,\text{A}$, FF= 0.606 and $\eta=12.8\%$.

Britt and *Ferekides* [34.21] reported the fabrication and characteristics of high-efficiency thin-film CdS/CdTe heterojunction solar cells. CdS films with 0.07–$0.10\,\mu\text{m}$ in thickness were deposited on a 0.5-μm SnO_2:F transparent conducting oxide layer by chemical-bath deposition and p-CdTe films with a thickness of $5\,\mu\text{m}$ were deposited on the CdS layer by using the CSS method. Prior to the deposition of CdTe, CdS/SnO$_2$:F/glass was annealed in a H$_2$ atmosphere at temperatures of 350–425 °C for 5–20 min. The characteristics of the CdS/CdTe solar cell were $V_{oc}= 843\,\text{mV}$, $J_{sc}= 25.1\,\text{mA/cm}^2$, FF=74.5%, corresponding to a total area greater than $1\,\text{cm}^2$ with an air mass 1.5 (AM1.5) conversion efficiency of 15.8%.

With modern growth techniques, high-efficiency CdS/CdTe solar cells have been developed using ultra-thin CdS films having a thickness of 50 nm [34.22]. Figure 34.5 shows the cross-sectional structure of a CdS/CdTe solar cell. CdS films were deposited on an ITO glass substrates by the metal organic chemical vapor deposition (MOCVD) technique, and CdTe films were subsequently deposited by the CSS technique (see Fig. 34.6) for the fabrication of CdS/CdTe solar cells. A thin-film CdS/CdTe solar cell with an area of $1.0\,\text{cm}^2$ under AM1.5 conditions showed $V_{oc}= 840\,\text{mV}$, $J_{sc}= 26.08\,\text{mA/cm}^2$ and FF= 73%. As a result, a high conversion efficiency of 16.0% was achieved. Furthermore, they made developments in improving the uniformity of thickness and film qualities of CdS in order to obtain more efficient solar cells and to realize film deposition on large-area substrates of $30\times 60\,\text{cm}$ or more [34.23]. They found clear differences in the electrical and optical properties between high- and low-quality CdS films. A photovoltaic conversion efficiency of 10.5% was achieved by a solar module with an area of $1376\,\text{cm}^2$ under AM1.5 measurement conditions.

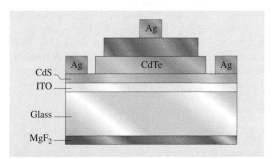

Fig. 34.5 The cross-sectional structure of a CdS/CdTe solar cell fabricating by the screen-printing method

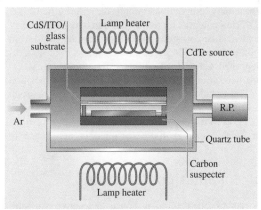

Fig. 34.6 Schematic of the close-spaced sublimation apparatus used for depositing CdTe films

In addition to dry processes such as screen printing, close-spaced sublimation etc., electrodeposition [34.24–27] has been also investigated for preparing polycrystalline CdTe layers for solar cells. *Awakura* and coworkers [34.25–27] reported the cathodic deposition behavior of CdTe thin layers under irradiation by visible light using ammoniacal basic aqueous solution as an electrolytic bath. Both deposition current density and current efficiency for the CdTe deposition were enhanced by irradiation. The deposited rate was over 10 times high than non-photoassisted electrodeposition. This was believed to represent significant progress in reducing the cost of Cd-based solar cells.

To date, solar cells with an n-CdS/p-CdTe heterojunction have been reported with efficiencies as high as 16% [34.22]. Recently, it was reported that thin-layered n-CdS/p-CdTe heterojunction solar cells have already been manufactured industrially [34.28]. n-CdS film with a thickness range of 500–1000 Å was deposited from an aqueous solution directly onto a transparent conductive oxide (TCO) substrate. After the CdS film was annealed for densification and grain growing, p-CdTe layer was formed by electrochemical deposition. The CdTe film was then annealed in air at 450 °C. Subsequently, monolithic TCO/CdS/CdTe was cut into discrete cells using an infrared laser. After other preparing process, a maximum efficiency of 10.6% for a CdTe 0.94-m^2 module with a power of 91.5 W was fabricated. The test results showed good stability. For practical applications, a 10-MW CdTe solar-cell manufacturing plant has been constructed [34.28].

34.2.4 CdZnTe Solar Cells

The ternary compound cadmium zinc telluride (CdZnTe, CZT) has potential for the preparation of high-efficiency tandem solar cell since its band gap can be tuned from 1.45 to 2.26 eV [34.29, 30]. *McCandless* et al. [34.29] deposited $Cd_{1-x}Zn_xTe$ films using the physical vapor deposition (PVD) and vapor transport deposition (VTD) techniques. The film composition was between 0.35 and 0.6, corresponding to a band gap from 1.7 to 1.9 eV. Post-deposition treatment of CdZnTe films in $ZnCl_2$ vapor at 400 °C resulted in no change to the alloy composition and caused recrystallization. Solar cells made from $Cd_{1-x}Zn_xTe$ films with $x \approx 0.35$ exhibited $V_{oc} = 0.78$ V and $J_{sc} < 10$ mA/cm^2. These results were similar to those obtained from CdS window layers. Analysis of the spectral response indicated that $Cd_{1-x}Zn_xTe$ with $x \approx 0.35$ has a band gap of about 1.7 eV. *Gidiputti* et al. [34.30] used two deposition technologies, co-sputtering from CdTe and ZnTe targets and co-close-spaced sublimation (CCSS) from CdTe and ZnTe powders, to prepare CdZnTe films. A structure similar to the CdTe superstrate configuration was initially utilized for cell fabrication: glass/ITO/CdS/CZT/graphite. Typical solar cell parameters obtained for CdZnTe/CdS (when E_g(CdZnTe) = 1.72 eV) solar cells were $V_{oc} = 720$ mV, and $J_{sc} = 2$ mA/cm^2. However, the spectral response indicated increasing loss of photocurrent at longer wavelengths. In order to improve the collection efficiency, CdZnTe devices were annealed in a H_2 atmosphere. This postprocessing treatment showed that J_{sc} increased to over 10 mA/cm^2. Study of CdZnTe solar cells is being carried out in various directions.

34.2.5 The Future of Cd-Based Solar Cells

From the development process of Cd-based solar cells, the study on the laboratory scale is being transferred to large-scale deposition and cell fabrication. Modules are being developed throughout the world, using the screen printing, evaporative deposition and close-space sublimation [34.31–33]

In order to make a commercial Cd-based photovoltaic cell with its full potential, a large-scale high-throughput manufacturing process is required. The process must possess excellent yields and produce high-efficiency devices with good long-term stability. In order to progress towards these goals, a pilot system for continuous, inline processing of CdS/CdTe devices has been developed [34.34–36]. High-quality low-cost thin-film CdTe modules with an average total area efficiency of 8% and cascaded production-line yield of > 70% have been manufactured, and technology-development programs will further increase production-line module efficiency to 13% within five years [34.36].

34.3 Radiation Detectors

A radiation detector is a device that converts a radiation ray into electrical signal. They can be divided into gas-filled detectors, scintillation detectors and semiconductor detectors. Since semiconductor radiation detectors have a high spectrometric performance, and can be made portable, they have been applied

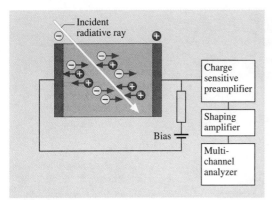

Fig. 34.7 The principle of operation of a semiconductor detector

in nuclear safeguards, medical physics and imaging diagnosis, industrial safety, nondestructive analysis, security and monitoring, nonproliferation and astrophysics [34.37, 38]. Meanwhile, they are also being gradually improved.

34.3.1 Basic Description of Semiconductor Radiation Detectors

The principle of operation of a semiconductor radiation detector is shown in Fig. 34.7. When high-energy photons (X- or γ-rays) radiate the detector, they lose the energy to induce electron–hole pairs in the semiconductor through photoelectric or Compton interactions, thereby increasing the conductivity of the material. In order to detect the change in conductivity, a bias voltage is required. An externally applied electric field separates the electron–hole pairs before they recombine, and electrons drift towards the anode, holes to the cathode. The charges are then collected by the electrodes. The collected charges produce a current pulse, whose integral equals the total charge generated by the incident photons, indicating the radiation intensity. The read-out goes through a charge-sensitive preamplifier, followed by a shaping amplifier. The multichannel analyzer is used for spectroscopic measurements of the radiation ray. This type of detector can produce both count and energy information. Typically, detectors are manufactured in a square or rectangular configuration to maintain a uniform bias-current distribution throughout the active region. The wavelength of the peak response depends on the material's band-gap energy.

There are several important parameters to describe a detector performance. Detection efficiency is defined as the percentage of radiation incident on a detector system that is actually detected. Detector efficiency depends on the detector size and shape (larger areas and volumes are more sensitive), radiation type, material density (atomic number), and the depth of the detection medium. Energy resolution is one of the most important characteristics of semiconductor detectors. It is usually defined as the full-width at half-maximum (FWHM) of a single-energy peak at a specific energy. High resolution enables the system to more clearly resolve the peaks within a spectrum. So the reliability of the detector is also high. Detector sensitivity implies minimal detectable counts. It is defined as the number of counts that can be distinguished from the background. The relative detector response factor expresses the sensitivity of a detector relative to a standard substance. If the relative detector response factor is expressed on an equal-mass (-weight) basis, the determined sensitivity values can be substituted for the peak area.

34.3.2 CdTe and CdZnTe Radiation Detectors

Semiconductor radiation detectors are fabricated from a variety of materials including: germanium (Ge), silicon (Si), mercuric iodide (HgI_2), CdTe, CdZnTe etc.. Typical detectors for a given application depend on several factors. Table 34.2 shows the features of some semiconductor detectors. From Table 34.2, Ge detectors have the best resolution, but require liquid-nitrogen cooling, which makes them impractical for portable applications.

Table 34.2 The features of some semiconductor detectors

	NaI	Si	Ge	CdTe	$Cd_{0.9}Zn_{0.2}Te$
Bandgap (eV) at 300 K	–	1.14	0.67	1.54	1.58
Density (g/cm^3)	3.67	2.33	5.35	6.2	5.78
Energy per e–h pair (eV)	–	3.61	2.96	4.43	4.64
Detection/absorption	High	Low	Medium	High	High
Operation at room temp.	Yes	Yes	No	Yes	Yes
Spectrometic performance	Medium	High	Very high	High	High

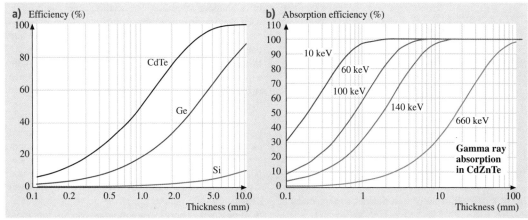

Fig. 34.8a,b γ-ray absorption in CdTe (**a**) and CdZnTe (**b**) detectors as a function of thickness

Si detectors also need cooling, and are inefficient in detecting photons with energies greater than a few tens of keV. CdTe and CdZnTe detectors possess their own advantages.

Single crystals of CdTe and CdZnTe are very important materials for the development of X- and γ-ray detectors. Currently Ge and Si detectors have to be used at liquid-nitrogen temperatures, and suffer from poor detection efficiency. Since the compound semiconductors CdTe and CdZnTe with high atomic number ($Z_{Cd} = 48$, $Z_{Zn} = 30$, $Z_{Te} = 52$) have a significantly higher photoelectric absorption efficiency in the -100–500 keV range, they can be fabricated into detectors that provide two advantages for use in portable instrumentation: (1) the large band-gap energy ($E_g > 1.54$ eV at room temperature) results in extremely small leakage current of a few nanoamps at room temperature. Therefore, these detectors have the potential to serve as useful detectors for radiation spectroscopy without the need for cryogenic cooling as required for more conventional silicon and hyper-pure germanium detectors; (2) the high density of the crystal provides excellent stopping power over a wide range of energies. The ability to operate at room temperature without the need for liquid-nitrogen cooling allows the construction of compact devices. CdTe and CdZnTe detectors can be fabricated into a variety of shapes and sizes, which makes it possible to produce detectors capable of meeting the requirements of a wide assortment of applications that are unsupported by other detector types. Their small size and relatively simple electronics allow them to open new areas of detector application. In many instances, CdTe and CdZnTe detectors can be substituted for other detector types in existing applications.

Figure 34.8 shows the γ-ray absorption efficiency in: (a) CdTe, and (b) CdZnTe detectors as a function of thickness. It was reported that, for a detector with a thickness of about 10 mm detecting 100-keV γ-rays, the detection efficiency for a CdTe detector is 100%, for a Ge detector 88%, for a Si detector only 12% [34.39]. In the case of a CdZnTe detector, CdZnTe with a thickness of only about 7 mm can absorb 100% of the γ-rays [34.40]. This efficiency assumes that the γ-ray is completely absorbed and the generated electron–hole pairs are completely collected. In fact, a large amount of charge loss occurs in CdTe and CdZnTe detector, depending on the quality of the materials.

34.3.3 Performance of CdTe and CdZnTe Detectors

Although CdTe and CdZnTe have many advantages over Si and Ge, including higher attenuation coefficients and lower leakage currents due to the wider bandgap, their disadvantages limit their application range [34.41]. The main disadvantage is that their crystals usually contain higher density defects. These defects are remaining impurities and structural defects such as mechanical cracks, twins, grain and tilt boundaries, and dislocations all lower the crystalline perfection of the materials.

In CdTe and CdZnTe detectors, the charge transport determines the collection efficiency of generated electron-hole pairs, and structural properties of materials control the uniformity of the charge transport. The problem of poor transport and collection of carriers, especially hole, can be overcome by applying high voltage to device and a range of electrode config. The former is related to high resistivity material, and later is related to the design of a detector. In addition, it was found that detected pulse signal intensity degraded with the time [34.42, 43]. This phenomenon was thought to be mainly related to the defects existing in bandgap. When the detector starts to work, these defects could act as slowly ionized acceptors. The low resistivity and poor quality material always exhibits such a polarization effect.

Summarizing above, improvement of a detector performance is equivalent to the improvement of the crystallinity of CdTe and CdZnTe. The key feature of all applications except substrate materials of CdTe and CdZnTe is the resistivity. This is because high resistivity can be obtained only by controlling native defects and impurity concentration. For many years, much effort was done in preparing high quality CdTe and CnZnTe single crystals.

Progresses in Crystal Growth of High Quality CdTe and CdZnTe

Various techniques have been applied to grow high-quality CdTe and CdZnTe single crystals. The Bridgman method [34.44, 45], the traveling-heater method [34.46, 47], growth from Te solvent [34.48], the gradient-freeze method [34.49, 50], and physical vapor transport [34.51, 52] are the most widely used. Crystals with applicable quality and size have become available, and have also fostered the rapid progress of research on CdTe.

As-grown CdTe single crystals commonly contain high concentration of both residual impurities and intrinsic defects. Preparation of high-purity high-resistivity CdTe and CdZnTe crystals has been widely attempted [34.53]. Early works was done by *Triboulet* and *Marfaing* [34.54] in obtaining high-purity lightly compensated zone-melting growth following synthesis by the Bridgman method. The room-temperature carrier concentrations range from 1 to 5×10^{13} cm^{-3}, and the resistivity from 100 to 400 Ωcm. The carrier mobility at 32 K reaches as high as 1.46×10^5 cm^2/Vs. The total concentration of electrically active centers was estimated to be about 10^{14} cm^{-3}.

Usually, as-grown CdTe crystal shows p-type conductivity owing to the existence of remaining acceptor impurities or native defects. For this reason, many results have been reported in preparing high-resistivity CdTe. Chlorine (Cl) is thought to be a suitable donor for the compensation of these remaining acceptors because Cl can be doped quite uniformly due to its very small segregation coefficient [34.55].

The growth of high-resistivity CdTe:Cl single crystals with device quality was successfully performed by the traveling-heater method (THM) [34.55]. Solvent alloys for THM growth were synthesized in ampoules filled with Te, CdTe and CdCl$_2$ so that the molar ratio of Te/Cd was the same as in the solvent zone during the growth. The Cl concentration in the grown crystal was 2 weight − ppm. According to the dynamics of the crystal growth, it is important to control the shape of the solid–liquid to grow high-quality single crystals. Therefore, the solvent volume was optimized. The use of a slightly tilted seed from <111>B was also effective in limiting the generation of twins with different directions. Single-crystal (111) wafers, larger than 30×30 mm^2 were successfully obtained from a grown crystal with a diameter of 50 mm. Pt/CdTe/In detectors with dimensions of $2 \times 2 \times 0.5$ mm^3 showed better energy resolution, because a higher electric field can be applied. The effective detector resistivity was estimated to be 10^{11} Ωcm.

CdTe doped with chlorine (Cl) or indium (In) with a resistivity of 3×10^9 Ωcm and CdZnTe with a resistivity of 5×10^{10} Ωcm were grown by the high-pressure Bridgman (HPB) technique [34.56]. The material was polycrystalline with large grains and twins. Although the crystalline quality of HPB CdTe and CdZnTe is poor, the grains are large enough to obtain volume detectors of several cm^3. Photoluminescence (PL) spectra at 4 K showed that the free exciton could be observed and the FWHM of the bound excitons is very low. Some impurity emissions were identified. The detectors were fabricated from HPB CdTe and CdZnTe. These detectors showed excellent performance. Gamma-ray spectra were presented with high-energy resolution in an energy range from 60 to 600 keV. Using a $10 \times 10 \times 2$ mm^2 HPB detector, at a bias of 300 V, the peak at 122 keV from ^{57}Co had a FWHM of 5.2 KeV. Detectors with high-energy resolution were fabricated.

Because of the difference in vapor pressure between Cd and Te, grown crystals contain a large number of Cd vacancies (V_{Cd}). These Cd vacancies manifest acceptor behaviour. Therefore, high-resistivity CdTe can also be obtained by controlling the concentration of Cd vacancies. However, a prerequisite is that the purity of the CdTe crystal is high enough that the remaining impurities do not play a substantial role in determining the conductivity. Recently, the preparation of ultra-high-purity CdTe single crystals was reported [34.57, 58]. In order to obtain a high purity the starting mater-

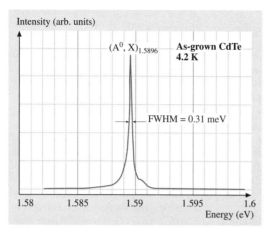

Fig. 34.9 A typical high-resolution PL spectrum of a high-purity CdTe sample

ials of Cd and Te, Cd were first purified by vacuum distillation (VD) and the overlap zone-melting (OZM) method [34.57], while Te was purified using the normal freezing method [34.58]. The results of glow-discharge mass spectroscopy (GDMS) and the measurement of the residual resistivity ratio (RRR) showed that refined Cd and Te had the purity of 6N-up. Using the refined Cd and Te as starting materials, extremely high-purity twin-free CdTe single crystals were prepared by the traditional vertical Bridgman technique. Figure 34.9 shows a typical high-resolution PL spectrum of a high-purity CdTe sample. Only a sharp and strong emission line of (A^0, X) (an exciton bound to a neutral acceptor) at 1.5896 eV was observed. The FWHM of the (A^0, X) line is 0.31 meV. This is the narrowest value ever reported for CdTe single crystals grown by the Bridgman method. All of the above indicate that the sample is of high purity and quality. For a good performance detector, the leakage current must not exceed a few nanoamps, which requires a material with high resistivity. When growing CdTe crystals, the CdTe contain large number of Cd vacancies due to the extreme volatility of Cd. For these reason, CdTe single crystals were annealed in a Cd atmosphere in order to obtain a high-resistivity crystal. Hall-measurement results showed that the conductive type changed from as-grown p-type to n-type at a Cd pressure around 1.5×10^{-2} atm, corresponding to a Cd source temperature of 500 °C. Samples annealed under these conditions showed a resistivity as high as 10^9 Ωcm.

Lachish [34.59] grew CdTe and CdZnTe crystals with excess cadmium in order to avoid the Te precipitates found in crystals grown by Bridgman method. During crystal growth one ampoule end was kept at a low temperature, which determines a constant, nearly atmospheric, vapor pressure in the system. The constant vapor pressure maintains a constant liquid composition and balanced amounts of cadmium and tellurium within the crystal. Although the highly donor-doped crystals are highly electrically conducting, successive annealing in Te vapor transforms the crystal wafers into a highly compensated state showing high electrical resistivity and high gamma sensitivity, depending on the doping level.

Improvements of Transport and Collection of Carriers in CdTe and CdZnTe Detectors

The transport and collection of carriers in CdTe and CdZnTe, especially holes, are the main problems that affect their performance. Poor carriers transport causes position-dependent charge collection and limits their capability as high-resolution spectrometers [34.60]. As a result, this limits the volume of detector. Therefore, the efficient detecting resolution and range are decreased. *Amman* et al. [34.61] studied the affect of nonuniform electron trapping on the performance of a γ-ray detector. An analysis of the induced charge signals indicated that regions of enhanced electron trapping are associated with inclusions, and that these regions extend beyond the physical size of the inclusions. Such regions introduce nonuniform electron trapping in the material that then degrades the spectroscopic performance of the material as a γ-ray detector. The measurements showed that the degree of nonuniformity that affects detector performance could be at the 1% level. Consequently, any useful characterization and analysis technique must be sensitive down to this level.

Detectors equipped with Ohmic contacts, and a grounded guard ring around the positive contact, have a fast charge-collection time. Conductivity adjustment, in detectors insensitive to hole trapping, optimizes the detector operation by a trade-off between electron lifetime and electrical resistance. Better hole collection has been achieved by designing detectors and electrodes of various geometries. *Nakazawa* et al. [34.62] improved the CdTe diode detectors by fabricating the guard-ring structure in the cathode face, and the leakage current of the detector decreased by more than an order of magnitude. The new CdTe detector was operated with a bias as high as 800 V at 20 °C and showed a good energy resolution (0.93 keV and 1 keV FWHM for 59.5 keV and 122 keV, respectively) and high stability for long-term operation at room temperature. Furthermore, they also developed a large-area (20×20 mm^2) detector with very high resolution. Owing to its high stopping power and high resolution at room temperature, pixel imagers have also been developed in the same group, and it is thought that this large diode detector could possibly be a substitute for scintillation detectors.

Since the carriers drift slow and have a short lifetime in CdTe detectors, the number of photons in the photopeak is reduced and the spectrum is distorted by a tail towards lower energies. For these reasons, CdTe detectors are usually made into Schottky diodes, because this structure can withstand much higher bias voltage with a leakage current orders of magnitude lower than detectors with Ohmic contacts. *Takahashi* et al. [34.63] adopted the configuration of Schottky CdTe diode, which due to the low leaking current, makes it possible to apply a much higher bias voltage to ensure complete charge collection in relatively thin (< 1 mm) devices. Both the improved charge-collection efficiency and the low leakage current lead to an energy resolution of better than 600 eV FWHM at 60 keV for a 2×2 mm^2 device without any electronics for charge-loss correction. Meanwhile, they also fabricated large-area detectors with dimensions of 21.5×21.5 mm^2, with a thickness of 0.5 mm and an energy resolution of 2.8 keV. Stacked detectors can measure the energies as high as 300 keV. Furthermore, a large-array detector, consisting of 1024 individual CdTe diodes, was also made. Every detector had a dimension of 1.2×5.0 mm^2. The total area, including the spaces between the detector elements, is 44×44 mm^2. This array detector is expected to be used in next-generation Compton telescopes.

34.3.4 Applications of CdTe and CdZnTe Detectors

Owing to the convenience of the smaller collimator, better resolution and temperature stability, CdTe and CZT detectors have been used in safeguard applications by the International Atomic Energy Agency (IAEA) and some countries for over ten years [34.64, 65]. With the gradual improvement in performance, CdTe and CdZnTe detectors are replacing NaI detectors used in spent-fuel attribute tests [34.66], though their sensitivity is still low compared to NaI and Ge detectors, although already sufficient for many applications. CdTe detectors have a sensitive volume of about 20–100 mm^3 and a probe diameter of 8–9 mm. CdZnTe detectors have a larger volume than CdTe detectors. The largest commercial CdZnTe detectors have a geometric volume of 1687 mm^3 ($15\times15\times7.5$ mm^3). Detectors are mainly of hemispheric design to obtain high carrier-collection efficiency. These large-volume detectors have been made into portable and hand-held isotope-identification devices and are being used to detect radioactive sources. They will become commercially available in the near future.

At present, conventional X-ray film or scintillator mammograms are used in medical diagnostics such as screening for breast cancer. However, these show a nonlinear response to X-ray intensity and the detection quantum efficiency is low. Room-temperature semiconductor detectors such as CdTe and CdZnTe have favorable physical characteristics for medical applications. From the start of these investigations in the 1980s, rapid progress has been achieved [34.67–72]. *Barber* [34.68] presented results concerning, first, a CdTe two-dimensional (2-D) imaging system (20×30 mm^2 with 400×600 pixels) for dental radiology and, second, a CdZnTe fast pulse-correction method applied to a $5\times5\times5$ mm^3 CdZnTe detector (energy resolution of 5% for a detection efficiency of 85% at 122 keV) for medical imaging. After that, a 2-mm-thick CdZnTe detector was fabricated for application to digital mammography [34.70]. The preliminary images showed high spatial resolution and efficiency. Furthermore, CdTe and CdZnTe detectors with a thickness of 0.15–0.2 mm were fabricated [34.72]. The detectors are indium-bump-bonded onto a small version of a chip. Their detection quantum efficiency was measured as 65%. This result showed that CdTe and CdZnTe detectors are superior to scintillator-based digital systems, whose quantum efficiency is typically around 30 –-40%. This showed that CdTe and CdZnTe detectors have potential applications in medical imaging, as well as industrially for nondestructive evaluation inspection.

In universe exploration, Cd-based detectors can be used in an advanced Compton telescope (ACT) planned as the next-generation space-based instrument devoted to observations of low/medium-energy γ-rays (≈ 0.2–30 MeV) and to the nonthermal energy exploration telescope (NeXT) [34.73]. In the universe, radiation rays have a wide energy range of 0.5–80 keV. In order to detect this wide range of radiation, *Takahashi* et al. [34.74–76] proposed a new focal-plane detector based on the idea of combining an X-ray charge-coupled device (CCD) and a CdTe pixel detector as the wide-band X-ray imager (WXI). The WXI consists of a soft-X-ray imager and a hard-X-ray imager. For the detection of soft X-rays (10–20 keV) with high positional resolution, a CCD with a very thin dead layer will be used. For hard X-rays, CdTe pixel detectors serve as absorbers. This study is now under way.

34.4 Conclusions

Devices fabricated from Cd-based compounds, such as solar cells and radiation detectors, are being applied in our daily life. In the case of solar cells, although recent success have improved their conversion efficiency and reduced the cost, many problems remain. Fundamental understanding of the CdTe-based solar-cell properties is limited, particularly as a result of their polycrystalline nature. Therefore, the fundamental electronic properties of polycrystalline Cd-based thin films should be studied deeply. Other challenges are to reduce the cost and to lengthen the operating life span. These problems are being studied and solved [34.77].

In the case of radiation detectors, CdTe and CdZnTe detectors have many advantages, such as room-temperature operation, high count rates, small size, and direct conversion of photons to charge, which make them attractive candidates for a wide variety of applications in industrial gauging and analysis, as well as medical instrumentation and other areas. These detectors are being made available commercially at present. However, they have severe problems such as polarization effects, long-term stability and their high price. Further efforts should still be focused on the preparation of high-quality materials and improvement of the stability and reliability of detectors. We are confident that radiation detectors made from Cd-based compounds will achieve more widespread application.

Definition of Terms

- **Cd-based compound semiconductor:** a semiconductor that contains the element Cd.
- **Heterojunction:** a junction between semiconductors that differ in their doping-level conductivities, and also in their atomic or alloy composition.
- **Band gap:** energy difference between the conduction band and the valence band.
- **n-type conductivity:** a semiconductor material, with electrons as the majority charge carriers, that is formed by doping with donor atoms.
- **p-type conductivity:** a semiconductor material in which the dopants create holes as the majority charge carrier, formed by doping with acceptor atoms.
- **Solar cell:** a semiconductor device that converts the energy of sunlight into electric energy. Also called a photovoltaic cell.
- **Semiconductor detector:** a device that converts the incident photons directly into an electrical pulse.
- **Conversion efficiency:** the ratio of incident photon energy and output electricity energy.
- **Detection efficiency:** percentage of radiation incident on a detector system that is actually detected.

References

34.1   E. Becquerel: Compt. Rend. Acad. Sci. (Paris) **9**, 561 (1839)

34.2   Yu. A. Vodakov, G. A. Lomakina, G. P. Naumov, Yu. P. Maslakovets: Sov. Phys. Solid State **2**, 1 (1960)

34.3   T. Aramoto, S. Kumazawa, H. Higuchi, T. Arita, S. Shibutani, T. Nishio, J. Nakajima, M. Tsuji, A. Hanafusa, T. Hibino, K. Omura, H. Ohyama, M. Murozono: Jpn. J. Appl. Phys. **36**, 6304 (1997)

34.4   B. Yang, Y. Ishikikawa, T. Miki, Y. Doumae, M. Isshiki: J. Cryst. Growth **179**, 410 (1997)

34.5   A. W. Brinkman: Properties of Narrow Gap Cadmium-Based Compounds. In: *Electronic Materials Information Services*, Vol. 10, ed. by P. Capper (IEE, London 1994) p. 591

34.6   R. W. Swank: Phys. Rev. **156**, 844 (1967)

34.7   D. A. Cusano: Solid State Electron. **6**, 217 (1963)

34.8   R. G. Little, M. J. Nowlan: Progress in Photovoltaics **5**, 309 (1997)

34.9   Y.-S. Tyan, E. A. Perez-Albuerne: In: *Proc. 16th IEEE Photovoltaic Specialists Conf.* (IEEE, New York 1982) p. 794

34.10 J. M. Woodcock, A. K. Turner, M. E. Özsan, J. G. Summers: In: *Proc. 22nd IEEE Photovoltaic Specialists Conf., Las Vegas* (IEEE, New York 1991) p. 842

34.11 K. Kuribayashi, H. Matsumoto, H. Uda, Y. Komatsu, A. Nakano, S. Ikegami: Jpn. J. Appl. Phys. **22**, 1828 (1993)

34.12 J. Britt, C. Ferekides: Appl. Phys. Lett **62**, 2851 (1993)

34.13 H. W. Schock, A. Shah: In: *Proc. 14th European Photovoltaics Solar Energy Conf.*, ed. by H. A. Ossenbrink, P. Helm, H. Ehmann (H. S. Stephens & Ass., Bedford, UK 1997) p. 2000

34.14 A. D. Compaan, A. Gupta, J. Drayton, S.-H. Lee, S. Wang: Phys. Stat. Solid B **241**, 779 (2004)

34.15 R. W. Dutton, R. S. Muller: Solid State Electron. **11**, 749 (1968)

34.16 K. W. Mitchell, A. L. Fahrenbruch, R. W. Bube: J. Appl. Phys **48**, 4365 (1977)
34.17 H. Uda, A. Nakano, K. Kuribayashi, Y. Komatsu, H. Matsumoto, S. Ikegami: Jpn. J. Appl. Phys. **22**, 1822 (1983)
34.18 N. Nakayama, H. Matsumoto, K. Yamaguchi, S. Ikegami, Y. Hioki: Jpn. J. Appl. Phys. **15**, 2281 (1976)
34.19 S. Ikegami, T. Yamashita: J. Electron. Mater. **8**, 705 (1979)
34.20 N. Nakayama, H. Matsumoto, A. Nakano, S. Ikegami, H. Uda, T. Yamashita: Jpn. J. Appl. Phys. **19**, 703 (1980)
34.21 J. Britt, C. Ferikides: Appl. Phys. Lett **62**, 2851 (1993)
34.22 T. Aramoto, S. Kumazawa, H. Higuchi, T. Arita, S. Shibutani, T. Nishio, J. Nakajima, M. Tsuji, A. Hanafusa, T. Hibino, K. Omura, H. Ohyama, M. Murozono: Jpn. J. Appl. Phys. **36**, 6304 (1997)
34.23 M. Tsuji, T. Aramoto, H. Ohyama, T. Hibino, K. Omura: Jpn. J. Appl. Phys. **39**, 3902 (2000)
34.24 M. P. R. Panicker, M. Knaster, F. A. Kröger: J. Electrochem. Soc. **125**, 566 (1978)
34.25 K. Murase, H. Uchida, T. Hirato, Y. Awakura: J. Electrochem. Soc. **146**, 531 (1999)
34.26 K. Murase, M. Matsui, M. Miyake, T. Hirato, Y. Awakura: J. Electrochem. Soc. **150**, 44 (2003)
34.27 M. Miyake, K. Murase, H. Inui, T. Hirato, Y. Awakura: J. Electrochem. Soc. **151**, 168 (2004)
34.28 D. W. Cunningham, M. Rubcich, D. Skinner: Prog. Photovoltaics **10**, 59 (2002)
34.29 B. McCandless, K. Dobson, S. Hegedus, P. Paulson: *NCPV and Solar Program Review Meeting Proceeding, March 24-26 2003, Denver, Colorado* (NREL, Golden, Colorado 2003) p. 401 Available in electronic form, NREL/CD-520-33586
34.30 G. Gidiputti, P. Mahawela, M. Ramalingan, G. Sivaraman, S. Subramanian, C. S. Ferekides, D. L. Morel: *NCPV and Solar Program Review Meeting Proceeding, March 24-26 2003, Denver, Colorado* (NREL, Golden, Colorado 2003) p. 896 Available in electronic form, NREL/CD-520-33586
34.31 P. D. Maycock: PV News **17**, 3 (1998)
34.32 R. C. Powell, U. Jayamaha, G. L. Dorer, H. McMaster: *Proc. NCPV Photovoltaics Program, Review*, ed. by M. Al-Jassim, J. P. Thornton, J. M. Gee (American Institute of Physics, New York 1995) p. 1456
34.33 D. Bonnet, H. Richter, K.-H. Jager: *Proc. 13th European Photovoltaic Solar Energy Conference*, ed. by W. Freiesleben, W. Palz, H. A. Ossenbrink, P. Helm (Stephens, Bedford, UK 1996) p. 1456
34.34 K. Zweibel, H. Ullal: *Proceeding of the 25th IEEE Photovoltaic Specialists Conference, Washington DC 1996* (IEEE, New York 1996) p. 745
34.35 K. L. Barth, R. A. Enzenroth, W. S. Sampath: *NCPV and solar Program Review Meeting Proceeding, March 24-26 2003, Denver, Colorado* (NREL, Golden, Colorado 2003) p. 904 Available in electronic form, NREL/CD-520-33586
34.36 A. Abken, C. Hambro, P. Meyers, R. Powell, S. Zafar: *NCPV and solar Program Review Meeting Proceeding, March 24-26 2003, Denver, Colorado* (NREL, Golden, Colorado 2003) p. 393 Available in electronic form, NREL/CD-520-33586
34.37 K. Zanio: , Vol. 13 (Academic, New York 1978) p. 164
34.38 R. Triboulet, Y. Marfaing, A. Cornet, P. Siffert: J. Appl. Phys. **45**, 2759 (1974)
34.39 G. Sato, T. Takahashi, M. Sugiho, M. Kouda, T. Mitani, K. Nakazawa, Y. Okada, S. Watanabe: IEEE Trans. Nucl. Sci **48**, 950 (2001)
34.40 C. Szeles: Phys. Stat. Solid B **241**, 783 (2004)
34.41 H. Yoon, J. M. Van Scyoc, T. S. Gilbert, M. S. Goorsky, B. A. Brunett, J. C. Lund, H. Hermon, M. Schieber, R. B. James: *Infrared Applications of Semiconductors II. Symposium, Boston, MA, USA, 1–4 Dec. 1997*, ed. by D. L. McDaniel Jr., M. O. Manasreh, R. H. Miles, S. Sivananthan, P. A. Warrendale (Materials Research Society, Pittsburgh, PA 1998) p. 241. USA: Mater. Res. Soc, 1998
34.42 R. O. Bell, G. Entine, H. B. Serreze: Nucl. Instrum. Methods **117**, 267 (1974)
34.43 P. Siffert, J. Berger, C. Scharager, A. Cornet, R. Stuck, R. O. Bell, H. B. Serreze, F. V. Wald: IEEE Trans. Nucl. Sci. **23**, 159 (1976)
34.44 R. K. Route, M. Woff, R. S. Feigelson: J. Cryst. Growth **70**, 379 (1984)
34.45 K. Y. Lay, D. Nichols, S. McDevitt, B. E. Dean, C. J. Johnson: J. Cryst. Growth **86**, 118 (1989)
34.46 R. O. Bell, N. Hemmat, F. Wald: Phys. Stat. Solid A **1**, 375 (1970)
34.47 R. Triboulet, Y. Mafaing, A. Cornet, P. Siffert: J. Appl. Phys. **45**, 375 (1970)
34.48 K. Zanio: J. Electron. Mat. **3**, 327 (1974)
34.49 M. Azoulay, A. Raizman, G. Gafni, M. Roth: J. Cryst. Growth **101**, 256 (1990)
34.50 A. Tanaka, Y. Masa, S. Seto, T. Kawasaki: Mater. Res. Soc. Symp. Proc. **90**, 111 (1987)
34.51 W. Akutagawa, K. Zanio: J. Cryst. Growth **11**, 191 (1971)
34.52 C. Ceibel, H. Maier, R. Schmitt: J. Cryst. Growth **86**, 386 (1988)
34.53 M. Isshiki: *Wide-gap II-VI Compounds for Opto-Electronic Applications* (Chapman Hall, London 1992) p. 3
34.54 R. Triboulet, Y. Mafaing: J. Electrochem. Soc. **120**, 1260 (1973)
34.55 M. Funaki, T. Ozaki, K. Satoh, R. Ohno: Nucl. Instr. Meth. A **322**, 120 (1999)
34.56 M. Fiederle, T. Feltgen, J. Meinhardt, M. Rogalla, K. W. Benz: J. Cryst. Growth **197**, 635 (1999)
34.57 B. Yang, Y. Ishikawa, Y. Doumae, T. Miki, T. Ohyama, M. Isshiki: J. Cryst. Growth **172**, 370 (1997)
34.58 S. H. Song, J. Wang, M. Isshiki: J. Cryst. Growth **236**, 165 (2002)
34.59 http://urila.tripod.com/crystal.htm

34.60 T. Takahashi, S. Watanabe: IEEE Trans. Nucl. Sci. **48**, 950 (2001)
34.61 M. Amman, J. S. Lee, P. N. Luke: J. Appl. Phys. **92**, 3198 (2002)
34.62 K. Nakazawa, K. Oonuki, T. Tanaka, Y. Kobayashi, K. Tamura, T. Mitani, G. Sato, S. Watanabe, T. Takahashi, R. Ohno, A. Kitajima, Y. Kuroda, M. Onishi: IEEE Trans. Nucl. Sci. **51**, 1881 (2004)
34.63 T. Takahashi, T. Mitani, Y. Kobayashi, M. Kouda, G. Sato, S. Watanabe, K. Nakazawa, Y. Okada, M. Funaki, R. Ohno, K. Mori: IEEE Trans. Nucl. Sci **49**, 1297 (2002)
34.64 R. Arlt, D. E. Rundquist: Nucl. Instr. Methods Phys. Res. A **380**, 455 (1996)
34.65 T. Prettyman: *2nd Workshop on Science and Modern Technology for Safeguards, Albuquerque, NM, U.S.A., 21–24 September 1998*, ed. by C. Foggi, E. Petraglia (European Commission, Albuquerque, NM 1998)
34.66 W. K. Yoon, Y. G. Lee, H. R. Cha, W. W. Na, S. S. Park: INMM J. Nucl. Mat. Manage. **27**, 19 (1999)
34.67 C. Scheiber, J. Chambron: Nucl. Instr. Meth. A **322**, 604 (1992)
34.68 H. B. Barber: J. Electron. Mater. **25**, 1232 (1996)
34.69 L. Verger, J. P. Bonnefoy, F. Glasser, P. Ouvrier-Buffet: J. Electron. Mater. **26**, 738 (1997)
34.70 S. Yin, T. O. Tümay, D. Maeding, J. Mainprize, G. Mawdsley, M. J. Yaffe, W. J. Hamilton: IEEE Trans. Nucl. Sci. **46**, 2093 (1999)
34.71 C. Scheiber: Nucl. Instr. Meth. A **448**, 513 (2000)
34.72 S. Yin, T. O. Tümay, D. Maeding, J. Mainprize, G. Mawdsley, M. J. Yaffe, E. E. Gordon, W. J. Hamilton: IEEE Trans. Nucl. Sci. **49**, 176 (2002)
34.73 T. Tanaka, T. Kobayashi, T. Mitani, K. Nakazawa, K. Oonuki, G. Sato, T. Takahashi, S. Watanabe: New Astron. Rev. **48**, 269 (2004)
34.74 T. Takahashi, B. Paul, K. Hirose, C. Matsumoto, R. Ohno, T. Ozaki, K. Mori, Y. Tomita: Nucl. Instr. Meth. A **436**, 111 (2000)
34.75 T. Takahashi, K. Nakazawa, T. Kamae, H. Tajima, Y. Fukazawa, M. Nomachi, M. Kokubun: SPI **4851**, 1228 (2002)
34.76 T. Takahashi, K. Makishima, Y. Fukazawa, M. Kokubun, K. Nakazawa, M. Nomachi, H. Tajima, M. Tashiro, Y. Terada: New Astron. Rev. **48**, 309 (2004)
34.77 V. K. Krishna, V. Dutta: J. Appl. Phys **96**, 3962 (2004)

35. Doping Aspects of Zn-Based Wide-Band-Gap Semiconductors

The present Chapter deals with the wide-band-gap (defined here as greater than 2 eV) Zn chalcogenides, i.e. ZnSe, ZnS, and ZnO (mainly in bulk form). However, since recent literature on ZnS is minimal, the main coverage is of ZnSe and ZnO. In addition $Zn_{1-x}Be_xSe$ ($x \leq 0.5$) is included, since Be is expected to reduce degradation (from light irradiation/emission) in ZnSe. The main emphasis for all these materials is on doping, in particular p-type doping, which has been a problem in all cases. In addition, the origin of light emission in ZnO is not yet well established, so this aspect is also briefly covered.

35.1	ZnSe	843
	35.1.1 Doping – Overview	843
	35.1.2 Results on p-Type Material with N as the Primary Dopant	845
35.2	ZnBeSe	848
35.3	ZnO	849
	35.3.1 Doping	849
	35.3.2 Optical Properties	850
References		851

The present Chapter treats the wide-band-gap (defined here as greater than 2 eV) Zn chalcogenides (as well as ZnBeSe), i.e., ZnSe, ZnS, and ZnO, with room-temperature band gaps of 2.7 eV, 3.7 eV, and 3.4 eV, respectively. We shall here concentrate mainly on bulk properties, since quantum dots and quantum wells are treated elsewhere in this Handbook except when these (or other nanostructures) are involved in bulk doping (Sect. 35.1.2). The primary emphasis will be on literature from 2000 to 2004. Moreover, since there have been few publications on ZnS in the last four years (our litarture search showed only seven publications) [35.1–7], the present review will effectively cover ZnSe, ZnBeSe, and ZnO.

It is well known that the primary interest in these materials is their ability to provide light emission and/or detection in the green and higher spectral ranges. One of the major problems for these materials is obtaining good bipolar doping, in particular good p-type doping for ZnO, ZnSe, and ZnBeSe with low fractions of Be; this problem has for instance been reviewed for ZnO by *Pearton* et al. [35.8] and by *Look* and *Claflin* [35.9] and for ZnSe by *Neumark* [35.10]. A second problem, especially for ZnSe-based devices, is that of degradation under photon irradiation, including those generated during light emission [35.11–14]. It is for this reason that ZnBeSe is of high interest, since Be is expected to *harden* ZnSe, i.e. to reduce defect formation and thus degradation [35.15–17].

ZnO is one of the most studied materials in the group of II–VI semiconductors because of its wide band gap (3.36 eV at room temperature) and its bulk exciton-binding energy (60 meV), which is larger than the room-temperature thermal energy. In addition to room-temperature ultraviolet (UV) optoelectronic devices, it can be used for magnetic [35.18] and biomedical applications [35.19] and references therein.

35.1 ZnSe

35.1.1 Doping – Overview

Despite many years of effort, p-type doping of ZnSe is still a problem. The main success to date has been achieved with nitrogen as the primary dopant. Of other dopants, Li diffuses extremely fast [35.20] and also self-compensates via interstitial Li [35.20, 21], Na has a predicted maximum equilibrium solubility of 5×10^{17} cm^{-3} [35.22] and also self-compensates (via interstitial Na), as shown by *Neumark* et al. [35.23], P and As give DX centers and thus give deep levels (as summarized for instance by *Neumark* [35.10]), and Sb to date has given net acceptor concentrations of only about 10^{16} cm^{-3} (see Table 35.1). Regarding, N doping, Table 35.2 lists recent results on concentrations of holes (p) or net acceptors ($[n_a - n_d]$), where n_a (n_d) is the accep-

Table 35.1 p-type doping of ZnSe with dopants other than N

Dopant	p (cm^{-3})	$n_a - n_d$ (cm^{-3})	E_a (meV)	Method	Reference
Sb		1.5×10^{16}	69	MOVPE	[35.25]
Sb	$\approx 10^{16}$			MOVPE	[35.26]
Sb			55 ± 5	MOVPE	[35.27]
Sb		$(7 \pm 3) \times 10^{16}$		PVT	[35.28]
K	9×10^{17}			Eximer laser	[35.29]
Na	5×10^{19}			Eximer laser	[35.29]
Co-doping Li, I		2×10^{16}		MOVPE	[35.30]
Co-doping Li, Cl		3.8×10^{16}		MBE	[35.31]
GaAs:Zn nano-cluster		1×10^{17}		MOMBE[a]	[35.32]

[a] metalorganic molecular beam epitaxy

Table 35.2 Doped ZnSe with p or $(n_a - n_d)$ above 10^{18} cm^{-3}

	Best p or $n_a - n_d$ (cm^{-3})	E_a (meV)	Comments on degradation	Reference
Sub-monolayer (N+Te) δ^3-doped	6×10^{18}	38–87	Expected to be minimal	[35.24, 33]
Li$_3$N diffusion	8×10^{18}			[35.34]
MOVPE-grown N-doped	1×10^{18}			[35.35]
ZnSe/ZnTe:N δ-doped superlattice	7×10^{18}	30	Expected to be high	[35.36]

Table 35.3 n-type doping of ZnSe

Dopant	n (cm^{-3})	Method	Comments	Reference
Cl			The PL is dominated by the Cl^{0}X line at 2.797 eV (10 K). Above 200 K, the intensity of the Cl^{0}X line decreases rapidly due to the presence of a nonradiative center with a thermal activation energy of ≈ 90 meV. The decrease of the Cl^{0}X line over the temperature range 10–200 K is due to the thermal activation of the Cl^{0}X line bound exciton to a free exciton with abactivation energy of ≈ 9.0 meV	[35.37]
Cl		MBE	At high ZnCl$_2$ beam intensity, crystallinity deteriorates due to excess Cl atom	[35.38]
Cl		MBE	At low T, the dominate PL is due to neutral donor-bound excitons; at high T, the dominate PL is due to free-to-bound recombination. At low T, two additional lines on the high-energy side are observed (light- and heavy-hole free-exciton transitions); one additional peak at the low-energy side (DAP[a] transition)	[35.39]
Al	$4.2 \times 10^{18} - 1.2 \times 10^{19}$	MBE	Three deep levels are reported: an acceptor-like state at 0.55 eV above VBM[b] and two donor states at 0.16 eV and 0.80 eV below CBM[c]	[35.40, 41]
Br	$1.4 - 4.1 \times 10^{17}$	Vertical sublimation		[35.42]
Br	4.0×10^{16}	PVT	Two deep electron traps with thermal activation energy 0.20 eV and 0.31 eV are reported	[35.43]
In		Dopant diffusion	A temperature range can be found where electron concentration decreases with an increase in temperature	[35.44]

[a] donor–acceptor pair; [b] valence band maximum; [c] conduction band maximum

tor (donor) concentration, in various approaches, where these are greater than 10^{18} cm^{-3}. We note in connection with Table 35.2 that degradation associated with N can be a severe problem [35.11, 12], and we also give some comments on degradation in the table. We shall discuss two N-doped systems in more detail below Sect. 35.1.2. One uses delta-doping with Te as co-dopant (for this system, a material used to help in incorporating the dopant); this system has given net acceptor concentrations up to 6×10^{18} cm^{-3} [35.24] with very low Te concentrations, so that minimal degradation is expected. The second system is that of Li$_3$N doping, with a report of carrier concentrations close to 10^{19} cm^{-3}. We list recent work on p-type dopants other than N in Table 35.1.

Interestingly, there are two reports that Sb gives quite low activation energies, one being 69 meV [35.25] and the other being 55 meV [35.27] (note that the activation energy for N is 111 meV [35.48]), with the former paper giving a net acceptor concentration of about 10^{16} cm^{-3}; in this connection it should still be noted that, as mentioned, As and P are generally believed to form DX centers and give deep levels (for a summary [35.10]). Other dopants used were K and Na, with doping carried out via excimer laser annealing; high doping levels were reported, but the excimer procedure would be expected to introduce high defect densities and resultant strong degradation (note that the maximum equilibrium solubility for Na was predicted to be about 5×10^{17} cm^{-3} by *Van de Walle* et al. [35.22]). A further approach was that of co-doping, where the term in this case means incorporation of both donors and acceptors; here, experimental tests were reported for Li with I in one case, and with Cl in another, but in both cases net acceptor concentrations were only in the 10^{16} cm^{-3} range. An additional method was to use planes of p-type GaAs (doped with Zn) to inject holes into ZnSe; net acceptor concentrations of 10^{17} cm^{-3} were reported in [35.32], where metalorganic molecular beam epitaxy (MOMBE) was used.

For completeness, we also list in Table 35.3 recent results on n-type doping.

35.1.2 Results on p-Type Material with N as the Primary Dopant

Recent methods for p-type doping with p or $[n_a - n_d]$ exceeding 10^{18} cm^{-3} have been listed in Table 35.2. Note that all of these use N as the primary dopant. As additional comments we note that growth by metalorganic vapor-phase epitaxy (MOVPE) is now relatively standard, and that a quite comprehensive discussion of this method has recently been given [35.49] (although it must be noted that the "hole concentration" of 3×10^{18} cm^{-3}, given in Table 1 of *Prete* et al. [35.49] from data given in *Fujita* et al. [35.50], is in fact the N concentration, with *Fujita* et al. [35.50] giving p as 8×10^{17} cm^{-3}); in view of this extensive recent paper, we do not discuss MOVPE here, but merely give in Table 35.4 some recent references (not in [35.49]). We further note that the use of a δ-ZnSe/ZnTe superlattice (SL) resulted in average Te concentrations of around 9%, which in turn increases the lattice mismatch between the GaAs substrate and the film, since the ZnTe lattice constant is larger than that of ZnSe. This is expected to lead to degradation problems [35.36].

A novel, interesting approach, which has given net acceptor concentrations up to 6×10^{18} cm^{-3}, is that of incorporating both N as a dopant and Te as a co-dopant into the δ-layer(s) with fractional ZnTe coverage, via molecular-beam epitaxy (MBE) [35.24]; as previously mentioned (Sect. 35.1.1), co-dopant here means a material which aids in the incorporation of the dopant, and it is well known that it is easy to obtain p-type ZnTe [35.51, 52]. Electrochemical capacitance–voltage (E–CV) profiling results for various samples are shown in Fig. 35.1 (Fig. 3 of [35.24]); it can be seen that good doping was obtained when three contiguous layers of N and Te were incor-

Table 35.4 Nitrogen-doped ZnSe grown by metalorganic chemical vapor deposition (MOCVD) or MOVPE

$n_a - n_d$ (cm^{-3})	E_a (meV)	Comments	Reference
6.7×10^{17}	109	ZnSe:N epilayers were grown on ZnSe substrates by low-pressure MOCVD at 830 K and annealed in Zn saturated vapor. The net acceptor concentration is enhanced	[35.45]
1.2×10^{18}		ZnSe:N grown on GaAs. A radio-frequency (RF) plasma nitrogen source was used for doping	[35.35]
		ZnSe:N grown by MOVPE with hydrazines as dopants. The acceptor concentration is limited by the residual impurities in the sources	[35.46]
		ZnSe:N was grown by photo-assisted MOVPE. Post-growth annealing is critical to reducing the hydrogen concentration (by a factor of 10)	[35.47]

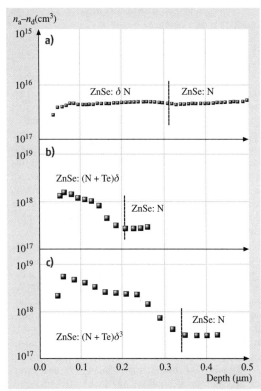

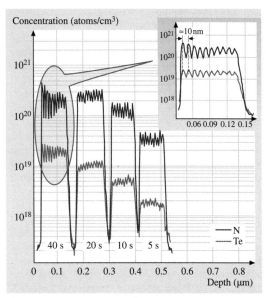

Fig. 35.1 (a) Depth-dependent $(n_a - n_d)$ of a conventional δ-doped sample with 5-ML spacer (nominal undoped ZnSe); **(b)** a (N + Te) δ-doped sample with 4-ML spacer; **(c)** a (N + Te) δ^3-doped sample with 7-ML. After [35.24]

Fig. 35.2 SIMS results on a δ^3-doped ZnSe:(Te, N) sample. The *upper line* represents the [Te] concentration, and the *lower line* represents the [N] concentration

porated (δ^3-doped). A very important aspect of this system, established by subsequent work [35.33], is that the N is preferentially located within ZnTe, which was shown conclusively [35.33] to be present in submonolayer quantities, without formation of a standard SL (considering a standard SL to require full monolayers). This result was established by transmission electron microscopy (TEM), secondary-ion emission spectroscopy (SIMS), and high-resolution X-ray diffraction (HRXRD). SIMS data were taken on a specially prepared sample, in which the spacer regions (undoped ZnSe separating the δ-layers) were thick enough (in view of the SIMS resolution) that the SIMS measurements effectively gave the N and Te concentrations in the delta region. The results for a triple-doped sample are shown in Fig. 35.2, with a Te concentration of about 5×10^{20} cm^{-3} and an N concentration of about 5×10^{19} cm^{-3} for a "standard" 5 s Te + N deposition time [35.24]. Thus, both are present at far less than monolayer quantities.

Results from HRXRD are shown in Fig. 35.3, which gives (004) $\theta - 2\theta$ (solid black line) of a triple delta ZnSe:(Te,N) sample; this sample was grown in the [001] direction with a 10 nm ZnSe buffer layer, spacers of 10 monolayer (ML, where we here assume 1 ML, in the [001] growth direction, to be half of the lattice constant), and 200 spacer/δ-region periods and was grown using a standard Te deposition time of 5 s [35.24]. The strongest peak, at $2\theta \approx 66.01°$, is from the GaAs substrate. In addition, satellite peaks associated with the periodic structure along the growth direction are observed. The result of a simulation using dynamical diffraction theory [35.53–55] is shown by the dashed line. This fit is obtained with a δ-layer and spacer thicknesses of 0.25 ML and 10.4 ML, respectively. These values are in excellent agreement with the nominal growth conditions. The average Te concentrations are $\approx 37\%$ and 2.2% in the δ-layers and the spacers, respectively. The low average Te coverage and its relatively high concentration within the δ^3-layers indicate that Te is not uniformly distributed within these layers, and, thus, forms ZnTe-rich nano-islands (such nano-islands have been observed by *Gu* et al. [35.56], optically, in similar samples grown without nitrogen). The relatively

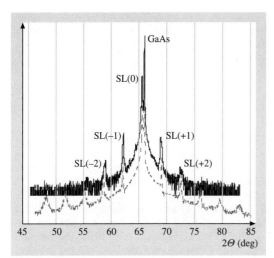

Fig. 35.3 Symmetric $\theta - 2\theta$ scan of a δ^3-doped ZnSe: (Te,N). The *solid black line* is the experimental result, and the *dashed line* is the result of simulation. For clarity, the curves are shifted vertically

high Te concentration in the δ-layers is consistent with doping results obtained for ZnSeTe alloys, where it was shown that high acceptor concentrations are observed only for Te concentrations exceeding 15% [35.57]. A further important result for understanding the doping mechanism in the present system was that the photoluminescence (PL) quenched, with increasing temperature, with quite a low activation energy [35.33]; results are shown in Table 35.5, where it can be seen that the activation energies are far lower than for N in ZnSe (111 meV [35.48]) and decrease with increasing Te concentration, down to 38 meV, which is within the range 30–65 meV reported for ZnTe [35.36, 58]. Thus, the N is associated primarily with ZnTe, i. e., the N is embedded primarily in Te-rich regions. It can also be noted from the SIMS results (Fig. 35.2) that N and Te are located in the same spatial region, and this is of course totally consistent with the view that N is embedded in Te-rich nano-islands.

We next consider the case of doping by diffusion of Li$_3$N into MOVPE-grown material [35.34]. The view has been expressed [35.59] that the resultant good doping was due to the incorporation of Li into Zn sites, and N into Se sites, with both such species being acceptors. In our view, this conjecture is unlikely. Thus, we note that in hard-to-dope wide-band-gap materials, strong compensation is expected [35.60]; since interstitial Li is a donor, it seems very likely that considerable Li is incorporated into the interstitial site after diffusion. Moreover, it is known [35.20] that Li diffuses very quickly. Thus we suggest that, during contact formation, even with minimal heating, a good fraction of the interstitial Li diffuses into precipitates, leaving the material p-type. We note that this view is reinforced by the work of *Strassburg* et al. [35.61], who show that this method works very well if doping is carried out by ion implantation; such implantation is expected to cause a high density of lattice defects, where defects would be expected to act as nucleation sites for precipitation of interstitial Li.

Last, but not least, no discussion of N doping would be complete without pointing out that it is now realized that N, to a greater or lesser extent (depending on conditions), does self-compensate, i. e. it does introduce donors. The nature of the donors will depend on the Fermi level and on the Zn (Se) and N chemical potentials, as shown in theoretical analyses by *Van de Walle* et al. [35.22], *Kwak* et al. [35.62] and *Cheong* et al. [35.63]. A discussion and comparison of these papers, as well as of the minimum requirements for reliable first-principles calculations, has been given by *Neumark* [35.10]. Additional work by *Faschinger* et al. [35.64] and *Gundel* and *Faschinger* [35.65] suggested, based on first-principles calculations, a complex between interstitial N (N$_i$) and a Se vacancy (V$_{Se}$), but no dependence on the Fermi level or chemical potentials was given.

Moreover, experimentally, *Kuskovsky* et al. [35.66] have reported a double N interstitial donor at high N doping, and *Desgardin* et al. [35.67] have reported a [V$_{Se}$N$_{Se}$] complex and a V$_{Zn}$ point defect. Furthermore, it has also been shown that the N$_i$ species (and probably complexes) contribute to degradation [35.11, 12, 68], but details of this process do not yet appear well understood, and we thus merely mention its existence. However,

Table 35.5 Sample parameters and photoluminescence properties of δ^3 (Te, N)-doped ZnSe

Te concentration (%)	$n_a - n_d$ (cm^{-3})	PL quenching activation energy (meV)
< 3	6.0×10^{18}	38
1.3	4.0×10^{18}	72
< 1	3.0×10^{18}	87

a point we do want to emphasize in this regard is that it might be highly advantageous to be able to use a dopant other than N. We thus note that, with the approach of Lin et al. [35.24], one can envision the use of a dopant other than N, since the acceptors are not located within ZnSe, but rather in a favorable ZnTe-rich environment. We thus point out that P and As are excellent p-dopants in ZnTe [35.69, 70].

35.2 ZnBeSe

As mentioned in Sect. 35.1, the best p-type dopant developed to date for ZnSe is nitrogen, and such N-doped material suffers from degradation problems. To alleviate this problem, the use of ZnBeSe has been suggested [35.15, 16]. BeSe is *harder* than ZnSe (Fig. 35.4 [35.71]) and, since it is expected that harder materials are less susceptible to defect formation (dislocations etc.), it is expected to be less susceptible to degradation [35.15, 16]. It has been shown that the hardness of ZnBeSe increases with increasing Be content, at least up to 60% Be, as shown in Fig. 35.4 [35.71], where it should be noted that the experimental error at the higher Be concentrations is quite large (moreover, the main interest in ZnBeSe is in the direct-band range, i.e. below 46% [35.72]).

It can be pointed out that two additional advantages of ZnBeSe over ZnSe are that one can adjust the lattice constant for better lattice-matching to various materials of interest (GaAs – the most frequently used substrate), and that one can obtain a wider band gap. For instance, $Zn_{0.028}Be_{0.972}Se$ is lattice-matched to GaAs [35.73] and $Zn_{0.55}Be_{0.45}Se$ [35.74] is lattice-matched to Si (assuming a BeSe lattice constant of ≈ 5.138 Å). The variation of the band gap has been studied in a number of recent papers [35.72, 75]. One result, over the entire concentration range, is shown in Fig. 35.5 [35.72]. It can be seen that the band gap becomes indirect for Be concentrations above 46%; thus, very high Be concentrations are not as interesting, since they cannot be used to give diode lasers.

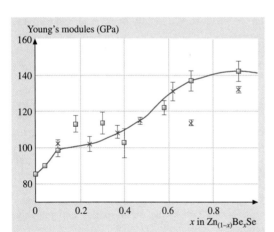

Fig. 35.4 Variation of elastic modulus E as a function of alloy composition x in $ZnBe_xSe_{1-x}$. The data points represent the average value. The *squares* – joined by the *full line* – show results obtained under peak loads of 1 mN for alloys grown onto GaAs. *Crosses* are for data obtained under 10 mN for alloys grown onto GaAs and *open circles* show data obtained under 1 mN for alloys grown onto GaP. We note that in general Young's modulus is related to material hardness [35.17]. After [35.71]

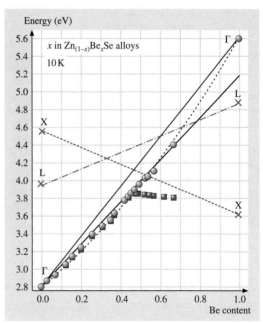

Fig. 35.5 Evolution of the direct band gap (○) and of the main PL peak (■) as a function of the Be content in $Zn_{1-x}Be_xSe$ alloys. After [35.72]

Table 35.6 XRD, EPD, and C–V results for undoped, N-doped, and (N + Te) δ-doped ZnBeSe epilayers [35.76]

	ZnBeSe	ZnBeSe : N	ZnBeSe : (N + Te)δ	ZnBeSe : (N + Te)δ^3
FWHM (arcs)	23	30	45	51
Be content (%)	3.1	2.6	2.6	2.5
Te content (%)	0	0	0.3	0.5
EPD (cm^{-2})	4×10^4	1×10^5	6×10^5	5×10^5
$n_a - n_d$ (cm^{-3})	–	2×10^{17}	3×10^{17}	1.5×10^{18}

A problem for ZnBeSe, as for ZnSe, is that of p-type doping. The highest bulk net acceptor concentration in p-type ZnBeSe does not exceed $\approx 2 \times 10^{17}$ cm^{-3} [35.76]. The best p-type results were again obtained via delta-doping [35.76], using the same method that *Lin* et al. [35.24] used for ZnSe. The results from *Guo* et al. [35.76] are shown in Table 35.6 (Table I from [35.76]).

35.3 ZnO

ZnO is a wide-band-gap (3.36 eV at room temperature) semiconductor with a bulk exciton-binding energy (60 meV), larger than the room-temperature thermal energy, which makes this material very suitable for a variety of applications (see recent reviews by *Pearton* et al. [35.18], *Heo* et al. [35.87] and *Djurišić* et al. [35.88]) in the UV spectral range. However, as for ZnSe and ZnBeSe, one of the major problems for ZnO is p-type doping, and we shall therefore emphasize this aspect.

35.3.1 Doping

ZnO can be grown by a wide range of techniques (some of which are listed in Table 35.7). As-grown ZnO is usually n-type, and heavily n-type ZnO is easily obtained by using group III elements. It is assumed that nominally undoped ZnO is n-type due to shallow native defects such as interstitial zinc (Zn$_i$) [35.89, 90] or, alternatively, due to the presence of hydrogen [35.91]. Experimentally, hydrogen in ZnO has been observed

Table 35.7 p-type doping of ZnO

Dopant	Growth method	Resistivity (Ωcm)	Dopant concentration (cm^{-3})	Carrier concentration (cm^{-3})	Mobility (cm^2/Vs)	Reference
As	Evaporation followed by sputtering	0.4	Mid 10^{19}	4×10^{18}	4	[35.77]
As	Hybrid beam deposition	2	3×10^{18}	4×10^{17}	35	[35.78]
P	RF sputtering followed by RTA	0.59–4.4		1.0×10^{17} –1.7×10^{19}	0.53–3.51	[35.79]
N	Ultrasonic spray pyrolysis	≈ 0.03		8.59×10^{18}	24.1	[35.80]
N	Thermal oxidization of Zn$_3$N$_2$ thin films		Up to 10^{21}	4.16×10^{17}		[35.81]
N	Implantation	10.11–15.3		Up to 7.3×10^{17}	2.51–6.02	[35.82]
N	MOCVD	3.02		1.97×10^{18}	1	[35.83]
N	CVD	17.3	Up to 3×10^{20}	1.06×10^{18}	0.34	[35.84]
N + Al	Direct-current (DC) reactive magnetron sputtering	57.3		2.25×10^{17}	0.43	[35.85]
N + In	Ultrasonic spray pyrolysis	0.017		2.44×10^{18}	155	[35.86]

via electron paramagnetic resonance (EPR), electron nuclear double resonance (ENDOR), optical, and IR absorption measurements [35.92–95]. The activation energy of the hydrogen donor is 35–46 meV [35.92, 96]. We note that sometimes oxygen vacancies (V_O) are cited as shallow donors [35.97]; however, *Zhang* et al. [35.90] estimated this species to be a relatively deep level. Also, *Vanheusden* et al. [35.98, 99] suggested that charged oxygen vacancies are responsible for the deep green luminescence in ZnO (see below).

Obtaining good p-type ZnO has however proven difficult. There is a good discussion and summary of growth methods as well as achieved resistivities in p-type ZnO up to 2003 in *Look* and *Claflin* [35.9] and *Look* et al. [35.102]. The latter publication also discusses background impurities in ZnO. We therefore present only some later results and give a short discussion of models proposed for p-type doping of ZnO.

Group V acceptors, based on theoretical arguments, are expected to form very deep substitutional acceptors; for instance, *Park* et al. [35.103] have calculated that the ionization energies of N, P, and As are 0.40 eV, 0.93 eV, and 1.15 eV, respectively. So successes (Table 35.7) in obtaining p-type ZnO with N, P, and As are surprising. Also group I (Li, Na, and K) impurities [35.103] have, in general, lower ionization energies, but these impurities are amphoteric and thus self-compensate. Experimentally, interstitial Li and Na donors were observed by *Orlinskii* et al. [35.104], and recent attempt to use Li_3N to dope ZnO to be p-type produced n-type conductivity instead [35.105].

To achieve p-type doping, *Wang* and *Zunger* [35.106] have proposed a cluster co-doping method using Ga or Al as co-dopants along with group V dopants; experimentally, p-type ZnO has been obtained using co-doping with Al and In (Table 35.7); N–Ga co-doping has been attempted [35.107] but no p-type conductivity has been observed via the Hall effect.

Recently, to explain p-type ZnO obtained via group V doping, *Limpijumnong* et al. [35.108] proposed, using first-principles calculations, that group V elements give shallow acceptors by forming complexes with native defects. Specifically, these authors proposed that ZnO:As and ZnO:Sb are p-type due to $As_{Zn} - 2V_{Zn}$ and $Sb_{Zn} - 2V_{Zn}$ complexes, which behave as shallow acceptors. These complexes have low formation energies (1.59 eV and 2.00 eV, respectively) as well as low ionization energies (0.15 eV and 0.16 eV, respectively). Experimentally, the activation energy for ZnO:As was reported to be between 0.12 [35.78] and 0.18 [35.109]. As for nitrogen, the most often used p-type dopant,

Look et al. [35.77] reported that the ionization energy was as low as 0.090 eV for heavily doped material (see also [35.110] and references therein).

Regarding doping using phosphorous, we note that *Kim* et al. [35.79] obtained p-type ZnO only after annealing at high temperatures using rapid thermal annealing (RTA), while as-grown material was n-type. The authors suggested that the annealing removes the compensating donors; however, we suggest that the formation of shallow acceptor complexes cannot be ruled out, especially in view of enhanced n-type behavior with increased P concentration [35.87].

Finally, we note that *Lee* and *Chang* [35.111] have proposed, theoretically, ways to use [group I – Hydrogen] complexes for p-type doping. These authors have found that an intentional co-doping with H impurities suppresses the formation of compensating interstitials and greatly enhances the solubility of Li and Na acceptors. This type of effect, in general, was clearly predicted by *Neumark* [35.112]. H atoms can be easily removed from ZnO by post-growth annealing at relatively low temperatures. Apparently, this method is similar to that used to obtain p-type GaN. These authors [35.111] also found, as did *Park* et al. [35.103], that Li and Na have lower ionization energies than substitutional group V dopants such as nitrogen.

35.3.2 Optical Properties

Finally, we shall briefly discuss some optical properties of ZnO. Low-temperature PL of undoped ZnO is dominated by near-band, edge emission, with up to 20 lines observed within the spectral range 3.34–3.38 eV [35.96]. Detailed studies of bound excitons (BX) and donor–acceptor pair luminescence have recently been published by *Meyer* et al. [35.96], so here

Table 35.8 Low-temperature bound-excitonic position and assignments

BX line energy (eV)	Assignment	Donor binding energy (meV)	References
≈ 3.3567	In	63.2	[35.96, 100]
≈ 3.3598	Ga	54.6	[35.96, 101]
≈ 3.3628	H	37 35 46.1	[35.92] [35.95] [35.96]
≈ 3.3608	Al	54.8 51.55	[35.96] [35.100]

we merely summarize some of their results and compare these with other reports.

An important conclusion was that all bound-excitonic lines are due to neutral donor-bound excitons, rather than, as previously suggested, to acceptor-bound excitons [35.113–115]. For instance, the ≈ 3.357 eV and ≈ 3.3608 lines are attributed to In and Al donors, respectively [35.113–115]. Similar conclusions were obtained by *Morhain* et al. [35.100] using PL and selective PL (SPL) on MBE-grown ZnO. It must be noted that such assignments were previously proposed by *Block* et al. [35.116] and *Gonzalez* et al. [35.117]. In Table 35.8 we summarize the BX lines that have been identified with a specific impurity or center. A line at 3.3631 eV, which is slightly above the 3.3628 eV line, was also assigned to hydrogen by *Look* et al. [35.95] who performed Hall-effect, PL, and EPR measurements on a series of ZnO samples annealed in air at various temperatures. The dominant donor had an activation energy of ≈ 37 meV, but disappeared after high-temperature annealing, and was replaced by a 67 meV donor [35.95]. The line at ≈ 3.3631 eV has been assigned to the 37 meV donor; the authors suggested, following *Hofmann* et al. [35.92], that this donor is hydrogen. This assignment has been confirmed by *Meyer* et al. [35.96] by SPL and by *Morhain* et al. [35.100] by magneto-optics, with *Meyer* et al. [35.96] reporting an H ionization energy of ≈ 46.1 meV.

Another important feature of bulk ZnO is a visible luminescence, often referred to as the green band. It is usually observed at (2.38 ± 0.04) eV [35.118–121]. The origin of this band, however, remains controversial: transitions associated with O_{Zn} antisites [35.119], oxygen vacancies [35.97–99, 122], zinc interstitials [35.123], ZnO antisites [35.124], donor–acceptor pairs [35.125], and Cu^{2+} ions [35.126] have all been suggested. It must be noted that the origin of the green luminescence could be different in ZnO prepared via different methods, since various defects and/or impurities can contribute to the emission [35.127]. We note that oxygen vacancies are the species most often suggested as the defect associated with the green luminescence. Oxygen vacancies can have three states – neutral, singly and doubly positively charged. The transition thus depends on the type of free carrier that is participating in recombination. *Vanheusden* et al. [35.98, 99] suggested that holes participate in this recombination while *Djurišić* et al. [35.88] (also references therein) suggested the involvement of electrons. We have recently shown that the green luminescence most likely involves holes rather than electrons [35.128], via studies of quantum ZnO wires. A similar conclusion was also recently obtained by *Kang* et al. [35.129] who investigated PLD-grown ZnO. Their conclusion was that the most likely channel for the green PL is through a deep donor (attributed to oxygen vacancies) and the holes in the valence bands. Lastly, a further suggestion by *Harada* and *Hashimoto* [35.130] is that surface states associated with oxygen vacancies could play a significant role in the emission within the visible spectral region in bulk ZnO.

References

35.1 L. Svob, C. Thiandourme, A. Lusson, M. Bouanani, Y. Marfaing, O. Gorochov: Appl. Phys. Lett. **76**, 1695 (2000)

35.2 S. Kishimoto, T. Hasegawa, H. Kinto, O. Matsumoto, S. Iida: J. Cryst. Growth **214/215**, 556 (2000)

35.3 S. Kishimoto, A. Kato, A. Naito, Y. Yakamato, S. Lida: Phys. Status Solidi B **229**, 391 (2002)

35.4 Y. Abiko, N. Nakayama, K. Akimoto, T. Yao: Phys. Status Solidi B **229**, 339 (2001)

35.5 S. Nakamura, J. Yamaguchi, S. Takagimoto, Y. Yamada, T. Taguchi: J. Cryst. Growth **237/239**, 1570 (2002)

35.6 S. Kohiki, T. Suzuka, M. Oku, T. Yamamoto, S. Kishimoto, S. Iida: J. Appl. Phys. **91**, 760 (2002)

35.7 K. Ichino, Y. Matsuki, S.T. Lee, T. Nishikawa, M. Kitagawa, H. Kobayashi: Phys. Status Solidi C **1**, 710 (2004)

35.8 S.J. Pearton, D.P. Norton, K. Lp, Y.W. Heo, T. Steiner: J. Vac. Sci. Technol. B **22**, 932 (2004)

35.9 D.C. Look, B. Claflin: Phys. Status Solidi B **241**, 624 (2004)

35.10 G.F. Neumark: Mater. Lett. **30**, 131 (1997)

35.11 D. Albert, J. Nürnberger, V. Hock, M. Ehinger, W. Faschinger, G. Landwehr: Appl. Phys. Lett. **74**, 1957 (1999)

35.12 V.N. Jmerik, S.V. Sorokin, T.V. Shubina, N.M. Shmidt, I.V. Sedova, D.L. Fedorov, S.V. Ivanov, P.S. Kop'ev: J. Cryst. Growth **214/215**, 502 (2000)

35.13 H. Ebe, B.-P. Zhang, F. Sakurai, Y. Segawa, K. Suto, J. Nishizawa: Phys. Status Solidi B **229**, 377 (2002)

35.14 K. Katayama, T. Nakamura: J. Appl. Phys. **95**, 3576 (2004)

35.15 A. Wagg, F. Fischer, H.-J. Lugauer, Th. Litz, T. Gerhard, J. Nürnberger, U. Lunz, U. Zehnder, W. Ossau, G. Landwehr, B. Roos, H. Richter: Mater. Sci. Eng. B **43**, 65 (1997)

35.16 C. Verie: J. Cryst. Growth **184/185**, 1061 (1998)

35.17 F. C. Peiris, U. Bindley, J. K. Furdyna, H. Kim, A. K. Raudas, M. Grimsditch: Appl. Phys. Lett. **79**, 473 (2001)

35.18 S. J. Pearton, C. R. Abernathy, M. E. Overberg, G. T. Thaler, D. P. Northon, N. Theodorpoulou, A. F. Hebard, Y. D. Park, F. Ren, J. Kim, L. A. Boatner: J. Appl. Phys. **93**, 1 (2003)

35.19 Y. W. Heo, D. P. Norton, L. C. Tien, Y. Kwon, B. S. Kang, F. Ren, S. J. Pearton, J. R. LaRoche: Mater. Sci. Eng. R **47**, 1 (2004)

35.20 M. A. Haase, H. Cheng, J. M. DePuydt, J. E. Potts: J. Appl. Phys. **67**, 448 (1990)

35.21 G. F. Neumark, S. P. Herko: J. Cryst. Growth **59**, 189 (1982)

35.22 C. G. Van de Walle, D. B. Laks, G. F. Neumark, S. T. Pantelides: Phys. Rev. B **47**, 9425 (1993)

35.23 G. F. Neumark, S. P. Herko, T. F. McGee III, B. J. Fitzpatrick: Phys. Rev. Lett. **53**, 604 (1984)

35.24 W. Lin, S. P. Guo, M. C. Tamargo, I. Kuskovsky, C. Tian, G. F. Neumark: Appl. Phys. Lett. **76**, 2205 (2000)

35.25 M. Takemura, H. Goto, T. Ido: Jpn. J. Appl. Phys. **36**, L540 (1997)

35.26 H. Goto, T. Ido, A. Takatsuka: J. Cryst. Growth **214/215**, 529 (2000)

35.27 H. Kalisch, H. Hamadeh, R. Rüland, J. Berntgen, A. Krysa, M. Hluken: J. Cryst. Growth **214/215**, 1163 (2000)

35.28 M. Prokesch, K. Irmscher, U. Rinas, H. Makino, T. Yao: J. Cryst. Growth **242**, 155 (2002)

35.29 Y. Hatanaka, M. Niraula, A. Nakamura, T. Aoki: Appl. Surf. Sci. **175/176**, 462 (2001)

35.30 I. Suemune, H. Ohsawa, T. Tawara, H. Machida, N. Shimoyama: J. Cryst. Growth **214/215**, 562 (2000)

35.31 M. Yoneta, H. Uechi, K. Nanami, M. Ohishi, H. Saito, K. Yoshino, K. Ohmori: Physica B **302**, 166 (2001)

35.32 J. Hirose, I. Suemune, A. Ueta, H. Machida, N. Shimoyama: J. Cryst. Growth **214/215**, 524 (2000)

35.33 I. L. Kuskovsky, Y. Gu, Y. Gong, H. F. Yan, J. Lau, G. F. Neumark, O. Maksimov, X. Zhou, M. C. Tamargo, V. Volkov, Y. Zhu, L. Wang: Phys. Rev. Lett. B **73**, 195306 (2006)

35.34 O. Schulz, M. Strassburg, T. Rissom, U. W. Pohl, D. Bimberg, M. Klude, D. Hommel: Appl. Phys. Lett. **81**, 4916 (2002)

35.35 E. D. Sim, Y. S. Joh, J. H. Song, H. L. Park, S. H. Lee, K. Jeong, S. K. Chang: Phys. Status Solidi B **229**, 213 (2002)

35.36 H. D. Jung, C. D. Song, S. Q. Wang, K. Arai, Y. H. Wu, Z. Zhu, T. Yao, H. Katayama-Yoshida: Appl. Phys. Lett. **70**, 1143 (1997)

35.37 S. Z. Wang, S. F. Yoon, L. He, X. C. Shen: J. Appl. Phys. **90**, 2314 (2001)

35.38 M. Yoneta, K. Nanami, H. Uechi, M. Ohishi, H. Saito, K. Yoshino: J. Cryst. Growth **237/239**, 1545 (2002)

35.39 Y. Gu, I. L. Kuskovsky, G. F. Neumark, X. Zhou, O. Maksimov, S. P. Guo, M. C. Tamargo: J. Lumin. **104**, 77 (2003)

35.40 D. C. Oh, J. S. Song, J. H. Chang, T. Takai, T. Handa, M. W. Cho, T. Yao: Mater. Sci. Semicond. Process. **6**, 567 (2003)

35.41 D. C. Oh, J. H. Chang, T. Takai, J. S. Song, K. Godo, Y. K. Park, K. Shindo, T. Yao: J. Cryst. Growth **251**, 607 (2003)

35.42 H. Kato, H. Udono, I. Kikuma: J. Cryst. Growth **229**, 79 (2001)

35.43 M. Yoneta, T. Kubo, H. Kato, K. Yoshino, M. Ohishi, H. Saito, K. Ohmori: Phys. Status Solidi B **229**, 291 (2002)

35.44 K. Lott, O. Volobujeva, A. Öpik, T. Nirk, L. Türn, M. Noges: Phys. Status Solidi C **0**, 618 (2003)

35.45 J. F. Wang, D. Masugata, C. B. Oh, A. Omino, S. Seto, M. Isshikim: Phys. Status Solidi A **193**, 251 (2002)

35.46 U. W. Pohl, J. Gottfriedsen, H. Schumann: J. Cryst. Growth **209**, 683 (2000)

35.47 M. U. Ahmed, S. J. C. Irvine: J. Electron. Mater. **29**, 169 (2000)

35.48 P. J. Dean, W. Stutius, G. F. Neumark, B. J. Fitzpatrick, R. N. Bhargava: Phys. Rev. B **27**, 2419 (1983)

35.49 P. Prete, N. Lovergine: Prog. Cryst. Growth Char. Mater. **44**, 1 (2002)

35.50 Y. Fujita, T. Terada, T. Suzuki: Jpn. J. Appl. Phys. **34**, L1034 (1995)

35.51 C. M. Rouleau, D. H. Lowndes, G. W. McCamy, J. D. Budai, D. B. Poker, D. B. Geohegan, A. A. Puretzky, S. Zhu: Appl. Phys. Lett. **67**, 2545 (1995)

35.52 T. Baron, K. Saminadayar, N. Magnea: J. Appl. Phys. **83**, 1354 (1998)

35.53 S. Takagi: Acta Crystallogr. **15**, 1311 (1962)

35.54 D. Taupin: Bull. Soc. Franc. Miner. Crystallogr. **88**, 469 (1964)

35.55 M. A. G. Halliwell, M. H. Lyons, M. J. Hill: J. Cryst. Growth **68**, 523 (1984)

35.56 Y. Gu, I. L. Kuskovsky, M. van der Voort, G. F. Neumark, X. Zhou, M. C. Tamargo: Phys. Rev. B **71**, 045340 (2005)

35.57 W. Lin, B. S. Yang, S. P. Guo, A. Elmoumni, F. Fernandez, M. C. Tamargo: Appl. Phys. Lett. **75**, 2608 (1999)

35.58 N. J. Duddles, K. A. Dhese, P. Devine, D. E. Ashenford, C. G. Scott, J. E. Nicholls, J. E. Lunn: J. Appl. Phys. **76**, 5214 (1994)

35.59 S. W. Lim, T. Honda, F. Koyama, K. Iga, K. Inoue, K. Yanashima, H. Munekata, H. Kukimoto: Appl. Phys. Lett. **65**, 2437 (1994)

35.60 G. F. Neumark, R. M. Park, J. M. Depudyt: Phys. Today **47 (6)**, 26 (1994)

35.61 M. Strassburg, O. Schulz, U. W. Pohl, D. Bimberg, S. Itoh, K. Nakano, A. Ishibashi, M. Klude, D. Hommel: IEEE J. Sel. Top. Quant. Electron. **7**, 371 (2001)

35.62 K. W. Kwak, R. D. King-Smith, D. Vanderbilt: Physica B **185**, 154 (1993)

35.63 B.-H. Cheong, C. H. Park, K. J. Chang: Phys. Rev. B **51**, 10610 (1995)

35.64 W. Faschinger, S. Gundel, J. Nürnberger, D. Albert: *Proc. Conf. Optoelectronic and Microelectronic Materials and Devices* (IEEE, Piscataway 2000) p. 41

35.65 S. Gundel, W. Faschinger: Phys. Rev. B **65**, 035208 (2001)

35.66 I. L. Kuskovsky, G. F. Neumark, J. G. Tischler, B. A. Weinstein: Phys. Rev. B **63**, 161201 (2001)

35.67 P. Desgardin, J. Oila, K. Sarrnen, P. Hautojärvi, E. Tournié, J.-P. Faurie, C. Corbel: Phys. Rev. B **62**, 15711 (2000)

35.68 S. Tomiya, S. Kijima, H. Okuyama, H. Tsukamoto, T. Hino, S. Taniguchi, H. Noguchi, E. Kato, A. Ishibashi: J. Appl. Phys. **86**, 3616 (1999)

35.69 F. El. Akkad: Semicond. Sci. Technol. **2**, 629 (1987)

35.70 A. Kamata, H. Yoshida: Jpn. J. Appl. Phys. (Pt. 2) **135**, L87 (1996)

35.71 S. E. Grillo, M. Ducarrori, M. Nadal, E. Tournié, J.-P. Faurie: J. Appl. Phys. D: Appl. Phys. **35**, 3015 (2002)

35.72 C. Chauvet, E. Tournié, J.-P. Faurie: Phys. Rev. B **61**, 5332 (2000)

35.73 V. Bousquet, E. Tournié, M. Laügt, P. Venéguès, J.-P. Faurie: Appl. Phys. Lett. **70**, 3564 (1997)

35.74 J. P. Faurie, V. Bousquet, P. Brunet, E. Tournié: J. Cryst. Growth **184/185**, 11 (1998)

35.75 M. Malinski, L. Bychto, S. Legowski, J. Szatkowski, J. Zakrzewski: Microelectron. J. **32**, 903 (2001)

35.76 S. P. Guo, W. Lin, X. Zhou, M. C. Tamargo, C. Tian, I. L. Kuskovsky, G. F. Neumark: J. Appl. Phys. **90**, 1725 (2001)

35.77 D. C. Look, G. M. Renlund, R. H. Burgener II, J. R. Sizelove: Appl. Phys. Lett. **85**, 5268 (2004)

35.78 Y. R. Ryu, T. S. Lee, H. W. White: Appl. Phys. Lett. **83**, 87 (2003)

35.79 K.-K. Kim, H. S. Kim, D.-K. Hwang, J.-H. Lim, S.-J. Park: Appl. Phys. Lett. **83**, 63 (2003)

35.80 J. M. Bian, X. M. Li, C. Y. Zhang, W. D. Yu, X. D. Gao: Appl. Phys. Lett. **85**, 4070 (2004)

35.81 B. S. Li, Y. C. Liu, Z. Z. Zhi, D. Z. Shen, Y. M. Lu, J. Y. Zhang, X. W. Fan, R. X. Mu, D. O. Henderson: J. Mater. Res. **18**, 8 (2003)

35.82 C. C. Lin, S. Y. Shen, S. Y. Cheng, H. Y. Li: Appl. Phys. Lett. **84**, 5040 (2004)

35.83 W. Xu, Z. Ye, T. Zhou, B. Zhao, L. Zhu, J. Huang: J. Cryst. Growth **265**, 133 (2004)

35.84 X. Li, Y. Yan, T. A. Gessert, C. L. Perkins, D. Young, C. DeHart, M. Young, T. J. Coutts: J. Vac. Sci. Technol. A **21**, 1342 (2003)

35.85 J. G. Lu, Z. Z. Ye, F. Zhuge, Y. J. Zeng, B. H. Zhao, L. P. Zhu: Appl. Phys. Lett. **85**, 3134 (2004)

35.86 J. M. Bian, X. M. Li, X. D. Gao, W. D. Yu, L. D. Chen: Appl. Phys. Lett. **84**, 541 (2004)

35.87 Y. W. Heo, K. Ip, S. J. Park, S. J. Peaton, D. P. Norton: Appl. Phys. A **78**, 53 (2004)

35.88 A. B. Djurišić, Y. Chan, E. H. Li: Mater. Sci. Eng. R **38**, 237 (2002)

35.89 D. C. Look, J. W. Hemsky, J. R. Sizelove: Phys. Rev. Lett. **82**, 2552 (1999)

35.90 S. B. Zhang, S.-H. Wei, A. Zunger: Phys. Rev. B **63**, 075205 (2001)

35.91 C. G. Van de Walle: Phys. Rev. Lett. **85**, 1012 (2000)

35.92 D. M. Hofmann, A. Hofstaetter, F. Leiter, H. Zhou, F. Henecker, B. K. Meyer, S. B. Orlinskii, J. Schmidt, P. G. Baranov: Phys. Rev. Lett. **88**, 045504 (2002)

35.93 E. V. Lavrov, J. Weber, F. Börnert, C. G. Van de Walle, R. Helbig: Phys. Rev. B **66**, 165205 (2001)

35.94 M. D. McCluskey, S. J. Jokela, K. K. Zhuravlev, P. J. Simpson, K. G. Lynn: Appl. Phys. Lett. **81**, 3807 (2002)

35.95 D. C. Look, R. L. Jones, J. R. Sizelove, N. Y. Garces, N. C. Giles, L. E. Halliburton: Phys. Status Solidi A **195**, 171 (2003)

35.96 B. K. Meyer, H. Alves, D. M. Hofmann, W. Kriegseis, D. Forster, F. Bertram, J. Christen, A. Hoffmann, M. Straßburg, M. Dworzak, U. Haboeck, A. V. Rodina: Phys. Status Solidi B **241**, 231 (2004)

35.97 F. A. Kroger, H. J. Vink: J. Chem. Phys. **22**, 250 (1954)

35.98 K. Vanheusden, C. H. Seager, W. L. Warren, D. R. Tallent, J. A. Voigt: Appl. Phys. Lett. **68**, 403 (1996)

35.99 K. Vanheusden, W. L. Warren, C. H. Seager, D. R. Tallent, J. A. Voight: J. Appl. Phys. **79**, 7983 (1996)

35.100 C. Morhain, M. Teisseire-Doninelli, S. Vézian, C. Deparis, P. Lorenzini, F. Raymond, J. Guion, G. Neu: Phys. Status Solidi B **241**, 631 (2004)

35.101 H. J. Ko, Y. F. Chen, S. K. Hong, H. Wenisch, T. Yao, D. C. Look: Appl. Phys. Lett. **77**, 3761 (2000)

35.102 D. C. Look, B. Claflin, Ya. I. Alivov, S. J. Park: Phys. Stat. Sol. A **201**, 2203 (2004)

35.103 C. H. Park, S. B. Zhang, S.-H. Wei: Phys. Rev. B **66**, 073202 (2002)

35.104 S. Orlinskii, J. Schmdit, P. G. Baranov, D. M. Hofmann, C. de M. Donegá, A. Meijerink: Phys. Rev. Lett. **92**, 047603 (2004)

35.105 H.-J. Ko, Y. Chen, S.-K. Hong, T. Yao: J. Cryst. Growth **251**, 628 (2003)

35.106 L. G. Wang, A. Zunger: Phys. Rev. Lett. **90**, 256401 (2003)

35.107 M. Sumiya, A. Tsukazaki, S. Fuke, A. Ohtomo, H. Koinuma, M. Kawasaki: Appl. Surf. Sci. **223**, 206 (2004)

35.108 S. Limpijumnong, S. B. Zhang, S.-H. Wei, C. H. Park: Phys. Rev. Lett. **92**, 155504 (2004)

35.109 C. Morhain, M. Teisseire, S. Vézian, F. Vigué, F. Raymond, P. Lorenzini, J. Guion, G. Neu, J.-P. Faurie: Phys. Status Solidi B **229**, 881 (2002)

35.110 D. C. Look, B. Claflin, Ya. I. Alivov, S. J. Park: Phys. Status Solidi A **201**, 2203 (2004)

35.111 E.-C. Lee, K. J. Chang: Phys. Rev. B **70**, 115210 (2004)

35.112 G. F. Neumark: Phys. Rev. Lett. **62**, 1800 (1989)

35.113 G. Blattner, C. Klingshirn, R. Helbig, R. Meinl: Phys. Status Solidi B **107**, 105 (1981)

35.114 C. Klingshirn, W. Maier, G. Blatter, P. J. Dean, G. Klobbe: J. Cryst. Growth **59**, 352 (1982)

35.115 J. Gutowski, N. Presser, I. Broser: Phys. Rev. B **38**, 9746 (1988)

35.116 D. Block, A. Hervé, R. T. Cox: Phys. Rev. B **25**, 6049 (1982)

35.117 C. Gonzalez, D. Block, R. T. Cox, A. Hervé: J. Cryst. Growth **59**, 357 (1982)

35.118 X. Liu, X. Wu, H. Cao, R. P. H. Chang: J. Appl. Phys. **95**, 3141 (2004)

35.119 B. Lin, Z. Fu, Y. Jia: Appl. Phys. Lett. **79**, 943 (2001)

35.120 D. Banejee, J. Y. Lao, D. Z. Wang, J. Y. Huang, Z. F. Ren, D. Steeves, B. Kimball, M. Sennett: Appl. Phys. Lett. **83**, 2061 (2003)

35.121 T.-B. Hur, G. S. Jeen, Y.-H. Hwang, H.-K. Kim: J. Appl. Phys. **94**, 5787 (2003)

35.122 P. H. Hasai: Phys. Rev. **130**, 989 (1963)

35.123 M. Liu, A. H. Kitai, P. Mascher: J. Lumin. **54**, 35 (1992)

35.124 D. C. Reynolds, S. C. Look, B. Jogai, H. Morkoc: Solid State Commun. **101**, 643 (1997)

35.125 D. C. Reynolds, S. C. Look, B. Jogai: J. Appl. Phys. **89**, 6189 (2001)

35.126 N. Y. Garces, L. Wang, L. Bai, N. C. Giles, I. E. Halliburton, G. Cantwell: Appl. Phys. Lett. **81**, 622 (2002)

35.127 D. Li, H. Leung, A. B. Djurišić, Z. T. Liu, M. H. Xie, S. L. Shi, S. J. Xu, W. K. Chan: Appl. Phys. Lett. **85**, 1601 (2004)

35.128 Y. Gu, I. L. Kuskovsky, M. Yin, S. O'Brien, G. F. Neumark: Appl. Phys. Lett. **85**, 3833 (2004)

35.129 H. S. Kang, J. S. Kang, J. W. Kim, S. Y. Lee: Phys. Status Solidi C **1**, 2550 (2004)

35.130 Y. Harada, S. Hashimoto: Phys. Rev. B **68**, 045421 (2003)

36. II–VI Narrow-Bandgap Semiconductors for Optoelectronics

The field of narrow-gap II–VI materials is dominated by the compound semiconductor mercury cadmium telluride, ($Hg_{1-x}Cd_xTe$ or MCT), which supports a large industry in infrared detectors, cameras and infrared systems. It is probably true to say that HgCdTe is the third most studied semiconductor after silicon and gallium arsenide. $Hg_{1-x}Cd_xTe$ is the material most widely used in high-performance infrared detectors at present. By changing the composition x the spectral response of the detector can be made to cover the range from 1 μm to beyond 17 μm. The advantages of this system arise from a number of features, notably: close lattice matching, high optical absorption coefficient, low carrier generation rate, high electron mobility and readily available doping techniques. These advantages mean that very sensitive infrared detectors can be produced at relatively high operating temperatures. $Hg_{1-x}Cd_xTe$ multilayers can be readily grown in vapor-phase epitaxial processes. This provides the device engineer with complex doping and composition profiles that can be used to further enhance the electro-optic performance, leading to low-cost, large-area detectors in the future. The main purpose of this chapter is to describe the applications, device physics and technology of II–VI narrow-bandgap devices, focusing on HgCdTe but also including $Hg_{1-x}Mn_xTe$ and $Hg_{1-x}Zn_xTe$. It concludes with a review of the research and development programs into third-generation infrared detector technology (so-called GEN III detectors) being performed in centers around the world.

36.0.1 Historical Perspective and Early Detectors 856
36.0.2 Introduction to HgCdTe 857
36.0.3 Introduction to Device Types 857

36.1 **Applications and Sensor Design** 858
36.2 **Photoconductive Detectors in HgCdTe and Related Alloys** 860
    36.2.1 Introduction to the Technology of Photoconductor Arrays 860
    36.2.2 Theoretical Fundamentals for Long-Wavelength Arrays 861
    36.2.3 Special Case of Medium-Wavelength Arrays.... 863
    36.2.4 Nonequilibrium Effects in Photoconductors 863
36.3 **SPRITE Detectors** 864
36.4 **Photoconductive Detectors in Closely Related Alloys** 866
36.5 **Conclusions on Photoconductive HgCdTe Detectors** ... 867
36.6 **Photovoltaic Devices in HgCdTe** 867
    36.6.1 Ideal Photovoltaic Devices 868
    36.6.2 Nonideal Behavior in HgCdTe Diodes....................... 869
    36.6.3 Theoretical Foundations of HgCdTe Array Technology 870
    36.6.4 Manufacturing Technology for HgCdTe Arrays 873
    36.6.5 HgCdTe 2-D Arrays for the 3–5 μm (MW) Band 878
    36.6.6 HgCdTe 2-D Arrays for the 8–12 μm (LW) Band 879
    36.6.7 HgCdTe 2-D Arrays for the 1–3 μm (SW) Band........... 879
    36.6.8 Towards "GEN III Detectors"........ 880
    36.6.9 Conclusions and Future Trends for Photovoltaic HgCdTe Arrays.... 882
36.7 **Emission Devices in II–VI Semiconductors** 882
36.8 **Potential for Reduced-Dimensionality HgTe–CdTe** 883
References 883

The main commercial application for narrow-bandgap semiconductors is in infrared radiation detection. There are very few elemental or compound semiconductors with the correct energy gap to sense photons within the infrared spectrum, particularly at longer wavelengths. It is also beneficial to accurately match the spectral sensitivity to certain atmospheric windows, so the ability to tailor the wavelength is very important. By using an alloy of two different compounds with widely separate energy gaps it is possible to synthesize crystals with an intermediate energy gap. II–VI compounds have crystal properties that make them very suitable for mixing and they have a range of bandgaps from near zero for the semimetal Hg compounds to more than 1 eV.

However, amongst the wider bandgap II–VI semiconductors, only Cd, Zn, Mn and Mg have been shown to open up the bandgap of the Hg-based semimetals HgTe and HgSe. By far the most developed alloy system is $Hg_{1-x}Cd_xTe$ (HgCdTe), which is a semiconductor formed from the semimetal HgTe and the wide-bandgap semiconductor CdTe. By adjusting the alloy composition "x", the properties (including the bandgap) can be varied smoothly between HgTe and CdTe.

There has been some interest in developing alternative ternary alloys to replace HgCdTe, since from theoretical considerations the already weak Hg–Te lattice bond is further destabilized by alloying with CdTe and there is potential for obtaining materials with increased hardness and detectors with better temperature stability. Most other compound combinations have crystal growth problems or doping limitations that make them unsuitable for device fabrication, with the possible exception of $Hg_{1-x}Mn_xTe$ and $Hg_{1-x}Zn_xTe$. Photoconductive detectors have been reported using $Hg_{1-x}Mn_xTe$, and photovoltaic detectors using both materials, but devices have not reached the maturity of $Hg_{1-x}Cd_xTe$. Also, the technology reported for these materials is very similar to that of HgCdTe, and so many of the detector design, technology and performance factors described in this chapter for HgCdTe are also relevant to other II–VI alloys.

The main purpose of this chapter is to describe the applications, device physics and technology of II–VI narrow-bandgap devices, focusing on HgCdTe. It concludes with a review of the research and development programs into third-generation infrared detector technology (so-called GEN III detectors) being performed in centers around the world.

36.0.1 Historical Perspective and Early Detectors

Passive thermal imaging is the term used to describe imaging of the natural thermal radiation emitted by all objects around us. The contrast in such images is due to temperature differences and changes in emissivity or spectral radiance of surfaces. The atmosphere is rather inconvenient for infrared imaging and is only transparent in certain wavelength "windows". There is a short-waveband (SW) window between 2.0 and 2.25 µm, a medium-waveband (MW) window between 3.0 and 5.0 µm and a long-waveband (LW) window between 7.5 and 14 µm. The spectral radiance curve for bodies around room temperature shows a peak that best matches the LW window and this is the preferred wavelength of operation for thermal imaging detectors. At first sight the MW window looks compromised by the photon flux, which is some two orders lower than the LW window. However, it is much easier to make detectors for this wavelength, and the photon flux disadvantage can be offset by using staring architectures that enable a longer integration (or stare) time, and so this band is very widely used. The SW window is rarely used for passive thermal imaging because of low flux, but active imaging using a source such as starlight or an infrared laser is of growing interest.

The first infrared photon detectors, based on thallous sulfide, were developed in the USA during World War I and were sensitive to about 1.4 µm. The next important developments occurred before and during World War II in Germany, with work on thin-film polycrystalline PbS devices with a response up to 2.5 µm, later extended further to the 3–5 µm region using PbSe and PbTe. During the 1950s and 1960s indium antimonide (InSb) detectors emerged that were capable of detecting wavelengths up to 5.5 µm, and during the same period impurity photoconductivity was studied in doped germanium and doped silicon. Germanium was favored at this time because it was available with fewer compensating impurities and gave better detector performance. Ge : Hg, with an impurity activation energy of 0.09 eV, provided the first practical detector to be used in real-time thermal imaging, employing linear arrays for the 8–13 µm region, although it required cooling to below 30 K. Because of the stringent cooling requirements, extrinsic silicon devices have not found favor for terrestrial applications.

The next important phase of semiconductor infrared detector research took place during the late 1960s and 1970s, when research efforts were directed towards an intrinsic detector for the 8–13 µm band that would operate more conveniently at around 80 K. Two alloy semiconductors were developed, $Hg_{1-x}Cd_xTe$ (HgCdTe) and $Pb_{1-x}Sn_xTe$ (LTT). The first report of the synthesis of the semimetal HgTe and the wide-bandgap semiconductor CdTe to form the semiconductor alloy HgCdTe was published by the Royal Radar Establishment in Malvern, UK [36.1]. This landmark paper reported both photoconductive as well as photovoltaic response at wavelengths extending out to 12 μm, and pointed out that this new alloy semiconductor showed promise for intrinsic infrared detectors. Soon after many centers around the world switched detector development for major thermal-imaging programs to the HgCdTe system. High-performance LW and MW linear arrays were produced for the first generation of thermal imaging equipment developed and manufactured in the 1970s through to the present. Many tens of thousands of detectors have been delivered.

36.0.2 Introduction to HgCdTe

$Hg_{1-x}Cd_xTe$ is the material most widely used for high-performance infrared detectors at present. By changing the composition x, the detector spectral response can be made to cover the range from 1 μm to beyond 17 μm. The advantages of this system arise from a number of features, notably: close lattice matching, high optical absorption coefficient, low carrier generation rate, high electron mobility and readily available doping techniques. These advantages mean that very sensitive infrared detectors can be produced at relatively high operating temperatures. HgCdTe continues to be developed as the material of choice for high-performance long-wavelength (8–12 μm) arrays and has an established market at the medium- (3–5 μm) and short-wavelength (1–3 μm) ranges.

In the LW band, the main competitive technologies are $Pb_{1-x}Sn_xTe$ (LTT) and multiple quantum well (MQW) detectors (usually using AlGaAs/GaAs technology). Work on LTT largely stopped in about the mid 1970s partly because the large dielectric constant made them unsuitable for photoconductors. However, there is more recent interest in large arrays of LTT photodiodes due to a potential cost advantage, but at present poor diode quality and excess noise makes them inferior to $Hg_{1-x}Cd_xTe$ on grounds of sensitivity alone. MQW arrays are essentially tuned to a wavelength, say 8.0 μm, and need deeper cooling to suppress thermal leakage currents. The imaging performance can be good due to the absence of low-frequency noise sources but they have a much lower ultimate sensitivity than HgCdTe.

36.0.3 Introduction to Device Types

HgCdTe Photoconductive Arrays
A photoconductor usually comprises a small slab or element of material with two contacts. The aim is to detect the change in resistance of the element when the photon flux is changed. The first HgCdTe devices were photoconductive because of the simplicity of the technology, and the relative ease of achieving near-ideal infrared performance and excellent reliability. HgCdTe photoconductive detectors have been in routine production since the early 1980s and are often called first-generation detectors. They are the key component of the US Common Module Thermal Imager and, in the form of the SPRITE detector, they feature in the UK Class II Common Module imager. Detectors for thermal imaging are most commonly fabricated with a peak response in the 8–12 μm atmospheric window region and are cooled to 80 K by means of a Joule–Thompson expansion cooler or, more recently, a Stirling engine refrigerator. They are also quite commonly made for operation in the 3–5 μm atmospheric window, and either use cooling to 80 K or employ Peltier (or thermoelectric) coolers to cool to around 200 K. Peltier-cooled MW detectors are used in many small handheld cameras and a whole host of heat sensing applications. The reliable performance, low levels of defects and easily understood physics has led to a long product life for photoconductive arrays. However the array size is limited and first-generation thermal imaging systems need to employ complex optics to scan the infrared image over the array to build up a scene. The main limitation to developing larger arrays arises from the difficulty involved with amplifying and multiplexing elements electronically on the focal plane. In consequence, there needs to be a separate electrical connection to each photoconductor so that they can be connected to low noise current amplifiers outside of the cryogenic encapsulation. For larger arrays of, say, more than several hundred elements the cryogenic encapsulations become cumbersome and expensive. Also, the power consumption can become a problem with too many elements, so this and cryogenics set a practical limit to the size of photoconductor arrays for commercial thermal imaging to a few hundred elements.

HgCdTe Photovoltaic Arrays
A photovoltaic device is essentially a light-sensitive diode. Photons absorbed in the semiconductor create electron–hole pairs, and the minority carriers diffuse to the p–n junction where they are "separated" and the voltage across the junction changes. Because there is an integration of carriers, the signal can be built up over a time called the integration time, and there is the potential for larger signals and better infrared sensitivity than can be achieved with photoconductors.

In the mid-1970s attention turned to the use of photovoltaic HgCdTe for thermal imaging applications. At that time it was seen that, in the future, many infrared applications would need higher radiometric performance and/or higher spatial resolution than could be achieved with first-generation photoconductive infrared detectors. Photovoltaic arrays consume very little power, and can be easily multiplexed using an on-focal plane silicon chip, so they are well-suited to long linear and large, two-dimensional infrared arrays. Systems based upon such focal planes can be made smaller and lighter, with lower power consumption, and can result in much higher performance than systems based on first-generation detectors. Photovoltaic detectors can also have less low frequency noise, faster response time, and can avoid the need for complex scanning optics. To some extent these advantages are offset by the more complex processes needed to fabricate photovoltaic detectors, and so their development and industrialization have been slower, particularly for large arrays. Another point is that, unlike photoconductive detectors, the field of photovoltaic HgCdTe arrays shows a large variety of different material growth methods and device structures, often unique to individual companies and research organizations. The collective wealth of data on photovoltaic devices is therefore spread over a wide field.

Large two-dimensional and long linear arrays have the common feature that they are all mass-connected to a custom-designed silicon integrated circuit called a multiplexer or ROIC (for readout integrated circuit). The multiplexer performs the function of integrating the infrared signal and scanning the array. The evolution of large arrays was delayed by the slow emergence of silicon integrated circuits large enough to perform this function economically. Another economic factor was matching the cost of first-generation infrared cameras which had a well-established market and had set a benchmark for display picture points and cost. Cost depends strongly on array size and infrared array pixels tend to be much larger than those commonly found in visible imaging arrays. Infrared arrays then tend to be physically large and, together with the extra complexity of the manufacturing process, relatively difficult to cost-reduce. The commercial viability of second-generation focal plane arrays has depended on matching the spatial resolution (number of elements in the array) and cost of existing first-generation systems, which was only really achieved in the 1990s.

HgCdTe Metal–Insulator–Semiconductor (MIS) Arrays

The MIS HgCdTe detector operates much like a silicon charge-coupled device (CCD). The MIS detector was the basic element in a family of "monolithic" HgCdTe arrays in which the detection, integration and multiplexing functions were all done within the HgCdTe material itself. Unlike the PV and PC detectors, the MIS device operates under strongly nonequilibrium conditions, with large electric fields in the deep depletion regions. This makes the MIS detector much more sensitive to material and process defects than the PV and PC detectors [36.2]. This sensitivity, particularly acute for LW devices, caused the monolithic approach to be abandoned in favor of various hybrid approaches where the IR photon detection is performed in HgCdTe, and the signal processing is restricted to the silicon multiplexer.

36.1 Applications and Sensor Design

Currently the most important market for HgCdTe arrays is in thermal imaging in the long (8–12 μm) and medium wavebands (3–5 μm). Often the detector is the performance-limiting component in the system, and it is necessary to use detectors with a sensitivity limited only by the random rate of arrival of photons from the scene (so-called background limited or BLIP detectors). In a narrow-gap semiconductor using both photoconductive and photovoltaic detectors, it is necessary to cool so that the thermal generation and associated excess noise are suppressed and the sensitivity becomes "BLIP-limited". The choice of cooler is therefore a key technology in an infrared detector, and the common options are described here. One of the points to emphasize is that the focal plane array is only one of a number of components that need to be optimized to preserve the BLIP performance. The optical design, cryogenics and signal processing are critical to maintaining the sensitivity of the focal plane array. The precision engineering involved in designing infrared detectors and coolers, particularly for survival in high shock and vibration environments, is still challenging.

The means of cooling depends on the detector type and application. For instance, photoconductive arrays often need a lot of cooling power because of the Joule heating on the focal plane and the need for many bond wires that add to the thermal load. Two-dimensional arrays, on the other hand, often consume little power

Fig. 36.1 Typical fast cooldown detector using Joule–Thompson cooling. (Courtesy of SELEX Infrared)

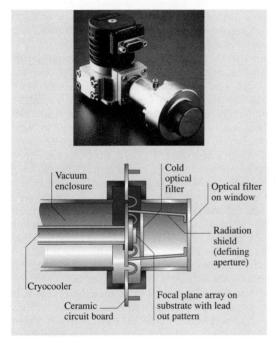

Fig. 36.2 Typical integrated detector: cooler assembly (IDCA). (Courtesy of SELEX Infrared)

Fig. 36.3 Typical thermoelectrically cooled detector. (Courtesy of SELEX Infrared)

and have few connections, allowing alternative cooling methods. Other considerations are cooldown time, power, weight, space-volume, robustness and life.

Historically the Joule–Thompson cooler was the first available method of cooling, relying on compressed gas bottles for power. The cooling power can be very high. Most first-generation detectors use Joule–Thompson cooling. In more recent times the high cooling power has been routinely exploited in fast-cooldown applications. Figure 36.1 illustrates a typical fast-cooldown product, which can be cooled in a few seconds to liquid argon temperature (92 K). HgCdTe arrays are well-suited to this type of application because, unlike other materials such as indium antimonide or multiple quantum wells, BLIP operation is easily achieved at relatively high temperatures.

For most modern thermal imaging applications the normal means of cooling is a cryocooler based on the Stirling cycle, which can achieve temperatures as low as 55 K but is normally regulated at a temperature in the range 80–140 K. Figure 36.2 shows a typical integrated detector–cooler assembly (IDCA). The cutaway in Fig. 36.2 shows the front end of the detector in more detail. The infrared array is bonded directly on the cold finger of the cooler. To ensure only photons from the scene reach the detector, a cold stop or radiation shield is bonded to the cold finger. The cold stop will have internal baffles and special coatings to stop internal reflections and is a vital component for achieving BLIP operation of the array. The detector will have an optical pass-band filter that controls the upper and lower wavelengths. This will vary with the application, but typical bands are 3.2 to 4.2 µm, 3.7 to 4.9 µm or 7.7 to 10.6 µm. Two methods are commonly used. The filter can be incorporated in the cold stop, and is therefore cooled, producing a very efficient block to stray radiation, say from the dewar itself. Almost as good performance can be achieved with filters on the front window, and a well-designed cold stop.

The third common means of cooling is Peltier cooling (usually based on bismuth telluride devices). A typical detector is illustrated in Fig. 36.3. State of the art coolers can achieve temperatures as low as 180 K, but in general they are not as efficient as Stirling cryocoolers, and the cooling performance can be poor in high ambient temperatures. Nevertheless the main strengths of long-term reliability, greater shock resistance and smaller weight and space-volume has created a large volume market, especially for small photoconductor arrays. The temperature of 180 K is not quite cold enough for BLIP thermal imaging, and the applications tend to be those that can tolerate lower performance infrared detectors.

36.2 Photoconductive Detectors in HgCdTe and Related Alloys

The physics and technology of HgCdTe photoconductive detectors are described here. It also serves as an introduction to SPRITE detectors described in Sect. 36.3 and related work on other II–VI compounds summarized in Sect. 36.4.

Being quite a simple device, the photoconductor provides a convenient introduction to the physics and technology of infrared detectors made from HgCdTe. Figure 36.4 shows a photomicrograph and scanning electron microscope view of a typical photoconductive element. It comprises a small slab of material, typically 50 μm square and 8 μm thick with two ohmic contacts. The element illustrated has been "labyrinthed" to improve its performance. The device is operated using a constant current bias, supplied by means of a voltage source, and a series resistance, which is large compared to the detector resistance. The signal is measured as a voltage change across the detector, so that the device is essentially a radiation-sensitive resistor.

36.2.1 Introduction to the Technology of Photoconductor Arrays

HgCdTe photoconductors were developed in the early 1970s when only bulk-grown material (from a Bridgman process or a solid state recrystallization process) was available. The simple device geometry allowed arrays to be manufactured by chemomechanical polishing of thick slices of material. Bulk material is still used because of the economic and convenient processes involved, but many workers [36.3–5] report arrays produced from LPE and MOVPE material. Because the electron mobility is many orders higher than the hole mobility in narrow-bandgap HgCdTe, it is advantageous to work with n-type material where the photoconductive gain is very high. n-type material can be manufactured directly, using the Bridgman growth process, or can be produced by two-temperature annealing. Chapter 14 describes the range of growth processes for HgCdTe. For photoconductors there is a requirement for a long minority carrier lifetime, which implies low levels of surface and bulk recombination centers. n-type material with a low doping level is ideal, because the fundamental carrier lifetime is more than adequate for most applications, and crystalline defects in n-type material appear to have a weak effect on the carrier lifetime. The quality of surfaces is critical because elements can be as thin as 4 μm. The technique most commonly used is to passivate the surfaces by anodic oxidation. The anodic oxide has a charge state that attracts electrons (accumulated surface) and provides an electric field which repels minority carriers, preventing them from interacting with surface recombination sites. The so-called surface recombination velocities are as low as 50 cm/s [36.6]. The individual elements are defined using a photomask and ion beam milling. As described later, ion beam milling results in a heavily doped n-type skin layer which also produces a low surface recombination velocity. Ion milling is also used to prepare the contact windows. The resulting highly doped n-type layer provides an excellent ohmic contact for electron flow, and a blocking contact to minority carrier holes. This condition enhances the minority carrier lifetime,

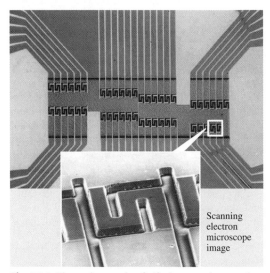

Fig. 36.4 Photomicrograph of 48-element photoconductive array using labyrinthed structures for enhanced responsivity. (Courtesy of SELEX Infrared)

Scanning electron microscope image

which leads to high signals and high sensitivity. As will be seen in the performance analysis, the device geometry and doping levels arising from the technology provide an excellent platform for producing sensitive detectors at relatively moderate cooling.

36.2.2 Theoretical Fundamentals for Long-Wavelength Arrays

Despite its apparent simplicity, a rigorous analysis of photoconductor operation is quite complex [36.7, 8]. Two-dimensional and three-dimensional treatments have been reported [36.9, 10], but the analysis here is restricted to a one-dimensional treatment.

For the purposes of simplicity, the analysis will concentrate on an n-type element, which is the most commonly used device structure. Consider a slab of HgCdTe of length l, width w and thickness t, having "ohmic" contacts formed to the end faces, and exposed to an infrared background flux of ϕ_b photons/cm^2.

Consider an electron–hole pair that is generated by a photon. Holes would drift towards the negative electrode with a velocity $\mu_a E$, where μ_a is the ambipolar mobility given by

$$\mu_a = (n - p)\mu_e \mu_h / (n\mu_e + p\mu_h) \,, \tag{36.1}$$

where n and p are the electron and hole densities, and μ_e and μ_h are the electron and hole mobilities. The ambipolar mobility approximates to the minority carrier mobility in most situations, except when the temperature or the background flux is very high. At low values of the applied field, the average drift length of a minority carrier, $\mu_a E \tau$, is very much less than the detector length l, and recombination occurs predominantly in the bulk of the sample. The minority carrier density is uniform along the length of the sample (neglecting diffusion), and in the presence of background radiation, is given by:

$$p = p_0 + p_b \tag{36.2}$$

where p_0 is the thermal equilibrium concentration of holes, and

$$p_b = \eta \phi_b \tau / t \tag{36.3}$$

is the excess carrier density, in equilibrium with the background radiation, where η is the quantum efficiency.

Equations for the device parameters are given below. In large n-type detectors where the minority carrier drift length, $\mu_a E \tau$, is small compared to l, then the responsivity in units of V/W is given by:

$$R_v = \frac{\eta V_0 \tau}{E_\lambda l w t n} \,, \quad \text{low bias} \tag{36.4}$$

where V_0 is the bias voltage and E_λ is the energy of a photon with wavelength λ.

At higher bias levels, the minority carrier lifetime in weakly doped n-type material can be made long enough to ensure that little recombination takes place before the carrier is swept to the negative contact (the so-called sweepout regime), such that $\mu_a E \tau > l$. The effective lifetime is reduced to a value τ_{eff} and, in the limit, to τ_α, where it is simply the average transit time. For simplicity we will assume that the equilibrium carrier densities are unchanged under bias. At the biases used in practical detectors, blocking contacts influence the carrier densities, and the effects must be taken into account.

The responsivity in sweepout is given by:

$$R_v(\text{max}) = \frac{\eta l}{2 E_\lambda w t n_0 \mu_h} \,, \quad \text{high bias}. \tag{36.5}$$

Taking $\eta = 0.6$, $l = w = 50\,\mu\text{m}$, $t = 8\,\mu\text{m}$, $n_0 = 5 \times 10^{14}\,\text{cm}^{-3}$, $E_\lambda = 0.1\,\text{eV}$ and $\mu_h = 450\,\text{cm}^2/\text{V s}$, a value

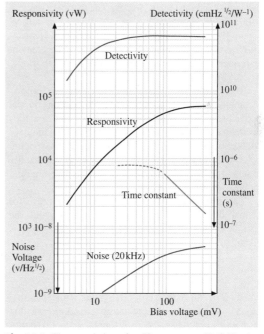

Fig. 36.5 Characteristics of a 50 μm-square HgCdTe detector, operated at 80 K, as a function of bias voltage. (After SELEX Infrared)

for the maximum responsivity of $\approx 1 \times 10^5$ V/W is obtained. The responsivity observed in practical devices is lower than the calculated value, typically by a factor of two, because the anodic oxidation technique used to produce low recombination velocity surfaces also enhances the surface conductance, shunting the bulk of the device and lowering its resistance, i.e., n_0 is effectively increased in (36.5). The behavior of a typical device, showing the saturation in responsivity with bias, is shown in Fig. 36.5.

Responsivities in photoconductors tend to be low and there have been a number of novel techniques reported to boost the signal. Several methods concentrate on prolonging the life of minority carriers. One technique [36.11, 12] is to use a remote negative contact and an opaque screen to define the sensitive area so that holes are swept under the screen and contribute to the signal for longer. Another technique [36.13, 14] is to use a low recombination velocity n^+n contact to increase the responsivity by a factor of five by blocking the recombination of holes. Even larger increases have been obtained [36.15] by using a heterojunction contact, i.e., an epitaxially grown double layer is employed with a high x region of n-type HgCdTe between the active region and the metal contacts. The theory of devices with blocking contacts has been published in several papers [36.14–16]. Significant increases in responsivity have also been obtained by increasing the element resistance by defining a "meander path" or "labyrinth" device (Fig. 36.4).

The principal noise sources are generation–recombination (g–r) noise due to fluctuations in the optically or thermally generated free-carrier densities and Johnson–Nyquist noise, associated with the finite resistance of the devices. In addition, noise with a spectral power density varying with frequency as $1/f$ is usually observed, which has its origin in surface trapping and mobility modulation effects.

The equation for root mean square g–r noise for a small device in strong sweep-out, and in which the background-induced carrier density $p_b \tau_a / 2\tau \gg p_0$, is given by:

$$V_{g-r} = \frac{1}{n_0 \mu_h t} \left(\frac{l^3 \eta \phi_b B}{w} \right)^{1/2}, \quad (36.6)$$

where B is the bandwidth. Using the same parameters as used above, and assuming a background flux of 5×10^{16} cm^{-2}/s ($f/2$ cold shield and 300 K ambient temperature), the value for V_{g-r} in saturation is 5×10^{-9} V/Hz$^{1/2}$. This corresponds to a noise-equivalent resistance of approximately 1.5 kΩ, and since the actual resistance of the device is $\approx 100\,\Omega$ the design of low-noise preamplifiers is relatively straightforward. A plot of the g–r noise for a practical device, with a peak response wavelength of 12 μm, exposed to 300 K background radiation in an $f/2$ field of view, is shown in Fig. 36.5. The saturation of the noise at high bias is clearly demonstrated.

A necessary condition for background-limited operation is that $p_b \gg p_0$ or

$$\frac{\eta \phi_b \tau}{t} \gg \frac{n_i^2}{n_0}, \quad (36.7)$$

where n_i is the intrinsic carrier density, which is approximately 10^{13} cm^{-3}. The lifetime is itself a function of the background flux, but in reduced field of view, values in excess of 2 μs are observed. Putting $\eta = 0.6$, $t = 10$ μm and $n_0 = 5 \times 10^{14}$ cm^{-3}, we see that background-limited operation should be observed when $\phi_b > 1 \times 10^{15}$ cm^{-2}/s, i.e., a field of view greater than about 5°, and for a background scene temperature of 300 K.

It is generally the case [36.17–19] that the lifetime in good-quality n-type HgCdTe, with a cut-off wavelength near 10 μm at 80 K, is determined by Auger 1 recombination, except possibly at very low donor concentrations near 1×10^{14} cm^{-3}, where Shockley–Read recombination may be significant. Auger 1 is essentially impact ionization by electrons in the high-energy tail of the Fermi–Dirac distribution, which have energies greater than the band gap E_g. An expression for Auger lifetime is:

$$\tau = 2\tau_{Ai1} \frac{n_i^2}{n(n+p)}, \quad (36.8)$$

where τ_{Ai1} is the intrinsic Auger 1 lifetime, which has a value of approximately 1×10^{-3} s.

The sensitivity of a detector is often described by the parameter called the detectivity. The detectivity (or D^*) is a signal-to-noise parameter normalized for area and bandwidth, and is used to compare photoconductor performance. Dividing the responsivity by the g–r noise expression gives

$$D_\lambda^* = \frac{\eta^{1/2}}{2E_\lambda} \left(\frac{1}{\phi_b} \right). \quad (36.9)$$

The limiting D^* in zero background flux is given by

$$D_\lambda^* = \frac{\eta}{2E_\lambda} \left(\frac{2\tau_{Ai1}}{n_0 t} \right)^{1/2}. \quad (36.10)$$

Using the values for the parameters as taken earlier

$$D_\lambda^* = 1 \times 10^{12} \text{ cmHz}^{1/2}/\text{W} \quad (36.11)$$

and experimental results close to this value have been reported [36.19].

In the high-bias sweep-out condition, photoconductive detectors can have a frequency response of several MHz, the response time being determined by the transit time of the holes between the electrodes, $l^2/\mu_h V_0$, rather than the excess carrier lifetime. The effect of bias on time constant is shown for a practical device in Fig. 36.5.

Detectors for space applications, with cut-off wavelengths of between 15 and 16 μm, have been reported with D^* of $3-4 \times 10^{11}$ cm Hz$^{1/2}$/W, when operated at 60 K in a background flux of 2×10^{15} photons/cm² s [36.18].

36.2.3 Special Case of Medium-Wavelength Arrays

Detectors operating in the 3–5 μm range, with an aperture of $f/1$ or faster, normally have near-BLIP performance. Figure 36.6 shows the D_λ^* versus cut-off wavelength for n-type photoconductive detectors. The theoretical maxima include Auger generation and radiative generation only. The experimental points are for 230 μm square n-type detectors, except where indicated. Small detectors in very low background flux conditions are expected to be Johnson noise-limited due to sweep-out. The limiting value of detectivity for n-type detectors in zero background flux [36.20, 21] is

$$D_\lambda^* = \frac{\eta l}{4 E_\lambda t^{1/2}} \left(\frac{q}{kT}\right)^{1/2} \left(\frac{b}{\mu_h n}\right)^{1/2}$$
$$\cong 1.7 \times 10^{19} \left(\frac{b}{\mu_h n}\right)^{1/2}, \qquad (36.12)$$

where the numerical value is obtained assuming $l = w = 100$ μm, $t = 10$ μm, $\eta = 0.7$ and $E_\lambda = 0.25$ eV.

36.2.4 Nonequilibrium Effects in Photoconductors

So far, the modeling of photoconductive devices has assumed equilibrium conditions for carrier densities. In reality, the application of bias fields results in departures from thermal equilibrium. An example occurs at the positive contact, where the highly doped buffer layer results in negligible hole emission (so-called excluding contact). When the bias is applied, no holes can be in-

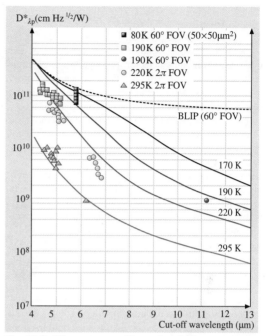

Fig. 36.6 D_λ^* versus cut-off wavelength for n-type photoconductive detectors. *Solid square* 80 K 60° FOV (50 × 50 μm²); *open square* 190 K 80° FOV; *solid circle* 190 K 60° FOV; *open circle* 220 K 2π FOV; *open triangle* 295 K 2π FOV; *solid line* Theoretical maximum in equilibrium (Auger 1, Auger 7 and radiative generation); *dotted line* BLIP (60° FOV, 30% reflection loss). (After SELEX Infrared, with permission)

jected into the material to sustain the flow of holes in the bulk away from the contact. As a consequence, p is reduced for some distance away from the contact. In order to maintain near space-charge neutrality, n also falls to a value close to the extrinsic value. The dominant thermal generation process in n-type material is Auger 1. For the example shown, n is reduced by a factor of about thirty, which would reduce the Auger 1 generation rate by the same factor, and increase the detectivity by a factor of more than five if other noise sources, such as Shockley–Read, were low. In devices with noninjecting contacts this effect results in a significant operating temperature advantage.

36.3 SPRITE Detectors

The SPRITE detector was invented in the early 1970s [36.20, 22, 23] and it forms the basis of the UK common module detector, produced in large numbers. A thorough description of the device is reported in [36.24]. It is a very elegant method of amplifying the sensitivity of a photoconductive device. It is also a very compact device, which enables highly efficient radiation shielding and economical use of material. SPRITE detectors are used in serial-parallel scan imaging systems, because they perform a time-delay and signal integration function, in addition to a detector function. The operating principle of the device is illustrated in Figs. 36.7a and 36.7b. It consists of a strip of n-type HgCdTe, typically 700 μm long, 60 μm wide and 10 μm thick, with three ohmic contacts. A constant-current circuit provides bias through the two end contacts. The third contact is a potential probe for signal read-out. There are two important points concerning the operation of the device. Firstly, the bias field must be sufficiently high to ensure that excess minority carriers, generated along the filament length, can reach the negative end contact before recombination occurs. Secondly, the bias field must be chosen so that the ambipolar drift velocity V_a is equal to the image scan velocity. SPRITE detectors have been commonly used in 8-row, 16-row and 24-row arrays and are commercially available in these forms.

Consider now an image feature scanned along the filament, as illustrated in Fig. 36.7a. The density of excess carriers in the filament, at a position corresponding to the illuminated element, increases during the scan, as illustrated in Fig. 36.7b. The carrier concentration increases during the scan as $[1 - \exp(-x/v_s \tau)]$, where x is the distance from the positive bias contact along the scan direction, and τ is the excess carrier lifetime. When the illuminated region enters the readout zone, the increased conductivity modulates the voltage on the contact and provides an output signal. Since the integration time of the signal flux, which is τ in a long element (or L/v_a in a shorter element), is greater than the dwell time on a conventional discrete element in a fast-scanned, serial system, a larger conductivity modulation and, hence, a larger output signal is observed.

Therefore, the SPRITE detector continuously performs a time-delay-integration function and the signal-to-noise ratio is improved as a result of coherent integration of the signal and incoherent integration of the noise. The net gain in the signal-to-noise ratio is proportional to the square root of the integration time. For an amplifier-limited or a Johnson noise-limited detector, the gain is proportional to the integration time.

The viability of the device depends on three aspects. The minority carrier lifetime must be maximized to give the longest possible integration length, the speed of the carriers (holes) must match a practical image scanning speed and the thermal spread of the carriers must be minimized as this will blur the image. In practice, this implies the need for large τ values to provide long integration times, and low minority carrier diffusion lengths Q_h to provide good spatial resolution. The material properties obtained for low carrier concentration n-type HgCdTe (as shown in Table 36.1) proves this to be a very suitable material for the device. Lifetime values of about 2 μs are obtained in alloy compositions suitable for 8–14 μm, 80 K operation, with an extrinsic carrier density of $\approx 10^{14}$ cm^{-3}. Recombination is principally by the Auger 1 process in these compositions, and longer

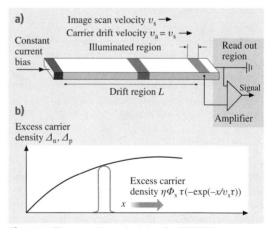

Fig. 36.7 The operating principle of a SPRITE detector

Table 36.1 Properties of n-type HgCdTe

Waveband (μm)	Operating temperature (K)	Lifetime (μs)	Hole mobility (μ_h) (cm^2/V/s)	D_h (cm^2/s)	Diffusion length Q_h (μm)
8–14	80	2–5	480	3.2	25
3–5	230	15–30	100	2.0	5

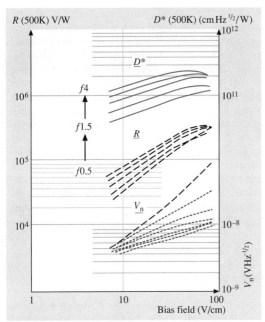

Fig. 36.8 Typical performance of a LW SPRITE detector ($\lambda_c = 11\,\mu\text{m}$, $T = 77\,\text{K}$). (After [36.25])

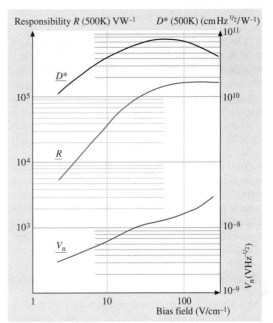

Fig. 36.9 Typical performance of a 3–5 μm SPRITE operating at 190 K. (After [36.25])

lifetimes, up to $\approx 10\,\mu\text{s}$ (achievable in low-background conditions), may be obtained from lower carrier density material. Lifetime values of 15–30 μs are obtained in compositions used for 3–5 μm operation at thermoelectric temperatures. The low hole mobility in HgCdTe results in short diffusion lengths, even with long lifetimes, yielding a spatial resolution in the 8–14 μm band SPRITE devices of $\approx 50\,\mu\text{m}$ and $\approx 100\,\mu\text{m}$ in 3–5 μm devices if the full integration is used.

The detectivity for a long, background-limited device in which $L \gg \mu_a E\tau$ and $\eta\phi_b\tau/t \gg p_0$ is

$$D_\lambda^* = \frac{\eta^{1/2}}{2E_\lambda}\left(\frac{1}{\phi_b w}\right)^{1/2}\left\{1 - \frac{\tau}{\tau_a}[1 - \exp(\tau_a/\tau)]\right\}^{1/2}, \tag{36.13}$$

where w is the element width, l is the length of the readout region, and $\tau_a = \ell/v_a$ is the transit time through the readout region. At sufficiently high scan speeds, such that $\tau_a \ll \tau$,

$$D_\lambda^* = (2\eta)^{1/2} D^* \,(\text{BLIP})\left(\frac{l}{w}\right)^{1/2}\left(\frac{\tau}{\tau_a}\right)^{1/2}. \tag{36.14}$$

It is convenient to express D_λ^* in terms of the pixel rate S which, for a nominal resolution size of $w \times w$, is v_s/w.

Thus, in the high scan speed limit

$$D_\lambda^* = (2\eta)^{1/2} D^* \,(\text{BLIP})\,(S\tau)^{1/2}. \tag{36.15}$$

The number of background-limited elements that would be required to provide the same performance in a serial array is

$$N_{\text{eq}}\,(\text{BLIP}) = 2S\tau. \tag{36.16}$$

For example, a 60 μm wide element scanned at a speed of 2×10^4 cm/s, and with τ equal to 2 μs, gives $N_{\text{eq}}\,(\text{BLIP}) = 13$.

The performance achieved in the 8–12 μm band is illustrated in Fig. 36.8. It may be seen that very high values of responsivity are obtained, as well as high values of D^*. The latter increases with the square root of the bias field, except at the highest fields, where Joule heating raises the element temperature. The detectivity increases with increasing cold-shield effective f/number to about $f/4$. It is, in practice, desirable to operate with efficient cold shielding with an f/number of ≥ 2 to avoid reductions of the carrier lifetime resulting from increased carrier density due to the background flux. An example of the results obtained from a SPRITE operating in the 3–5 μm band is shown in Fig. 36.9. Useful performance in this band can be obtained at temperatures up to about 240 K.

The spatial resolution of the SPRITE detector when the scan velocity and the carrier velocity are matched throughout the device length, is determined by the diffusive spread of the photogenerated carriers and the spatial averaging in the readout zone. This may be expressed through the modulation transfer function (MTF):

$$\mathrm{MTF} = \left(\frac{1}{1+k_s^2 Q_h^3}\right)\left[\frac{2\sin(k_s l/2)}{k_s l}\right], \qquad (36.17)$$

where k_s is the spatial frequency. The behavior of noise in the device at high frequencies is described in several papers [36.26, 27]. Some optimization of the shape of the filament has been carried out, involving a tapering of the read-out zone and a slight tapering of the main body of the element [36.28]. Following these modifications, very good agreement with (36.17) was obtained. In practice, the length of the read-out zone can be chosen sufficiently small that the diffusion term sets the limit. It was believed initially that this limit was fundamental, and that the only way in which it could be reduced was to restrict the integration by means of a shorter device. In fact, several different methods have been discussed [36.29] by means of which better spatial resolution can be obtained at the expense of additional power dissipation on the focal plane. The technique that has found most favor with system designers has been to increase the focal length of the detector lens in an afocal system, usually by employing anamorphic optics [36.30]. This allows the diffusion spot radius to be reduced below the diffraction spot radius.

36.4 Photoconductive Detectors in Closely Related Alloys

There has been some interest in developing alternative ternary alloys to replace HgCdTe, as from theoretical considerations it has been shown that the already weak Hg−Te lattice bond is further destabilized by alloying with CdTe [36.31]. It is predicted that the Hg−Te bond may be more stable in alternative alloys, leading to materials with increased hardness and detectors with better temperature stability. Apart from Cd and Zn, the other elements capable of opening up a bandgap in the semimetals HgTe and HgSe are Mn and Mg. In the case of Mg, little reinforcement of the Hg−Te bond occurs at compositions appropriate to 10 μm detectors, and in the case of $Hg_{1-x}Cd_x Se$, difficulties have been experienced in obtaining type-conversion and lightly doped n-type material. Most of the attention, therefore, has been focused on $Hg_{1-x}Zn_x Te$ and $Hg_{1-x}Mn_x Te$. As mentioned in the introduction, the technology and general properties of these materials are very similar to HgCdTe and much of the device processing is common to both. The status of the research is described by [36.32].

The hardness of $Hg_{1-x}Zn_x Te$ is about a factor of two better than HgCdTe and interdiffusion about one tenth, so this compound offers promise of competing with HgCdTe. However $Hg_{1-x}Zn_x Te$ presents more difficult material problems than HgCdTe. For instance, the separation of the liquidus and solidus curves is large and leads to high segregation coefficients. Also, the solidus lines are flat, leading to a strong composition dependence on growth temperature, and a very high mercury overpressure is needed for bulk crystal growth. The best crystals have been grown by the traveling heater method (THM). A D_λ^* value of 8×10^9 cm Hz$^{1/2}$/W has been quoted [36.33] for photoconductive detectors at 80 K prepared from LPE material, a value substantially lower than that typical of HgCdTe devices. 10.6 μm laser detectors with D_λ^* values in the range $3-6.5 \times 10^7$ cm Hz$^{1/2}$/W have been reported [36.34], comparable to those observed in equilibrium devices in HgCdTe. The most promising results, however, are those for long-wavelength detectors, of interest for space applications, with cut-off wavelengths as long as 17 μm [36.35]. The D_λ^* values were $8-10 \times 10^{10}$ cm Hz$^{1/2}$/W, measured at 65 K. Photodiodes have been produced by all of the established techniques, including ion implantation into THM material [36.36] and Hg diffusion into isothermal vapor phase material [36.37]. Both report comparable performance to HgCdTe diodes. The maximum annealing temperature is reported to be 10 to 20° better than HgCdTe. Challenges for the future will be to match the uniformity and defect levels of HgCdTe.

$Hg_{1-x}Mn_x Te$ is a semimagnetic narrow-gap semiconductor, but strictly it is not a II–VI ternary compound because Mn is not a group II element but is included here for completeness. $Hg_{1-x}Mn_x Te$ also has material disadvantages compared to HgCdTe and HgMnTe crystals need to be much more uniform in composition to achieve the same wavelength uniformity. For epitaxial growth the strong variation of lattice parameter with composition is also thought to be a disadvantage compared with HgCdTe. Photodiodes have been produced by Hg diffusion into THM material and VPE material grown on CdMnTe substrates, but the RoA values were low [36.38] and not yet competitive with other materials.

36.5 Conclusions on Photoconductive HgCdTe Detectors

Photoconductive HgCdTe detectors have been very successful in producing arrays of up to a few hundred elements or the equivalent in SPRITEs, for use in first-generation thermal imaging systems. The fundamental properties of HgCdTe have been found to be near-ideal for the fabrication of high-responsivity single elements, and for the fabrication of SPRITEs with very high responsivity and detectivity together with good spatial resolution. Frequency response into the MHz region is obtained from both types of device due to sweep-out effects. The limitations of photoconductive detectors are apparent when very large focal plane arrays are required. The low impedance of the photoconductor makes it unsuitable for coupling to direct injection gates of silicon charge transfer devices or MOS field effect transistors, therefore each element requires a lead-out through the vacuum encapsulation to an off-focal plane amplifier. The complexity of the dewar, therefore, limits the size of the array. This problem is considerably reduced in the SPRITE, but in this case, the effect of the Joule heating in the elements on the heat load of the cooler limits the maximum array size. For comprehensive reviews see [36.8, 39–41].

36.6 Photovoltaic Devices in HgCdTe

Photovoltaic arrays have inherently low power consumption and can be easily connected to a silicon integrated circuit to produce a retina-like focal plane array. Such arrays are key to so-called second-generation thermal imaging cameras which break through the performance limits imposed by photoconductive arrays in first-generation cameras. Photovoltaic devices tend to fall into two categories: long linear arrays of long-wavelength (8–12 μm) diodes and matrix arrays of medium-wavelength (3–5 μm) diodes.

Long linear arrays are required for systems with just one dimension of scanning. Here HgCdTe is the sensor material of choice because the stringent sensitivity requirements of these systems demands a high quantum efficiency and near BLIP-limited performance detectors. Development work in the 1980s and 1990s in the USA and Europe has led to a family of detectors for long-waveband thermal imaging, such as the US SADA II system based on a 480×6 format or the UK STAIRS "C" system based on a 768×8 array format (producing a high-definition 1250×768 image). In the detector examples quoted there are six or eight rows of diodes and the signal from each row is delayed in time and added to provide an enhancement. This is called time delay and integration (TDI). An important advantage of this scheme is that defective pixels can be deselected, so typically the STAIRS system will only use the best six of

Fig. 36.10 Medium-wavelength infrared image from a 384×288 element array (OSPREY) and the Kenis System. (Courtesy Denel Ltd)

the eight diodes in one channel. Consequently the detector performance can be relatively insensitive to defects in the HgCdTe, and this is the reason why LW long linear arrays have reached maturity much earlier than LW matrix arrays.

Long linear arrays can also use the MW band, but the flux levels are over an order less than the LW band and the performance is compromised by the short integration time in a scanned system. MW detectors then tend to use matrix arrays where the integration time can approach the frame time. The problem of point defects is considerably reduced in the medium waveband and large 2-D arrays are more practical. The emphasis for MW detectors is on developing wafer-scale processes to provide large, economical 2-D arrays for staring thermal imaging cameras. The MW band thermal imaging market is concentrated on arrays using half-TV (320×256 or 384×288) or full-TV formats (640×480 or 640×512), and is divided between indium antimonide (InSb), HgCdTe and platinum silicide. Broadly, manufacturing companies from the USA and Israel tend to prefer InSb, and European and some US companies have specialized in HgCdTe. Currently, infrared cameras based on HgCdTe arrays are in production and are producing remarkable sensitivities, over an order better than first-generation cameras. Figure 36.10 shows a representative image from a state of the art HgCdTe 2-D array.

Unlike photoconductive arrays, manufacturers use a variety of different technologies for photovoltaic arrays. Sections 36.6.1 to 36.6.3 describe the theory of photovoltaic detectors and the fundamental principles behind HgCdTe device technology. These principles help to explain the various approaches used by manufacturers to produce detectors in both the LW and MW bands described in Sects. 36.6.4 to 36.6.7. Section 36.6.8 describes the research and development progress in so-called Gen III (third-generation thermal imaging equipment) programs aimed at advanced infrared detectors.

36.6.1 Ideal Photovoltaic Devices

The current–voltage characteristic for an ideal diffusion-limited diode exposed to a photon flux ϕ_λ is given by:

$$I_d = I_s \left(\exp \frac{qV_d}{kT} - 1 \right) - qA\eta\phi_\lambda , \quad (36.18)$$

where I_d is the diode current, I_s is the diffusion current, V_d is the diode bias voltage (taking negative values in reverse-bias), η is the quantum efficiency and A is the detector area. The second term in (36.18) is the photocurrent, I_{ph}. The quantum efficiency is given by:

$$\eta = (1 - R)[1 - \exp(-\alpha_\lambda t)]F , \quad (36.19)$$

where R is the reflectivity of the front surface, normally minimized by an antireflection coating, α_λ is the absorption coefficient, t is the sample thickness, which is typically around $6-10\,\mu m$ to give adequate absorption in HgCdTe detectors, and F is a geometry factor which describes the number of photogenerated carriers within the pixel which reach the junction before recombining.

The condition that the detector be background limited is crucial for high-performance systems, and to meet this condition the internal thermal generation must be much less than the photon generation in the lowest flux case, i.e. $I_s \ll I_{ph}$. By differentiating (36.18), I_s is given by kT/qR_0, where R_0 is the zero bias resistance. The condition for background-limited performance is therefore often written as $R_0 \gg kT/qI_{ph}$, and this is the origin of the commonly used figure of merit, R_0A. As an example, in the case of a long waveband system, say $10\,\mu m$ cut-off, with an F2 optic and a lowest background scene temperature of $-40\,°C$, an R_0A of greater than $30\,\Omega\,cm^2$ is required. Detectors need to be cooled sufficiently to suppress thermally generated currents, but there are always pressures to avoid very low temperatures to give savings in power and cooldown time and better engine reliability. This is particularly the case for cut-off wavelengths longer than $10\,\mu m$, and detector technologies have emerged that provide high R_0A as a first priority.

The white noise current at low frequencies is given by:

$$i_n^2 = 2qI_s \left[1 + \exp\left(\frac{qV_d}{kT}\right) \right] B + 2q(qA\eta\phi_\lambda)B , \quad (36.20)$$

where B is the bandwidth in Hz. In the absence of photocurrent, the noise at zero bias is equal to the Johnson noise, $4kTB/R_0$. In reverse bias it tends to the normal expression for shot noise, $2q(I_s + I_{ph})B$. Note that the mean-square shot noise in reverse bias is half that of the Johnson noise at zero bias.

The current responsivity is $\eta q/E_\lambda$, where E_λ is the photon energy, and this leads to the general expression for detectivity:

$$D_\lambda^* = \frac{\eta q}{E_\lambda} \sqrt{\frac{A}{i_n^2}} . \quad (36.21)$$

When the photocurrent exceeds the diffusion current, the device is said to be background limited (BLIP), and the first term in (36.20) can be ignored. The background-limited detectivity is then given by:

$$D_\lambda^* \text{ (BLIP)} = \frac{1}{E_\lambda}\sqrt{\frac{\eta}{2\phi_\lambda}}. \tag{36.22}$$

Note that, in comparison with (36.9), the background-limited detectivity of a photodiode is a factor of $\sqrt{2}$ better than that of a photoconductor.

36.6.2 Nonideal Behavior in HgCdTe Diodes

In medium-waveband arrays the current–voltage (I–V) characteristic is usually close to ideal, but at longer wavelength a number of leakage currents can impact on the I–V characteristics and degrade the performance of the detector. Figure 36.11 illustrates a typical I–V characteristic for a LW diode, along with two types of HgCdTe photodiode, a planar diode and a via-hole diode. The diagram illustrates how factors such as thermally generated current, shunt resistance and breakdown currents at high bias can effect the diode behavior. The performance can be degraded directly; for example, extra currents produce white noise which may be significant if the leakage currents are of the same order as the photocurrent. In addition, some leakage current mechanisms are associated with trapping or tunneling and often produce excess, low-frequency noise (often called $1/f$ noise due to its typical trend with frequency). However an additional constraint arises from the need to inject photocurrent into a silicon multiplexer. The simplest method is called direct injection and literally means injection of the photocurrent into a common gate MOSFET. A capacitor on the drain performs the integration. Good injection efficiency is only achieved if the diode has a dynamic resistance higher than the input impedance of the MOSFET. Unless the dynamic impedance of the diode is high enough, the signal can be attenuated, additional noise added and nonlinearity generated in the response. Electronic correction for nonuniformity in the array is essential in IR detectors and nonlinearity can add spatial noise to the image after nonuniformity correction.

Achieving a high dynamic resistance has been a driver in the development of most infrared technologies. Very roughly there is a distinction between medium-wavelength (MW) arrays with cut-offs up to 6 μm and long-wavelength (LW) arrays. In MW arrays high injection efficiencies are routinely seen because the photodiodes are near-ideal. In the long waveband there will be a limit to the upper cut-off wavelength and/or flux levels that routinely allow high injection efficiencies using direct injection.

In order to inject the photocurrent signal efficiently into the input MOSFET, the input impedance of the MOSFET must be much lower than the internal dynamic resistance of the photodiode. The input impedance of a MOSFET is a function of the source-drain current (in this case the total diode current) and is usually expressed in terms of the transconductance, g_m, given by $qI_d/(nKT)$ for low injected currents (n is an ideality factor that can vary with technology usually in the range 1–2). The expression for injection efficiency ε given approximately by:

$$\varepsilon = \frac{R_V I_d}{R_V I_d + \frac{nkT}{q}}, \tag{36.23}$$

where R_V is the dynamic impedance of the diode and I_d is the total injected diode current equal to I_{ph} in the background-limited case.

In the background-limited case, (36.23) becomes a simple function of $I_{ph}R_V$. For a very high performance system requiring an efficiency of 99.9%, the value of $I_{ph}R_V$ approaches 10 V, and in systems with slow

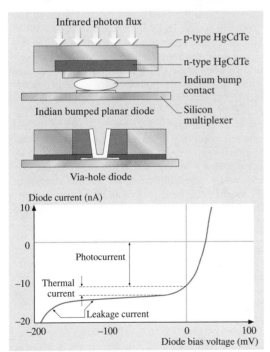

Fig. 36.11 Types of HgCdTe diode and a typical current–voltage characteristic for a LW diode showing main nonideal features

optics and low background temperatures, the requirement for R_v can exceed 10^9 Ω. In the long waveband this can present a technological challenge. There are more complex injection circuits that effectively reduce the input impedance and allow lower HgCdTe resistances to be used, but these require more silicon area and a higher power consumption, which may not be practical. The choice of buffer circuit depends on the application, and good comprehensive summaries are presented by [36.42, 43].

For most applications the detector performance depends on operating the diode in a small reverse bias where the dynamic resistance is at a maximum. It is then necessary to minimize extraneous leakage currents. The control of these leakage currents and the associated low-frequency noise is therefore of crucial interest to the device engineer. The origin of the leakage currents and the measures that are taken to minimize them are explained here in detail.

36.6.3 Theoretical Foundations of HgCdTe Array Technology

Good detailed summaries of photodiode fundamentals have been provided by [36.44, 45] and the reference lists contained therein. The purpose of this section is to highlight the key issues that control the thermal generation, leakage currents and quantum efficiency in detectors. From this analysis it is easier to understand the measures taken by detector manufacturers to make practical high-performance detectors.

Thermal Diffusion Currents in HgCdTe
Expressions for the diffusion current in photovoltaic devices have been derived [36.44] and the fundamental R_0A expression is given in (36.24):

$$R_0 A(\text{n-side}) = \frac{N_d}{q n_i^2} \sqrt{\frac{kT\tau_h}{q\mu_h}}, \qquad (36.24)$$

where N_d is the net donor concentration on the n-side, n_i is the intrinsic carrier concentration, which dominates the temperature dependence, and τ_h and μ_h are the minority carrier lifetime and mobility, respectively.

A similar expression describes the contribution of the p-side. In practice, the diffusion length L in normally doped HgCdTe is often larger than the 6–10 μm needed for effective infrared radiation absorption. In this case, the volume available for the generation of diffusion current is restricted, and a suitable modified expression is:

$$R_0 A(\text{n-side}) = \frac{kT N_d \tau_h}{q^2 n_i^2 \tau_n}, \qquad (36.25)$$

where t_n is the thickness of the n-type material.

In the case of L being greater than t_n, the surfaces and contacts can act as sources of extra diffusion current if the surface recombination velocity is greater than the diffusion velocity D/L. It is essential to ensure properly passivated surfaces and for contacts employ a minority carrier barrier, such as a higher doped or wider band gap layer under the metal contact.

The behavior of (36.25) with doping depends upon the dominant recombination process, i.e. radiative, Auger or Shockley-Read (S-R). Auger recombination in HgCdTe is a phenomenon involving the interaction of three carriers. The Auger 1 lifetime in n-type material is due to the interaction of two electrons and a hole and is generally minimized by using a low carrier concentration on the n-side. Fortunately it is relatively easy to produce low carrier concentrations in n-type material ($< 5 \times 10^{14}$ cm^{-3}).

In p-type material Auger recombination involves two holes and an electron and is referred to as Auger 7. The lifetime in p-type HgCdTe is reported [36.46–48] to show an inverse, linear dependence on doping, and this is attributed to the Shockley-Read process. S-R recombination is often associated with the Hg vacancy and can be modeled by a strong donor level located ≈ 30 mV from the conduction band, which appears to be independent of composition in the $x = 0.2$ to 0.3 range. The density of these donor S-R centers is said to be proportional to the Hg vacancy concentration, but lower by a factor of ≈ 20. If Hg vacancies are replaced by acceptor atoms such as Cu, Ag, Na or Au, the lifetime can be increased by over an order of magnitude. The use of extrinsic doping is an effective way to reduce thermal currents from the p-side in homojunction devices.

In order to engineer a detector with low thermal diffusion current (or high R_0A) it is important to recognize that it is only necessary to use one side of the junction for collecting photocurrent; the other side, in principle, can be made with a wider bandgap, thereby minimizing the thermal current contribution. Devices with layers of different band gaps are called heterostructures. The most common design is to use a wider bandgap on the p-side to reduce the thermal diffusion current and use the n-side as an absorber taking advantage of the long minority carrier diffusion lengths to maximize the quantum efficiency.

Thermal Generation Through Traps in the Depletion Region

The thermal generation rate g_{dep} within the depletion region via traps is given by the usual Shockley-Read expression:

$$g_{dep} = n_i^2 (n_1 \tau_{n0} + p_1 \tau_{p0})^{-1}, \qquad (36.26)$$

where n_1 and p_1 are the electron and hole concentrations which would be obtained if the Fermi energy was at the trap energy, and τ_{n0} and τ_{p0} are the lifetimes in the strongly n-type and p-type regions.

The leakage current is often known as g–r current. Normally, one of the terms in the denominator of (36.26) will dominate, and for the case of a trap at the intrinsic level, n_1 and $p_1 = n_i$, giving $g_{dep} = n_i/\tau$. It is the weaker dependence on n_i, and therefore on temperature, that distinguishes generation within the depletion layer from thermal diffusion current.

Where the depletion region intercepts the surface, there is often enhanced generation due to the presence of a high density of interface states. This can be exacerbated if the surface passivation is not properly optimized to give a flat band potential at the junction. In an extreme case, the surface on one side of the junction may become inverted, creating an extension of the depletion layer along the surface, and leading to high generation currents. A practical solution to this mechanism is to widen the band gap in the material where the junction intercepts the surface, so-called heteropassivation. This can be achieved by using a thin film of CdTe together with a low-temperature anneal [36.49], and this is the commonest passivation technique used by manufacturers.

Interband Tunneling

Due to the very low effective mass of the electron in HgCdTe, direct band-to-band tunneling can occur from filled states in the valence band to empty states at the same energy in the conduction band. An expression for the current due to this process has been developed by [36.50]. The tunneling current increases very rapidly as the applied voltage or doping is increased, or the band gap or temperature is decreased. It is the normal modern practice to use low doping on one side of the junction to minimize interband tunneling under normal operating conditions, but nevertheless, at low temperature or for very long wavelength devices, interband tunneling can become dominant.

Trap-Assisted Tunneling

Trap-assisted tunneling is generally accepted to be one of the main causes of leakage current and excess noise in LW diode arrays, but it is not a fundamental limitation, often being associated with impurities or structural defects within the depletion region. The definition of a trap is a center with a capture coefficient for minority carriers many times larger than that for majority carriers (otherwise it is a generation center). The role played by traps in the depletion region is very complicated, allowing for three possible two-step processes: thermal-tunnel, tunnel-thermal and tunnel-tunnel. The nonthermal step can include tunneling of electrons from the valence band to traps, and tunneling from the traps to the conduction band.

The formulation for trap-assisted tunneling in HgCdTe has been developed [36.51, 52] based upon original work in silicon [36.53]. Models based on a thermal-tunnel process can explain the bias-dependent behavior of reverse current in long-wavelength diodes [36.54–56].

A trap-assisted tunneling process can explain the observed behavior of n^+-p diodes [36.57]. The process involves a thermally excited, bulk Shockley-Read center, modeled for the special case of the trap residing at the Fermi level. The physical picture for this assumption is that there is a uniform distribution of trapping centers throughout the bandgap and the barrier for tunneling is lowest at the uppermost center that is still occupied. Hence, the occupied trapping center that coincides with the Fermi level has the highest transition probability and plays the dominant role in the thermal trap-assisted tunneling process. This model helps to explain several unusual properties, such as an observed reduction in leakage current when the temperature is reduced or the doping level increased. Note that this is in contrast to direct band-to-band tunneling. Also, there is a much weaker dependence on diode bias voltage. The model also predicts the commonly observed "ohmic" region illustrated in Fig. 36.11 at a bias of around -100 mV.

The physical origin of the trap has not been established yet but it appears to be an acceptor-type impurity or defect within the depletion region. The population of such traps is likely to be dependent on the HgCdTe material and the junction forming technology and so the quality of LW diodes is highly process-specific.

Impact Ionization

Underlying trap-assisted tunneling, a more fundamental source of leakage current has been proposed, which is called impact ionization.

The reverse bias characteristics of homojunction arrays often show behavior that is not easily explained by conventional trapping mechanisms. For instance, the product of p-side diffusion currents (including photocurrent) and reverse bias resistance is often observed to be insensitive to temperature and cut-off wavelength over a wide range (at least 4–11 μm s), and the current increases much more slowly with reverse bias than tunneling models would predict. A model based upon an impact ionization effect within the depletion layer gives a good fit to these observations [36.21]. The effect arises because in HgCdTe the electron scattering mechanisms tend to be weak and hot electrons can penetrate deep into the conduction band where they readily avalanche. Leakage current arises because extra electron-hole pairs are created within the depletion region due to impact ionization by minority carrier electrons from the p-side. The leakage mechanism has been confirmed to have a linear relationship with optically injected minority carriers over a wide temperature range [36.58].

Calculations have been performed on the effect of impact ionization on homojunction performance as a function of the doping levels [36.59]. These predict that to achieve a high dynamic resistance and therefore a high injection efficiency, the n-side doping must be very low. Routinely achieving low carrier concentrations ($< 5 \times 10^{14}$ cm^{-3}) is an important aim for homojunction technology. Most manufacturers introduce a donor, such as indium, to the crystal to control the n-type level to around $3-5\,\text{e}^{14}$ cm^{-3}. The carrier concentration in the p-region has a second-order effect compared with that of the n-region for the range of concentrations normally used.

Photocurrent and Quantum Efficiency

HgCdTe has a strong optical absorption coefficient and only thin layers are needed to produce high quantum efficiency. Typically in MW detectors the absorber need only be 4–5 μm thick and about twice this in LW detectors. Ideally the absorption should occur well within a diffusion length of the p–n junction to avoid signal loss due to recombination. A long carrier lifetime is nearly always observed in n-type material with low carrier concentration. Device engineers tend to favor using n-type absorbers for the best quantum efficiency and try to minimize the volume of the p-region for lower thermal leakage currents.

Excess Noise Sources in HgCdTe Diodes

There are many potential sources of excess noise in infrared detectors, and manufacturers strive to optimize obvious areas, such as the surface passivation (to limit surface leakage currents) and contacts. There are two other sources that have been reported in depth. The first is linked to tunneling currents through traps, and the second is associated with crystal defects.

Many authors have reported an empirical relationship between tunneling leakage current and $1/f$ noise. The scatter within databases is usually large and the noise depends strongly on the technology used, but the noise trend is roughly given as αi_L^β, where α and β are variables depending on the device and leakage current mechanism. Leakage current from the trap-assisted tunneling mechanism results in β values close to 0.5, with α taking a value of 1×10^6 A$^{0.5}$ [36.60]. With band-to-band processes the value of β moves towards 1. The nature of the trap has not been identified yet and it is not possible to exclude a variety of crystallographic defects. The physics of crystallographic defects in HgCdTe is complex because dislocations distort the local band structure and, via strain fields, the local bandgap, and this probably accounts for the variability in data in the literature.

The electrically active nature of dislocations and other crystal defects in HgCdTe is well reported [36.61, 62]. Dislocations can appear as n-type pipes and are associated with active defect centers. Many workers [36.63–68] have found that the reverse bias characteristics and $1/f$ noise of HgCdTe diodes depend strongly on the density of dislocations intercepting the junction. For HgCdTe this is an important observation because dislocations can easily be introduced during materials growth and the device fabrication process. Dislocations could increase the g–r current linearly in p-on-n heterostructures, along with a corresponding increase in $1/f$ noise current density [36.66]. Tunnel currents are also strongly associated with crystal defects, particularly at low temperatures.

The nature of defects in HgCdTe LW arrays has been studied in detail [36.58] in homojunction via-hole arrays (loophole arrays) made using high-quality LPE material, and a model has been proposed for the noise-generating mechanism. Threading dislocations that originate from the CdZnTe substrate and rise vertically through the layer can cause strong leakage in reverse bias if they intercept the junction, possibly because these dislocations can become randomly decorated during growth. Consequently, the threading dislocation density in the substrate is very important for controlling defect levels. Process-induced dislocations have a weaker effect on the junction properties resulting in an effective shunt resistance of about 40 MΩ, so an accumulation of this type of dislocation can result in a defect.

The current trend is to move towards lower temperature growth processes and growth on substrates with poor lattice match, and this has refocused attention on grown-in dislocations. Many of these processes are not yet suitable for LW arrays. More data is needed for the VPE processes because the nature of the misfit dislocations and the geometry of the absorber will influence the magnitude of the excess noise.

36.6.4 Manufacturing Technology for HgCdTe Arrays

Considerable progress has been made over the last two decades in the epitaxial growth of HgCdTe. Bulk growth methods are still used to provide good quality material for photoconductor arrays, but for photovoltaic arrays there are problems associated with crystal grain boundaries, which are electrically active, and cause lines of defects. Also, there are limitations in the boule size, which makes it suitable for small arrays only.

Several epitaxial growth techniques are in use today. Manufacturers will select a technique that suits their device technology and the type of detectors they are trying to make. For instance, high-performance LW arrays will call for the best possible crystal quality, whereas large-area MW arrays can probably accept poorer material but must have large, uniform wafers. It is the aim of most manufacturers to produce high-quality layers in large areas at low cost, but this ideal has been elusive. At the present time the best structural quality material is grown using liquid phase epitaxy, LPE, onto lattice-matched crystals of CdZnTe, and this has been used successfully in homojunction technologies where the photosensitive junction is diffused into a homogenous monolith of material. There is a trend to move away from expensive CdZnTe substrates and both LPE and vapor-phase epitaxy (VPE) are now used on a variety of alternative substrates. Many groups favor VPE because the composition and doping profile can be easily controlled to produce complex devices, such as two-color detectors. The main HgCdTe growth processes are described here.

Summary of Growth Using Liquid-Phase Epitaxy (LPE)

Liquid-phase epitaxy (LPE) of HgCdTe at present provides the lowest crystal defect levels, and very good short and long-range uniformity. LPE layers are grown using an isothermal supersaturation or programmed cooling technique or some combination. A detailed knowledge of the solid-liquid-vapor phase relation is essential to control the growth particularly in view of the high Hg pressure. Challenges include: the compositional uniformity through the layer, the surface morphology, the incorporation of dopants and the specifications for thickness, wavelength, etc. A common component leading to high structural quality is the use of lattice-matched substrates of CdZnTe. These are grown by a horizontal Bridgman process and can supply layers as large as 6×4 cm. The CdZnTe substrates must be of the highest quality, and often this is a significant cost driver for the process.

Two different technical approaches are used: growth from a Hg-rich solution, and growth from a Te-rich solution. Advantages of the Hg-rich route include: excellent surface morphology, a low liquidus temperature, which makes cap layer growth more feasible, and the ease of incorporation of dopants. Also, large melts can provide for very good compositional and thickness uniformity in large layers and give consistent growth characteristics over a long period of time. Growth from Te-rich solutions can use three techniques to wipe the melt onto the substrate: dipping, tipping and sliding. Sliding boat uses small melt volumes and is very flexible for changing composition, thickness and doping. Tipping and dipping can be scaled up easily and can provide thick, uniform layers but the large melts limit flexibility. Double layers are also more difficult to grow.

Most manufacturers have taken their chosen growth system and tailored it to provide optimum material for their device technology. In particular the use of dopants and the deliberate introduction of compositional grades are very specific to the device structure. A crucial figure of merit however is the dislocation count, that controls the number of defects in 2-D arrays. Etch pit densities of $3-7 \times 10^4$ cm^{-2} are typically seen in the Te-rich sliding boat process reproducing, the substructure of the CdZnTe substrate [36.69]. The etch pits are associated with threading dislocations which appear to be normal or near normal to the layer surface. The substrate defect level can be as low as mid-10^3 cm^{-2} in some horizontal Bridgman CdZnTe, but this is not easy to reproduce. Similar defect densities are found in CdTeSe material but the impurity levels have proved difficult to control in the past. Device processing therefore must expect to cope with defect levels in the mid-10^4 cm^{-2} range for routine CdHgTe epilayers, and this will set the ultimate limit on the number of defects in HgCdTe 2-D arrays. Several groups have used the LPE process on low-cost substrates, including: CdZnTe or CdTe on GaAs/Si wafers [36.70] or the PACE technology on sapphire described in detail in Sect. 36.6.4.

Summary of Growth Using Metalorganic Vapor-Phase Epitaxy (MOVPE)

A detailed summary of the state of the art for MOVPE technology has been produced [36.71], but this section summarizes the main points.

MOVPE growth depends on transporting the elements Cd and Te (and dopants In and As) at room temperature as volatile organometallics. They react along with Hg vapor in the hot gas stream above the substrate or catalytically on the substrate surface. The drive to lower temperatures and hence lower Hg equilibrium pressures has resulted in the adoption of the Te precursor di-isopropyl telluride, which is used for growth in the 350–400 °C range. A key step in the success of this process is to separate the CdTe and HgTe growth so that they can be independently optimized. This is called the IMP process (interdiffused multilayer process) [36.72]. IMP results in a stack of alternating CdTe and HgTe layers and relies on the fast interdiffusion coefficients in the pseudobinary to homogenize the structure at the growth temperature. Doping is straightforward using Group III metals (acceptors) and Group VII halogens (donors). For instance, ethyl iodide is used for iodine doping.

The main morphological problem for MOVPE are macro defects called hillocks, which are caused by preferred 111 growth, nucleated from a particle or polishing defect. Hillocks can cause clusters of defects in arrays. Orientations 3–4° off 100 are used primarily to reduce both the size and density of hillocks.

A variety of device structures with layers have been reported [36.73] based on MOVPE-grown HgCdTe layers on 75 mm-diameter GaAs on silicon.

Summary of Growth Using Molecular Beam Epitaxy (MBE)

A good detailed summary of the state of the art for MBE technology has been reported [36.74].

MBE offers the lowest temperature growth under an ultrahigh vacuum environment, and, in common with MOVPE, in situ doping and control of the composition and interfacial profiles. These are essential for the growth of advanced and novel device structures. Typically growth is carried out at 180–190 °C on 211 CdZnTe substrates. Effusion cells of CdTe, Te and Hg are commonly used. Hg is incorporated into the film only by reacting with free Te, and so the composition depends on the Te to CdTe flux ratio. The structural perfection depends strongly on the Hg to Te flux ratio and growth is usually restricted to a tight temperature range. Indium is the most widely used n-dopant and is well activated. p-type dopants are less conveniently incorporated in situ but manufacturers have devised a number of processes to force As onto the proper Te site. Again the Hg to Te ratio and growth temperature is crucial to achieving good activation. In general, reproducibility seems more difficult to achieve than MOVPE. MBE structural problems center mainly on pinholes or voids. Some very good EPD levels have been reported, but in general the EPD levels are an order or more higher than the best quality LPE.

Junction Forming Techniques

For n–p devices, crystal growers can obtain the desired p-type level by controlling the density of acceptor-like mercury vacancies within a carrier concentration range of, say, 10^{16} to 10^{17} cm^{-3}. Neutralizing the Hg vacancies and relying on a background level of donors to give the n-type conversion creates the photodiode junctions. Mercury can be introduced by thermal diffusion from a variety of sources, but high-temperature processes are not very compatible with HgCdTe at the device level. However, type-conversion can be readily achieved by processes such as ion beam milling [36.75] and ion implantation [36.76–81]. Type conversion can also be achieved using plasma-enhanced milling in the VIAP process [36.81] and also using H_2/CH_4 plasmas [36.82]. The common feature is that the conversion depth is much deeper than would be expected from the implantation range alone.

The current knowledge on type conversion using ion beam milling is described by [36.58] and the current knowledge for ion implantation is summarized by [36.83]. The explanation for the behavior of HgCdTe under ion beam bombardment involves a number of physical mechanisms. Firstly, the low binding energies, ionic bond nature and open lattice of HgCdTe encourages the liberation of free mercury at the surface and subsequent injection by the ion beam. The injection mechanism probably involves a recoil implantation process. Once the Hg interstitial is injected, the mobility is apparently extremely high and there is some evidence that this is stimulated by the ion beam in a process related to the anomalously high diffusion rates of impurities often observed in SIMS analysis. Another factor is the movement of dislocations under the influence of the ion beam and possibly stimulated by strain fields. A number of workers report that the n-type carrier concentration in the converted region is very low, and in fact fast-diffusing impurities such as the Group IB elements Cu, Ag and Au and the Group IA elements Na, K and Li, which reside on the metal sublattice, are swept out of the n$^-$ region of the diode by the flux of Hg interstitials. This impurity sweep-out effect is serendipitous because it means that in ion beam generated junctions the n-regions are very pure, free of S-R centers and have weak doping, creating an ideal structure for high-performance detectors. The detailed atomic level processes taking place including the role of mercury interstitials, dislocations and ion bombardment in the junction-forming process are complex and not well-understood in detail. Despite the complex physics involved, manufacturers have achieved good phenomenological control of the junction depth and n-dopant profiles with a variety of processes.

Via-Hole Technologies

Via-hole devices share many of the process stages of photoconductive arrays and have been commercially produced since 1980. Arrays made using this process are known as loophole devices [36.84] and HDVIP^{TM} devices [36.81]. By using a thin monolith of HgCdTe bonded rigidly to the silicon, the thermal expansion mismatch problem is overcome because the strain is taken up elastically. This makes the devices mechanically and electrically very robust and suitable for very long linear arrays, such as a 512×2 array of over 16 mm in length [36.69].

The process flow of the via-hole device using LPE material is illustrated in Fig. 36.12 [36.59, 69]. The process has two simple masking stages. The first defines a photoresist film with a matrix of holes of, say, 5 μm diameter. Using ion beam milling, the HgCdTe is eroded away in the holes until the contact pads are exposed. The holes are then backfilled with a conductor, to form the bridge between the walls of the hole and the underlying metal pad. The junction is formed around the hole during the ion beam milling process, as described in Sect. 36.6.4. The second masking stage enables the p-side contact to be applied around the array. Figure 36.12 shows a photomicrograph of one corner of the hybrid, illustrating the membrane-like nature of the HgCdTe on the silicon. The device structure inherently produces low crosstalk in small pixels.

The HDVIP^{TM} process [36.81] results in a similar structure but uses a plasma etching stage to cut the via-hole and an ion implantation stage to create a stable junction and damage region near the contact. In order to achieve higher lifetimes and lower thermal currents, Cu is introduced at the LPE growth stage. This is swept out during the diode formation and resides selectively in the p-region, partially neutralizing the S-R centers associated with Hg vacancies and achieving dark currents approaching those of fully doped heterostructures.

In via-hole processes the cylindrical shape of the junction minimizes the intersection with threading dislocations (Sect. 36.6.3). This has been demonstrated by the difference in yields when the HDVIP^{TM} device is extended by a planar junction [36.81].

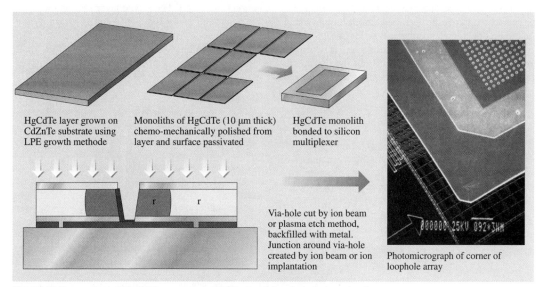

Fig. 36.12 Processing of HgCdTe arrays using LPE material and via-hole technology

Planar Device Structures Using LPE

The planar device structure illustrated in Fig. 36.13 is the simplest device structure currently used. It is consistent with a number of junction forming processes, e.g. ion implantation, diffusion and ion milling. The matrix of junctions is mass connected to an underlying silicon multiplexer using indium bumps. The strength of the process is the simplicity and the compatibility with epitaxially grown materials. A process based on high-quality LPE material and ion-implanted junctions is conducive to good-quality detectors and volume production [36.85]. In the simplest form, the process needs three masking stages for the junction and pixel contact, and one for the contact to the p-side. The device is backside-illuminated, i.e. it is illuminated through the substrate, and so this must be of high optical quality. Careful control of the junction geometry is needed to avoid crosstalk due to the diffusion of minority carriers into adjacent pixels, especially in the case of small pixel sizes. The thermal expansion mismatch between the HgCdTe/CdZnTe substrate combination and the silicon multiplexer is another important consideration in this device structure and this can restrict the practical size of the array unless the CdZnTe substrate is thinned.

Double Layer Heterojunction Devices (DLHJ)

Thermal leakage currents in HgCdTe devices tend to arise from the p-type side, and most advanced technologies strive to improve the operating temperature by using the n-type side as the absorber. n-type HgCdTe is easier to control at low carrier concentrations and is relatively free of Schottky-Read centers that limit the lifetime in p-type material. The p-type side then is often minimized in volume or uses a wider bandgap to reduce the thermal generation. For heterostructures that are grown in situ, it is necessary to completely isolate the junction, forming a so-called mesa device. This is usually performed using a chemical etch because of the electrical side effects of using dry processing techniques.

When a device uses different compositions or doping levels, a nomenclature is adopted to describe the structure. The definition used in this chapter is as follows. An n-type diode formed in a p-type layer is described as an n–p device, so the first letter is the layer nearest the contact and photons are absorbed in the p-type layer. Wider bandgap material is indicated by using a capital letter. A superscript "−" or "+" denotes particularly low or high doping.

Double-layer heterojunction devices have been developed mainly in the US for long-wavelength detectors with low thermal leakage currents (or high R_0A values). A number of elegant device structures have been reported with R_0A values that are an order of magnitude higher than those of via-hole or planar diodes. The back-illuminated mesa P^+-n heterojunction, illustrated in Fig. 36.14, is a widely used device, and has been reported from both LPE and MBE material. This structure makes good use of the n-type absorber to give a high fill factor. Devices have been made using a vertical dipper LPE process from a Hg rich solution [36.86] and using a horizontal slider LPE process from a Te-rich solution [36.87–89]. The p-type layer is doped with arsenic to around $1-4 \times 10^{17}$ cm^{-3} and is grown by the vertical dipper LPE process from a Hg-rich solution. An n^--P^+ structure has been reportedly grown from MBE which has the advantage of very low crosstalk because of the complete electrical isolation of the absorber [36.90].

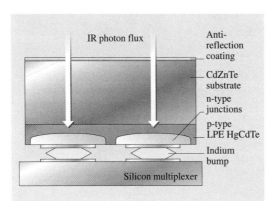

Fig. 36.13 Schematic of planar indium-bumped hybrid

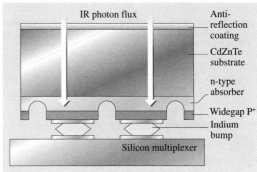

Fig. 36.14 Schematic of double-layer heterojunction (DLHJ)

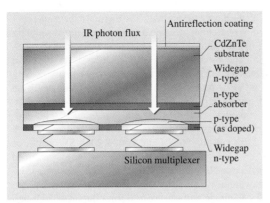

Fig. 36.15 Schematic of double-layer planar heterostructure (DLPH)

Other workers have produced the conventional P^+-n^- structure using an MBE process [36.91].

Mesa heterostructures have a technologically difficult passivation stage on the sidewalls of the mesa, which can lead to reverse bias leakage and uniformity problems if not optimum. In order to solve the sidewall problem, a clever diffused heterostructure has been reported [36.92], called a double-layer planar heterostructure (DLPH). This is a back-illuminated arsenic-implanted p-n-N planar buried junction structure [36.93], as illustrated in Fig. 36.15. The junction is formed by arsenic implantation into a three-layer N-n-N film grown in situ by MBE onto a CdZnTe substrate. The unique feature of this structure is that the junction is buried beneath the top wide-bandgap n-type layer. The junction intercepts the surface in wider gap CMT, thereby reducing the generation rates for any surface defects that may be present. The arsenic doped region must extend deep enough into the narrow gap absorber layer in order to collect photocarriers. This necessitates that the wide gap layer be thin (on the order of $0.5\,\mu m$) and that the compositional interdiffusion between the wide gap and absorber be minimal. Growth by MBE at low temperature (175 °C) satisfies these requirements. Another feature of this structure is the wide-gap n-type buffer layer between the absorber and the substrate. This buffer layer keeps carriers that are photogenerated in the absorber away from the film–substrate interface where there can be interface recombination. The DLPH process has been used to produce arrays of up to 1024×1024 on $18.5\,\mu m$ pitch (HAWAII multiplexers) with high quantum efficiency [36.94].

Wafer-Scale Processes Using Vapor Phase Epitaxy on Low-Cost Substrates

There is a dual purpose in developing vapor phase epitaxy technologies. Firstly, it opens up the possibility of growing complex multilayer materials for a whole range of new device structures, and secondly it allows growth on lower cost substrates. Certainly the elegance of this technique is that the composition, thicknesses and doping levels can be programmed, in principle, to grow any required detector structure. For instance, multilayer, fully doped heterostructures have been grown which demonstrate low thermal leakage current by a process called Auger exclusion [36.95]. Also, two-color detector structures can be grown as described in Sect. 36.6.8. The main commercial drive though is to cost-reduce the manufacture of large-area arrays and avoid expensive CdZnTe substrates. Silicon, gallium arsenide and sapphire are the commonest substrates used because they are available commercially in large-area wafers at relatively low cost. Figure 36.16 illustrates a typical process flow.

One of the main considerations when designing large arrays is the thermal mismatch problem between the various layers in the device and carrier, which can cause

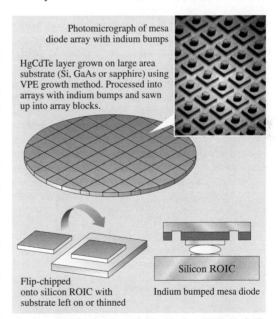

Fig. 36.16 Wafer-scale processing using VPE HgCdTe material, mesa diodes and indium bump interconnect technology

bowing or cracking at low temperature. The silicon multiplexer has a very low thermal expansion coefficient compared to most materials used and a number of different techniques are used to cope with this.

Some groups focus on the growth of HgCdTe on silicon substrates so that, when bump-bonded, the substrate and multiplexer are balanced. Silicon is also mechanically strong, available in large wafers at low cost, and avoids the recurring problem of impurity out-diffusion in CdZnTe substrates. A 75 mm-diameter silicon wafer can produce about twelve 640×512 arrays. For HgCdTe there is a 19% lattice constant mismatch with silicon, and this can potentially provide high levels of misfit dislocations, resulting in degraded performance, particularly at longer cut-off wavelengths. However at short and medium wavelengths the effect of misfit dislocations is reduced enough to make these wafer-scale processes viable at the present time. Most of the MOVPE work on silicon has used a GaAs layer to buffer the lattice mismatch between silicon and HgCdTe [36.73], and a variety of device structures are reported using MOVPE layers grown on 75 mm-diameter GaAs on silicon layers. MBE growth on silicon substrates of 100 mm in diameter has been used to grow MW HgCdTe layers [36.96]. The MBE work has used buffer layers of CdZnTe and CdTe to minimize the density of misfit dislocations. Both MBE and MOVPE growth has been used to produce LW arrays on misorientated silicon substrates [36.97]. An MBE CdTe/ZnTe layer is first grown and this acts as a buffer for the HgCdTe growth using an MOVPE process.

Another approach, known as PACE, has been described [36.98]. The process uses a 75 mm-diameter substrate of highly polished sapphire, which remains with the device throughout. The thermal expansion problem is overcome by using a carrier substrate of sapphire so that the silicon multiplexer is sandwiched in the middle. The process starts with a buffer layer of CdTe deposited using an organometallic vapor-phase epitaxy (OMVPE) process to 8.5 µm thickness followed by a 10 µm-thick p-type HgCdTe layer grown by LPE. The n–p diodes are formed by ion implantation. This process has been used to fabricate 1024×1024 arrays with 3.2 µm cut-off [36.42] and 2048×2048 SW arrays of 16 cm^2 in area [36.99]. The 2048×2048 array, produced by Rockwell Scientific Company LLC for an astronomy application, is shown in Fig. 36.17. The arrays are 37×37 mm in size and are backside-illuminated through the sapphire substrate.

36.6.5 HgCdTe 2-D Arrays for the 3–5 µm (MW) Band

In the MW band, the development of array technology has reached a stage where all of the reported device processes will result in essentially background-limited performance for a wide range of applications. Individual arrays differ from one another in array size and pixel size and in imaging artefacts, such as uniformity, crosstalk and blooming. Some mention needs to be made of the factors controlling sensitivity in 2-D arrays, because this does not depend primarily on the HgCdTe but the silicon multiplexer.

Both charge-coupled devices (CCDs) and standard foundry CMOS devices have been used for multiplexing infrared 2-D arrays, but CMOS is now the preferred choice and CMOS circuits will operate well at low temperatures. The function of the multiplexer is to integrate the photocurrent from the photodiode array, perform a limited amount of signal processing and sequentially readout the signals. Most manufacturers produce their own multiplexer designs because these often have to be tailored to the application. For background-limited detectors, the NETD (noise-equivalent temperature difference) can be shown to be inversely proportional to the square root of the number of integrated photogenerated electrons [36.100]. High sensitivity can therefore only be achieved if a large number of electrons are integrated and this requires the integration capacitance in each pixel to be fairly high. Normally the capacitor is created using a MOSFET structure with a thin gate oxide dielectric and capacitance densities as high as 3 fF µm^2 are typical. Nevertheless, the capacitance is restricted to about 1 pF in pixels of around 25 µm square, and so the best NETD that can be expected is about 10 mK per

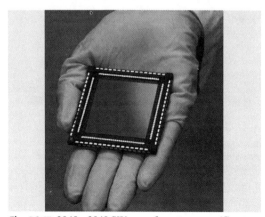

Fig. 36.17 2048×2048 SW array for astronomy. (Courtesy of Rockwell Scientific Company)

frame. It is important to note that this NETD limitation has nothing to do with the HgCdTe but only depends on the multiplexer design. Furthermore, as the pixel size is reduced the capacitance is decreased disproportionately, so the NETD cannot be maintained in small-pitch arrays.

Some general comments about operating temperature can be made. Standard vacancy-doped HgCdTe diodes, such as planar or via-hole homojunctions, need cooling to 120–140 K for BLIP performance in typical thermal imaging applications. The standard HDVIP^{TM}, which is extrinsically doped with many fewer vacancies, can operate up to 155 K. HgCdTe diodes fabricated with controlled doping and bandgap engineering can achieve BLIP operation near to ≈ 200 K. Operating at a higher temperature can prolong the life of cryocoolers and results in lower power consumption.

HgCdTe offers better spatial resolution (often quantified as the modulation transfer function, MTF) than some competing technologies such as planar InSb processes. Fully defined MOVPE or MBE diode structures will have the best MTF performance, but via-hole technologies (loophole and HDVIP) closely approach them because of their concentric junction shape, and this permits true resolution improvement in small pixels. DLHJ structures and planar structures rely on lateral collection to give very high fill factor, and high quantum efficiencies, and they need to have restricted sideways diffusion to limit blurring. Blooming due to an intense optical highlight follows a similar argument.

36.6.6 HgCdTe 2-D Arrays for the 8–12 μm (LW) Band

LW HgCdTe detectors were originally developed for long linear arrays and the processes can be extended to 2-D arrays with one additional consideration. The earth plane of the array must carry the sum of the photocurrent from all of the pixels in the array, and if there is insufficient conductance a voltage will appear between the center and edge of the array. This is called "substrate debiasing", and in the extreme case will inhibit injection in the center diodes. In any event it provides a potential long-range crosstalk mechanism and detector designers will try to minimize the resistance of the earth plane layer. In homojunction arrays, the p-type common layer cannot be doped high enough and a special metal grid is needed, as shown in the DRS Technologies array in Fig. 36.18. In other technologies the common layer must be doped high enough to prevent long-range crosstalk and use junctions with high dynamic resistance.

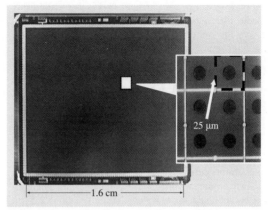

Fig. 36.18 LW HDVIP array with metal grid for substrate debiasing control. (Courtesy DRS Technologies)

36.6.7 HgCdTe 2-D Arrays for the 1–3 μm (SW) Band

Important applications in the SW (1–3 μm) waveband include: thermal imaging (using nightglow), spectroscopy and active imaging using lasers. Other materials can be used for SW detectors but products such as InGaAs tend to have high noise beyond 1.7 μm due to defects arising from the lattice mismatch with the InP substrate. Most HgCdTe technologies can be extended to SW with little change to the processing. Carrier lifetimes should be dominated by radiative recombination, but in fact S-R centers and technologically related limits probably apply in practice. SW detectors can produce good quality diodes in the presence of fairly high levels of misfit dislocations and so can be made using some of the newer technologies described in Sect. 36.6.4. For imaging and spectroscopy applications, typical operating temperatures are around 200 K and thermoelectric coolers are often offered as standard products.

A growing area is active imaging using eyesafe 1.54 μm lasers. The shorter wavelength results in higher resolution for a given optical aperture size and is particularly beneficial in long-range sensor systems. The very short pulse nature of the laser can be used to give range information and is often used with gated detectors to produce imaging in a certain range zone (so-called laser shape profiling). When range information is produced it is called laser range profiling and, in the limit, 3-D imaging. At present the field is in its early development stage and workers are using relatively small arrays to perform initial research. Detectors for this field need to be very fast and use a wide, weakly doped absorber region to give low capacitance. This is a so-called PiN diode, where the "i" stands for intrinsic. It also needs a very low series resistance to withstand the shock of the laser return, and this lends itself to metal grids and heavily doped heterostructures.

In most SW applications the photon flux is low and it is difficult to achieve reasonable signal-to-noise performance. However, it is relatively easy to enhance the signal by providing some avalanche gain in the device. Electron avalanching in MW HgCdTe via-hole diodes has been described [36.101, 102]. The electron and hole ionization rates are very different in HgCdTe, and this allows almost pure exponential, noise-free avalanche gain at fairly moderate voltages. So avalanche gain can readily be achieved for wavelengths above about $2.5\,\mu m$ when the absorber region is p-type and electrons are the minority carriers. A gain of 10 is typically observed for around 5 V at a cut-off of $5\,\mu m$. An alternative structure uses SW material ($1.6\,\mu m$) and a resonant enhancement of the hole impact ionization rate when the bandgap equals the spin-orbit split-off energy [36.103]. Gains of 30–40 have been seen with voltages of 85–90 V.

SW laser-gated imaging systems using avalanche gain in HgCdTe are now being reported for use in long-range identification applications [36.104]. Here the combination of sophisticated ROICs and high quality HgCdTe device processing is producing 320×256 arrays with a sensitivity down to 10 photons rms.

36.6.8 Towards "GEN III Detectors"

"GEN III" is a commonly used term which stems from the abbreviation for third-generation infrared detector. The definition of a "GEN III" detector can differ between different nations but the general guideline is any detector that offers an imaging advantage over conventional first- and second-generation systems. Common agreed examples include megapixel arrays with high density, dual color or even multispectral arrays, higher operating temperature, fast readout rates, very low NETD due to pixel-level signal compression and retina-level signal processing. Some of the progress made in such research is described here in more detail.

Two-Color Array Technology

Resolving the spectral signature can enhance the identification of objects in a thermal image. As a stepping stone to true multispectral arrays, many manufacturers are developing two-color detectors with simultaneous readout of flux levels in two separate infrared bands. A good example of a field requirement is to separate a sun glint from a hot thermally emitting object. The field is still in its early development stages, but some good results have been achieved.

Two-color arrays use two layers of HgCdTe with the longer wavelength layer underneath. The technological challenge is to make contact with the top layer without obstructing the sensitive area of the bottom layer. Ideal devices, where the integration takes place simultaneously and the sensitive area is co-spatial, are practically difficult to make on normal pixel sizes of say less than $50\,\mu m$. Most dual-color devices using indium bumps need two contacts within each pixel and make contact to the top diode by etching a hole in the bottom layer. The silicon needs to be custom-designed because the flux levels in the two bands may be markedly different. The polarity of the input MOSFET and the gain within the silicon must be matched to the technology and application.

Several successful arrays have been reported including a 64×64 simultaneous MW/LW dual-band HgCdTe array on $75\,\mu m$ pitch [36.105]. The array is fabricated from a four-layer P-n-N-P film grown in situ by MOVPE on CdZnTe and mounted on a custom ROIC. This approach has been applied to the double-layer planar heterostructure, DLPH, in a process called SUMIT, which stands for simultaneous unipolar multispectral integrated technology [36.106]. The DLPH structure can be turned into a two-color device by employing two absorbers with different bandgaps separated by a wide bandgap barrier. An etch step is used to expose the lower, shorter wavelength absorber and to form an arsenic doped junction. A pixel size of $40\,\mu m$ for each color is used in 128×128 demonstrators for MW/MW arrays. A variant of this basic process has also been presented by AIM [36.107] and they have achieved two colors in a concentric arrangement in 192×192 and 256×256 arrays with $56\,\mu m$ pitch, again using a full custom ROIC.

Two-color arrays have been fabricated using a via-hole type structure in which each pixel has two via-holes [36.81]. One is for connecting the longer wavelength array to the silicon in the normal way and the other is isolated from the longer wavelength material and connects the shorter wavelength to the silicon, as is illustrated schematically in Fig. 36.19. The spectral response of the two colors is shown in Fig. 36.20 for the two different doublets. The two-color arrays utilize standard 640×480, $25\,\mu m$ pitch ROICs (as used in LW focal plane arrays). The unit cell design in these ROICs incorporates one buffer amplifier circuit for every two HgCdTe pixels. The two-color array actually uses four

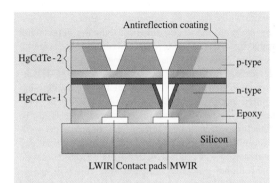

Fig. 36.19 MW/LW HDVIP FPA two-color composite. (Courtesy DRS Technologies)

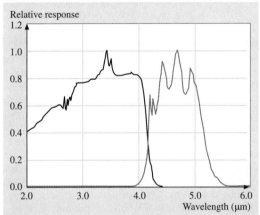

Fig. 36.20 Spectral response of MW/MW two-color HD-VIP HgCdTe FPA. (Courtesy DRS Technologies)

mono-color pixels to form a two-color unit cell, thus resulting in a 50 μm × 50 μm geometry, with two buffer amplifiers per unit cell, one for each color.

The shrink potential of dual-color technologies is limited by the need for two contacts per pixel. An alternative approach is to use a so-called bias-selectable detector [36.108–110], which needs only one bump and is compatible with small pixel sizes. This is an elegant technique but has the operational disadvantage of nonsimultaneous integration, and for this reason these detectors are often called sequential mode, dual-band detectors. Typical devices use a p-n-N-P or n-p-P-N sandwich structure with each p–n junction in different composition material. The wavelength can be selected by the bias polarity. There is a potential problem with parasitic transistor action and this needs to be suppressed by appropriate barriers. The silicon circuit requires a PMOS input for one polarity and an NMOS for the other and these must be switched for each color.

Higher Operating Temperature (HOT) Device Structures

The general principles for designing HOT HgCdTe detectors have been proposed over the past two decades, primarily as a result of research in the UK. Good summaries have been reported in [36.111, 112]. The advantages afforded by HOT detectors are in the area of cost of ownership and portability. LW arrays operating at only a few tens of degrees higher can significantly improve the mean time before failure (mtbf) of a Stirling engine cooler. In the MW band BLIP detectors on thermoelectric coolers could offer a much cheaper, smaller and more robust product, ideally suited to continuously operating security-type applications. Operation at 220 K has been demonstrated for a 128 × 128 array using MOVPE heterodiode technology [36.113], and the key step was designing a multiplexer that enabled very accurate biasing of the diodes at zero volts to suppress the dark current and associated $1/f$ noise.

The possibility of operating detectors at temperatures near room temperature has been proposed [36.114]. The basic structure is a P–p$^-$–N, as illustrated in Fig. 36.21. The active volume of the absorber is small relative to a minority carrier diffusion length, and it is operated in strong nonequilibrium by reverse biasing the minority carrier contact to completely extract all of the intrinsically generated minority carriers. To preserve space-charge neutrality, the majority carrier concentration drops to the background dopant concentration and Auger generation is effectively suppressed. Remaining components are associated with S-R centers and possibly injection from the contact and surface regions. The realization of HOT LW arrays will depend on the suppression of high $1/f$ noise associated with the reverse bias operation. In MW arrays it is also important to

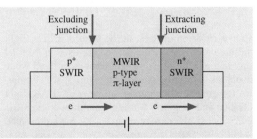

Fig. 36.21 Schematic of HOT device structure

suppress competing recombination mechanisms. Nevertheless, the device concept has enormous importance as a route towards high-performance infrared detectors with minimal cooling.

Retina-Level Processing

In very large arrays operating at high frame rates, the downloading of signal data and the subsequent signal processing can be daunting and is called the data processing bottleneck problem. The human retina presents an example of how evolution has dealt with the problem. The eye performs a number of image processing operations in the "z-plane", including, in order: logarithmic photon sensing, spatial filtering, temporal filtering, motion sensitivity and data decomposition. An essential function is the Difference of Gaussian or DoG filter which is used for both edge and contrast enhancement [36.115].

Focal plane arrays with neuromorphic processing are under development at a number of centers [36.116–118]. A good example is the MIRIADS programme (miniature infrared imaging applications development system). MIRIADS uses a neuromorphic FPA, with temporal high-pass filtering, frame co-adding and a Difference of Gaussians operation to detect motion, enhance edges and reject ambient light levels. The current reported array is a 64×64.

36.6.9 Conclusions and Future Trends for Photovoltaic HgCdTe Arrays

Photovoltaic 2-D HgCdTe detectors are serving many applications worldwide and are supplanting photoconductive detectors in second-generation systems for military, commercial and scientific applications, offering improved temperature sensitivity, lower power consumption, weight and volume. The continuing development work within manufacturing centers, supported by the wider scientific community, is aimed at increasing the radiometric performance, reducing defects and reducing manufacturing costs. Medium-term research is being directed at the best advanced materials and device structures for the next generation of HgCdTe arrays, which will combine state of the art background limited performance with low-cost wafer-scale processing. In the future it is envisaged that HgCdTe detectors will be grown directly on silicon or even silicon multiplexers for very low cost detectors. Bandgap engineering will produce heterostructure detectors with much higher operating temperatures. This will enable background-limited operation in MW arrays at near room temperature and operation of LW detectors on thermoelectric coolers or low-power Stirling engines. Device structures will be extended to produce bispectral and multispectral capability. Advanced detectors will critically depend on the future development of multiplexers. Challenges for the future include: arrays with larger physical size, smaller pixels, higher sensitivity, faster frame rate and even perhaps retina-type processing to reduce the need for external signal processing. The ultimate performance potential of HgCdTe will ensure that it is the material of choice for all high-performance infrared systems. In MW arrays it will compete with InSb offering better imaging characteristics and higher operating temperature. Active imaging will drive a major growth in SW array technology. The top performing thermal imaging cameras currently depend on LW, HgCdTe, long linear arrays offering high-definition TV images: the only technology currently capable of doing this economically. The market for smaller LW 2-D arrays is dominated by uncooled (thermal) detectors and multiple quantum well detectors, but HgCdTe offers considerably improved sensitivity, and once the cost becomes competitive it should compete with these technologies for many applications.

36.7 Emission Devices in II–VI Semiconductors

Infrared emission devices, in the form of LEDs (light emitting diodes) and laser diodes, are used in a broad range of applications such as gas sensors, infrared scene simulators, molecular spectroscopy, free space communications, fiber optic communications and LIDAR (light detection and ranging). A useful summary of emission device theory and current progress has been presented by [36.119]. There is competition from lead salt semiconductors (in the 3 to 34 μm range) and narrow gap III–V compounds. HgCdTe LEDs use a forward-biased junction to inject minority carriers into an n-type region, typically with low doping to encourage radiative recombination. The external photon efficiency is limited by factors such as competing recombination mechanisms, reabsorption of the photons and internal reflection. Nevertheless, efficiencies as high as 5% have been achieved at 4 μm, using an immersion lens to overcome internal reflection [36.120].

One of the disadvantages of the LED is that the spectrum is often too broad for many applications, and lasers are needed. Lasers also offer greater output power and higher spatial coherence, but they are more difficult to make and have additional noise mechanisms. Infrared lasers covering the 3–5 μm band have been demonstrated with pulsed operation at 80 K. It is only recently that the complex double heterojunction structures needed for lasing have been available from low-temperature VPE processes, so the field is relatively young. Nevertheless, peak powers of up to several milliwatts have been recorded in diode pumped lasers and over 1 W in optically pumped lasers.

Using a structure similar to that reported in Sect. 36.6.8 and illustrated in Fig. 36.21, the carrier densities, in reverse bias, can be lowered to well below the thermal equilibrium values. Under this condition, the absorption of photons can exceed the emission rate and the surface of the device can look cold. This is called negative luminescence [36.121] and devices have been developed for use as temperature references in thermal imaging cameras.

Infrared LEDs have also been reported [36.122] in the dilute magnetic semiconductors $Hg_{1-x}Mn_xTe$ and $Hg_{1-x-y}Cd_xMn_yTe$, where, in principle, the bandgap could be tuned by the applied magnetic field.

36.8 Potential for Reduced-Dimensionality HgTe–CdTe

The emergence of low-temperature growth processes will enable HgTe–CdTe structures to be grown with dimensions small enough to show quantum confinement effects. At present this research is in its early stages, and there are no active programs to make devices, but this could change in future. HgTe–CdTe superlattices offer many benefits for detectors. The minority carrier lifetime could be enhanced beyond that of bulk material, providing better quantum efficiency. The optical absorption coefficient can be much larger than bulk material, allowing LW detectors to use absorbers as thin as a few microns. Furthermore, tunneling currents can be reduced by orders of magnitude, permitting operation at much longer wavelengths. The state of the art of reduced dimensionality material is summarized in [36.123].

Quantum well lasers are also known to offer significant advantages over double heterojunction devices with bulk active regions. For instance, the maximum operating temperatures for pulsed and CW lasers could be increased by over 100 K, and this would have important consequences for the attractiveness of narrow-gap II–VI emitters.

References

36.1 W. D. Lawson, S. Nielsen, E. H. Putley, Y. S. Young: J. Phys. Chem. Solids **9**, 325–329 (1959)
36.2 M. A. Kinch: Mater. Res. Soc. Symp. Proc. **90**, 15 (1987)
36.3 N. Duy, D. Lorans: Semicond. Sci. Technol. **6(12)**, C93 (1991)
36.4 S. Oguz, R. J. Olson, D. L. Lee et al.: Proc. SPIE **1307**, 560 (1990)
36.5 T. Tanaka, K. Ozaki, K. Yamamoto et al.: J. Cryst. Growth **117**, 24 (1992)
36.6 M. A. Kinch, S. R. Borello: Infrared Phys. **15**, 111 (1975)
36.7 C. T. Elliott: *Handbook on Semiconductors*, 1st edn., ed. by C. Hilsum (North-Holland, Amsterdam 1981) p. 727
36.8 C. T. Elliott, N. T. Gordon: *Handbook on Semiconductors*, 2nd edn., ed. by C. Hilsum (North-Holland, Amsterdam 1993) p. 841
36.9 A. Kolodny, I. Kidron: Infrared Phys. **22**, 9 (1992)
36.10 N. Oda: Proc. SPIE **915**, 20 (1988)
36.11 M. A. Kinch, S. R. Borello, A. Simmons: Infrared Phys. **17**, 127 (1977)
36.12 D. L. Smith: J. Appl. Phys. **54**, 5441 (1983)
36.13 T. Ashley, C. T. Elliott: Infrared Phys. **22**, 367 (1982)
36.14 D. L. Smith, D. K. Arch, R. A. Wood, M. W. Scott: Appl. Phys. Lett. **45(1)**, 83 (1984)
36.15 C. A. Musca, J. F. Siliquini, B. D. Nener, L. Faraone: IEEE Trans. Electron. Dev. **44(2)**, 239 (1997)
36.16 R. Kumar, S. Gupta, V. Gopal, K. C. Chabra: Infrared Phys. **31(1)**, 101 (1991)
36.17 M. A. Kinch, S. R. Borello, B. H. Breazale, A. Simmons: Infrared Phys. **16**, 137 (1977)
36.18 M. B. Reine, E. E. Krueger, P. O'Dette et al.: Proc. SPIE **2816**, 120 (1996)
36.19 I. M. Baker, F. A. Capocci, D. E. Charlton, J. T. M. Wotherspoon: Solid-State Electron. **21**, 1475 (1978)
36.20 C. T. Elliott: Electron. Lett. **17**, 312 (1981)
36.21 C. T. Elliott, N. T. Gordon, R. S. Hall, G. J. Crimes: J. Vac. Sci. Technol. A **8**, 1251 (1990)
36.22 C. T. Elliott: UK Patent 1488, p. 258 (1977)
36.23 C. T. Elliott, D. Day, D. J. Wilson: Infrared Phys. **22**, 31 (1982)

36.24 C. T. Elliott, C. L. Jones: *Narrow-Gap II–VI Compounds for Optoelectronic and Electromagnetic Applications* (Chapman Hall, New York 1997) Chap. 16
36.25 A. Blackburn, M. V. Blackman et al.: Infrared Phys. **22**, 57 (1982)
36.26 D. J. Day, T. J. Shepherd: Solid-State Electron. **25(6)**, 707 (1982)
36.27 T. J. Shepherd, D. J. Day: Solid-State Electron. **25(6)**, 713 (1982)
36.28 T. Ashley, C. T. Elliott, A. M. White et al.: Infrared Phys. **24(1)**, 25 (1984)
36.29 C. T. Elliott: Proc. SPIE **1038**, 2 (1989)
36.30 A. Campbell, C. T. Elliott, A. M. White: Infrared Phys. **27(2)**, 125 (1987)
36.31 A. Sher, A. B. Chen, W. E. Spicer, C. K. Shih: J. Vac. Sci. Technol. A **3**, 105 (1985)
36.32 A. Rogalski: *Infrared Detectors and Emitters: Materials and Devices*, Electron. Mater. Vol. 8 (Kluwer Academic, Dordrecht 2001) Chap. 12
36.33 E. J. Smith, T. Tung, S. Sen et al.: J. Vac. Sci. Technol. A **5**, 3043 (1987)
36.34 J. Piotrowski, T. Niedziela: Infrared Phys. **30**, 113 (1990)
36.35 E. A. Patten, M. H. Kalisher, G. R. Chapman et al.: J. Vac. Sci. Technol. B **9**, 1746 (1991)
36.36 J. Ameurlaine, A. Rousseau, T. Nguyen-Duy, R. Triboulet: Proc. SPIE **929**, 14 (1988)
36.37 D. L. Kaiser, P. Becla: Mater. Res. Soc. Symp. Proc. **90**, 397 (1987)
36.38 P. Becla: J. Vac. Sci. Technol. A **4**, 2014 (1986)
36.39 R. M. Broudy, V. J. Mazurczyk: Semicond. Semimet., 18 (1991)
36.40 M. B. Reine: Proc. SPIE **443**, 2 (1983)
36.41 M. B. Reine, K. R. Maschoff, S. B. Tobin et al.: Semicond. Sci. Technol. **8**, 788 (1993)
36.42 L. J. Kozlowski: Proc. SPIE **2745**, 2 (1996)
36.43 L. J. Kozlowski, J. Montroy, K. Vural, W. E. Kleinhans: Proc. SPIE **3436**, 162 (1998)
36.44 M. B. Reine, A. K. Sood, T. J. Tredwell et al.: *Semiconductors and Semimetals*, Vol. 18, ed. by R. K. Willardson, A. C. Beer (Academic, New York 1981) Chap. 6
36.45 M. B. Reine: *Infrared Detectors and Emitters: Materials and Devices*, Electron. Mater. Vol. 8 (Kluwer Academic, Dordrecht 2001) Chap. 12, p. 8
36.46 D. E. Lacklison, P. Capper et al.: Semicond. Sci. Technol. **2**, 33 (1987)
36.47 P. L. Polla, R. L. Aggarwal, D. A. Nelson et al.: Appl. Phys. Lett. **43**, 941 (1983)
36.48 O. K. Wu, G. S. Kamath, W. A. Radford et al.: J. Vac. Sci. Technol. A **8**(2), 1034 (1990)
36.49 O. P. Agnihotri, C. A. Musca, L. Faraone: Semicond. Sci. Technol. **13**, 839–845 (1998)
36.50 W. W. Anderson: Infrared Phys. **20**, 353 (1980)
36.51 J. Y. Wong: IEEE Trans. Electron. Dev. **27**, 48 (1980)
36.52 W. W. Anderson, K. J. Hoffman: J. Appl. Phys. **53**, 9130 (1982)
36.53 C. T. Sah: Phys. Rev. **123**, 1594 (1961)
36.54 R. E. DeWames, J. G. Pasko, E. S. Yao, A. H. B. Vanderwyck, G. M. Williams: J. Vac. Sci. Technol. **A6**, 2655 (1988)
36.55 Y. Nemirovski, D. Rosenfeld, R. Adar, A. Kornfeld: J. Vac. Sci. Technol. **A7**, 528 (1989)
36.56 D. Rosenfeld, G. Bahir: IEEE Trans. Electron. Dev. **39**, 1638–45 (1992)
36.57 Y. Nemirovsky, R. Fastow, M. Meyassed, A. Unikovsky: J. Vac. Sci. Technol. **B9**(3), 1829 (1991)
36.58 I. M. Baker, C. D. Maxey: J. Electron. Mater. **30**(6), 682 (2003)
36.59 I. M. Baker, G. J. Crimes, C. K. Ard et al.: IEE Conf. Pub. **321**, 78 (1990)
36.60 Y. Nemirovsky, A. Unikovsky: J. Vac. Sci. Technol. **B10**, 1602 (1992)
36.61 J. H. Tregilgas: J. Vac. Sci. Technol. **21**, 208 (1982)
36.62 J. P. Hirth, H. Ehrenreich: J. Vac. Sci. Technol. **A3**, 367 (1985)
36.63 A. Szilagyi, M. N. Grimbergen: J. Cryst. Growth **86**, 912 (1988)
36.64 A. J. Syllaios, L. Colombo: *Proc. IEDM Conf.* (IEEE, New York 1982) p. p137
36.65 B. Pelliciari, G. Baret: J. Appl. Phys. **62**, 3986 (1987)
36.66 S. M. Johnson, D. R. Rhiger, J. P. Rosberg et al.: J. Vac. Sci. Technol. **B10**, 1499 (1992)
36.67 P. W. Norton, A. P. Erwin: J. Vac. Sci. Technol. **A7**, 503 (1989)
36.68 P. S. Wijewarnasuriya, M. Zandian, D. B. Young et al.: J. Electron. Mater. **28**, 649–53 (1999)
36.69 I. M. Baker, G. J. Crimes, J. E. Parsons, E. S. O'Keefe: Proc. SPIE **2269**, 636 (1994)
36.70 S. M. Johnson, J. A. Vigil, J. B. James et al.: J. Electron. Mater. **22**, 835 (1993)
36.71 S. J. C. Irvine: *Narrow-gap II–IV Compounds for Optoelectronic and Electromagnetic Applications* (Chapman and Hall, New York 1997) Chap. 3
36.72 J. Tunnicliffe, S. J. Irvine, S. Dosser, J. Mullin: J. Cryst. Growth **68**, 245 (1984)
36.73 C. D. Maxey, J. P. Camplin, I. T. Guilfoy et al.: J. Electron. Mater. **32**(7), p656 (2003)
36.74 O. K. Wu, T. J. deLyon, R. D. Rajavel, J. E. Jensen: *Narrow-Gap II–IV Compounds for Optoelectronic and Electromagnetic Applications*, Part 1 (Chapman and Hall, New York 1997) Chap. 4
36.75 M. V. Blackman et al.: Elec. Lett. **23**, 978 (1987)
36.76 S. Margalit, Y. Nemirovsky, I. Rotstein: J. Appl. Phys. **50**, 6386 (1979)
36.77 A. Kolodny, I. Kidron: IEEE Trans. Electron. Dev. **ED-27**, 37 (1980)
36.78 L. O. Bubulac, W. E. Tennant, R. A. Riedel et al.: J. Vac. Sci. Technol. **21**, 251 (1982)
36.79 L. O. Bubulac, W. E. Tennant et al.: Appl. Phys. Lett. **51**, 355 (1987)
36.80 J. Syz, J. D. Beck, T. W. Orient, H. F. Schaake: J. Vac. Sci. Technol. **A7**, 396 (1989)
36.81 M. A. Kinch: Proc. SPIE **4369**, 566 (2001)
36.82 J. White et al.: J. Electron. Mater. **30**(6), 762 (2001)

36.83 L. O. Bubulac, C. R. Viswanathan et al.: J. Cryst. Growth **123**, 555 (1992)
36.84 I. M. Baker, R. A. Ballingall: Proc. SPIE **510**, 210 (1985)
36.85 P. Tribulet, J-P. Chatard, P. Costa, S. Paltrier: J. Electron. Mater. **30**(6), 574 (2001)
36.86 T. Tung, M. H. Kalisher, M. H. Stevens et al.: Mater. Res. Soc. Symp. Proc. **90**, 321 (1987)
36.87 C. C. Wang: J. Vac. Sci. Technol. **B9**, 740 (1991)
36.88 G. N. Pulz, P. W. Norton, E. E. Krueger, M. B. Reine: J. Vac. Sci. Technol. **B9**, 1724 (1991)
36.89 P. W. Norton, P. LoVecchio, G. N. Pultz et al.: Proc. SPIE **2228**, 73 (1994)
36.90 T. Tung: J. Cryst. Growth **86**, 161 (1988)
36.91 J. Arias, M. Zandian, J. G. Pasko et al.: J. Appl. Phys. **69**, 2143 (1991)
36.92 J. M. Arias, J. G. Pasko, M. Zandian et al.: Appl. Phys. Letts. **62**, 976 (1993)
36.93 J. Bajaj: Proc. SPIE **3948**, 42 (2000)
36.94 K. W. Hodapp, J. K. Hora, D. N. B. Hall et al.: New Astronomy **1**, 177 (1996)
36.95 C. D. Maxey, C. J. Jones, N. Metcalf et al.: Proc. SPIE **3122**, 453 (1996)
36.96 J. B. Varesi, R. E. Bornfreund, A. C. Childs et al.: J. Electron. Mater. **30**(6), 56698 (2001)
36.97 D. J. Hall, L. Buckle, N. T. Gordon et al.: Proc. SPIE **5406**, 317 (2004)
36.98 G. Bostrup, K. L. Hess, J. Ellsworth, D. Cooper, R. Haines: J. Electron. Mater. **30**(6), 560 (2001)
36.99 K. Vural, L. J. Kozlowski, D. E. Cooper et al.: Proc. SPIE **3698**, 24 (1999)
36.100 N. T. Gordon, I. M. Baker: *Infrared Detectors and Emitters: Materials and Devices*, Electron. Mater. Vol. 8 (Kluwer Academic, Dordrecht 2001) Chap. 2, p. 23
36.101 J. D. Beck, C.-F. Wan, M. A. Kinch, J. E. Robinson: Proc. SPIE **4454**, 188 (2001)
36.102 M. A. Kinch, J. D. Beck, C.-F. Wan et al.: J. Electron. Mater. **33**(6), 630 (2003)
36.103 T. J. de Lyon, J. E. Jenson, M. D. Gordwitz et al.: J. Electron. Mater. **28**, 705 (1999)
36.104 I. M. Baker, S. S. Duncan, J. W. Copley: Proc. SPIE **5406**, 133 (2004)
36.105 M. B. Reine, A. Hairston, P. O'Dette et al.: Proc. SPIE **3379**, 200 (1998)
36.106 W. E. Tennant, M. Thomas, L. J. Kozlowski et al.: J. Electron. Mater. **30**(6), 590 (2001)
36.107 W. Cabanski, R. Brieter, R. Koch et al.: Proc. SPIE **4369**, 547 (2001)
36.108 J. M. Arias, M. Zandian, G. M. Williams: J. Appl. Phys. **70**(8), 4620 (1991)
36.109 R. D. Rajavel, D. M. Jamba, O. K. Wu et al.: J. Electron. Mater. **26**, 476 (1997)
36.110 R. D. Rajavel, D. M. Jamba, O. K. Wu et al.: J. Electron. Mater. **27**, 747 (1998)
36.111 C. T. Elliott, N. T. Gordon, A. M. White: Appl. Phys. Lett. **74**, 2881 (1999)
36.112 C. T. Elliott: *Infrared Detectors and Emitters: Materials and Devices*, Electron. Mater. Vol. 8 (Kluwer Academic, Dordrecht 2001) Chap. 11
36.113 N. T. Gordon, C. L. Jones, D. J. Lees et al.: Proc. SPIE **5406**, 145 (2004)
36.114 C. T. Elliott, T. Ashley: Electron. Lett. **21**, 451 (1985)
36.115 D. Marr: *Vision* (W. H. Freeman, San Francisco 1982)
36.116 M. Masie, P. McCarley, J. P. Curzan: Proc. SPIE **1961**, 17 (1993)
36.117 P. McCarley: Proc. SPIE **3698**, 716 (1999)
36.118 C. R. Baxter, M. A. Massie, P. L. McCarley, M. E. Couture: Proc. SPIE **4369**, 129 (2001)
36.119 N. T. Gordon: *Narrow-Gap II-IV Compounds for Optoelectronic and Electromagnetic Applications* (Chapman and Hall, New York 1997) Chap. 17
36.120 P. Bouchut, G. Destefanis, J. P. Chamonal et al.: J. Vac. Sci. Technol. B **9**, 1794 (1991)
36.121 T. Ashley, C. T. Elliott, N. T. Gordon et al.: Infrared Phys. Technol. **36**, 1037 (1995)
36.122 R. Zucca, J. Bajaj, E. R. Blazewski: J. Vac. Sci. Technol. A **6**, 2725 (1988)
36.123 J. R. Meyer, I. Vurgaftman: *Infrared Detectors and Emitters: Materials and Devices*, Electron. Mater. Vol. 8 (Kluwer Academic, Dordrecht 2001) Chap. 14

37. Optoelectronic Devices and Materials

Unlike the majority of electronic devices, which are silicon based, optoelectronic devices are predominantly made using III–V semiconductor compounds such as GaAs, InP, GaN and GaSb and their alloys due to their direct band gap. Understanding the properties of these materials has been of vital importance in the development of optoelectronic devices. Since the first demonstration of a semiconductor laser in the early 1960s, optoelectronic devices have been produced in their millions, pervading our everyday lives in communications, computing, entertainment, lighting and medicine. It is perhaps their use in optical-fibre communications that has had the greatest impact on humankind, enabling high-quality and inexpensive voice and data transmission across the globe. Optical communications spawned a number of developments in optoelectronics, leading to devices such as vertical-cavity surface-emitting lasers, semiconductor optical amplifiers, optical modulators and avalanche photodiodes. In this chapter we discuss the underlying theory of operation of the most important optoelectronic devices. The influence of carrier–photon interactions is discussed in the context of producing efficient emitters and detectors. Finally we discuss how the semiconductor band structure can be manipulated to enhance device properties using quantum confinement and strain effects, and how the addition of dilute amounts of elements such as nitrogen is having a profound effect on the next generation of optoelectronic devices.

37.1 Introduction to Optoelectronic Devices .. 888
    37.1.1 Historical Perspective................. 888
37.2 **Light-Emitting Diodes and Semiconductor Lasers** 890

37.2.1 Carrier–Photon Interactions in Semiconductors...................... 890
37.2.2 Direct- and Indirect-Gap Semiconductors 890
37.2.3 Emission and Absorption Rates and the Einstein Relations 891
37.2.4 Population Inversion 892
37.2.5 Gain in Semiconductors 892
37.2.6 Density of States 893
37.2.7 Optical Feedback in a Fabry–Perot Laser Cavity...... 896
37.2.8 Wave-Guiding 897
37.2.9 Carrier Confinement 898
37.2.10 Current Confinement 898
37.2.11 Laser Threshold and Efficiency 899
37.2.12 Carrier Recombination Processes . 900
37.2.13 Temperature Sensitivity and T_0 ... 903

37.3 **Single-Mode Lasers**.............................. 904
    37.3.1 DFB lasers 904
    37.3.2 VCSELs 905

37.4 **Optical Amplifiers** 906
    37.4.1 An Introduction to Optical Amplification 906
    37.4.2 Semiconductor Optical Amplifiers (SOAs) 907

37.5 **Modulators** 907
    37.5.1 Modulator Theory...................... 907
    37.5.2 Polarisation-Insensitive Modulators................................. 909
    37.5.3 High-Speed High-Power QCSE Modulators..................... 910
    37.5.4 The Electro-Optic Effect.............. 911

37.6 **Photodetectors**.................................... 911
    37.6.1 Photodetector Requirements 912
    37.6.2 Photodetection Theory............... 912
    37.6.3 Detectors with Internal Gain....... 913
    37.6.4 Avalanche Photodetectors 913

37.7 **Conclusions**... 914

References 915

37.1 Introduction to Optoelectronic Devices

In this chapter we introduce the underlying theory and operating principles of semiconductor optoelectronic devices. There exist today a plethora of optoelectronic devices, which are used in a multitude of applications. These devices include sources such as light-emitting diodes (LEDs) and laser diodes, photodetectors, optical amplifiers and optical modulators. With such devices, one can generate, modulate, detect and switch photons in an analogous way to electrons in an electrical circuit. We begin this chapter by considering the underlying physical interactions between electrons and photons that occur in semiconductors and how they may be harnessed to produce a wide variety of devices. At the time of writing, optoelectronic devices have found their way into many different aspects of modern life whether it be the ubiquitous indicator LEDs on hi-fi systems, televisions, computers, solid-state lighting and countless other items or in the bar-code scanning systems at the supermarket, the compact disc (CD) player, CD-ROM/CD±R/CD±RW or digital versatile disk (DVD), DVD-ROM/DVD±R/DVD±RW/DVD-RAM at home, the laser printer in the office, or when using a telephone or watching cable television. Over the past decade or so there has been an information explosion whereby information from all over the world can be quickly accessed by anyone equipped with a computer and access to the internet. In all of these applications it is a semiconductor-based optoelectronic device that forms an essential part of the system.

One of the major advantages of semiconductor devices is their small size. For example, a typical edge-emitting laser measures approximately $500\,\mu m$ long by $250\,\mu m$ wide with a thickness of $100\,\mu m$. Several thousand such devices can be made from a single wafer. Thus, even when packaged, these form very compact sources of coherent radiation. Other types of laser, such as the gas laser, simply cannot compare with the semiconductor laser in terms of size, modulation rates, flexibility of application and power consumption. Furthermore, semiconductor devices can be tailored to meet the exacting requirements of an application by simply altering the composition of the various layers forming the structure.

Of the uses of semiconductor devices outlined above, their use in telecommunications stands out as having the largest impact on modern life. Digital-based data transmission allows information to be transmitted over large distances with a much lower degradation in signal quality compared with older, analogue-based systems. Optical telecommunications are ideal for use in digital systems, enabling data transmission rates in excess of $10\,Gbit/s$ using short optical pulses of $< 100\,ps$ in duration. Even at this exceptionally fast rate of data transmission, bit error rates of better than 1 in every 10^9 bits can be achieved. Such capabilities are a direct consequence of significant research and development work that has gone into producing semiconductor devices for the emission and detection of light. In the next section, we give a brief history of this development before proceeding to discuss the key elements of physics and technology related to the device operation.

37.1.1 Historical Perspective

The development of optoelectronic devices began in the early 1960s with the development of the light-emitting diode and soon thereafter, the semiconductor laser. *Holonyak* had been experimenting with the alloy GaAsP to produce visible light via spontaneous emission and was successful in producing visible (red) light-emitting diodes [37.1]. Stimulated emission was first predicted by *Albert Einstein* in his famous 1917 paper, *Zur Quantentheorie der Strahlung* (On the quantum theory of radiation), [37.2]. In 1961 the possibility of obtaining stimulated emission in semiconductors was discussed by *Bernard* and *Duraffourg* who, for the first time, derived the condition for lasing action in semiconductor materials [37.3]. This became known as the Bernard–Duraffourg condition, as discussed later in this chapter. In the following year, the first reports of lasing action in semiconductor materials were published by four independent groups, *Nathan* and co-workers [37.4], *Hall* and co-workers [37.5], *Quist* and co-workers [37.6] and *Holonyak* and co-workers [37.1]. These lasers were based upon GaAs which, with its direct band gap, made it suitable for use as an optical source. The first devices consisted of simple p–n homo-junctions which have a single interface between the n- and p-doped regions. With a wavelength of near to $900\,nm$, these lasers emitted in the near-infrared region of the electromagnetic spectrum. Following on from his success with LEDs, Holonyak produced visible (red) semiconductor lasers. The first semiconductor lasers had very high threshold current densities, J_{th} (defined later in this chapter), and could only be operated under pulsed conditions. This made such devices impractical but started a period of intensive research into producing the first continuous-wave (CW) room-temperature semiconductor-based laser. After much investment into developing the growth technology of liquid-phase epitaxy (LPE), it became possible to produce a high-quality double heterostructure consisting of GaAs sandwiched between higher-band-gap $Al_xGa_{1-x}As$ layers. The double heterostructure brought two key advantages over the homojunction; firstly, the lower-band-gap GaAs region formed a reservoir for the carriers (carrier confinement) where they could recombine across the band gap, and secondly, the higher refractive index of GaAs with respect to $Al_xGa_{1-x}As$ provided better confinement of the optical field. These two improvements resulted in a significant reduction in J_{th} of two orders of magnitude and enabled CW operation at room temperature to be achieved for the first time [37.7, 8]. Almost 30 years later, in 2000, Alferov and Kroemer were co-recipients of the Nobel prize for physics for their pioneering work on the development of semiconductor heterostructures. In the late 1970s further improvements in semiconductor growth technology led to the development of molecular-beam epitaxy (MBE) and vapour-phase epitaxy (VPE) which enabled increasingly thin layers to be grown reproducibly. Layer thicknesses of the order of less than 100 Å became achievable and introduced the regime in which quantum-confinement effects could be harnessed. These quantum well (QW) structures [37.9] brought about further improvements in laser performance, including a further increase in carrier confinement, narrower line width and extended wavelength tunability for a given material composition. The splitting of the degeneracy of the valence band also resulted in a reduced density of states at the top of the valence band. These improvements led to a substantially reduced J_{th}. In 1986, *Adams*, and also *Yablonovitch* and *Kane*, independently predicted that the introduction of biaxial strain into the active region of quantum wells would result in a further improvement in the laser characteristics [37.10, 11]. The proposed benefits of strain included a further reduction in the density of states at the top of the valence band (for compressive strain) and the ability to tailor the symmetry of the carrier distribution to that of the laser beam. The introduction of tensile strain made it possible to produce a semiconductor laser with transverse magnetic (TM) polarisation, something that was not previously possible. Strained-layer quantum wells are now included in the majority of commercially available semiconductor laser products. Recent material developments include quantum dot lasers, in which carriers are restricted in all three dimensions. There is a major effort to develop quantum dot (QD) lasers with better performance than existing quantum well lasers. More than 20 years ago the first theoretical prediction [37.12] showed that using three-dimensionally (3D)-confined structures with an atomic-like discrete density of states in the active region of semiconductor laser should allow the development of devices with low threshold current density and very high thermal stability. Since this time, quantum dot lasers have been demonstrated with record low room-temperature threshold current densities, $< 20 A/cm^2$ [37.13]. Work is ongoing to produce temperature-insensitive quantum dot lasers, particularly at 1.3 μm.

The late 1970s and early 1980s also gave rise to the development of single-mode lasers such as the distributed feedback (DFB) laser which have made long-haul optical communications possible [37.14]. The development of such advanced semiconductor laser devices spawned research into other semiconductor optoelectronic devices such as monolithic tunable lasers (discussed in detail by [37.15]), the semiconductor optical amplifier (SOA), optical modulators and advanced photodetectors. These devices are discussed in the latter part of this chapter.

In parallel with the developments mentioned above, other technologies were being investigated such as the vertical-cavity surface-emitting laser (VCSEL) originally proposed by *Iga* in 1977 and demonstrated by his group in 1979 (for a comprehensive review see [37.16]). VCSELs promise low-cost single-mode lasers which can easily be made into arrays. There remains intense activity to develop VCSELs for operation at the important telecommunications wavelength of 1.55 μm. VCSELs are discussed in later in this chapter. Other devices of note include the quantum cascade laser (QCL) developed by *Capasso* and co-workers in 1994 [37.17]. The development of QCLs was driven by the need to produce lasers emitting in the mid-infrared (mid-IR) ($\approx 3-12$ μm) for gas sensing and environmental monitoring applications. For conventional interband lasers as discussed above, this requires narrow-band-gap materials which are much less developed than materials such as GaAs and InP. QCLs get around this problem by utilising *intra*-band transitions involving only one type of carrier, e.g. electrons (thus they are unipolar devices), and can be grown using conventional GaAs- or InP-based materials. In such devices, the carrier cascades through several quantum wells, giving rise to many photons per injected electron, quite unlike conventional interband devices. Clearly, QCLs offer great possibilities for the future development of mid-IR lasers and perhaps, with suitable materials such as GaN, they may allow emission in the near-IR (for a review see [37.18]).

Unfortunately, it is impossible to go into detail about the developments of the full range of optoelectronic devices in a single book chapter. However, we hope that this chapter provides a basic grounding on the relevant physics of the most important optoelectronic devices allowing the interested reader to investigate further, starting with the bibliography.

37.2 Light-Emitting Diodes and Semiconductor Lasers

37.2.1 Carrier–Photon Interactions in Semiconductors

In its simplest form, a semiconductor can be considered as a system containing two energy bands populated with electrons. In such a system there exist three possible processes by which the electrons can move between the two bands: the spontaneous emission, absorption and stimulated emission processes, as shown schematically in Fig. 37.1. Spontaneous emission occurs when an electron in the conduction band with energy E_2 recombines with a hole in the valence band at a lower energy E_1. The difference in energy, $(E_2 - E_1)$ is emitted in the form of a photon. This process, as its name implies, is random and photons may be emitted in any direction with arbitrary polarisation. The probability of spontaneous emission is proportional to the density of electrons in the conduction band (n) and to the density of holes in the valence band (p). In the absorption process, an electron in the valence band is promoted to the conduction band by the absorption of a photon with an energy equal to the optical band gap. The probability of absorption is proportional to the photon density, the density of electrons in the valence band and the density of empty states in the conduction band. The stimulated emission process occurs when a photon of energy $E_2 - E_1$ interacts with an electron in the conduction band, causing it to recombine with a hole in the valence band thereby generating a photon. The photon released by this process has both identical energy, phase and momentum to that of the incident photon. The probability of stimulated emission is therefore proportional to n, p and the photon density. In contrast to spontaneous emission, stimulated emission produces photons that are essentially identical and therefore very pure monochromatic light is generated.

At room temperature in an undoped semiconductor, there are far fewer electrons in the conduction band than in the valence band. As a consequence, absorption is far more likely than emission. To achieve optical gain, the probability of stimulated emission must exceed that of absorption. By applying either an electrical current (electrical pumping) or by injecting light with photon energies greater than the transition energy (optical pumping), it is possible to inject electrons into the conduction band and holes into the valence band. Electrical pumping is used for practical devices.

37.2.2 Direct- and Indirect-Gap Semiconductors

Semiconductors may exist in two basic forms where the band gap is either direct or indirect. Whether or not the band gap is direct or indirect has a profound influence on their suitability for use in optoelectronic devices. Figure 37.2 illustrates the difference between a direct- (e.g. GaAs) and an indirect-band-gap (e.g. Ge) semiconductor. It can be seen that for the direct-gap material, the conduction-band (CB) energy minimum occurs at the same k-value as the valence-band (VB) maximum (direct). In contrast, for the indirect-gap material the CB minimum lies at a different k-value to the VB maximum. For any electron transition, the total energy and momentum must be conserved. The photon wavevector has a magnitude given by $2\pi/\lambda$, where λ

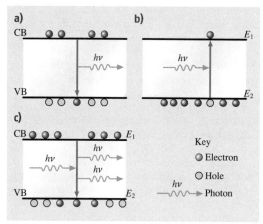

Fig. 37.1a–c The three electron–photon interactions in semiconductors: (a) spontaneous emission, (b) stimulated absorption, and (c) stimulated emission

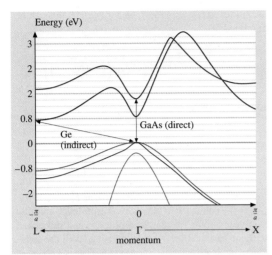

Fig. 37.2 Band structure of a direct-gap semiconductor (GaAs) and an indirect semiconductor (Ge)

is typically $\approx 1\,\mu\text{m}$, whilst the magnitude of the electron wavevector ranges between $-\pi/a$ and π/a within the first Brillouin zone, where a is the crystal lattice spacing, typically $\approx 1\,\text{Å}$. Clearly therefore, the photon wavevector is much smaller than the possible electron wavevectors and hence, if a photon interacts with an electron, the transition must occur with virtually no change in wavevector, hence only vertical transitions are allowed. It is clear that such transitions can occur in the direct-band-gap semiconductor. However, in the case of indirect semiconductors, such transitions are only possible via the interaction of a phonon. Since this becomes a three-particle interaction, the transition probability is significantly reduced. Stimulated emission is inhibited in indirect-band-gap semiconductors and it is for this reason that direct-band-gap semiconductors (such as GaAs, InP and related alloys) are primarily used to produce photoemissive devices such as LEDs and lasers. Indirect semiconductors, such as elemental germanium or silicon are therefore impractical for the production of such devices. However, these are very effective semiconductor materials for detection, as will be discussed later in this chapter.

37.2.3 Emission and Absorption Rates and the Einstein Relations

The populations of electrons and holes in the conduction and valence bands respectively are governed by both the densities of states and the Fermi–Dirac occupation probabilities. In a semiconductor, electron–electron and hole–hole scattering rates are typically $\approx 100\,\text{fs}$ whilst electron–hole recombination times are $\approx 1\,\text{ns}$. We can therefore assume that the electrons and holes are in thermal equilibrium with themselves and can be described using a Fermi–Dirac distribution. It is useful to define two quasi-Fermi levels. These are referred to as F_c and F_v, corresponding to the energy at which the occupation probability equals $\frac{1}{2}$ for *electrons* in the CB and VB, respectively. In thermal equilibrium, $F_c = F_v$. The corresponding energy-dependent Fermi functions are

$$f_c(E) = \frac{1}{1+\exp\left(\frac{E-F_c}{k_B T}\right)} \quad \text{and}$$

$$f_v(E) = \frac{1}{1+\exp\left(\frac{E-F_v}{k_B T}\right)}, \quad (37.1)$$

where k_B is the Boltzmann constant and T is the absolute temperature.

By considering each of the processes illustrated in Fig. 37.1 one can obtain mathematical expressions relating each of the processes to the another. If we label states in the VB as "1" and states in the CB as "2", the *absorption rate*, r_{12}, for photons of energy, $h\nu$ is given by

$$r_{12} = B_{12}\rho_v f_v (1-f_c)\rho_c P(h\nu) \quad (37.2)$$

where ρ_c and ρ_v are the CB and VB densities of states, respectively. $P(h\nu)$ is the photon density at energy $h\nu$ and B_{12} is the Einstein coefficient for the absorption process. In a similar way, the stimulated emission rate, r_{21}, may be written as

$$r_{21} = B_{21}\rho_c f_c (1-f_v)\rho_v P(h\nu) \quad (37.3)$$

where B_{21} is the Einstein coefficient for the stimulated emission process. For the *spontaneous* emission rate r_{21}^{spon}

$$r_{21}^{\text{spon}} = A_{21}\rho_c f_c (1-f_v)\rho_v . \quad (37.4)$$

A_{21} is the Einstein coefficient for the spontaneous emission process. Note that, since spontaneous emission does not require a photon to initiate the process, r_{21}^{spon} does not depend on $P(h\nu)$. Under steady-state conditions, the total upward transition rate must equal the total downward transition rate, thus

$$r_{12} = r_{21} + r_{21}^{\text{spon}} . \quad (37.5)$$

Thus, by combining equations (37.2)–(37.4), we obtain

$$B_{12}\rho_v f_v (1-f_c)\rho_c P(h\nu)$$
$$= B_{21}\rho_c f_c (1-f_v)\rho_v P(h\nu)$$
$$+ A_{21}\rho_c f_c (1-f_v)\rho_v . \quad (37.6)$$

In thermal equilibrium ($F_c = F_v$),

$$P(h\nu) = \frac{A_{21}}{B_{12}\exp\left(\frac{h\nu}{k_B T}\right) - B_{21}}. \quad (37.7)$$

The standard expression for black-body radiation is given by Planck's law as

$$P(h\nu) = \frac{8\pi^3 n^3 (h\nu)^2}{(hc)^3} \frac{1}{\exp\left(\frac{h\nu}{k_B T}\right) - 1}. \quad (37.8)$$

Here, n is the refractive index of the semiconductor. For simplicity, here we assume that the medium is nondispersive. From these two expressions of $P(h\nu)$ we obtain the result that

$$B_{12} = B_{21} = B \quad (37.9)$$

and

$$A_{21} = B \frac{8\pi^3}{(hc)^3} n^3 (h\nu)^2. \quad (37.10)$$

37.2.4 Population Inversion

The quantity of greatest importance to semiconductor laser operation is the *net* stimulated emission rate, which may simply be calculated as $r_{stim} = r_{21} - r_{12}$. Positive values of r_{stim} mean that an optical wave will grow in intensity as it travels through the semiconductor, whilst a negative r_{stim} implies that the optical wave would be reduced in intensity. From (37.2) and (37.3), r_{stim} can be written as

$$\begin{aligned}r_{stim} &= r_{21} - r_{12} \\ &= B_{21}\rho_c f_c (1-f_v)\rho_v P(h\nu) \\ &\quad - B_{12}\rho_v f_v (1-f_c)\rho_c P(h\nu)\end{aligned} \quad (37.11)$$

therefore

$$r_{stim} = r_{21} - r_{12} = B\rho_c \rho_v (f_c - f_v) P(h\nu). \quad (37.12)$$

The ratio of spontaneous emission to net stimulated emission is given by

$$\begin{aligned}\frac{r_{spon}}{r_{stim}} &= \frac{A_{21}\rho_c\rho_v f_c(1-f_v)}{B\rho_c\rho_v f_c (f_c - f_v) P(h\nu)} \\ &= \frac{A_{21} f_c (1-f_v)}{BP(h\nu)(f_c - f_v)}.\end{aligned} \quad (37.13)$$

Combining (37.13) with (37.10) leads to

$$\frac{r_{spon}}{r_{stim}} = \frac{8\pi^3 n^3 (h\nu)^2}{(hc)^3 P(h\nu)} \frac{f_c(1-f_v)}{(f_c - f_v)}. \quad (37.14)$$

Using the expressions for f_c and f_v (37.1), this can be simplified to

$$\frac{r_{spon}}{r_{stim}} = \frac{8\pi^3 n^3 (h\nu)^2}{(hc)^3 P(h\nu)} \frac{1}{\left[1 - \exp\left(\frac{h\nu - (F_c - F_v)}{k_B T}\right)\right]} \quad (37.15)$$

and consequently

$$\begin{aligned}r_{stim} = r_{spon} &\frac{(hc)^3 P(h\nu)}{8\pi^3 n^3 (h\nu)^2} \\ &\times \left[1 - \exp\left(\frac{h\nu - (F_c - F_v)}{k_B T}\right)\right].\end{aligned} \quad (37.16)$$

From this expression we see that, when $F_c - F_v = h\nu$, $r_{stim} = 0$. Thus, when the quasi-Fermi-level splitting equals the photon energy, the absorption and stimulated emission rates cancel. At this injection level, the semiconductor is effectively transparent. This is known as the Bernard–Duraffourg condition. For photons of energy $h\nu > F_c - F_v$, there will be absorption. However, for photons of energy $h\nu < F_c - F_v$ there is gain. This defines the condition necessary to achieve population inversion in the semiconductor and lasing action can occur when $F_c - F_v > E_g$.

37.2.5 Gain in Semiconductors

If $\alpha(h\nu)$ is the rate of gain/loss per unit length and $P(h\nu)$ is the photon density, then the rate at which the photon density increases/decreases per unit length is simply $\alpha(h\nu)P(h\nu)$. The corresponding rate of increase of photon density per unit time is then

$$r_{stim} = P(h\nu)\alpha(h\nu)\frac{c}{n} \quad (37.17)$$

where c/n is the speed of light within a semiconductor of refractive index, n. By equating our two expressions for r_{stim} [(37.16) and (37.17)] we find that

$$\begin{aligned}\alpha(h\nu) = r_{spon} &\frac{h^3 c^2}{8\pi^3 n^2 (h\nu)^2} \\ &\times \left[1 - \exp\left(\frac{h\nu - (F_c - F_v)}{k_B T}\right)\right].\end{aligned} \quad (37.18)$$

Under conditions of low pumping or at high photon energies ($h\nu \gg F_c - F_v$), (37.18) can be approximated to

$$\begin{aligned}\alpha(h\nu) = -r_{spon} &\frac{h^3 c^2}{8\pi^3 n^2 (h\nu)^2} \\ &\times \exp\left(\frac{h\nu - (F_c - F_v)}{k_B T}\right)\end{aligned} \quad (37.19)$$

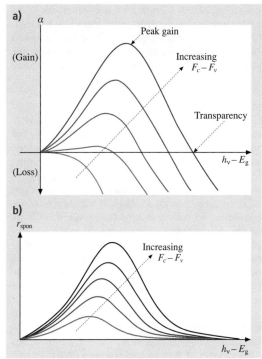

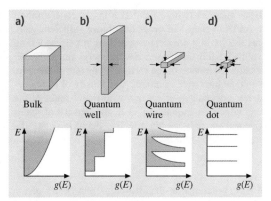

Fig. 37.4a–d Density of states in systems of differing dimensionality: (**a**) bulk, (**b**) quantum well, (**c**) quantum wire, and (**d**) quantum dot

Fig. 37.3 (**a**) gain/loss and (**b**) spontaneous emission spectra as a function of increasing quasi-Fermi-level splitting

from which we have

$$r_{\text{spon}} = -\alpha(h\nu)\frac{8\pi^3 n^2 (h\nu)^2}{h^3 c^2}$$
$$\times \exp\left(\frac{(F_c - F_v) - h\nu}{k_B T}\right) \quad (37.20)$$

Typical plots of the absorption and spontaneous emission rates can be seen in Fig. 37.3. From Fig. 37.3a it can be seen that, with increasing pumping level (increasing quasi-Fermi-level splitting), the semiconductor moves from purely absorption and becomes a gain medium. With increasing pumping, the spread of photon energies over which gain occurs also increases, with the peak gain moving towards higher photon energy. In Fig. 37.3b, the spontaneous emission rate is plotted as a function of photon energy. From this is can be seen that the emission spectrum is broad and corresponds to the output spectrum of a light-emitting diode (LED). With increasing pumping level, the integrated spontaneous emission increases whilst the peak shifts to higher photon energies. In practical devices, current-induced Joule heating

may lead to a slight decrease in the band gap, which reduces the overall blue shift.

37.2.6 Density of States

As a semiconductor laser is put under forward bias, the quasi-Fermi levels in the conduction and valence bands move towards higher energies. The separation of the quasi-Fermi levels is dependent on the density of states $\rho(E)$ in the conduction and valence bands. At laser transparency, where the quasi-Fermi-level separation equals the photon energy, the transparency carrier density, n_{tr} and hence the threshold carrier density, n_{th} depend upon the number of states in a given energy range. By reducing the number of available states, n_{th} and consequently I_{th} (J_{th}), can be lowered. The density of states can be modified by changing the band structure. This can be achieved by reducing the dimensionality, thereby limiting the motion of the carriers, and also by introducing strain into the active region of the laser, as shall be described shortly.

Dimensionality

In conventional bulk semiconductor materials, the carriers have unrestricted motion in all directions and can be described by a 3D density of states,

$$\rho_{\text{3-D}} \propto m_{\text{eff}}^{3/2} \sqrt{E} \quad (37.21)$$

where m_{eff} is the effective mass. This is illustrated in Fig. 37.4a, where the shaded region indicates the carrier population as a function of energy. In such (bulk) devices there are relatively few states near to the band edges and hence the quasi-Fermi levels move towards higher energies (where there is a larger concentration of available states) in order to achieve sufficient gain. This results in a broad output spectrum corresponding to radiative transitions over a wide energy range. By restricting the motion of the carriers along one dimension, as in quantum well structures, the density of states becomes independent of energy, with many states at the band edges. In a quantum well, the density of states is dependent only on the effective mass, m_{eff} and the quantum well width, L_z as:

$$\rho_{2\text{-D}} \propto \frac{m_{\text{eff}}}{L_z} \,. \tag{37.22}$$

In bulk structures, at the Brillouin zone centre of the valence band, the light- and heavy-hole bands are degenerate. Consequently, as the laser is pumped, states that do not contribute to the gain of the lasing mode also become filled. In a quantum well, the energy states form discrete levels. For an infinite square well, the confinement energy of the nth energy level E_n is dependent on the quantum well width L_z and the growth-direction effective mass m_{eff}^z, and at the zone centre can be simply expressed analytically as

$$E_n = \frac{\hbar^2}{2m_{\text{eff}}^z}\left(\frac{n\pi}{L_z}\right)^2 \,. \tag{37.23}$$

Thus, in a quantum well, the degenerate heavy- and light-hole bands become split (due to the difference in m_{eff}^z), resulting in a further reduction in the density of states at the valence-band Brillouin zone centre. Therefore in a quantum well laser, a higher proportion of the filled states can contribute to the gain of the lasing mode.

Further improvements in carrier confinement can be achieved by moving towards quantum wire or even quantum dot structures. In quantum wire lasers, the carriers are further confined, allowing freedom of movement along one direction only (Fig. 37.4c). In principle, such structures can be realised, e.g. by growth along lithographically defined V-grooves and lasers have been successfully produced in this manner [37.19] although the complexity of the processing has meant that they have yet to become a commercial product. In a quantum dot laser the carriers are confined still further so that they are restricted in all three dimensions (Fig. 37.4d). Thus, an ideal quantum dot is analogous to an atom. By completely restricting the carrier motion, one may theoretically expect to achieve temperature-insensitive operation [37.12]. Initial attempts to produce quantum dot lasers concentrated on lithographic processes but it was not until the successful demonstration of self-assembled growth (the Stranski–Krastanov technique) that quantum dot lasers became a reality. Record low room-temperature threshold current densities ($< 20\,\text{A/cm}^2$) have been achieved using quantum dot lasers (e.g. [37.13]). However, due to the low gain, long lasers are required which increases the threshold current.

The influence of strain

A detailed discussion of strain in semiconductors is given in Chapt. 20 of this book. However, due to the enormous importance of strain in optoelectronic devices, in this section, we discuss the influence of strain on device characteristics. For many years, manufacturers of semiconductor lasers struggled to produce dislocation- and strain-free material. In 1986, it was suggested independently by *Adams* and by *Yablonovitch* and *Kane* that the incorporation of strain into the active layer of semiconductor lasers would lead to improved performance [37.10, 11]. Many of these benefits have since been demonstrated [37.20–22]. The majority of the lasers currently manufactured in the world today are based upon strained layers.

In a strained-layer structure, the lattice constant of the epilayer material a_{epi} is different from the lattice constant of the substrate a_{sub} on which it is grown. When a thin epilayer of this material is grown onto the much thicker substrate, the lattice constant of the grown epilayer becomes equal to that of the substrate, a_{sub}, and the epilayer is in a state of biaxial stress. The resulting net strain in the $(x-y)$ plane of the layer ε is given by

$$\varepsilon_\| = \varepsilon_{xx} = \varepsilon_{yy} = (a_{\text{sub}} - a_{\text{epi}})/a_{\text{epi}} \,, \tag{37.24}$$

where it is assumed that the substrate is much thicker than the epilayer. Hence $a_{\text{epi}} > a_{\text{sub}}$ results in compressive strain while $a_{\text{epi}} < a_{\text{sub}}$ produces tensile strain.

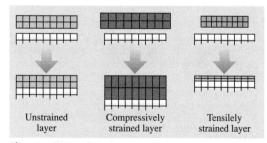

Fig. 37.5 Illustration of unstrained, compressively strained, and tensile-strained layers. For the strained layers the epilayer expands (contracts) in the plane whilst expanding (contracting) in the growth direction, corresponding to compressive (tensile) strain respectively

Figure 37.5 shows the three possible types of epilayer growth: (a) compressively strained, (b) tensile strained and (c) unstrained. The sign convention used here is that positive strain indicates that the layer is under biaxial compression whilst a negative strain corresponds to a layer being in a state of biaxial tension. For compressive strain, when the lattice constant of the epilayer is reduced in the plane, the lattice constant in the growth direction, $a_\perp$, is increased. The strain in the growth direction $\varepsilon_\perp$ $(=\varepsilon_{zz})$ is governed by Poisson's ratio σ and is given by

$$\varepsilon_\perp = \varepsilon_{zz} = -2\sigma\varepsilon_\parallel/(1-\sigma) \,. \tag{37.25}$$

In tetrahedral semiconductors, $\sigma \approx 1/3$ therefore, upon substituting this into (37.25), it can be seen that $\varepsilon_\perp \approx -\varepsilon_\parallel$. In order to understand how strain affects the band structure, it is useful to split the strain into an axial component ε_{ax}

$$\varepsilon_{ax} = \varepsilon_\perp - \varepsilon_\parallel \approx -2\varepsilon_\parallel \tag{37.26}$$

and a hydrostatic component, ε_{vol} $(=\Delta V/V$, the fractional change in volume of the epilayer)

$$\varepsilon_{vol} = \varepsilon_{xx} + \varepsilon_{yy} + \varepsilon_{zz} \approx \varepsilon_\parallel \,. \tag{37.27}$$

The axial and hydrostatic strain components affect the band structure as described in the next section. In direct-band-gap semiconductors, whilst strain has only a small effect on the conduction band, it greatly alters the valence-band structure.

The maximum layer thickness that can be grown under a state of strain is determined by the critical thickness h_c. Above this thickness, it becomes energetically favourable to relieve the strain energy through the formation of dislocations. Thus, in order to grow high-quality strained material it is necessary to keep the thickness of each layer below h_c. As the strain in a layer is increased, the corresponding critical thickness decreases. The maximum strain–thickness product for InGaAs/GaAs materials is ≈ 200 Å% [37.23]. It is however possible to use alternating layers of compressive and tensile strain to produce zero-net-strain structures that can prevent the formation of dislocations. Strained-layer semiconductor lasers are often fabricated with strain-compensated wells and barriers when there are many strained quantum wells, thereby providing a large strain in the wells whilst reducing the overall strain in the structure.

The introduction of strain into the active region of semiconductor diode lasers brings about several changes in the band structure. The initial predictions of *Adams* and *Yablonovitch* and *Kane* concentrated on

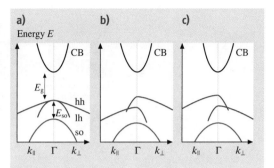

Fig. 37.6a–c Effects of strain for a bulk-like direct-band-gap semiconductor for (**a**) unstrained, (**b**) tensile strain, and (**c**) compressive strain

the benefits of compressive strain. It has since been shown that tensile strain can also enhance laser performance [37.24–26]. In the previous section, it was shown that the strain in an epilayer can be split into purely axial and hydrostatic components, ε_{ax} and ε_{vol}, respectively. The axial component of the strain splits the cubic symmetry of the semiconductor and hence the degeneracy of the heavy- and light-hole valence bands at the Brillouin zone centre is lifted (it should be noted that the valence band will already be nondegenerate in a quantum well due to the effects of quantum confinement). In addition, the valence-band dispersion becomes anisotropic. In the case of compressive strain the heavy-hole band is moved to higher energy with respect to the light-hole band, while the opposite is true for tensile strain. Also, for compressive strain the *in-plane* (x–y) mass becomes lighter than the mass in the growth (z) direction. Again, the opposite is true for tensile strain. The hydrostatic component of the strain causes an increase in the mean band gap for compressive strain and a decrease in the mean band gap for tensile strain. These effects on the band structure of a strained *bulk-like* material are shown schematically in Fig. 37.6.

Due to the much reduced in-plane highest hole band mass, the density of states in a compressively strained laser is lower than that for an unstrained device. As a result, n_{tr} is decreased, which not only lowers the amount of unwanted spontaneous emission [37.27] but also reduces carrier-density-dependent nonradiative processes such as Auger recombination [37.28]. Also, due to the increased heavy- and light-hole band splitting, carrier spill-over into higher sub-bands can be substantially reduced. Consequently, compressively strained lasers have been shown to have much lower threshold current densities than similar unstrained devices [37.20, 29].

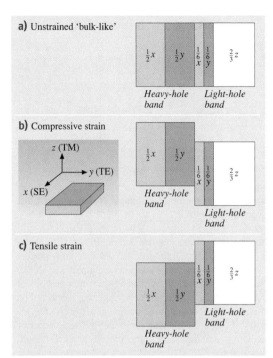

Fig. 37.7a–c Schematic illustration of the valence-band alignment for (**a**) unstrained bulk-like, (**b**) compressive strain, and (**c**) tensile strain

To understand the improvement in performance of tensile-strained lasers it is necessary to examine the character of the valence band. The heavy-hole band consists of 1/2 x-like states and 1/2 y-like states but contains no states with z-like character. In contrast, the light-hole band consists of 1/6 x-like states, 1/6 y-like states and 2/3 z-like states. y-like states contribute to transverse electric (TE) gain where the electric field oscillates across the facet (y-axis) and z-like states contribute to transverse magnetic (TM) gain, where the electric field oscillates along the growth direction. x-like states contribute only to spontaneous emission (SE), as the electric field vector lies in the propagation direction of the lasing mode. This is illustrated in Fig. 37.7. In bulk, unstrained quantum well and compressively strained quantum well lasers the lasing mode is always TE-polarised as $g_{th}(TE) < g_{th}(TM)$. With the inclusion of tensile strain in the quantum wells of semiconductor lasers it became possible for the first time to produce a device with TM polarisation [37.25].

From Fig. 37.6 it can be seen that the in-plane mass of a tensile-strained laser is relatively large, resulting in an increased density of states. However, from Fig. 37.7 it can be seen that the top of the valence band in a tensile-strained device contains 2/3 z-like states, which can contribute to TM gain. Only 1/3 of the injected carriers will *not* produce gain in the lasing mode, compared with 2/3 in unstrained (bulk-like) and 1/2 in compressively strained QW devices. Hence, tensile-strained lasers can also exhibit a lower threshold current [37.25]. However, if only a moderate amount of tensile strain is incorporated into the quantum well, the threshold current will increase due to the heavy- and light-hole sub-bands becoming degenerate.

37.2.7 Optical Feedback in a Fabry–Perot Laser Cavity

The probability of stimulated emission depends not only on maintaining population inversion, but also on the photon density. In a laser, the photon density is sustained by the use of optical feedback. In a gas laser, optical feedback is realised by the use of partially metallised mirrors, which form a Fabry–Perot cavity around the gain medium. When the light reaches the end of the cavity, a certain fraction (R) is reflected back into the gain medium, where it again becomes available to stimulate the emission of further photons. In a standard edge-emitting semiconductor laser, the same principle is used but the mirrors are formed by cleaving the semiconductor along a crystallographic plane, (most commonly the [110] plane) forming laser facets. This is illustrated in Fig. 37.8. By using as-cleaved facets the resulting reflectivity, R is typically ≈ 30% due to the refractive-index step between the laser material and (usually) air. This can be either increased or decreased with suitable facet

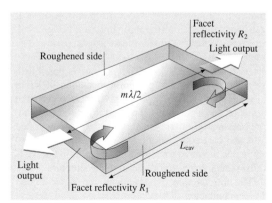

Fig. 37.8 Schematic diagram of a Fabry–Perot edge-emitting laser

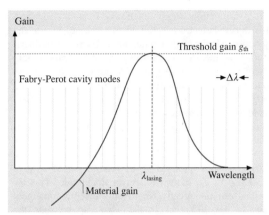

Fig. 37.9 Illustration of gain in a semiconductor laser. The cavity mode that lies closest to the material gain peak at threshold will lase

the peak of the material gain curve at threshold will lase (Fig. 37.9) although other modes near to the gain peak may also be present in the output spectrum of an edge-emitting Fabry–Perot laser. With increasing bias however, the relative strengths of these modes will decrease with respect to the dominant lasing mode. This produces a very narrow and spectrally pure output, in stark contrast with the spontaneous emission spectrum of an LED.

37.2.8 Wave-Guiding

As discussed earlier in this chapter, the stimulated emission process is not only dependent on the carrier density, but also on the photon density within the laser cavity. In particular, it is a requirement that the photon density is maintained as high as possible so that the probability of stimulated emission is maximised whilst still allowing sufficient radiation to escape. As the optical field propagates along the laser cavity, it can extend in both the growth (z) and transverse (y) directions. Hence it is necessary for the optical field to be confined along two directions. In a semiconductor laser, wave-guiding can be accomplished by two methods: index-guiding and gain-guiding. Index-guiding is achieved by surrounding the active region with a material of larger band gap and hence, lower refractive index. In the *growth* direction of an edge-emitting double-heterostructure semiconductor laser the optical field is always index-guided. Figure 37.10 shows an example of index-guiding in a typical double-heterostructure laser together with the associated refractive-index profile in the growth direction. Index-guiding confines most of the optical field within the active region due to total internal reflection at the cladding heterojunctions with only a small fraction

coatings. Semiconductor lasers usually have roughened sides as a result of the fabrication process. This has the desirable effect of reducing the reflectivity of the sides of the laser and hence reduces the probability of generating unwanted transverse modes.

The application of optical feedback using a Fabry–Perot cavity has the effect of producing Fabry–Perot resonances or modes in the output spectrum of the laser. From Fig. 37.8 it can be seen that the optical field will form a standing wave when the cavity length is equal to an integer number of half wavelengths, given by

$$\frac{m\lambda}{2} = nL_{\text{cav}} \quad m = 1, 2, 3\ldots, \quad (37.28)$$

where m is the (integer) number of half-wavelengths, λ is the lasing wavelength (as measured), n is the refractive index of the laser cavity and L_{cav} is the cavity length (typically $250\,\mu\text{m} \leq L_{\text{cav}} \leq 1500\,\mu\text{m}$). It can be seen that the maximum wavelength that the cavity can support is $\lambda = 2nL_{\text{cav}}$. If (37.21) is rearranged for λ and differentiated with respect to m, the mode spacing, $\Delta\lambda$ is found to be

$$\Delta\lambda = \frac{\lambda^2}{2nL_{\text{cav}}}. \quad (37.29)$$

Thus, the possible modes of a semiconductor laser consist of a comb of many hundreds of delta functions spaced $\Delta\lambda$ apart, as illustrated in Fig. 37.9. For a typical 1.5-µm laser with $L_{\text{cav}} = 500\,\mu\text{m}$ and $n \approx 3.2$, $\Delta\lambda \approx 0.7\,\text{nm}$. The wavelength at which lasing occurs is therefore determined by both the material gain curve and the mode spacing. The mode which lies closest to

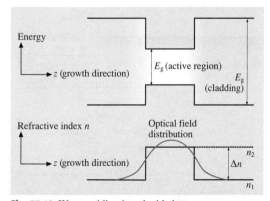

Fig. 37.10 Wave-guiding in a double heterostructure

of the field evanescently decaying into the higher-bandgap layers. The degree of confinement depends upon the refractive-index step between the active region and the surrounding layers.

In the transverse direction, the optical field can be confined by either index- or gain-guiding mechanisms. Gain-guiding arises from a weak current-induced change in refractive index across the active region caused by the nonuniform carrier distribution. Index-guided structures can be considered as either strongly index-guided or weakly index-guided, corresponding to either a large or small refractive-index step in the transverse direction. Gain-guided structures are easier to fabricate than index-guided structures but are less effective in confining the optical field when compared with index-guided structures [37.30].

37.2.9 Carrier Confinement

In addition to confining the optical mode, laser characteristics can be greatly enhanced by improving the confinement of the electrons and holes, thereby increasing the probability of radiative recombination. The first laser structures [37.1, 4–6] were based upon single p–n junctions (homojunctions). In such structures, the injected electrons and holes recombine close to the junction. However, due to the nonuniform carrier distribution, as the electrons diffuse out of the gain region, there is a variation in gain with position. In addition, due to the small difference in refractive index at the edge of the gain region, a large amount of the optical field spreads into absorbing regions. Thus, the combination of poor carrier and field confinement gives rise to large threshold current densities $\approx 10^5$ A/cm^2 in homojunction lasers. The development of the double-heterostructure laser [37.7, 8] brought about a large decrease in J_{th} of two orders of magnitude. This results from the greatly improved carrier confinement due to the difference in the band gap between the active and cladding layers, which forms a potential barrier to the carriers. A further improvement in carrier confinement is obtained with quantum well structures. In a quantum well laser the injected carriers rapidly diffuse into the thin (≈ 25–150 Å) quantum well layers [37.9]. Due to the large potential barriers between the well and barrier material it is difficult for the carriers to subsequently escape. In addition to improved carrier confinement, quantum well lasers offer many other benefits over bulk lasers as discussed later in this chapter. Figure 37.11 illustrates the improvement in carrier confinement obtainable on going from homojunctions to double heterostructures and quantum well laser structures.

37.2.10 Current Confinement

The threshold current of semiconductor lasers can be further decreased by reducing the area of the device that is being electrically pumped. Reducing the threshold current can also suppress current-heating effects and extend the operating lifetime of the laser. Current confinement can also help to produce a stable fundamental mode. There are several fabrication techniques and associated laser structures designed to increase current confinement [37.30]. The ridge structure Fig. 37.12a is a weakly index-guided type of laser. In this structure, following the epitaxial growth of the complete layer structure, the wafer is etched to produce a ridge that is ≈ 2–10 μm wide. The wafer is then coated with a dielectric (usually SiO$_2$) and metallised. The weak index-guiding is provided by the large refractive-index step between the active region material and the dielectric. The current is directly injected into the ridge and therefore travels through a relatively narrow region of the device. The two main other structures are strongly index-guided buried-

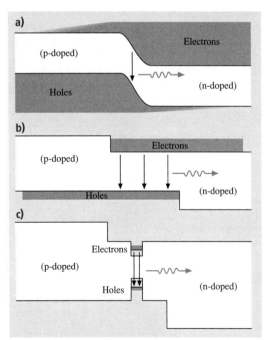

Fig. 37.11a–c Advances in carrier confinement (**a**) homojunction, (**b**) double heterostructure, and (**c**) quantum well

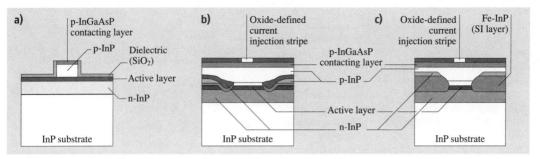

Fig. 37.12a–c Schematic diagrams of typical long-wavelength InP lasers (**a**) ridge structure, (**b**) double-channel planar buried heterostructure (DCPBH), and (**c**) semi-insulating planar buried heterostructure (SIPBH)

heterostructure (BH) devices. In double-channel planar buried heterostructure (DCPBH) lasers (Fig. 37.12b), following the growth of the layer structure, two channels, $\approx 3\,\mu\text{m}$ deep and $\approx 10\,\mu\text{m}$ wide, are chemically etched, forming a mesa. After cleaning, n- and p-doped current blocking layers are grown followed by the cladding and contacting layers. The current confinement is provided by the reverse-biased p–n junctions that surround the mesa.

The final structure considered is the semi-insulating planar buried heterostructure (SIPBH), as can be seen in Fig. 37.12c. In a similar way to the DCPBH devices, SIPBH lasers are based upon an etched mesa. In an SIPBH device the current confinement is provided by semi-insulating layers that surround the mesa. In these structures, after wet-chemically etching the wafer to form the mesa (typically $1-2\,\mu\text{m}$ wide and $2-3\,\mu\text{m}$ high) an Fe-doped layer is grown, followed by an n-doped anti-diffusion layer, p-doped cladding and contacting layers. The resistivity of the Fe-doped regions surrounding the mesa is typically $\approx 10^8\,\Omega\,\text{cm}$ thereby ensuring that the current only flows through the active region. Buried heterostructure lasers tend to have lower threshold currents than ridge structures due to the combination of excellent current and optical field confinement. For high-speed applications such as telecommunications, SIPBH devices outperform DCPBH lasers due to their much lower parasitic capacitance.

37.2.11 Laser Threshold and Efficiency

In Sect. 37.2.4, the condition for material transparency was derived. In a laser structure the threshold is reached when the total gain is equal to the optical losses as the wave travels along the cavity plus the (useful) loss of light from the facets (mirror loss). If we consider the laser cavity shown in Fig. 37.8, the threshold gain condition, g_th is given by

$$g_\text{th} = \frac{1}{\Gamma}\left[\alpha_\text{m} + \alpha_\text{i}\right]$$
$$= \frac{1}{\Gamma}\left[\frac{1}{2L_\text{cav}} \ln\left(\frac{1}{R_1 R_2}\right) + \alpha_\text{i}\right], \quad (37.30)$$

where R_1 and R_2 are the facet reflectivities, α_m is the mirror loss and α_i is the internal loss per unit length in the *entire* structure including scattering losses, free-carrier absorption, and inter-valence-band absorption [37.31]. Γ is the optical confinement factor, which is equal to the fraction of the optical field intensity that overlaps the active region. This therefore accounts for the fact that not all of the optical field will give rise to gain. Note that g_th is the threshold gain per unit length of the device. For a typical 1000-μm-long 1.5-μm quantum well device ($\Gamma \approx 0.02$) with as-cleaved facets ($R_1 = R_2 \approx 0.3$) and $\alpha_\text{i} = 10\,\text{cm}^{-1}$, $g_\text{th} \approx 1100\,\text{cm}^{-1}$.

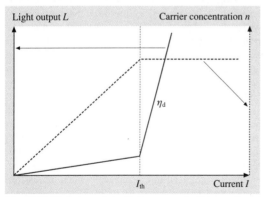

Fig. 37.13 Variation of the light output (*solid curve*) and carrier density (*dashed curve*) as a function of the injection current. Above threshold, the light output increases abruptly whilst the carrier density becomes pinned

The carrier density required to reach g_{th} is termed n_{th}, and the corresponding current (density) is termed I_{th} (J_{th}). Figure 37.13 shows how the light output and carrier density increase as current is injected into the laser structure. Below threshold, increasing the current causes the Fermi-level splitting to increase, thereby increasing the electron and hole (carrier) densities in the active region of the laser. In this regime, the laser is acting as an LED and spontaneous emission is produced. However, the electrical-to-optical conversion efficiency is relatively low ($\approx 1\%$). In practice nonradiative recombination processes (as discussed in the next section) cause the carrier density to have a nonlinear dependence on current below the laser threshold. At threshold, the stimulated lifetime decreases, causing each additional injected carrier to undergo stimulated emission. These carriers rapidly transfer to the valence band with the production of a photon via stimulated emission giving rise to an abrupt increase in the light output. Thus, at threshold the carrier density becomes pinned at n_{th} since every additional carrier quickly undergoes stimulated emission. Hence, above threshold the intrinsic *differential* quantum efficiency, η_d, of the lasing process can be as high as 100% in an ideal laser. However, effects including carrier leakage, absorption and self-heating can reduce the overall measured differential quantum efficiency.

The differential quantum efficiency, η_d, is defined as the ratio of the incremental number of photons emitted from the facets to the incremental number of carriers injected into the laser (above threshold), thus

$$\eta_d = \frac{dL}{dI} \frac{e}{h\nu}, \qquad (37.31)$$

where L is the optical power emitted from both facets, e is the electronic charge and $h\nu$ is the photon energy (the lasing energy). Thus, η_d can simply be determined from the light–current characteristic (providing that all of the stimulated light is collected). η_d can more generally be defined from

$$\eta_d = \eta_i \left(\frac{\alpha_m}{\alpha_i + \alpha_m} \right); \qquad (37.32)$$

η_i is the *internal* quantum efficiency and accounts for the fact that, in a real laser, not all of the injected carriers will result in the production of a photon. This may be due to inefficient injection of carriers into the active region of the laser (dependent on both the laser materials and the geometry of the laser). It is convenient to rewrite (37.32) in terms of the inverse efficiency so that

$$\frac{1}{\eta_d} = \frac{1}{\eta_i} \left(\frac{\alpha_i + \alpha_m}{\alpha_m} \right) = \frac{1}{\eta_i} \left(\frac{\alpha_i}{\alpha_m} + 1 \right)$$
$$= \frac{1}{\eta_i} \left[\frac{2\alpha_i L_{cav}}{\ln\left(\frac{1}{R_1 R_2}\right)} + 1 \right]. \qquad (37.33)$$

If R_1 and R_2 are known (easily calculated), by measuring η_d for lasers with several different cavity length taken from the same wafer and plotting a graph of $1/\eta_d$ versus L_{cav} one can obtain a value for both η_i (from the intercept) and α_i (from the slope/intercept).

From Fig. 37.13 it is clear that the *overall* efficiency of the laser will depend on both the differential quantum efficiency η_d, which we wish to maximise, *and* the threshold current I_{th}, which we would like to minimise. In the discussion thus far, the threshold current has been described only in terms of a spontaneous emission current. In practice there exist many other recombination paths that contribute to the laser threshold. These are discussed in the next section.

37.2.12 Carrier Recombination Processes

When carriers are injected into a semiconductor laser there are many possible recombination paths. In an ideal semiconductor laser, above threshold, the carrier density pins, as illustrated in Fig. 37.13. Therefore, above threshold, any recombination path which depends upon n will also become pinned. However, the threshold current itself is determined by the different mechanisms by which carriers recombine in a real laser structure. In this section, the important radiative and nonradiative recombination mechanisms are briefly discussed. The threshold current of a semiconductor laser is defined as the current required to provide enough gain to reach the threshold gain level g_{th}. Due to differences between laser structures, it is frequently more useful to compare the threshold current density J_{th} between devices, which accounts for the overall area of the device that is being pumped, thereby allowing a useful comparison to be made between different device structures. In this chapter, the two are used interchangeably when discussing laser characteristics.

At threshold, the total current I can be expressed as the sum of the current paths as

$$I = eV(An + Bn^2 + Cn^3) + I_{leak} \qquad (37.34)$$

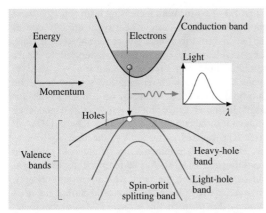

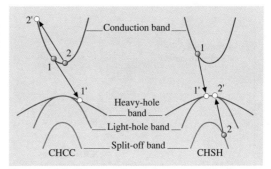

Fig. 37.15a,b The two primary Auger recombination processes: (**a**) CHCC producing hot electrons and (**b**) CHSH producing hot holes

Fig. 37.14 Spontaneous emission in a bulk semiconductor

where e is the electron charge, V is the active region volume and n is the carrier density (where we assume that the electron (n) and hole (p) densities are equal, $n = p$).

The An term is due to recombination at defects and is considered to be negligible in high-quality material. It does however make significant contributions to some laser materials, most notably InGaAsN/GaAs where it can account for up to 50% of I_{th} even in the best 1.3-μm devices [37.32].

The Bn^2 term corresponds to the radiative recombination of electrons and holes giving rise to spontaneous emission, as observed from LEDs, and semiconductor lasers operated below threshold. Radiative recombination occurs (for interband processes) when an electron in the conduction band recombines with a hole in the valence-band, resulting in the spontaneous emission of a photon as illustrated in Fig. 37.14. For lasing to occur it is necessary to have spontaneous recombination to initiate the stimulated emission process. However, because spontaneous emission is a random process, emitting photons in all directions over a broad energy range, only a small fraction of the photons can couple to the laser gain. Hence for efficient laser operation it is desirable that the amount of unwanted spontaneous emission is reduced.

The Cn^3 term describes *nonradiative* Auger recombination. In an Auger recombination process, the energy of a recombining electron–hole pair that would normally produce a photon, is instead given to a third carrier (electron or hole) which is excited further into its respective band. Thus, in an Auger recombination process, three carriers are involved, the initial electron–hole pair plus an additional electron or hole.

The two main Auger recombination processes are illustrated in Fig. 37.15, where the closed circles represent electrons and open circles represent holes. In the conduction–hole–conduction–conduction (CHCC) process, the energy and momentum of a recombining conduction-band electron and valence-band hole are transferred directly to a second conduction-band electron, which is then excited into a higher-energy conduction-band state. In the direct conduction–hole–spin–hole (CHSH) process, the energy and momentum are transferred to a second valence-band hole, which is then excited into the spin split-off band. In addition to these direct Auger recombination processes, there may be phonon-assisted processes whereby the conservation

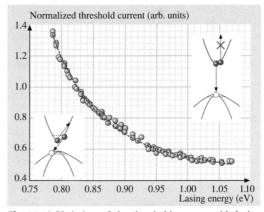

Fig. 37.16 Variation of the threshold current with lasing energy over the wavelength range 1.2–1.6 μm. With increasing wavelength (decreasing energy), the threshold current increases swiftly due to the increased probability of Auger recombination, as illustrated in the *insets*

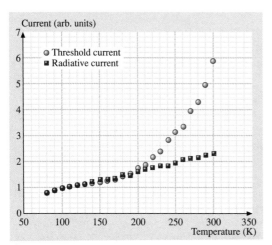

Fig. 37.17 Threshold current (*circles*) and radiative current (*squares*) versus temperature for a 1.3-μm InGaAsP laser. The threshold current increases more strongly with temperature than the radiative current due to the thermally induced onset of Auger recombination

of momentum and energy of the carriers is relaxed by the emission or absorption of a phonon. These are however of less importance in 1.3-μm and 1.5-μm lasers [37.33]. It should be noted that Auger-generated carriers are very energetic and initially lie many $k_B T$ more than E_g above their respective band edges. Whilst these carriers quickly scatter back to thermal equilibrium, there is a finite probability that they will scatter from the active region into the cladding layers, thus forming an additional Auger-generated leakage current.

Auger recombination processes are very sensitive to the band gap, E_g, and are severely detrimental to laser performance. For emission wavelengths below 1 μm, the probability of an Auger recombination process occurring is small due to the near-vertical transitions that are required on an energy–momentum diagram, which decreases the probability of the process occurring (inset Fig. 37.16). However, as E_g decreases, the transitions become less vertical and the probability of finding a state for the *hot* Auger carrier increases. Auger recombination therefore begins to influence the performance of semiconductor lasers and by 1.5 μm it dominates their behaviour at room temperature, forming ≈ 80% of the threshold current [37.28]. This is illustrated in Fig. 37.16 where the normalised variation of the measured threshold current with lasing energy was determined by performing high-pressure measurements on several lasers over the nominal wavelength range 1.3–1.6 μm. With increasing lasing energy (band gap), the threshold current decreases strongly due to the lowered contribution of Auger recombination. Thus, long-wavelength lasers are more susceptible to Auger recombination. Auger recombination is also largely responsible for the high-temperature sensitivity of the threshold current of 1.3-μm and 1.5-μm semiconductor lasers. In Fig. 37.17 we plot the temperature dependence of the threshold current (circles) for a 1.3-μm InGaAsP laser. The squares correspond to the measured radiative current as determined from spontaneous emission measurements [37.28]. At low temperature, the threshold current follows the radiative current very closely, however, above ≈ 160 K, the threshold current increases strongly due to thermally activated Auger recombination. This is primarily due to the n^3 dependence of the Auger current but also because the Auger coefficient C is thermally activated [37.34].

I_{leak} is associated with heterobarrier leakage and current spreading around the active region. With increasing temperature there is an increasing probability that electrons in the conduction band and holes in the valence band will have enough thermal energy to escape from the quantum wells. This is termed heterobarrier leakage. Here these carriers may recombine radiatively (at shorter wavelengths) or nonradiatively. If they have sufficient energy, they may even escape into the cladding layers. The probability of carriers escaping into the cladding layers depends upon the position of the quasi-Fermi levels with respect to the cladding band edge, as illustrated in Fig. 37.18. With increasing temperature, the thermal spread of carriers broadens and the population of unconfined carriers at threshold increases, thereby increasing the threshold current. For diffusion-driven leakage of electrons, the leakage current I_{leak} can be described via an activation energy E_a, equal to the difference in en-

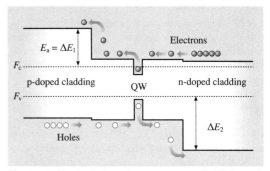

Fig. 37.18 Carrier leakage in a quantum well (QW) semiconductor laser

ergy between F_c and the lowest (Γ, L or X) CB edge of the cladding layer. Thus

$$I_{\text{leak}} = I_0 \exp\left(-\frac{E_a}{k_B T}\right), \qquad (37.35)$$

where k_B is the Boltzmann constant, T is the temperature and I_0 is independent of temperature. As with Auger recombination, carrier leakage is strongly temperature-dependent and can lead to a large change in the threshold current with temperature. For this reason, much effort has been put into designing lasers which minimise carrier leakage. In III–V semiconductors, due to the large hole effective mass, F_v moves towards the valence band at a much slower rate than F_c moves towards the conduction-band edge. For InGaAs(P)-based materials as used in 1.3-μm and 1.5-μm lasers, the quantum well to barrier/SCH band offset in the conduction band is much lower than the valence-band offset. Additionally, holes have a much lower mobility than electrons. Consequently, the overflow of electrons is generally considered to be more significant compared to the overflow of holes in this materials systems. Carrier leakage is a particular problem for visible semiconductor lasers for which the Al-containing layers may be close to, or indeed are, indirect, whereby the conduction-band X-minima are at a lower energy than the conduction band Γ-minimum. Thus, electrons may thermally escape from the Separate confinement Heterostructure (SCH) into the cladding, forming a leakage current. An example of this is shown in Fig. 37.19 where the threshold current is plotted as a function of temperature for a 670-nm Al(GaInP) laser.

The strong increase in threshold current with temperature is due to the thermal leakage of electrons from the GaInP QW via the SCH region into the X-minima of the AlGaInP barrier layers. At room temperature, leakage accounts for $\approx 20\%$ I_{th} at room temperature, rising to $\approx 70\%$ at 80 °C at this wavelength [37.35]. Due to the smaller activation energy, shorter-wavelength devices suffer still further from carrier leakage and this, in practice, sets the lower wavelength limit for lasers produced using this material system.

37.2.13 Temperature Sensitivity and T_0

As described in the previous section, the threshold current of lasers operating in the visible and in the 1.3-μm to 1.5-μm range is very sensitive to the ambient temperature. From a commercial perspective, this is very unsatisfactory as these lasers are required to provide a constant light output for a given drive current above threshold all year round. However, throughout the year, the ambient temperature can typically vary significantly. Hence, further cost results from having to use expensive temperature-control electronics to stabilise the temperature and hence light output of these lasers. This can be seen in Fig. 37.20 where the power–current curves are plotted for a 1.3-μm laser over the temperature range 20–70 °C. Clearly, by operating the laser at a constant bias current, the light output decreases considerably as the temperature decreases. In extreme cases the current may no longer be sufficient to reach laser threshold, at which point the device will no longer lase and the output would only be spontaneous emission (as in Fig. 37.20,

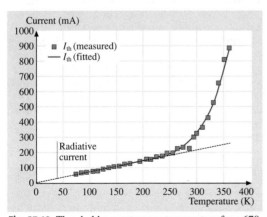

Fig. 37.19 Threshold current versus temperature for a 670-nm GaInP quantum well (QW) laser. The strong temperature sensitivity above 250 K is due to electron leakage into the X-minima of the AlGaInP barrier layers

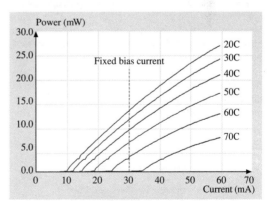

Fig. 37.20 Typical curves of power versus current for a 1.3-μm laser as a function of temperature. The large increase in threshold current with increasing temperature gives rise to a strong loss of power for a fixed bias current

for the 70 °C curve). In practical situations this therefore necessitates the use of temperature-control feedback electronics. These add considerable cost (many times the cost of the laser itself) to the system. This has become known as the T_0 problem. Around room temperature and over a limited range of temperature, the threshold current of many semiconductor lasers was observed to increase approximately exponentially with temperature [37.36]. Hence it is usual to write that $I_{th} = I_0 \exp(T/T_0)$, where I_0 is a constant and T_0 is the characteristic temperature of the threshold current given by

$$\frac{1}{T_0} = \frac{1}{I_{th}} \frac{dI_{th}}{dT} = \frac{d\ln(I_{th})}{dT}, \qquad (37.36)$$

T_0 is usually expressed in units of Kelvin. It follows that a low value of T_0 corresponds to a high temperature sensitivity and vice versa. Hence, a high T_0 is desirable. Experimentally, T_0 can be determined from measurements of I_{th} from

$$T_0 = \frac{T_2 - T_1}{\ln\left[\frac{I_{th}(T_2)}{I_{th}(T_1)}\right]}, \qquad (37.37)$$

where $I_{th}(T_1)$ and $I_{th}(T_2)$ are the threshold currents at T_1 and T_2 respectively and $T_2 > T_1$. In practice, over a wider temperature range, T_0 is *itself* temperature-dependent due to the temperature dependence of the recombination mechanisms that occur in the lasers. However, T_0 is a parameter often used to investigate the temperature dependence of the threshold current. For a typical 1.3-µm or 1.5-µm QW laser, $T_0 \approx 50$–60 K. This is due to Auger recombination, as discussed above. It can be shown that, for an ideal QW laser, the maximum $T_0 = 300$ K around room temperature, where the threshold current is entirely due to radiative recombination [37.37]. Furthermore, by going to quantum dot lasers, due to the 3D-confinement effect, an infinite T_0 has been predicted [37.12]. Recently reported experimental work has shown that this may be possible by using p-doped quantum dot layers, which exhibit temperature-insensitive operation around room temperature [37.38].

37.3 Single-Mode Lasers

37.3.1 DFB lasers

In an ideal laser, the output spectrum would be very narrow with no side modes. In practice, the output spectrum of Fabry–Perot devices such as those described earlier in this chapter consists of several competing modes. This is due to the comb of Fabry–Perot modes, which overlap with the laser gain spectrum (shown earlier in Fig. 37.9). There are a number of laser devices which can be operated with high spectral purity. Perhaps the most important of these is the distributed feedback (DFB) laser. In a Fabry–Perot laser, the end facets provide optical feedback to give rise to provide sufficient gain to achieve threshold. Because the reflectivity of the end facets is only slightly wavelength-dependent (for a given polarisation), this gives rise to several longitudinal modes. These are clearly visible in the output amplified spontaneous-emission spectrum of a Fabry–Perot laser, indeed this is one way in which the gain can be measured [37.39]. In contrast, in a DFB laser, there is an additional layer called the guiding layer (Fig. 37.21) grown close to the active layer. The guiding layer is essentially a grating consisting of a corrugated layer of a dissimilar-refractive-index semiconductor defined using lithographic and etching techniques. The guiding layer is usually overgrown although surface gratings are also possible and simpler to produce. The corrugations in the material provide a periodic variation in the refractive index along the laser cavity and hence the optical feedback is *distributed* along the length of the cavity. The forward and backward propagating waves will only interfere constructively if their frequency is related to the pitch of the grating. It can be shown that the correspondingly allowed modes lie symmetrically either side of the Bragg frequency of the grating. In practice, due to nonuniformity as a result of processing, or by artificially introducing a quarter-wavelength shift in the Bragg reflector, one of the modes may lase preferentially. DFB lasers are the mainstay of optical communications systems due to their single-mode behaviour with

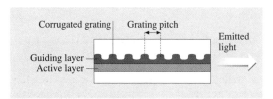

Fig. 37.21 Simplified diagram of a distributed feedback (DFB) laser

high side-mode suppression and narrow line widths (≈ 0.1 nm). In an optical-fibre communication system, this allows for the simultaneous transmission of several data channels with closely spaced wavelengths around the fibre attenuation minimum of 1.55 µm. this is known as dense wavelength-division multiplexing (DWDM).

An added benefit of the grating in a DFB laser is that it may be used to artificially reduce the temperature sensitivity of the laser threshold current and lasing wavelength. Unlike Fabry–Perot lasers, which primarily emit at the mode closest to the gain peak, as discussed above DFB lasers can be made to emit at one fixed mode. The peak of the gain spectrum moves approximately at the same rate as that of the band gap (≈ 0.5 nm/K) whilst the temperature dependence of the DFB mode follows the thermally induced change in the refractive index (≈ 0.1 nm/K) which is much less temperature sensitive. Thus, DFB lasers are far less susceptible to thermally-induced mode hops. Furthermore, by carefully designing the peak of the gain spectrum to lie at a shorter wavelength than the DFB wavelength, one can effectively *tune* the DFB laser with increasing temperature resulting in, over a limited temperature range, a relatively temperature-independent threshold current. This has been demonstrated as a means of achieving cooler-less 1.3-µm lasers, albeit at the expensive of higher threshold currents.

37.3.2 VCSELs

Vertical-cavity surface-emitting lasers (VCSELs) differ from conventional semiconductor lasers due to the fact that the optical cavity lies in the growth direction, in the same direction as the current flow. The output of the laser is therefore emitted from the surface. There are several major advantages of VCSELs compared with edge-emitting lasers, which have led to a considerable amount of research being undertaken to produce VCSELs at a variety of wavelengths. Unlike edge-emitting lasers, once the semiconductor wafer has been metallised, it is very easy to test the devices prior to dicing them into individual chips. This represents a considerable cost saving. Forming laser arrays is obviously therefore much more straightforward with VCSELs and has led to the development of parallel data links based upon VCSELs and high-power VCSEL arrays. The small size of VCSELs (aperture diameter <10 µm) also gives rise to both single-mode behaviour and low threshold currents. Furthermore, due to the fact that the light is (usually) emitted from a circular geometry contact, the beam profile is circular, making fibre coupling much simpler. In contrast, edge-emitting lasers have an elliptical beam profile due to diffraction of the laser light from the end facet, requiring expensive corrective optics and active alignment with optical fibres.

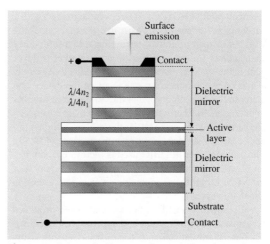

Fig. 37.22 Schematic diagram of a vertical-cavity surface-emitting laser (VCSEL)

Because the VCSEL emits from its surface, the optical cavity over which the optical field may experience gain is very short. Consequently, in order to reach the threshold gain level, the optical field needs to make several passes through the active region. This requires highly reflective end mirrors. At a typical III–V semiconductor–air interface, the refractive-index contrast gives rise to a reflectivity of $\approx 30\%$. This is enough to achieve sufficient gain to reach threshold in an edge-emitting laser, which is typically many hundreds of microns long. However, in a VCSEL, the cavity length is typically ≈ 0.1 µm, requiring mirror reflectivities $>99\%$. This is achieved by sandwiching the active region between two multilayer dielectric mirrors (Fig. 37.22). These so called distributed Bragg reflectors (DBRs) consist of alternating high- and low-refractive-index quarter-wave-thick pairs. In a similar way to the DFB lasers discussed above, the DBR mirrors in VCSELs provide a high degree of wavelength-selective reflectance, providing that the sum of the optical thicknesses of each layer pair is equal to a half wavelength (i.e. $n_1 d_1 + n_2 d_2 = \lambda/2$, where n_i and d_i are the refractive index and thickness of each layer respectively), thereby giving rise to constructive interference at the desired wavelength. Typically, there may be 30 or so layers in each DBR to achieve sufficiently high reflectivity. It is useful to note here that a related device, the resonant-cavity LED (RCLED), is similar to the VCSEL but contains far fewer DBR pairs in order that it does not lase, but instead produces narrow-band high-efficiency spontaneous emission. Whilst a large number of layer pairs in the VCSEL gives rise to a high reflectivity it may also decrease the quantum efficiency of the device and increase the operating voltage due to the large number of interfaces. Furthermore, due to free-carrier absorption and carrier leakage in the DBRs themselves, a high number of DBRs can degrade device performance [37.40]. The exact number of DBRs used is therefore a compromise which depends upon the operating wavelength of the VCSEL and the semiconductor material system on which it is based.

High-quality VCSELs have already been produced for emission at 650 nm and 850 nm (for plastic and silica fibre communications), for oxygen sensing using 760 nm VCSELs and high-power devices for emission at 980 nm [37.41]. However, there has been considerable difficulty in producing VCSELs at the telecommunications wavelengths of 1.3 μm and 1.55 μm. This is primarily due to the fact that GaAs/AlGaAs is the preferred system for making highly reflective DBRs due to the high refractive-index contrast between GaAs and AlAs. Such DBRs are therefore compatible with the GaAs substrates as used in devices below $\approx 1\,\mu$m. However, for longer-wavelength devices, which are primarily based on InP substrates it is difficult to form highly reflective DBRs using InGaAsP/InP due to their low refractive-index contrast. Alternative methods of producing long-wavelength VCSELs are inherently process-intensive, using techniques such as wafer fusion in which the active layer and InGaAsP/InP DBR is first grown on an InP substrate whilst a GaAs/AlAs DBR is grown separately on a GaAs substrate. The active layer is then fused onto the GaAs/AlAs DBR and the InP substrate is then removed (see for example, [37.42]). The wafer fusion process is understandably both costly and difficult, and therefore methods of growing monolithic 1.3-μm and 1.55-μm VCSEL structures using only a single GaAs substrate are particularly attractive. This has stimulated a great deal of research into producing GaAs-based active regions emitting at these wavelengths. In recent years there has been some success in achieving this by two approaches. The first approach is to use InAs/GaAs quantum dots as the active region. Quantum dot lasers with very low threshold current density ($< 20\,\text{A/cm}^2$) emitting at 1.3 μm have already been realised [37.13]. However, there are few reports in the literature on quantum-dot-based VCSELs. This is largely due to the difficulty in achieving sufficient gain from the quantum dots and can result in the need for several layers of quantum dots. In spite of this, quantum dots are showing promise for use in temperature-insensitive edge-emitting lasers and high-power lasers.

The second approach has been in the use of the so called dilute nitrides. In 1997, *Kondow* and co-workers originally proposed the use of InGaAsN/GaAs as an active material to achieve long-wavelength emission on GaAs [37.43]. This is due to the unusual band-gap bowing that occurs when small concentrations (≈ 2–5%) of nitrogen are added to GaAs. By growing the alloy InGaAsN, one may achieve long-wavelength emission whilst maintaining a low or zero strain. Although material quality remains an issue, there have been several successful demonstrations of edge-emitting lasers and VCSELs based upon this material emitting at 1.3 μm [37.44–46]. The push towards 1.55-μm emission has been largely hampered by material quality issues although at the time of writing this chapter, there have been the first reports of low-threshold 1.5-μm edge-emitting lasers based upon InGaAsN/GaAs [37.47]. Other researchers have produced close-to-1.5-μm edge-emitting and VCSEL devices with the pentenary material InGaAsNSb/GaAs [37.48]. It is speculated that antimony may act as a surfactant to improve the quality of growth. The dilute nitride approach does at present appear to offer the best possibility of obtaining VCSELs emitting at the technologically important wavelength of 1.55 μm.

37.4 Optical Amplifiers

37.4.1 An Introduction to Optical Amplification

As a light pulse propagates through an optical communication system, the pulse becomes attenuated until eventually it is necessary to regenerate the pulse to keep the signal above the background noise level. This was initially achieved by detecting the pulse using a photodetector, and then using the detected electronic pulse to trigger a laser giving a fresh output pulse. An alternative technique is to use direct optical amplification. This has been achieved for instance by doping optical fibres with rare-earth elements such as erbium, forming an erbium-doped fibre amplifier (EDFA). Population inversion is achieved by exciting the erbium atoms at 980 nm or 1.48 µm, where the excited carriers decay to a level from which they can recombine to give stimulated emission at 1.55 µm, amplifying weak signals in the fibre. Amplification can also be achieved within conventional silica fibres using the Raman effect; by pumping the fibre with a high-power pump laser (total power ≈ 1 W) at a fixed frequency above the signal frequency, Raman scattering gives rise to gain at a lower frequency (the difference is the phonon energy). Thus, by pumping an optical fibre at 1.45 µm one may achieve gain at 1.55 µm. The main advantage of Raman amplification compared with EDFAs is the fact that the gain curve may be dynamically tuned by judiciously using different wavelength pump lasers. The disadvantages of Raman amplification is that the overall gain is generally lower than an EDFA and cross-amplification effects can be a problem. For some applications, semiconductor optical amplifiers (SOAs) may be preferred.

37.4.2 Semiconductor Optical Amplifiers (SOAs)

The structure of a SOA is very similar to that of a laser, but with one significant difference: the reflectivity of the end facets, $R \approx 0$. This is achieved through the use of multilayer antireflection coatings together with angled facets or cavities. This suppresses lasing within the cavity, and eliminates reflected signals in the optical system.

The signal emerging from an optical fibre is, in general, randomly polarised and therefore it is very desirable that any optical amplifier provides gain that is independent of the direction of polarisation of the light. In a bulk heterostructure device, the TE and TM material gain are equal. However, the optical confinement factor Γ for TE light polarised in the plane of the heterostructure is slightly larger than for TM light polarised perpendicular to the heterostructure plane, leading to a larger overall TE gain. As discussed earlier in this chapter, in a normal quantum well system the quantum confinement brings the heavy-hole band to the top of the valence band and light polarised in the TE mode is amplified considerably more strongly than light polarised in the TM mode. This problem can be overcome using strained-layer techniques. It is possible to grow the wells or the barriers with a small amount of tensile strain. This raises the light-hole band and the strain can be adjusted to increase the device TM gain until it is just equal to that of the TE gain. Work to date indicates that the relative gains in the two modes are sensitive to the magnitude of the amplifier current and so the gains are equal over a limited range.

Strained SOAs
Another more promising approach is to introduce both compressive- and tensile-strained wells alternately into the active region. The compressive wells provide gain predominantly to the TE mode and the tensile wells predominantly to the TM mode. The structure can then be designed to make the two gain modes equal. The initial work by *Tiemeijer* et al. from Philips concentrated on 1.3-µm devices [37.49]. They found that a combination of four compressive wells and three tensile wells each with 1% strain gave TE and TM gain within 1 dB of each other over a wide wavelength band and over an order of magnitude change in amplifier current. This structure is an excellent example of the flexibility in device design that is afforded by the introduction of strained-layer techniques. Other applications of such a structure include two-polarisation or two-frequency lasers, polarisation control elements and nonabsorbing strain-overcompensated mirrors in high-power lasers.

37.5 Modulators

37.5.1 Modulator Theory

Although it is possible to modulate the output from a semiconductor laser by directly modulating the injected current, in some applications requiring high-speed or low-power switching, it is advantageous to run the laser source CW and modulate the light subsequently [37.50]. This is basically because, unlike lasers, semiconductor modulators are diodes operated in reverse bias. They therefore draw little current and, because there is no carrier injection, there is much less chirp.

Electroabsorption modulators make use of the fact that the presence of a large electric field adds a perturbation to the Schrödinger equation, which leads to a decrease in the band-gap energy E_g. Thus photons with

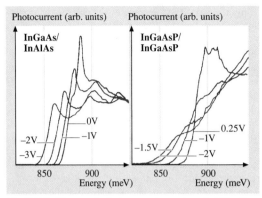

Fig. 37.23 Measured photocurrent spectra at 2 K of an InGaAs/InGaAlAs (*left*) and an InGaAsP/InGaAsP (*right*) electroabsorption modulator at several bias voltages. (After [37.50] with permission)

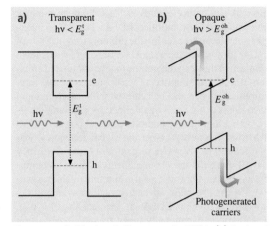

Fig. 37.24a,b The band alignment of a QW in (**a**) flat band (zero field) where the optical gap equals $E_g^t > h\nu$ (*transparent*) and (**b**) for a large field across the QW, where the optical gap equals $E_g^0 < h\nu$ (*opaque*)

energies just below the band gap become more strongly absorbed when a field is applied, leading to a decrease in light intensity. Electroabsorption modulators can benefit enormously from the use of multiple quantum wells, where the electroabsorption effect is more than one order of magnitude greater than that in comparable bulk structures.

In quantum wells the modulation of the energy levels by an electric field applied perpendicular to the plane of the wells is called the quantum-confined Stark effect (QCSE). For a given applied voltage, it causes more change, $\Delta\alpha$, to the absorption coefficient α close to the band edge than in equivalent bulk structures for three main reasons. Due to the quantum confinement, there is a step-like increase in the density of states at the band edges, as described above, and so a faster rate of increase in absorption coefficient with energy. Thus, $\Delta\alpha/\Delta E_g$, the change in absorption coefficient with band gap, is larger. Secondly, the steep rise in the absorption coefficient with energy at the band edge is further enhanced in quantum well structures due to excitonic effects. An exciton consists of an electron–hole pair bound together by electrostatic attraction and has associated with it a sharp line absorption spectrum which is superimposed on the quantum well absorption. Figure 37.23 shows the measured photocurrent spectra at low temperature as a function of the applied reversed bias for two long-wavelength modulator structures. The exciton peak is clearly visible at the absorption edge, particularly for low bias values. Excitons also exist in bulk material at low temperatures but not at room temperature. In quantum wells however, the electrons and holes are confined into the same region of space by the well and so the electrostatic binding energy is sufficiently large that they are still able to exist at room temperature. Finally, the actual energy of the band edge is largely determined by the effect of quantum confinement, which depends not only on the width of the quantum well but also on its shape. Since the shape of the well can be strongly modified by an applied electric field, so can the electronic energies at the band edges, as can be seen from the shift in absorption edge with applied bias in Fig. 37.23. Thus ΔE_g is enhanced by quantum confinement.

The effect of an electric field applied perpendicular to the plane of a quantum well is to change the shape of the well from the rectangular well shown in Fig. 37.24a to the triangular shape shown in Fig. 37.24b. Solutions to the Schrödinger equation show that the lowest confined states are closer together in the triangular situation than the rectangular one and so the effective absorption edge is moved to lower energies. Because the electron ground state is shifted towards one side of the quantum well, and the hole ground state towards the opposite side, the overlap of the electron and hole wavefunctions is decreased by the application of an electric field. The result is to decrease the absolute value of the absorption coefficient and to decrease the exciton absorption peak, as seen in Fig. 37.23. However, if one considers an energy just below the band-edge exciton energy at zero field, at about 850 meV in Fig. 37.23, the absorption coefficient at that energy can be greatly increased by the application of an electric field.

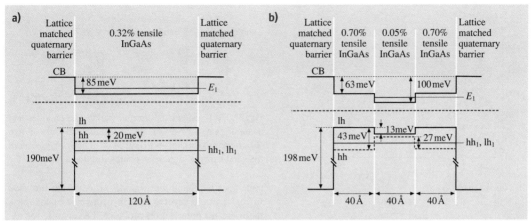

Fig. 37.25a,b Calculated conduction- and valence-band profiles for (**a**) conventional tensile QW and (**b**) multi-strain-stepped QW (After [37.51])

For a modulator, a very important parameter is the ratio of the light intensity transmitted in the *on* state to that transmitted in the *off* state when the field is applied. This is given simply by

$$P = \exp(\Delta\alpha L), \quad (37.38)$$

where L is the distance the light travels through the absorbing medium. In quantum well structures, although $\Delta\alpha$ is relatively large, if the light travels in a direction perpendicular to the plane of the wells, the length L is small; of the order of tens of nm. One way to overcome this problem is to use the QCSE modulator in the waveguide configuration; i. e. to pass the light parallel to the plane of the quantum wells in a waveguide structure similar to that for a laser as described previously. In this case the effective absorption per unit length will be decreased by the optical confinement factor (Γ) to $\Gamma\alpha$, but L can easily be increased much more; to tens or even hundreds of μm. Such a waveguide configuration can achieve a large on/off ratio but it has two potential problems. Firstly, it can have a large insertion loss because it is difficult to launch the light into the waveguide unless it is integrated with the laser source and also there can be light scattering and absorption along the length of the guide even in the on state. Secondly, the modulator will be dependent on the polarisation of the light beam travelling along the waveguide, as described below.

37.5.2 Polarisation-Insensitive Modulators

As was explained when considering strained-layer electronic properties, the degeneracy of the heavy- and light-hole bands, which exists in bulk material, is split by confinement effects in a quantum well. This is because the quantum confinement energy is proportional to $1/m^*$ and so the first light-hole valence band lh_l lies below the first heavy-hole valence band hh_1. As a consequence, the wavefunctions that make up the edge of the valence band are derived from the heavy-hole band and so only interact with TE-polarised photons as described above. Thus, while photons of energy close to the band edge and polarised in the TE mode are modulated as the band edges move together with applied electric field, photons of the same energy but polarised in the TM mode, are relatively unaffected. This is in general unacceptable if the photons are arriving from a system, such as an optical fibre, which may deliver photons of any polarisation in a manner which is sensitive to extraneous influences.

The problem of polarisation sensitivity in modulators, as with the polarisation-insensitive amplifier, may be dealt with by engineering the band structure with a judicious use of in-built strain within the quantum wells. As described above, the light-hole band may be raised in energy with respect to the heavy-hole band by the introduction of tensile strain. Thus, with tensile strain it is possible to shift the heavy- and light-hole bands until the confined states are once again degenerate, as shown in Fig. 37.25a. Such a system, although it may be polarisation insensitive at zero net field, can again become polarisation sensitive as a net field is applied, as is shown by the movement of the band edges in Fig. 37.26a. This effect arises since the heavy-hole confined level, being closer to its band edge, is more sensitive to the distortion of that band edge by the applied field. Once again, how-

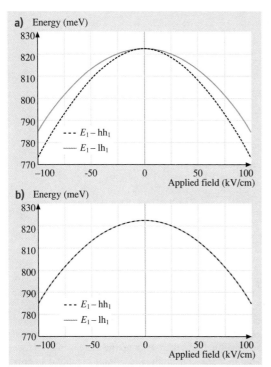

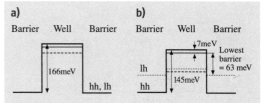

Fig. 37.27a,b Calculated valence-band edge energies and band alignments for (**a**) lattice-matched modulator and (**b**) strained modulator, as in Fig. 37.28. The *thick horizontal lines* indicate the highest hh confined state

Fig. 37.26a,b Calculated quantum confined Stark shifts of the E_1–hh_1 (*dashed lines*) and E_1–lh_1 (*solid lines*) transitions for structures (**a**) and (**b**) in Fig. 37.25. (After [37.51])

ever, it is predicted that this effect can be compensated by further band-structure refinements brought about by employing three zones of different strain within the same well. This is illustrated in Fig. 37.25b. In this multi-strain well, the tensile strain is larger in the outer thirds of the well. Thus, when a field is applied and the hole wavefunctions are moved towards the edge of the well, the heavy-hole level sees a smaller effective well width and its change in energy can be made the same as for the light hole. Figure 37.26b shows that excellent matching of the movements of the heavy- and light-hole bands can be achieved with the three-zone well over the full range of practical electric fields, up to $100\,kV/cm$.

37.5.3 High-Speed High-Power QCSE Modulators

Strained-layer QCSE modulators have been successfully integrated with strained-layer lasers to produce high-speed low-chirp sources operating at $1.55\,\mu m$ for optical communications. The laser and the modulator consist of the same wave-guide structure but are used in forward and reverse bias respectively. In this case there are no polarisation problems because if, for example, compressively strained wells are used the laser will produce the TE-polarised output, which is efficiently modulated by the modulator. One problem that is encountered, however, is that while the modulator is in the absorbing state a significant density of photogenerated carriers may accumulate in the quantum wells. This has two effects. Firstly, the carriers tend to fill the band-edge states, thus reducing the field in the well, and in addition bleaching the absorption process. Secondly, they may take a long time to escape, so that efficiency of switching to the nonabsorbing state is dependent on how long the modulator was absorbing and what carrier density had accumulated. In practice it is found that the escape rate of photogenerated holes from the quantum wells is slower than that for electrons and is the limiting process. This problem may be addressed by changing from the standard InGaAsP material system to InGaAlAs, which has a much smaller valence-band offset than in the equivalent InGaAsP layers [37.50]. Alternatively, the problem may be reduced by band-structure engineering in InGaAsP using strain.

Figure 37.27a shows the valence-band structure of a lattice-matched InGaAsP-based electroabsorption modulator designed to operate around $1.5\,\mu m$. The heavy-hole well was calculated to be $166\,meV$ deep, and the structure showed power saturation, as illustrated by the open squares in Fig. 37.28. However, when a 1% compressive strain was introduced into the well and 1% tensile strain into the barriers between the wells, the calculated excitation energy from the heavy-hole states in the well to light-hole states in the barrier was reduced to $63\,meV$, as illustrated in Fig. 37.27b. Since the strain in the barrier has little effect on the well depth for heavy holes, the heavy-hole wavefunctions and quan-

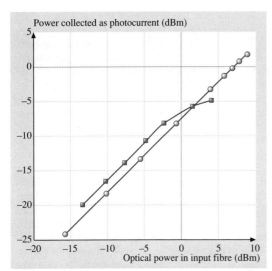

Fig. 37.28 Total power absorbed in lattice-matched (*open squares*) and strained (*closed squares*) electroabsorption modulator as a function of power in the input fibre. (After [37.52] with permission)

tum confined Stark effect should be comparable in both structures, but now thermal excitation out of the well by phonon scattering to the light-hole level in the barrier is greatly enhanced. The improvement in the power-saturation characteristics of the strained modulator are shown by the solid squares in Fig. 37.28; a considerable improvement in the 10-GHz switching performance was also observed.

37.5.4 The Electro-Optic Effect

The real part of the refractive index can be described in terms of virtual transitions between the valence and conduction band and is therefore related to the imaginary part of the refractive index. Hence, when an electric field applied to a semiconductor leads to a change in the absorption characteristics as described above, there is also

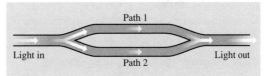

Fig. 37.29 Schematic of a Mach–Zender interferometer

an accompanying change in the real part of the refractive index. This results in a phase change of the light reaching the end of the active region of the modulator. Whilst this causes unwanted chirp in the electroabsorption modulators described above, it can be used to advantage in modulators and switches that employ interference effects. A good example of such a device is the modulator based on a Mach–Zender interferometer. In this structure the incoming wave is divided into two equal components at a Y-junction and then recombined at a similar junction, as shown in Fig. 37.29. If the two arms of the interferometer are exactly equivalent then the two beams will arrive in-phase and constructively recombine at the output. However, if the refractive index for light travelling down one path is increased by the electro-optic effect so that its phase at the output is changed by π, then the two beams will destructively interfere at the output and no light will be transmitted. Band-structure considerations for the electro-optic effect are ones with which we are already familiar. The line-width enhancement factor α_l is defined by

$$\alpha_l = -\frac{4\pi}{\lambda}\left(\frac{\frac{dN}{dg}}{\frac{dg}{dn}}\right) \quad (37.39)$$

and provides a measure of the ratio of the change in the real part N of the refractive index to the change in the imaginary part g of the refractive index, each as a function of carrier density n. Thus, for modulators based on the electro-optic effect, we would wish to have a band structure that results in a large α_l, while for electroabsorption modulators we require α_l to be as small as possible.

37.6 Photodetectors

In the first part of this chapter we dealt with optoelectronic devices designed to produce light for use in applications including displays, indicators, data storage and communications. The latter two of these applications also require the ability to detect the light once it has been created. Much of the explosion of interest in semiconductor optoelectronic devices over the past 20 or so years has been particularly aimed at the communications market. This has put high demands on the ability to detect small optical signals being transmitted at very fast data rates (> 40 Gbit/s). In this section, we consider the important aspects for detectors and describe some of the approaches that have been used.

37.6.1 Photodetector Requirements

There are many attributes important to the design of photodetectors. These can broadly be split into the following categories: sensitivity, speed, noise, physical size (footprint), reliability, temperature sensitivity, ease of use and finally cost. When deciding on a particular photodetector one must decide which of these factors is of most importance to the particular application of interest.

37.6.2 Photodetection Theory

The primary requirement of the band structure of semiconductor optical detectors is that the band gap of the active region is less than the photon energy to be absorbed. The electron–hole pair formed when the photon is absorbed is then separated by an applied or built-in field that exists in the active region. If, under the influence of the field, the electron and hole move apart by a distance, x, then the charge induced to flow in the external circuit connected to the detector is x/l. In a simple photoconductive detector consisting of a bulk semiconductor with conducting electrodes, l is the distance between the electrodes. In a p–n or p–i–n diode detector, l is the width of the depletion region.

Several factors need to be taken into account when considering the band structure of a detector. Firstly, the absorption depth of the light to be detected must be matched to the width of the active region, so that there is a high probability that the photogenerated carriers will be produced within the active region or less than the minority-carrier diffusion length from it. Figure 37.30 shows the absorption coefficient as a function of energy for direct-gap GaAs and for indirect-gap Si. An absorption coefficient of 10^3 cm^{-1}, typical for Si, corresponds to an absorption depth of $10\,\mu\text{m}$, while 10^4 cm^{-1} for GaAs corresponds to a 10-μm absorption depth. One must also consider the effect of the band structure on the dark current in the device, since this sets the lower limit on the strength of the signal that can be detected. There are three important effects that can give rise to the generation of electron–hole pairs even when no light is falling on the device. These are: a) thermal generation, b) band-to-band tunnelling and c) avalanche breakdown. Processes (a) and (b) are illustrated in Fig. 37.31. Thermal generation, which occurs even at zero electric field, can itself be conveniently divided into two parts. Firstly there is the part i_{diff} due to the diffusion of minority carriers into the field region. Secondly there is the current i_{gr} due to thermal generation within the field region. The thermal generation current increases strongly with temperature but is relatively independent of applied field, although i_{gr} increases slightly with reverse bias since this causes the depletion width to increase. Process (b), band-to-band tunnelling, is indicated by the horizontal arrow i_{t} in Fig. 37.31. At large reverse fields, electrons at the top of the valence band see a triangular-shaped potential barrier through which it is possible for them to tunnel into the conduction band. This quantum-mechanical tunnelling is relatively independent of temperature but is strongly dependent on the applied field. Process (c), avalanche breakdown, which occurs at very large electric fields, has to be avoided except in avalanche photodetectors, which will be described in some detail below. Avalanche breakdown involves carriers with initial ki-

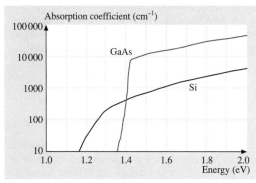

Fig. 37.30 Absorption coefficient as a function of energy for direct-gap GaAs and indirect-gap Si

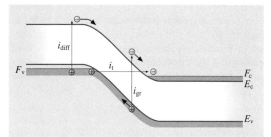

Fig. 37.31 Mechanisms contributing to the dark current in a reverse-biased p–i–n diode. Thermal generation has contributions from diffusion of minority carriers into the field region i_{diff} and from thermal generation within the field region i_{gr}. Zener (band-to-band) tunnelling is indicated by the horizontal arrow i_{t}

netic energies greater than the band-gap energy E_g away from the conduction-band minimum. This means that electrons in wide-band-gap III–V semiconductors will be at energies such that they are scattering between the central Γ minimum, the three X minima at the edges of the Brillouin zone in the $\langle 100 \rangle$ directions and the four L minima at the edges of the Brillouin zone in the $\langle 111 \rangle$ directions. Hydrostatic pressure measurements [37.53] showed that the effective ionisation energy $\langle E \rangle$ could be related to the average conduction-band energy simply by

$$\langle E \rangle = \frac{1}{8}(E_\Gamma + 3E_X + 4E_L) \quad (37.40)$$

where E_Γ, E_X and E_L are the energies of the Γ, X and L minima above the top of the valence band, respectively.

The first two effects giving rise to a dark current, described in (a) and (b) above, increase almost exponentially with decreasing band gap E_g. Therefore there exists an optimum value for the band gap E_g to be used to detect photons of a particular energy E_{ph}. Considering E_g to be the adjustable parameter, which can be determined by the alloy composition, as E_g is reduced below E_{ph} the absorption coefficient increases and light becomes more efficiently absorbed within the active region of the diode. However, once all the photons are absorbed, reducing E_g further adds nothing to the photocurrent but does continue to increase the dark current. Therefore, it is clear that the optimum band gap is just below that of the photon energy to be detected. The exact difference $E_{ph} - E_g$ depends upon the shape of the absorption edge of the material, the thickness of the active region and its position within the device.

37.6.3 Detectors with Internal Gain

In a photoconductive detector, gain can actually be achieved by adding impurities that trap one type of carrier, forming a centre with a low capture cross section for the other type of carrier. Assuming the trapped carrier is a hole, as the free electron reaches the anode another will be injected from the cathode and this process will continue until an electron recombines with the trapped hole. Under these circumstances x exceeds l and a gain of x/l is achieved: $x/l = \tau_r/t_t$, where t_t is the transit time of the electron between the electrodes and τ_r is the electron–hole recombination lifetime. The response time of such a system is $\tau_r = t_t x/l$ so that, as the gain increases, the speed of the device decreases. A related effect is obtained in a phototransistor. If, for example, we consider a photon being absorbed in the base or col-

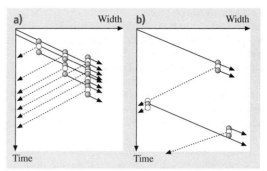

Fig. 37.32a,b Avalanche multiplication as a function of position and time where (**a**) the electron multiplication rate (*solid line*) is much greater than that for holes (*dashed line*). (**b**) when the electron and hole multiplication rates are equal, the current pulse continues for a long time

lector region of an n–p–n transistor then the presence of the hole, which will be effectively trapped in the base region, will cause the transit of electrons between the emitter and collector for the recombination lifetime of the hole. The gain of the system is then equal to the gain of the transistor, which is related to τ_r/t_t, where t_t is now the transit time of the electron from emitter to collector.

37.6.4 Avalanche Photodetectors

Interestingly, silicon as well as direct-band-gap materials can be used to produce efficient detectors, but for different reasons. In the direct-band-gap materials, photons with energies greater than the band gap are strongly absorbed by the creation of electron–hole pairs, leading to the generation in the detector of a primary photocurrent or photovoltage. In silicon, photons with energies just above the band gap are relatively weakly absorbed but the electrons generated can be greatly amplified by the process of avalanche multiplication. This occurs when a large reverse voltage is applied to the p–n or p–i–n junction of the detecting diode. Electrons drifting in the field can gain kinetic energy in excess of the band-gap energy E_g. It is then possible for the hot carrier to give up its energy and momentum in an impact ionisation process, resulting in the generation of an electron–hole pair. The generated carriers can, in turn, drift in the electric field and themselves generate further carriers. The band structure of silicon is particularly well suited for avalanche multiplication since impact ionisation can occur for hot electrons with conservation of energy and momentum when the electron's kinetic energy is almost exactly equal to E_g [37.54]. By contrast a hot hole in silicon requires an energy close to $3E_g/2$ before it can generate an electron–hole pair. Thus, in silicon, a large avalanche multiplication can be achieved with each electron that is optically generated and swept across the avalanche region. This situation is illustrated in Fig. 37.32a. Very large gains can be obtained but all of the carriers are swept out of the active region in a time equal or less than $t_e + t_h$, where t_e and t_h are the transit times taken by electrons and holes, respectively, to cross the active region.

In contrast, it is found that, in the majority of direct-band-gap materials, holes and electrons are able to produce impact ionisation at the same electric field. As a result, above this field, avalanching occurs for both types of carrier as they move in opposite directions and the carrier density escalates uncontrollably, leading to complete avalanche breakdown. This situation is illustrated in Fig. 37.32b. Gain, but also breakdown, occurs when the field is large enough to create an impact ionisation process just as the electron is about to leave the active region. If exactly the same happens for the hole, the situation can continue ad infinitum, leading to infinite gain but a current pulse that does not switch off. Of course, in fact, impact ionisation is a statistical process and the current pulse might stop after one electron transit or run away as described, leading to an unstable noisy detection system.

Silicon is therefore clearly a very useful material for the production of avalanche photodetectors (APDs). Unfortunately however, the detection bandwidth of silicon limits it from the visible spectrum up to $\approx 1\,\mu m$ in the near-infrared (IR). It is of significant technological importance to have fast high-gain low-noise detectors above this wavelength, in particular in the range $1.3–1.6\,\mu m$ for optical communications. The mid-infrared is also of importance due to the large number of gas absorption bands in this range. Due to the effects discussed above, the performance of APDs in the near-mid-infrared region is poor when compared with silicon APDs. Recently, *Adams* proposed that APDs based upon InGaAsN may give rise to improved characteristics for detection in the near- and mid-IR [37.55]. As discussed earlier in this chapter, InGaAsN is already showing great promise for long-wavelength lasers due to the unusually large band-gap bowing brought about by small concentrations of nitrogen in GaAs. This clearly makes it a potential candidate for the manufacture of long-wavelength detectors. However, in addition to the large band-gap bowing, InGaAsN has an unusually large conduction-band effective mass derived from the dilute amount of nitrogen in the lattice. This has several effects of benefit to APDs. Firstly, the high effective mass inhibits electrons from gaining kinetic energy under an applied field. Secondly, the increased mass, and hence increased density of states, will increase the probability of electron scattering. As the electrons heat up and their energy approaches that of the N levels, they will be strongly scattered. Furthermore, if any electrons manage to reach the top of the E^- band, they have to be scattered across, or tunnel through, the energy gap to the next conduction (E^+) band before they can gain more energy. All of these effects prevent electrons from avalanching. However, since the valence band is largely unaffected by the presence of nitrogen, the *holes* may be accelerated by the field and avalanche. It has been proposed that such devices will have a multiplication of $\approx 10^4$ compared with < 10 in the best currently available APDs at these wavelengths. Furthermore, it is also predicted that the noise factor of these devices could be reduced by two orders of magnitude. Although yet to be demonstrated in practice, InGaAsN may in the future provide both high-quality detectors as well as emitters in the infrared.

37.7 Conclusions

This chapter introduced some of the most important optoelectronic devices in use today. It has discussed how these devices were developed based upon highly innovative work which has transformed semiconductor growth technology over the past 40 years. The requirements for lasing action in semiconductors were discussed as well as the ways in which low-dimensional structures and strain may be used to improve the properties of devices such as lasers (both edge-emitting and surface-emitting), modulators, amplifiers and detectors. The chapter considered current state-of-the-art semiconductor laser devices and discussed emerging future directions in semiconductor optoelectronics.

References

37.1 N. Jnr. Holonyak, S. F. Bevacqua: Appl. Phys. Lett. **1**, 82 (1962)

37.2 A. Einstein: Phys. Z. **18**, 121 (1917)

37.3 M. G. A. Bernard, G. Duraffourg: Phys. Status Solidi **1**, 699 (1961)

37.4 M. I. Nathan, W. P. Dumke, G. Burns, F. H. Dill Jr., G. Lasher: Appl. Phys. Lett. **1**, 62 (1962)

37.5 R. N. Hall, G. E. Fenner, J. D. Kingsley, T. J. Soltys, R. O. Carlson: Phys. Rev. Lett. **9**, 366 (1962)

37.6 T. M. Quist, R. H. Rediker, R. J. Keyes, W. E. Krag, B. Lax, A. L. McWhorter, H. J. Zeiger: Appl. Phys. Lett. **1**, 91 (1962)

37.7 H. Nelson, J. I. Pankove, F. Hawrylo, G. C. Dousmanis: Proc. IEEE **52**, 1360 (1964)

37.8 Zh. I. Alferov, V. M. Andreev, D. Z. Garbuzov, Yu. V. Zhilyaev, E. P. Morozov, E. L. Portnoi, V. G. Trofim: Sov. Phys. Semicond. **4**, 1573 (1971)

37.9 R. Dingle, C. H. Henry: US Patent 3982207 (1976)

37.10 A. R. Adams: Electron. Lett. **22**, 249 (1986)

37.11 E. Yablonovitch, E. O. Kane: J. Light. Technol. **LT-4**, 504 (1986)

37.12 Y. Arakawa, H. Sakaki: Appl. Phys. Lett. **40**, 939 (1982)

37.13 G. Liu, A. Stintz, H. Li, K. J. Malloy, L. F. Lester: Electron. Lett. **35**, 1163 (1999)

37.14 Y. Sakakibara, K. Furuya, K. Utaka, Y. Suematsu: Electron. Lett. **16**, 456 (1980)

37.15 M.-C. Amann, J. Buus: *Tunable Laser Diodes* (Artech House, Boston 1998)

37.16 K. Iga: IEEE J. Sel. Top. Quant. Electron. **6**, 1201 (2000)

37.17 J. Faist, F. Capasso, D. L. Sivco, C. Sirtori, A. L. Hutchinson, A. Y. Cho: Science **264**, 553 (1994)

37.18 F. Capasso, C. Gmachl, R. Paiella, A. Tredicucci, A. L. Hutchinson, D. L. Sivco, J. N. Baillargeon, A. Y. Cho, H. C. Liu: IEEE J. Sel. Top. Quant. Electron. **6**, 931 (2000)

37.19 T. Kojima, M. Tamura, H. Nakaya, S. Tanaka, S. Tamura, S. Arai: Jpn. J. Appl. Phys. **37**, 4792 (1998)

37.20 P. J. A. Thijs, L. F. Tiemeijer, J. J. Binsma, T. van Dongen: IEEE J. Quant. Electron. **30**, 477 (1994)

37.21 G. Jones, A. Ghiti, M. Silver, E. P. O'Reilly, A. R. Adams: IEE Proc. J. **140**, 85 (1993)

37.22 A. Valster, A. T. Meney, J. R. Downes, D. A. Faux, A. R. Adams, A. A. Brouwer, A. J. Corbijn: IEEE J. Sel. Top. Quant. Electron. **3**, 180 (1997)

37.23 E. P. O'Reilly: Semicond. Sci. Technol. **4**, 121 (1989)

37.24 T. Yamamoto, H. Nobuhara, K. Tanaka, T. Odagawa, M. Sugawara, T. Fujii, K. Wakao: IEEE J. Quant. Electron. **29**, 1560 (1993)

37.25 M. P. C. M. Krijn, G. W. 't Hooft, M. J. B. Boermans, P. J. A. Thijs, T. van Dongen, J. J. M. Binsma, L. F. Tiemeijer: Appl. Phys. Lett. **61**, 1772 (1992)

37.26 G. Jones, A. D. Smith, E. P. O'Reilly, M. Silver, A. T. R. Briggs, M. J. Fice, A. R. Adams, P. D. Greene, K. Scarrott, A. Vranic: IEEE J. Quant. Electron. **34**, 822 (1998)

37.27 A. Ghiti, M. Silver, E. P. O'Reilly: J. Appl. Phys. **71**, 4626 (1992)

37.28 S. J. Sweeney, A. F. Phillips, A. R. Adams, E. P. O'Reilly, P. J. A. Thijs: IEEE Phot. Tech. Lett. **10**, 1076 (1998)

37.29 Y. Zou, J. S. Osinski, P. Godzinski, P. D. Dapkus, W. Rideout, W. F. Sharfin, R. A. Logan: IEEE J. Quant. Elec. **29**, 1565 (1993)

37.30 G. P. Agrawal, N. K. Dutta: *Long-Wavelength Semiconductor Lasers* (Van Nostrand, New York 1986)

37.31 A. R. Adams, M. Asada, Y. Suematsu, S. Arai: Jap. J. Appl. Phys. **19**, L621 (1980)

37.32 R. Fehse, S. Jin, S. J. Sweeney, A. R. Adams, E. P. O'Reilly, H. Riechert, S. Illek, A. Yu. Egorov: Electron. Lett. **37**, 1518 (2001)

37.33 S. J. Sweeney, A. R. Adams, E. P. O'Reilly, M. Silver, P. J. A. Thijs: *Conference on Lasers and Electro-Optics* (IEEE, San Francisco 2000)

37.34 W. W. Lui, T. Yamanaka, Y. Yoshikuni, S. Seki, K. Yokoyama: Phys. Rev. B. **48**, 8814 (1993)

37.35 S. J. Sweeney, G. Knowles, T. E. Sale, A. R. Adams: Phys. Status Solidi B. **223**, 567 (2001)

37.36 J. I. Pankove: IEEE J. Quant. Electron. **QE-4**, 119 (1968)

37.37 E. P. O'Reilly, M. Silver: Appl. Phys. Lett. **63**, 3318 (1993)

37.38 N. Hatori, K. Otsubo, M. Ishida, T. Akiyama, Y. Nakata, H. Ebe, S. Okumura, T. Yamamoto, M. Sugawara, Y. Arakawa: *30th European Conference on Optical Communication* (IEEE, Stockholm 2004)

37.39 B. W. Hakki, T. I. Paoli: J. Appl. Phys. **46**, 1299 (1975)

37.40 A. I. Onischenko, T. E. Sale, E. P. O'Reilly, A. R. Adams, S. M. Pinches, J. E. F. Frost, J. Woodhead: IEE Proc. Optoelectron. **147**, 15 (2000)

37.41 T. E. Sale: *Vertical Cavity Surface Emitting Lasers* (Res. Stud., London 1995)

37.42 Y. Ohiso, C. Amano, Y. Itoh, H. Takenouchi, T. Kurokawa: IEEE J. Quant. Electron. **34**, 1904 (1998)

37.43 M. Kondow, S. Kitatani, S. Nakatsuka, M. C. Larson, K. Nakahara, Y. Yazawa, M. Okai, K. Uomi: IEEE J. Sel. Top. Quant. Electr. **3**, 719 (1997)

37.44 T. Kitatani, K. Nakahara, M. Kondow, K. Uomi, T. Tanaka: Jpn. J. Appl. Phys. **39**, L86 (2000)

37.45 K. D. Choquette, J. F. Klem, A. J. Fischer, O. Blum, A. A. Allerman, I. J. Fritz, S. R. Kurtz, W. G. Breiland, R. Sieg, K. M. Geib, J. W. Scott, R. L. Naone: Electron. Lett. **36**, 1388 (2000)

37.46 G. Steinle, H. Riechert, A. Yu. Egorov: Electr. Lett. **37**, 93 (2001)

37.47　R. Averbeck, G. Jaschke, L. Geelhaar, H. Riechert: *19th International Semiconductor Laser Conference 2004, Conference Digest* (IEEE, Piscataway, NJ 2004)

37.48　M. A. Wistey, S. R. Bank, H. B. Yuen, L. L. Goddard, J. S. Harris: J. Vac. Sci. Tech. B **22**, 1562 (2004)

37.49　L. F. Tiemeijer, P. J. A. Thijs, T. van Dongen, R. W. M. Slootweg, J. M. M. van der Heijden, J. J. M. Binsma, M. P. C. M. Krijn: Appl. Phys. Lett. **62**, 826 (1993)

37.50　F. Devaux, S. Chelles, A. Ougazzaden, A. Mircea, J. C. Harmand: Semicond. Sci. Technol. **10**, 887 (1995)

37.51　M. Silver, P. D. Greene, A. R. Adams: Appl. Phys. Lett. **67**, 2904 (1995)

37.52　I. K. Czajkowski, M. A. Gibbon, G. H. B. Thompson, P. D. Greene, A. D. Smith, M. Silver: Electron. Lett. **30**, 900 (1994)

37.53　J. Allam, A. R. Adams, M. A. Pate, J. S. Roberts: Inst. Phys. Conf. Ser. **112**, 375 (1990)

37.54　A. R. Adams, J. Allam, I. K. Czajkowski, A. Ghiti, E. P. O'Reilly, W. S. Ring: Strained-layer lasers and avalanche photodetectors. In: *Condensed Systems of Low Dimensionality*, ed. by J. L. Beeby (Plenum, New York 1991) p. 623

37.55　A. R. Adams: Electr. Lett. **40**, 1086 (2004)

38. Liquid Crystals

This chapter outlines the basic physics, chemical nature and properties of liquid crystals. These materials are important in the electronics industry as the electro-optic component of flat-panel liquid-crystal displays, which increasingly dominate the information display market.

Liquid crystals are intermediate states of matter which flow like liquids, but have anisotropic properties like solid crystals. The formation of a liquid-crystal phase and its properties are determined by the shape of the constituent molecules and the interactions between them. While many types of liquid-crystal phase have been identified, this Chapter focuses on those liquid crystals which are important for modern displays.

The electro-optical response of a liquid crystal display depends on the alignment of a liquid-crystal film, its material properties and the cell configuration. Fundamentals of the physics of liquid crystals are explained and a number of different displays are described.

In the context of materials, the relationship between the physical properties of liquid crystals and their chemical composition is of vital importance. Materials for displays are mixtures of many liquid-crystal compounds carefully tailored to optimise the operational behaviour of the display. Our current understanding of how chemical structure determines the physical properties is outlined, and data for typical liquid-crystal compounds are tabulated. Some key

38.1	**Introduction to Liquid Crystals**...............	917
	38.1.1 Calamitic Liquid Crystals.............	919
	38.1.2 Chiral Liquid Crystals..................	921
	38.1.3 Discotic Liquid Crystals...............	923
38.2	**The Basic Physics of Liquid Crystals**........	924
	38.2.1 Orientational Order	924
	38.2.2 Director Alignment	925
	38.2.3 Elasticity	926
	38.2.4 Flexoelectricity..........................	928
	38.2.5 Viscosity...................................	929
38.3	**Liquid-Crystal Devices**	931
	38.3.1 A Model Liquid-Crystal Display: Electrically Controlled Birefringence Mode (ECB)	932
	38.3.2 High-Volume Commercial Displays: The Twisted Nematic (TN) and Super-Twisted Nematic (STN) Displays...........................	935
	38.3.3 Complex LC Displays and Other Cell Configurations	937
38.4	**Materials for Displays**...........................	940
	38.4.1 Chemical Structure and Liquid-Crystal Phase Behaviour ..	942
	38.4.2 The Formulation of Liquid-Crystal Display Mixtures	942
	38.4.3 Relationships Between Physical Properties and Chemical Structures of Mesogens	943
References		949

references are given, but reference is also made to more extensive reviews where additional data are available.

38.1 Introduction to Liquid Crystals

Liquid crystals have been known for almost 120 years but it is only in the last 30 years or so that their unique application in display devices has been recognised. Now they are seen as extremely important materials having made possible the development of thin screens for use with personal computers (PCs) and in televisions. In fact a wide range of different liquid-crystal (LC) display devices has been developed. The common feature for each of these is that the optical characteristics of the display are changed on application of an electric field across a thin liquid-crystal film. The process causing this change is associated with a variation in the macroscopic organisation of the liquid crystal within the cell. The liquid crystal is, therefore, strictly behaving as a molecular material and not an electronic one. Nonetheless the display itself is closely integrated with electronic components. Since liquid crystals may be unfamiliar to those concerned with conventional electronic materials, this section begins with an introduction to liquid crystals and the compounds that form them. The following section describes the basic physics for liquid crystals which are needed to understand their use in display devices. The functioning of the most important displays is described in Sect. 38.3, which makes contact with the basic physics outlined in Sect. 38.2. The liquid crystal materials used in display devices are discussed in the final section, where the necessary optimisation of a wide range of properties is addressed.

The majority of chemical compounds can exist in three states of matter, namely crystal, liquid or gas, each with its defining characteristics. There is a fourth state known as a liquid crystal and, as the name suggests, this state has characteristics of both crystals and liquids. Thus a liquid crystal flows when subject to a stress, like a liquid, but certain of its properties are anisotropic, like a crystal. This macroscopic behaviour, often used to identify the phase, implies that at the microscopic or molecular level the material has an element of long-range orientational order together with some translational disorder at long range. It is this combination of order and disorder that makes liquid crystals so fascinating and gives them their potential for applications, especially in the field of electro-optic displays.

A variety of different classes of materials are known to form liquid crystals at some point on their phase diagram [38.1]. These include organic materials where the liquid crystal is formed, on heating, between the crystal and isotropic liquid phases. Such materials are known as thermotropic liquid crystals and are the subject of this Chapter. Another class is formed by amphiphilic organic materials in which part of the constituent molecules favours one solvent, normally water, while the other part does not. When the amphiphile is dissolved in the water, the molecules form aggregates which then interact to give the liquid-crystal phase, the formation of which is largely controlled by the concentration of amphiphile. These are known as lyotropic liquid crystals; they underpin much of the surfactant industry, although they are not used in displays and so will not be considered further. Colloidal dispersions of inorganic materials such as clays can also form liquid-crystal phases depending on the concentration of the colloidal particles. Solutions of certain organic polymers also exhibit liquid-crystal phases and, like the colloidal systems, the solvent acts to increase the separation between polymer chains but does not significantly affect their state of aggregation. A prime example of such a system is the structural polymer Kevlar, which for the same weight is stronger than steel; it is formed by processing a nematic solution of the polymer.

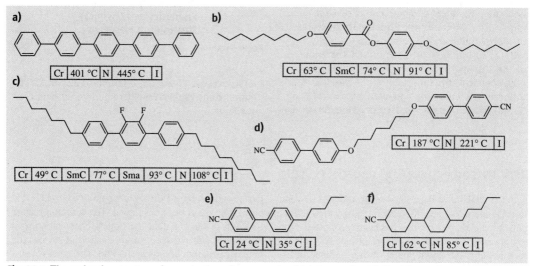

Fig. 38.1 The molecular structures for a selection of compounds which form calamitic liquid crystals

The following sections return to thermotropic liquid crystals and describe the molecular organisation within the phases, mention some of their properties and briefly indicate the relationship between the phase and molecular structures.

38.1.1 Calamitic Liquid Crystals

In view of the anisotropic properties of liquid crystals, it seems reasonable that a key requirement for their formation is that the molecules are also anisotropic. This is certainly the case, with the majority of liquid crystals having rod-like molecules, such as those shown in Fig. 38.1. One of the simplest nematogenic rod-like molecules is *p*-quinquephenyl (Fig. 38.1a), which is essentially rigid. However, flexible alkyl chains can also be attached at one or both ends of the molecule (Fig. 38.1b and c) or indeed in the centre of the molecule (Fig. 38.1d), and rigid polar groups such as a cyano (Fig. 38.1d–f) may be attached at the end of the molecule. The rigid part is usually constructed from planar phenyl rings (Fig. 38.1a–e) but they can be replaced by alicyclic rings (Fig. 38.1f) which enhance the liquid crystallinity of the compound. The term calamitic, meaning rod-like, is applied to the phases that they form. There are, in fact, many different calamitic liquid-crystal phases but we shall only describe those which are of particular relevance to display applications.

At an organisational level the simplest liquid crystal is called the nematic and in this phase the rod-like molecules are orientationally ordered, but there is no long-range translational order. A picture showing this molecular organisation, obtained from a computer simulation of a Gay–Berne mesogen [38.2] is given in Fig. 38.2b. The molecular shape is ellipsoidal and the symmetry axes of the ellipsoids tend to be parallel to each other and to a particular direction known as the director.

In contrast there is no ordering of the molecular centres of mass, except at short range. The essential difference between the nematic and isotropic phases (Fig. 38.2a) is the orientational order, which is only short range in the isotropic liquid. At a macroscopic level the nematic phase is characterised by its high fluidity and by anisotropy in properties such as the refractive index. The anisotropic properties for a nematic have cylindrical symmetry about the director, which provides an operational definition of this unique axis and is the optic axis for the phase. The anisotropy in the refractive index combined with the random director distribution results in the turbidity of the phase, which contrasts with the transparency of the isotropic liquid. This on its own would not be sufficient to identify the liquid crystal as a nematic phase but identification is possible from the optical texture observed under a polarising microscope. These textures act as fingerprints for the different liquid-crystal phases. An example of such a texture for a nematic phase is shown in Fig. 38.3a; it is created by the anisotropy or birefringence in the refractive index combined with a characteristic distribution of the director in the sample.

The next level of order within liquid crystal phases is found for the smectic A phase. Now, in addition to the long-range orientational order, there is translational order in one dimension, giving the layered structure shown in Fig. 38.2c [38.2]. The director associated with the orientational order is normal to the layers. Within a layer there is only short-range translational order as in a conventional liquid. In this structure the layer spacing is seen to be comparable but slightly less than the molecular length, as found experimentally for many smectic phases.

At a macroscopic level the layer structure means that the fluidity of a smectic A phase is considerably less than for a nematic phase. The properties are anisotropic and the birefringence is responsible for the turbidity of the phase, as found for a nematic. However, under a polar-

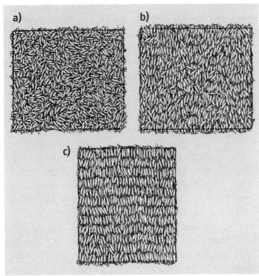

Fig. 38.2a–c The molecular organisation in (**a**) the isotropic phase (**b**) the nematic phase and (**c**) the smectic A phase obtained from the simulation of a Gay–Berne calamitic mesogen

Fig. 38.3a–c Typical optical textures observed with a polarising microscope for (**a**) nematic, (**b**) smectic A and (**c**) columnar liquid-crystal phases

ising microscope the optical texture is quite different to that of a nematic, as is apparent from the focal conic fan texture shown in Fig. 38.3b.

A variant on the smectic A is the smectic C phase. The essential difference to the smectic A phase is that the director in that smectic C phase is tilted with respect to the layer normal. The defining characteristic of the smectic C phase is then the tilt angle, which is taken as the angle between the director and the layer normal. This tilt in the structure reduces the symmetry of the phase to the point group C_{2h} in contrast to $D_{\infty h}$ for nematic and smectic A phases. This lowering in symmetry naturally influences the symmetry of the properties. The fluidity of the smectic C phase is comparable to that of a smectic A phase. However, the optical texture can be quite different and it has elements similar to a nematic phase and to a smectic A; the focal conic fan structure is less well defined and is said to be broken. The nematic-like features result from the fact that the tilt direction of the director is not correlated between the smectic layers and so it adopts a distribution analogous to that of the director in a nematic phase.

The molecular factors which influence the ability of a compound to form a liquid-crystal phase have been well studied both experimentally [38.3] and theoretically [38.4]. Consider the simple nematic as formed, for example, by *p*-quinquephenyl (Fig. 38.1a); this melts at 401 °C to form the nematic phase, which then undergoes a transition to the isotropic phase at 445 °C; the transition temperatures are denoted by T_{CrN} and T_{NI}, respectively. The very high value of T_{NI}, which is a measure of the stability of the nematic phase, is attributed to the large length-to-breadth ratio of *p*-quinquephenyl. In contrast *p*-quaterphenyl, formed by the removal of just one of the five phenyl rings, does not exhibit a liquid-crystal phase at atmospheric pressure, even though its shape anisotropy is still relatively large. This occurs because, on cooling, the isotropic liquid freezes before the transition to the nematic phase can occur. Indeed many compounds with anisotropic molecules might be expected to form liquid-crystal phases, but do not because of their high melting points. As a consequence the molecular design of liquid crystals needs to focus not only on increasing the temperature at which the liquid crystal–isotropic transition occurs but also on lowering the melting point. One way by which this can be achieved is to attach flexible alkyl chains to the end of the rigid core (Fig. 38.1). In the crystal phase the chain adopts a single conformation but in a liquid phase there is considerable conformational disorder, and it is the release of conformational entropy on melting that lowers the melting point. The addition of the chain also affects the nematic–isotropic transition temperature, which alternates as the number of atoms in the chain passes from odd to even. This odd–even effect is especially dramatic when the flexible chain links two mesogenic groups (Fig. 38.1d) to give what is known as a liquid-crystal dimer [38.5]. The odd–even effect is particularly marked because the molecular shapes for the dimers with odd and even spacers differ significantly on average, being bent and linear, respectively.

The attachment of alkyl chains to the rigid core of a mesogenic molecule has another important consequence, as it tends to promote the formation of smectic phases (Fig. 38.1b and c). The reason that the chains lead to the formation of such layered structures is that, both energetically and entropically, the flexible chains prefer not to mix with the rigid core, and so by forming a layer structure they are able to keep apart. Indeed it is known that biphenyl (a rigid rod-like structure) is not very soluble in octane (a flexible chain). The lack of compatibility of the core and the chain increases with the chain length and so along a homologous series it is those members with long alkyl chains that form smectic phases. For example, 4-pentyl-4′-cyanobiphenyl (Fig. 38.1e) forms only a nematic phase whereas the longer-chain homologue, 4-decyl-4′-cyanobiphenyl only exhibits a smectic A phase. To obtain a tilted smectic phase such as a smectic C there clearly needs to be a molecular interaction which favours an arrangement for a pair of parallel molecules that is tilted with respect to the intermolecular vector. Such a tilted structure can be stabilised by electrostatic interactions; for example by off-axis electric dipoles (Fig. 38.1b and c) [38.6] or with a quadrupolar charge distribution.

The ability of a compound to form a liquid crystal is not restricted to just one phase. The delicate balance of the intermolecular interactions responsible for the various liquid-crystal phases means that transitions between them can result from modest variations in temperature. This is apparent for the 4,4′-dialkyl-2′,3′-difluoroterphenyl shown in Fig. 38.1c, which on cooling the isotropic liquid forms nematic, smectic A and smectic C phases; such a compound is said to be polymorphic. Materials which form even more liquid-crystal phases are known [38.1]. The occurrence of several liquid-crystal phases in a single system can be of value in processing the material for display applications.

38.1.2 Chiral Liquid Crystals

The mesogenic molecules that have been considered so far are achiral in the sense that the molecule is superimposable on its mirror image. Molecules may also be

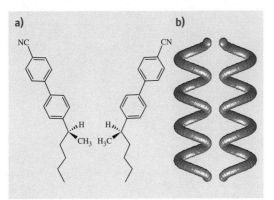

Fig. 38.4 (a) A chiral mesogenic molecule, (R) 2-[4′-cyano-4-biphenyl]-hexane, together with its mirror image, (S) 2-[4′-cyano-4-biphenyl]-hexane. (b) The left- and right-handed helical organisation of the director for a chiral nematic

chiral in that they are not superimposable on their mirror images; this chirality can result from the tetrahedral arrangement of four different groups around a single carbon atom. This is illustrated in Fig. 38.4a which shows such an arrangement together with its mirror image; these are known as enantiomers. The presence of a chiral centre will certainly change the nature of the interactions between the molecules and it is relevant to see whether the chiral interactions might not influence the structure of the liquid-crystal phases exhibited by the material. From a formal point of view it might be expected that the molecular chirality of a mesogen should be expressed through the symmetry of the liquid-crystal structure. This proves to be the case, provided no other interactions oppose the chiral deformation of the liquid-crystal phase. In fact the first liquid crystal to be discovered [38.7] was chiral; this was cholesteryl benzoate where the cholesteryl moiety contains many chiral centres. The structure of the liquid crystal is nematic-like in that there is no long-range translational order but there is long-range orientational order. However, the difference between this phase and a nematic formed from achiral materials is that the director is twisted into a helix. The helix may twist in a left-handed or a right-handed sense and these structures, shown in Fig. 38.4b, are mirror images of each other. The phase structure is certainly chiral and so is known as a chiral nematic, although originally it was called a cholesteric phase. The symbol for the phase is N*, where the asterisk indicates that the phase has a chiral structure.

The helical structure is characterised by the pitch of the helix p which is the distance along the helix axis needed for the director to rotate by 2π. Since the directions parallel and antiparallel to the director are equivalent the periodicity of the chiral nematic is $p/2$. For many chiral nematics the helical pitch is comparable to the wavelength of visible light. This, together with the periodic structure of the phase, means that Bragg reflection from a chiral nematic will be in the visible region of the spectrum and so this phase will appear coloured with the wavelength of the reflected light being related to the pitch of the helix. This pitch is sensitive to temperature, especially when the chiral nematic phase is followed by a smectic A. This sensitivity has been exploited in the thermochromic application of chiral nematics, where the reflected colour of the phase changes with temperature [38.1].

The chirality of cholesteryl benzoate clearly results from the chiral centres present in the mesogenic molecule. However, the chirality can also be introduced indirectly to a mesogen by simply adding a chiral dopant, which does not need to be mesogenic. The mixture will be chiral and this is sufficient to lead to a chiral nematic. The pitch of this mixture depends on the amount of the dopant and the inverse pitch, p^{-1}, proves to be proportional to its concentration. The handedness of the helix induced by the dopant will depend on its stereochemical conformation and will be opposite for the two enantiomers. Accordingly, if both enantiomers are present in equal amounts, i.e. as a racemic mixture, then doping a nematic with this will not convert it to a chiral nematic.

Chiral smectic phases are also known in which the director adopts a helical structure as a result of introducing molecular chirality into the material either as a dopant or as an intrinsic part of the mesogenic molecule. The chiral smectic C phase, denoted SmC*, provides an appropriate example with which to illustrate the structure of such phases. In an achiral smectic C phase the tilt direction of the director changes randomly from layer to layer, analogous to the random director orientation in an achiral nematic. For the chiral smectic C phase, as might be anticipated, the tilt direction of the director rotates in a given sense, left-handed or right-handed, and by a small, fixed amount from layer to layer. Other structural features of the smectic C phase remain unchanged. Thus, the director of the chiral smectic C phase has a helical structure with the helix axis parallel to the layer normal. The pitch of the helix is somewhat smaller than that of the associated chiral nematic phase. The magnitude of the pitch is inversely related to the tilt angle of the smectic C, and since this grows with decreasing temperature so the pitch decreases. The reduced symmetry, C_2, of the SmC* phase leads to the introduction of a macroscopic electrical polarisation [38.8]. This is of potential importance for the creation of fast-switching displays.

The ability of the SmC* phase to adopt a helical structure results from the fact that the tilt direction for the director acts in an analogous manner to the director in a nematic, and importantly that the layer spacing is preserved in the helical structure. In marked contrast there are strong forces inhibiting the creation of a twisted structure for a smectic A composed of chiral molecules. The director is normal to the layers and so the creation of a twisted structure would require a variation in the layer thickness but this has a high energy penalty associated with it. Accordingly many of the smectic A phases formed from chiral molecules have the same structure as those composed of achiral molecules. There are, however, exceptions and these occur when the chiral interactions are especially strong and, presumably, the translational order of the layers is small. Under such

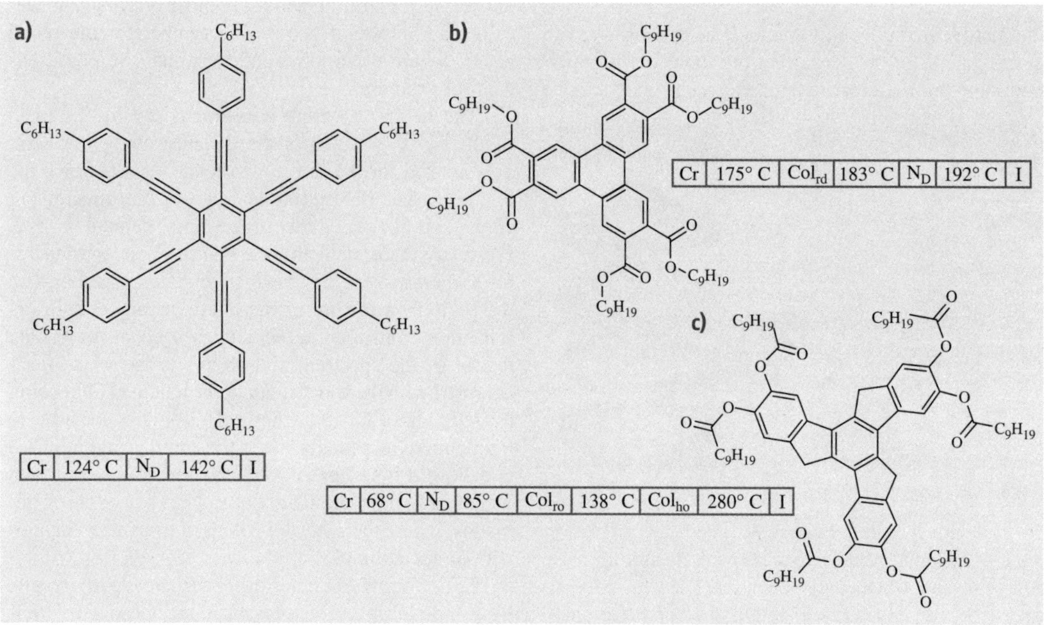

Fig. 38.5 A selection of molecular structures for compounds that form discotic liquid crystals

conditions the smectic A structure is partially destroyed, creating small SmA-like blocks about 1000 Å wide, separated by screw dislocations [38.9, 10]. These defects in the organisation allow the directions for the blocks to rotate coherently to give a chiral helical structure; the pitch of the helix is found to be larger than that in an analogous chiral nematic phase. This chiral phase is just one example of a liquid-crystal structure stabilised by defects; it is known as a twist grain-boundary phase and denoted by TGBA*. The letter A indicates that the director is normal to the layers in the small smectic blocks; there is a comparable phase in which the director is tilted, denoted by TGBC*.

38.1.3 Discotic Liquid Crystals

The key requirement for the formation of a liquid crystal is an anisotropic molecule, as exemplified by the rod-like molecules described in the previous sections. However, there is no reason why disc-like molecules should not also exhibit liquid-crystal phases. Nonetheless, it was not until 1977 that the first example of a thermotropic liquid crystal formed from disc-like molecules was reported [38.11]. Since that time several liquid-crystal phases have been identified and these phases are known collectively as discotic liquid crystals. The range of compounds that exhibits these phases is now extensive and continues to grow [38.12], although it does not match the number that form calamitic liquid crystals. The molecular structures of three compounds which form discotic liquid crystals are shown in Fig. 38.5.

As for rod-like molecules, the simplest liquid-crystal phase formed by disc-like molecules is the nematic, usually denoted N_D, where the D indicates the disc-like nature of the molecules. Within the nematic phase, shown in Fig. 38.6a, the molecular centres of mass are randomly distributed and the molecular symmetry axes are orientationally correlated. The nematic structure is the same as for that formed from rod-like molecules, the only difference being that the symmetry axes which are orientationally ordered are the short axes for the discs and the long axes for the rods. The point symmetry of the discotic and calamitic nematic is the same, namely $D_{\infty h}$. The discotic nematic is recognised in the same way as the calamitic nematic; that is it flows like a normal fluid and its anisotropy is revealed by a characteristic optical texture analogous to that shown in Fig. 38.3a. In fact, the refractive index of a discotic nematic along the director is smaller than that perpendicular to it, which is the opposite to that for a calamitic nematic. It has been suggested that this difference may be of value in display devices [38.13] but this concept has not as yet been commercialised.

The other class of discotic liquid crystals possesses some element of long-range translational order and these are known as columnar phases, two examples of which are sketched in Fig. 38.6b and c. The disc-like molecules are stacked face-to-face into columns. A single column has a one-dimensional structure, and as such is not expected to exhibit long-range translational order, although this can result from interactions between neighbouring columns in the liquid-crystal phase. The column–column interactions will result in the columns being aligned parallel to each other; these interactions will also determine how the columns are packed. When the discs are orthogonal to the column axis the cross section is essentially circular and so the columns pack hexagonally, as shown in Fig. 38.6b. The symbol given to this phase is Col_{hd}, where h denotes hexagonal packing of the columns and d indicates that the arrangement along the column is disordered. The point group symmetry of this phase is D_{6h}. The disc-like molecules may also be tilted with respect to the column axis, giving

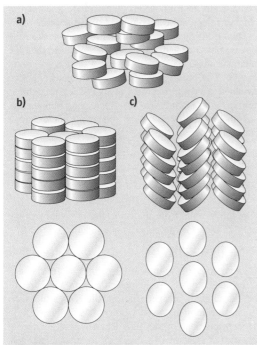

Fig. 38.6a–c The molecular organisation in discotic liquid-crystal phases, (**a**) nematic, (**b**) hexagonal columnar and (**c**) rectangular columnar

an elliptical cross section to the columns. As a result the columns are packed on a rectangular lattice; there are four possible arrangements and just one of these is indicated in Fig. 38.6c. In general the mnemonic used to indicate a rectangular columnar phase is Col_{rd}. The point group symmetry of the rectangular columnar phase is D_{2h} and the extent to which the structure deviates from that of the Col_{hd} phase will depend on the magnitude of the tilt angle within the column. The columnar phases can be identified from their optical textures and an example of one is shown in Fig. 38.3c.

The columnar phases have potential electronic applications because of the inhomogeneity of the molecules that constitute them; i.e. the central part is aromatic while the outer part is aliphatic. As a result of the overlap between the π-orbitals on the centres of neighbouring discs it should be possible for electrical conduction to take place along the core of the column. This should occur without leakage into adjacent columns because of the insulation provided by the alkyl chains. It should also be possible to anneal these molecular wires because of their liquid-crystal properties [38.14]. This and the ability to avoid defects in the columns which can prevent electronic conduction in crystals mean that the columnar phase has many potential advantages over non-mesogenic materials. In addition, discotic systems are also used as compensating films to improve the optical characteristics for some liquid-crystal displays.

At a molecular level the factors that are responsible for the formation of the discotic liquid-crystal phases are similar to those for calamitic systems. Thus, the molecular design should aim to increase the liquid crystal–isotropic transition temperature while decreasing the melting point. The latter is certainly achieved by attaching flexible alkyl chains to the perimeter of the rigid disc. The creation of the columnar phases should be relatively straightforward provided the central core is both planar and large. Then, because of the strong attractive forces between the many atoms in the rigid core the molecules will wish to stack face-to-face in a column. The formation of the columns will also be facilitated by the flexible alkyl chains attached to the core. Clearly then it may prove to be difficult to create the nematic phase before the columnar phase is formed unless the disc–disc interactions can be weakened. One way in which this can be achieved is by destroying the planarity of the core, for example, by using phenyl rings attached to the molecular centre so that they can rotate out of the plane (Fig. 38.5a). It is to be expected that the columnar phases should occur below the nematic phase, corresponding to an increase in order with decreasing temperature. This is usually observed, for example, for the hexasubstituted triphenylenes (Fig. 38.5b). However, the truxene derivatives, with long alkyl chains on the perimeter, (Fig. 38.5c) exhibit quite unusual behaviour. For these compounds the crystal melts to form a discotic nematic and then at a higher temperature a columnar phase appears. This deviates from the expected sequence, and because the nematic phase appears at a lower temperature than the columnar phase it is usually referred to as a re-entrant nematic. The occurrence of a re-entrant phase is often attributed to a conformational change which strengthens the molecular interactions with increasing temperature thus making the more ordered phase appear at higher temperatures.

38.2 The Basic Physics of Liquid Crystals

38.2.1 Orientational Order

The defining characteristic of a liquid crystal is the long-range order of its constituent molecules. That is, for rod-like molecules, the molecular long axes tend to align parallel to each other even when separated by large distances. The molecules tend to be aligned parallel to a particular direction known as the director and denoted by $\bm{n}$. This is an apolar vector, that is $\bm{n} = -\bm{n}$, because the nematic does not possess long-range ferroelectric order. The properties of the nematic phase are cylindrically symmetric about the director, which provides a macroscopic definition of this. The anisotropy of the properties results from the orientational order and the extent of this is commonly defined [38.15] by

$$S = \left\langle (3\cos^2\beta - 1)/2 \right\rangle , \qquad (38.1)$$

although other definitions are possible [38.16]. Here β is the angle made by a molecule with the director and the angular brackets indicate the ensemble average. In the limit of perfect order S is unity while in the isotropic phase S vanishes. The temperature dependence of S is shown in Fig. 38.7 for the nematogen, 4,4′-dimethoxyazoxybenzene; this behaviour is typical of most nematic liquid crystals. At low temperatures S is about 0.6 and then decreases with increasing

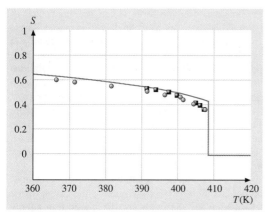

Fig. 38.7 The temperature variation of the orientational order parameter, S, for 4,4′-dimethoxyazoxybenzene; the different symbols indicate results determined with different techniques

temperature, reaching about 0.3 before it vanishes discontinuously at the nematic–isotropic transition, in keeping with the first-order nature of this transition. It is also found, both experimentally and theoretically, that the orientational order of different nematic liquid crystals is approximately the same provided they are compared at corresponding temperatures, either the reduced, T/T_{NI}, or shifted, $T_{\text{NI}} - T$, temperatures [38.16]. Since many properties of liquid crystals are related to the long-range orientational order these also vary with temperature especially in the vicinity of the transition to the isotropic phase.

38.2.2 Director Alignment

The director in a bulk liquid crystal is distributed randomly unless some constraint is applied to the system; a variety of constraints can be employed and two of these are of special significance for display applications. One of them is an electric field and because of the inherent anisotropy in the dielectric permittivity of the liquid crystal the director will be aligned. The electric energy density controlling the alignment is given by [38.17]

$$U_{\text{elec}} = -\varepsilon_0 \Delta\varepsilon (\boldsymbol{n} \cdot \boldsymbol{E})^2 / 2 \, . \tag{38.2}$$

Here ε_0 is the permittivity of a vacuum and the scalar product $\boldsymbol{n} \cdot \boldsymbol{E}$ is $E \cos\theta$, where E is the magnitude of the field $\boldsymbol{E}$ and θ is the angle between the director and the field, $\Delta\varepsilon$ is the anisotropy in the dielectric tensor

$$\Delta\varepsilon = \varepsilon_\| - \varepsilon_\perp \, , \tag{38.3}$$

where the subscripts denote the values parallel ($\|$) and perpendicular ($\perp$) to the director. If the dielectric anisotropy is positive then the director will be aligned parallel to the electric field and, conversely, if $\Delta\varepsilon$ is negative, then the director is aligned orthogonal to the field. The molecular factors that control the sign of $\Delta\varepsilon$ will be discussed in Sect. 38.4. Of course, intense electric fields are also able to align the molecules in an isotropic phase but what is remarkable about a nematic liquid crystal is the very low value of the field needed to achieve complete alignment of the director. Thus for a bulk nematic free of other constraints the electric field necessary to align the director is typically about 30 kV/m, although the value does clearly depend on the magnitude of the dielectric anisotropy. This relatively small value results because of the long-range orientational correlations which mean that the field acts, in effect, on the entire ensemble of molecules and not just single molecules.

The other constraint, essential for display devices, is the interaction between the director and the surface of the container [38.18]. At the surface there are two extreme arrangements for the director. One is with the director orthogonal to the surface, the so-called homeotropic alignment. In the other the director is parallel to a particular direction in the surface; this is known as the uniform planar alignment. The type of alignment depends on the way in which the surface has been treated. For example, for a glass surface coated with silanol groups a polar liquid crystal will be aligned homeotropically, while to achieve this alignment for a non-polar nematic the surface should be covered with long alkyl chains. In these examples the direct interaction of a mesogenic molecule with the surface produces an orthogonal alignment which is then propagated by the long-range order into the bulk. To achieve uniform planar alignment of the director the surface is coated with a polymer, such as a polyimide, which on its own would result in planar alignment. To force the director to be parallel to a particular direction in the surface the polymer is rubbed which aligns the director parallel to the direction of rubbing. There is still some uncertainty about the mechanism responsible for uniform planar alignment. It might result from alignment of the polymer combined with anisotropic intermolecular attractions with the mesogenic molecules, although it had been thought [38.19] to have its origins in surface groves and the elastic interactions which are described later.

The energy of interaction between the surface and the director clearly depends on the nature of the surface treatment and the particular nematic. *Rapini* and

Popular [38.20] have suggested the following simple form for the surface energy density

$$U_S = -A(\mathbf{n} \cdot \mathbf{e})^2/2, \qquad (38.4)$$

where A is the anchoring energy and $\mathbf{e}$ is the easy axis or direction along which the director is aligned. Clearly it has an analogous form to that for the anisotropic interaction between the nematic and an electric field (38.2) but is essentially phenomenological. The anchoring energy is determined to be in the range 10^{-7}–10^{-5} J/m^2 [38.21]. The upper value corresponds to strong anchoring in that typical values of the electric field would not change the director orientation at the surface. In contrast the lower value is associated with weak anchoring and here the director orientation at the surface can be changed by the field.

38.2.3 Elasticity

As the name suggests, a liquid crystal has some properties typical of crystals and others of liquids. The elastic properties of crystals should, therefore, be reflected in the behaviour of liquid crystals. Here it is the director orientation which is the analogue of the atomic positions in a crystal. In the ground state of a nematic liquid crystal the director is uniformly aligned. However, the elastic torques, responsible for this uniform ground state, are weak and, at temperatures within the nematic range, the thermal energy is sufficient to perturb the director configuration in the bulk. This perturbed state can take various forms depending on a combination of factors but, whatever the form, it can be represented as a sum of just three fundamental distortion modes. These are illustrated in Fig. 38.8 and are the splay, twist and bend deformations [38.22]. At a more formal level these modes are also shown in terms of the small deviations of the director from its aligned state at the origin as the location from this is varied. For example, for the twist deformation away from the origin along the x-axis there is a change in the y-component of the director, n_y, and the displacement along the y-axis causes a change in n_x. The magnitude of the twist deformation is measured by the gradients $\partial n_y/\partial x$ and $\partial n_x/\partial y$.

The energy needed to stabilise a given distortion of the director field is clearly related to the extent of the deformation via the gradients. This distortion energy density for a bulk nematic liquid crystal is given by continuum theory [38.22] as

$$f = \left[K_1(\nabla \cdot \mathbf{n})^2 + K_2(\mathbf{n} \cdot \nabla \times \mathbf{n})^2 + K_3(\mathbf{n} \times \nabla \times \mathbf{n})^2 \right]/2, \qquad (38.5)$$

where the terms $\nabla \cdot \mathbf{n}$, $\mathbf{n} \cdot \nabla \times \mathbf{n}$ and $\mathbf{n} \times \nabla \times \mathbf{n}$ correspond to the splay, twist and bend deformations, respectively. The contribution each makes to the free energy is determined by the proportionality constants K_1, K_2 and K_3, which are usually known as the Frank elastic constants for splay, twist and bend, respectively. They are small, typically 5×10^{-12} N, and their small magnitude explains why the thermal energy is able to distort the uniform director arrangement so readily. The elastic constants are not in fact constant but vary with temperature

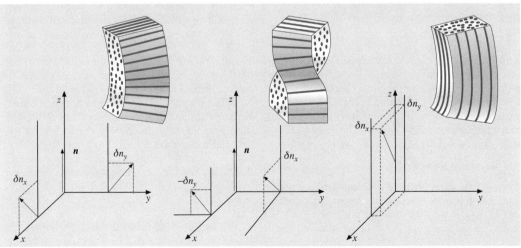

Fig. 38.8a–c The three fundamental deformations for the nematic director (**a**) splay, (**b**) twist and (**c**) bend

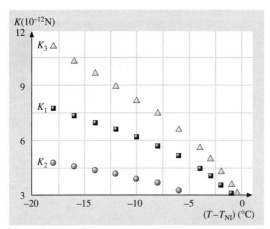

Fig. 38.9 The temperature dependence of the three elastic constants, K_1, K_2 and K_3, for the nematogen 5CB

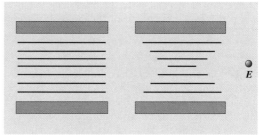

Fig. 38.10a,b The geometry for the Freedericksz experiment used to determine the twist elastic constant, K_2; (**a**) in zero field and (**b**) above the threshold value

and nor are they equal as the results for 4-pentyl-4′-cyanobiphenyl (5CB) shown in Fig. 38.9 demonstrate. The twist elastic constant is seen to be the smallest while the largest is the bend elastic constant. This means that it is easiest to induce a twist deformation in a nematic while a bend deformation is the most difficult to create. All three elastic constants decrease with increasing temperature in keeping with the decreasing order as the transition to the isotropic phase is approached. Like the orientational order the elastic constants vanish discontinuously at the first-order nematic–isotropic transition.

The continuum theory is especially valuable in predicting the behaviour of display devices and this is illustrated by considering one of the ingenious experiments devised by *Fréedericksz* to determine the elastic constants [38.23]. In these a thin slab of nematic is confined between two glass plates with a particular director configuration, either uniform planar or homeotropic produced by surface forces. These forces control the director orientation just at the two surfaces and the alignment across the slab is propagated by the elastic interactions. A field is then applied which will move the director away from its original orientation and the variation of the director orientation with the field strength provides the elastic constant. To employ the continuum theory in order to describe the experiment it is necessary to add the field energy density (38.2) to the elastic free-energy density in (38.5). The director configuration is then obtained by integrating the free-energy density over the volume of the sample, and minimising this, subject to the surface constraints. The geometry of the experiment is shown in Fig. 38.10 and provided the dielectric anisotropy, $\Delta\varepsilon$,

is positive the director will move from being orthogonal to the field to being parallel to it. However, the extent of this twist deformation will vary across the cell, being greatest at the centre and zero at the surfaces, in the limit of strong anchoring. The dependence of the director orientation, at the centre of the slab, with respect to the electric field is shown in Fig. 38.11. As the field strength is increased from zero the director orientation remains unchanged until a threshold value, E_{th}, is reached when the angle between the director and the field starts to decrease continuously. At very high values of the field the director at the centre of the slab tends to be parallel to the field. This behaviour can be understood in the following simple terms. Below the threshold field the elastic energy exceeds the electrical energy and so the director retains its uniform planar alignment. Above the threshold field strength the elastic energy is less than

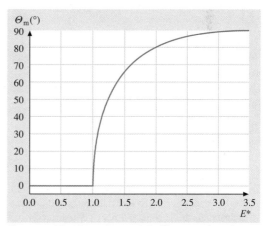

Fig. 38.11 The dependence of the director orientation θ_m at the centre of a nematic slab on the scaled field strength, $E* = E/E_{\text{th}}$

the electrical energy and so the director begins to move to be parallel to the electric field. The threshold electric field is predicted to be

$$E_{\text{th}} = (\pi/l)\sqrt{K_2/\varepsilon_0 \Delta\varepsilon} \,, \tag{38.6}$$

where l is the slab thickness. Once the threshold field has been measured and provided the dielectric anisotropy is known, the twist elastic constant would be available. Other Fréedericksz experiments with analogous expressions for the threshold field lead to the determination of the splay and bend elastic constants.

For real display applications (Sect. 38.3) the director alignment at the surface deviates from either uniform planar or homeotropic alignment. This deviation is known as a surface pre-tilt and is illustrated in Fig. 38.6a for near-uniform planar alignment with the tilt direction on the two surfaces differing by 180°. The cell with this arrangement has what is known as antiparallel alignment, so named because of the difference in the tilt direction caused by the direction of rubbing on the surfaces being antiparallel. In zero field, therefore, the director is uniformly aligned across the cell but tilted with respect to the x-axis set in the surface. A continuum theory calculation analogous to the case when the pre-tilt angle, θ_0, is zero allows the dependence of the director orientation θ_m in the centre of the cell to be determined as a function of the strength of the field applied across the cell. The results of these calculations,

for a nematic with positive $\Delta\varepsilon$, are shown in Fig. 38.12b as a function of the scaled field strength, E/E_{th}, where E_{th} is the threshold field for zero pre-tilt. The theoretical dependence for a tilt angle of zero is analogous to that considered for the twist deformation. In other words, below the threshold field the director is parallel to the easy axis and then above this threshold the director moves to become increasingly parallel to the field. When there is a surface pre-tilt the behaviour is quite different and this is especially apparent for a pre-tilt angle of 10° (Fig. 38.12b). In zero field the angle θ_m made by the director with the x-axis is 10° and, as the field increases, so does θ_m. The rate of increase grows as the threshold field is approached and then is reduced as E/E_{th} increases beyond unity. Comparison of this behaviour with the conventional Fréedericksz experiment (Fig. 38.12b) shows that a pre-tilt of 10° has a significant effect on the way in which the director orientation changes with the field strength. Indeed, even for a pre-tilt of just 2°, there is a pronounced difference in behaviour in the vicinity of the threshold field. This clearly has important implications for the accurate determination of the elastic constants [38.20]. It also shows how unique the behaviour is when the surface pre-tilt angle is zero.

Chiral nematics are often employed in liquid-crystal display devices. Locally, their structure is analogous to a nematic but the director is twisted into a helical structure. The continuum theory for the chiral nematic must, therefore, be consistent with the helical ground state structure of the phase. To achieve this, a constant is added to the twist term in the elastic free-energy density. Thus (38.5) for a nematic becomes

$$f = \Big[K_1 (\nabla \cdot \boldsymbol{n})^2 + K_2 (\boldsymbol{n} \cdot \nabla \times \boldsymbol{n} - 2\pi/p)^2 \\ + K_3 (\boldsymbol{n} \times \nabla \times \boldsymbol{n})^2 \Big]/2 \tag{38.7}$$

for a chiral nematic, where p is the pitch of the helix.

38.2.4 Flexoelectricity

Another property of solids which is mimicked by liquid crystals is piezoelectricity. For solids this is the generation of a macroscopic electrical polarisation as a result of the deformation of certain ionic materials. It is to be expected, therefore, that deformation of the director distribution for a liquid crystal will create a macroscopic polarisation; this proves to be the case [38.24] and the phenomenon is known as flexoelectricity. The origin of flexoelectricity can be understood in the following way. It is generally assumed that mesogenic molecules are

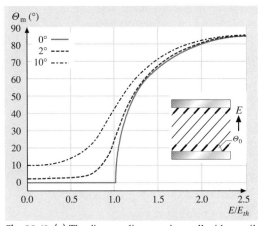

Fig. 38.12 (a) The director alignment in a cell with pre-tilt at the two surfaces assembled so that the rubbing directions are antiparallel. **(b)** The electric field dependence of the director orientation θ_m at the centre of the cell for values of the pre-tilt angle of 0° (*solid line*), 2° (*dashed line*) and 10° (*dash-dotted line*)

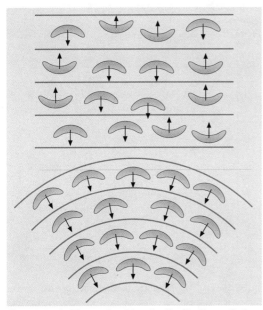

Fig. 38.13a,b The orientational distribution of bent molecules in (**a**) a uniformly aligned nematic and (**b**) one subject to a bend deformation; the x-axes are denoted by *arrows*

cylindrically symmetric but examination of the molecular structure of real mesogens shows that this is not the case. The molecules are asymmetric and at one extreme can be thought of as bent, as shown in Fig. 38.13; here z is the molecular long axis and x is the axis bisecting the bond angle. When the director is uniformly aligned the z-axis will be parallel to the director and the x-axis will be randomly arranged orthogonal to it (Fig. 38.13). If the director is now subject to a bend deformation the molecular long axis will still tend to be parallel to the local director, however, the x-axis will tend to align parallel to a direction in the plane formed by the bent director. This change in the distribution function for the x-axis will introduce polar order into the system and if there is an electrical dipole moment along the x-axis then the deformed nematic will exhibit a macroscopic polarisation, P. A similar argument shows that, if the molecule is wedge-shaped, a splay deformation of the director distribution will also induce a polarisation in the nematic [38.24].

The magnitude of P, the induced dipole per unit volume, clearly depends on the extent of the deformation in the director distribution. In the linear response regime the induced polarisation for mesogenic molecules of arbitrary asymmetry is given by [38.25]

$$P = e_1 n \nabla \cdot n + e_3 n \times \nabla \times n , \qquad (38.8)$$

where the vectors $n\nabla \cdot n$ and $n \times \nabla \times n$ represent the splay and bend deformations, respectively. Since the polarisation per unit volume is also a vector then the proportionality constants, e_1 and e_3, are scalars. These are known as the splay and bend flexoelectric coefficients, respectively, and their dimensions are Cm^{-1}. The determination of individual flexoelectric coefficients is challenging [38.26], however, it seems that $(e_1 + e_3)$ is of the order of 10^{-12}–10^{-11} Cm^{-1}, although from the previous discussion their magnitude should vary amongst the mesogens because of their dependence on the molecular asymmetry and the size and location of the dipole moment. The molecular model proposed to understand the flexoelectricity of nematics suggests that, for rod-like molecules, devoid of asymmetry, the flexoelectric coefficients should vanish, but that does not seem to be the case. Indeed, it has been proposed that polarisation can result even for rod-like molecules if they possess an electrostatic quadrupole moment [38.27]. This is not inconsistent with the polarisation predicted by (38.8) which follows from the reduced symmetry of a nematic with splay and bend deformations of the director [38.24]. Although the existence of flexoelectricity is of considerable fundamental interest the inverse effect in which the director is deformed from a uniform state by the application of an electric field is of relevance for liquid-crystal displays (Sect. 38.3). This deformation occurs because the polarisation induced by the director deformation can couple with the applied electric field and so stabilise the deformation [38.26]. Since the coupling is linear in the electric field then reversal of the field will simply reverse the deformation; a novel bistable device based solely on this reversal is described in Sect. 38.3.

38.2.5 Viscosity

For a nematogen the fluidity of the nematic phase is comparable to that of the isotropic phase appearing at a higher temperature. It is this fluidity, similar to that of a liquid, which is responsible, in part, for the display applications of nematics. An indication of the fluidity of conventional liquids is provided by a single viscosity coefficient, η, measured from the flow of the liquid subject to an applied stress in a viscometer. The flow behaviour of a nematic is made more complex by its defining long-range orientational order and the resultant anisotropy of the phase.

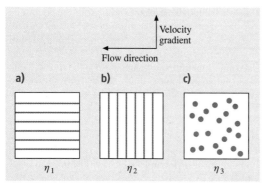

Fig. 38.14a–c The principal flow geometries of the Miesowicz experiments with the director pinned (**a**) parallel to the flow direction, (**b**) parallel to the velocity gradient and (**c**) orthogonal to both the flow direction and velocity gradient

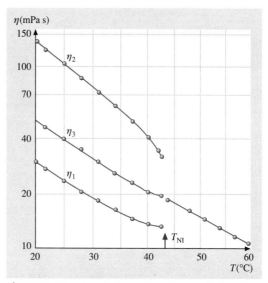

Fig. 38.15 The temperature dependence of the three Miesowicz viscosity coefficients, η_1, η_2, and η_3, for the nematic phase of MBBA; the viscosity coefficient for the isotropic phase is also shown

This complexity can be appreciated at a practical level in terms of the Miesowicz experiments to determine the viscosity coefficients of a nematic [38.28]. In these experiments it is helpful to consider flow in a viscometer with a square cross section such that there is a velocity gradient orthogonal to the direction of flow. For a nematic, flow through the viscometer will now depend on the orientation of the director with respect to these two axes. A magnetic field is employed to align the director along a particular axis and it must be sufficiently strong that flow does not perturb the director alignment. There are three relatively simple flow geometries and these are shown in Fig. 38.14: (a) with the director parallel to the flow direction, (b) with the director parallel to the velocity gradient, and (c) with the director orthogonal to both the flow direction and velocity gradient. The viscosity coefficients were denoted by *Miesowicz* as (a) η_1, (b) η_2, and (c) η_3, although other notation has been proposed in which η_1 and η_2 are interchanged [38.29]. The three viscosity coefficients are clearly expected to differ given the anisotropy of the nematic phase and these differences are found to be large. They are illustrated for the room-temperature nematogen 4-methoxybenzylidene-4′-butylaniline (MBBA) [38.30] in Fig. 38.15 where the viscosities are plotted against temperature.

It is immediately apparent that flow is easiest when the director is parallel to the direction of flow, as might have been anticipated. Conversely flow of the nematic is most difficult when the director is orthogonal to the flow direction but parallel to the velocity gradient. The intermediate viscosity, η_3, occurs when the director is orthogonal to both the flow direction and the velocity gradient. Here it is seen that this third viscosity coefficient for the nematic is very similar to the viscosity coefficient for the isotropic phase extrapolated to lower temperatures.

The viscosity coefficient $\eta(\theta, \phi)$ for an arbitrary orientation (θ, ϕ) of the director with respect to the flow direction and velocity gradient can be related to the director orientation. It might be expected that this orientation dependence would involve just the three Miesowicz viscosity coefficients, rather like the transformation of a second-rank tensor from its principal axis system. In fact this is not quite the case and a fourth viscosity coefficient η_{12} needs to be introduced. The orientation dependence is then [38.28]

$$\eta(\theta, \phi) = \eta_1 \cos^2 \theta + \left(\eta_2 + \eta_{12} \cos^2 \theta\right)$$
$$\times \sin^2 \theta \cos^2 \phi + \eta_3 \sin^2 \theta \sin^2 \phi \, ; \quad (38.9)$$

we see that for the principal orientations $(0, \pi/2)$, $(\pi/2, 0)$ and $(\pi/2, \pi/2)$ the expression gives η_1, η_2, and η_3, as required. The optimum director orientation with which to determine η_{12} is with the director at 45° to both the flow direction and the velocity gradient; the viscosity coefficient for this is

$$\eta(\pi/4, 0) = \eta_{12}/4 + (\eta_1 + \eta_2)/2 \, . \quad (38.10)$$

The value of η_{12} determined for MBBA proves to be significantly smaller than the other three viscosity coefficients.

The four viscosity coefficients have been defined at a practical level in terms of flow in which the director orientation is held fixed. The converse of these experiments, in which the director orientation is changed in the absence of flow, allows the definition of the fifth and final viscosity coefficient. This is known as the rotational viscosity coefficient; it is denoted by the symbol γ_1 and plays a major role in determining the response times of display devices (Sect. 38.4). To appreciate the significance of γ_1, it is helpful, as for the other four viscosity coefficients, to consider an experiment with which to measure it. In this an electric field is suddenly applied an angle θ to a uniformly aligned director, then providing the dielectric anisotropy is positive the director orientation will be changed and rotates towards the field direction. The electric torque responsible for the alignment is given by

$$\Gamma_{\text{elec}} = -\varepsilon_0 \Delta\varepsilon \sin 2\theta/2 \,, \tag{38.11}$$

which is the derivative of the electric energy in (38.2). The rotation of the director is opposed by the viscous torque

$$\Gamma_{\text{visc}} = \gamma_1 \, d\theta/dt \,, \tag{38.12}$$

which is linear in the rate at which the director orientation changes, with the proportionality constant being γ_1. Provided the only constraint on the director is the electric field and provided, $\theta_0 \leq 45°$, the director will move as a monodomain so that the elastic terms vanish. The inertial term for a nematic is small and so the movement of the director is governed by the equation in which the two torques are balanced. That is, the electric torque causing rotation is balanced by the viscous torque opposing it; this gives

$$\gamma_1 \, d\theta/dt = -\varepsilon_0 \Delta\varepsilon E^2 \sin^2\theta/2 \,. \tag{38.13}$$

The solution to this differential equation is

$$\tan\theta = \tan\theta_0 \exp(-t/\tau) \,, \tag{38.14}$$

where θ_0 is the initial orientation of the director with respect to the electric field and τ is the relaxation time

$$\tau = \gamma_1/\varepsilon_0 \Delta\varepsilon E^2 \,. \tag{38.15}$$

Measurement of the time-dependent director orientation allows τ to be determined and from this γ_1, given values of $\Delta\varepsilon$. For MBBA the rotational viscosity coefficient is found to be slightly less than η_2 and to parallel its temperature dependence [38.31].

The five independent viscosity coefficients necessary to describe the viscous behaviour of a nematic have been introduced in a pragmatic manner by appealing to experiments employed to measure these coefficients. However, the viscosity coefficients can be introduced in a more formal way as has been shown by Ericksen and then by Leslie in their development of the theory for nematodynamics [38.32]. The Leslie–Ericksen theory in its original form contained six viscosity coefficients, but subsequently Parodi has shown, using the Onsager relations, that there is a further equation linking the viscosity coefficients, thus reducing the number of independent coefficients to five [38.33]. These five coefficients are linearly related to those introduced by reference to specific experiments.

38.3 Liquid-Crystal Devices

The idea to use liquid crystals as electro-optic devices goes back to the early days of liquid-crystal research. In 1918 Björnståhl, a Swedish physicist, demonstrated that the intensity of light transmitted by a liquid crystal could be varied by application of an electric field [38.34]. As optical devices of various types became established in the first decades of the 20th century, mainly in the entertainment industry, ways of controlling light intensity became important to the developing technologies. One device that soon found commercial application was the Kerr cell shutter, in which an electric field caused the contained fluid (usually nitrobenzene) to become birefringent. Placing such a cell between crossed polarisers enabled a beam of light to be switched on and off very rapidly. The first report of liquid crystals being of interest for electro-optic devices was in 1936, when the Marconi Company filed a patent [38.35] which exploited the high birefringence of nematic liquid crystals in an electro-optic shutter. However, it was another 35 years before commercial devices became available which used the electro-optic properties of liquid crystals. The long interruption to the development of liquid-crystal devices can be attributed to the lack of suitable materials. We shall see in this Section how the physical properties of liquid crystals determine the performance of devices.

The optical properties of liquid crystals are exploited in displays, although the operational characteristics of such devices also depend crucially on many other physical properties (Sect. 38.2). Since the devices to be described all depend on the application of an electric field, their operation will be influenced by electrical properties such as dielectric permittivity and electrical conductivity. There is a range of electro-optic effects that can be used in devices, and the precise manner in which the properties of the liquid-crystal materials affect the device behaviour depends on the effect and the configuration of the cell. Thus there is not a single set of ideal properties that can define the best liquid-crystal material, rather the material properties have to be optimised for a particular application. In this section, different devices will be described, and their dependence on different properties will be outlined. The way in which the material properties can be adjusted for any application will be discussed in the final section of this Chapter.

38.3.1 A Model Liquid-Crystal Display: Electrically Controlled Birefringence Mode (ECB)

If an electric field is applied across a film of planar-aligned liquid crystal having a positive dielectric anisotropy, then the director of the liquid crystal will tend to align along the electric field. Thus the director will rotate into the field direction, and the optical retardation or birefringence of the film will change.

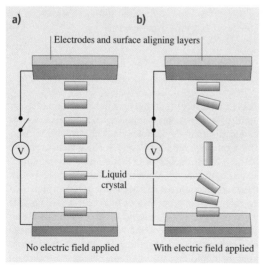

Fig. 38.16 Electrically controlled birefringence (ECB) cell

If the film is placed between crossed polarisers, then the change in optical retardation will be observed as a change in the intensity of transmitted light. The configuration for an ECB-mode display is schematically illustrated in Fig. 38.16. This illustrates the principal components of a liquid-crystal display. The liquid crystal is contained between two glass plates that have been coated with a transparent conducting layer, usually an indium–tin oxide alloy. Such treatment allows the application of an electric field across the liquid-crystal film, which can then be viewed along the field direction. For most applications, the surfaces of the electrodes are treated so that a particular director orientation is defined at the surface. This can be achieved in a variety of ways, depending on the desired surface director orientation. For the ECB-mode display under consideration, the director alignment should be parallel to the glass substrates and along a defined direction, which is the same on both surfaces. The standard technique to produce this alignment is to coat the glass plates with a thin (0.5 nm) layer of polyimide, which is mechanically rubbed in a particular direction. The cell is then placed between crossed polarisers, the extinction directions of which make angles of $\pm 45°$ with the surface-defined director orientation.

The unperturbed state of the liquid-crystal film will be determined by the surfaces which contain it. If these have been treated in such a way that the liquid-crystal director is parallel to the surface along a particular direction, then the film will act as an optical retardation plate. Thus incident polarised light will, in general, emerge elliptically polarised, and there will be a phase retardation between components of the light wave parallel to the fast and slow axes of the retardation plate. For electromagnetic waves polarised at $\pm 45°$ to the director the phase retardation ϕ, in radians, will be determined by the intrinsic birefringence of the liquid crystal ($\Delta n = n_\mathrm{e} - n_\mathrm{o}$), the film thickness ℓ and the wavelength of the light λ:

$$\phi = \frac{2\pi\ell}{\lambda}(n_\mathrm{e} - n_\mathrm{o}) ; \qquad (38.16)$$

n_e and n_o are respectively the extraordinary (slow) and ordinary (fast) refractive indices of the liquid crystal (assuming that Δn is positive). If the emergent elliptically polarised light passes through a second polariser, crossed with respect to the incident polarisation direction, only a proportion of the incident intensity will be transmitted. The normalised intensity of light transmitted by a pair of crossed polarisers having a birefringent element between them, the axis of which is at $\pm 45°$ to the extinction directions of the polarisers is given by

$$T = \left(\frac{1}{2}\right)\sin^2\frac{\phi}{2} \qquad (38.17)$$

(the factor of one half appears for incident unpolarised light – if the light is polarised, as from a laser, then the factor is one).

Thus the initial appearance of the cell will be brightest if the cell thickness, birefringence and wavelength are chosen to give ϕ equal to π, 3π, 5π etc. It is normal to select the cell thickness to give $\phi = \pi$, and under these conditions the display is known as *normally white*. It is possible, though less satisfactory, to configure the display so that it operates in a *normally black* state, corresponding to a phase retardation of a multiple of 2π.

Application of an electric field causes the director orientation to change such that the optical retardation of the cell decreases to zero, and hence the cell becomes non-transmitting, at least in the normally white configuration. Under these circumstances the optical retardation across the cell becomes

$$\phi = \frac{2\pi}{\lambda}\int_0^\ell \{n_e[\theta(z)] - n_0\}\,dz, \qquad (38.18)$$

where the effective extraordinary refractive index $n_e[\theta(z)]$ depends on the angle $[90° - \theta(z)]$ between the director and the field and is a function of position z in the cell. This effective index is given by

$$\frac{1}{n_e^2[\theta(z)]} = \frac{\sin^2[\theta(z)]}{n_o^2} + \frac{\cos^2[\theta(z)]}{n_e^2}; \qquad (38.19)$$

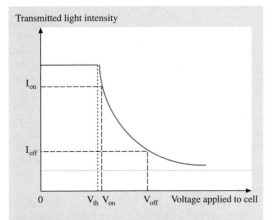

Fig. 38.17 Variation of optical transmission with voltage for an ECB cell between crossed polarisers

when the director is along the field direction $\theta = 90°$, so $n_e(90°) = n_o$, and $\phi = 0$.

The orientational distribution of the director in the cell in the presence of an applied electric field is determined by the strength of the field, the electric permittivity and elastic constants of the liquid crystal, and most importantly by the properties at the interface between the liquid crystal and the aligning surfaces. If the director at the surface satisfies the *strong anchoring* condition i.e. it is unaffected by the applied electric field, and the surface director is strictly perpendicular to the electric field, then the reorientation of the director exhibits a threshold response, known as a Fréedericksz transition. The change in transmitted intensity as a function of voltage can be calculated using continuum theory and simple optics [38.36], and a typical transmission curve for an ECB cell is illustrated schematically in Fig. 38.17.

The threshold voltage for director reorientation is independent of cell thickness, and is given by

$$V_{th} = \pi\sqrt{\frac{K_1}{\varepsilon_0 \Delta\varepsilon}}, \qquad (38.20)$$

where $\Delta\varepsilon$ is the anisotropy in the dielectric permittivity and K_1 is the splay elastic constant. However, it is clear from (38.17) and (38.18) that the intensity of light transmitted by the ECB cell depends on the cell thickness and the wavelength of light. This undermines the usefulness of displays based on the ECB mode, since they require cells of uniform thickness and also they will tend to show colouration in white light. Another important operating characteristic of displays is the *angle of view*, i.e. how the image contrast changes as the angle of incidence moves away from 90° with respect to the plane of the cell. This is clear from Fig. 38.16, where the optical paths in the distorted state for observation to the left or right of the perpendicular to the electrodes are clearly different.

In order to construct a useful display from a simple on/off shutter, it is necessary to consider how image data will be transferred to the display. This is known as *addressing*, and to a large extent it is determined by the circuitry that drives the display. However, we shall see that certain properties of liquid crystals also contribute to the effectiveness of different types of addressing. The simplest method of displaying images on a liquid-crystal display is to form an array of separate cells of the type illustrated by the ECB cell, each having a separate connection for the application of an electric field. Images can then be created by switching on, or off, those cells required to form the image. This technique is known as *direct addressing*, and can be illustrated by the seven-segment displays used in watches and numerical instrument displays (Fig. 38.18).

To create complex displays, a large number of separate cells, known as picture elements or pixels, must be fabricated, usually in the form of a matrix. Providing separate electrical connections for these pixel arrays (e.g. 640×480 for a standard visual graphics array (VGA) computer screen, is impossible, and so other methods of addressing have had to be developed. Historically, the first was the technique known as *passive matrix addressing*, in which the array of cells are identified in rows and columns, and connections are only made to the rows and columns: for an array of $n \times m$ pixels, only $n + m$ connections are made instead of $n \times m$, as required for direct addressing. Each row is activated in turn (sequentially), and appropriate voltage pulses applied to the columns. Only those pixel elements for which the sum of column and row voltages exceeds a *threshold* are switched to an on-state. However, the problem with this method is that many unwanted pixels in the off-state still have a voltage applied, and may be partially activated in the display: so-called *crosstalk*. The time-sharing of activating signals is known as multiplexing, and it relies on the addressed pixels holding their signal while other pixels which make up the image are activated. Provided that the multiplexing is on a time scale of milliseconds, any fluctuation in the image goes unnoticed. However, if the multiplexing becomes too slow, the image starts to flicker. In fact there is a limit to the number of rows of the matrix which can be addressed (n_{max}), which is related to the ratio of the on-voltage to off-voltage by a result due to *Alt* and *Pleshko* [38.37]:

$$\frac{V_{on}}{V_{off}} = \sqrt{\frac{n_{max}^{1/2} + 1}{n_{max}^{1/2} - 1}}. \tag{38.21}$$

The cell characteristic which determines the values of V_{on} and V_{off} is the optical transmittance curve, as illustrated in Fig. 38.17. Depending on the desired contrast ratio I_{on}/I_{off} for a pixel, then the voltages V_{on}/V_{off} are determined. Thus, in the operation of a passively addressed matrix display, there is a tradeoff between the contrast ratio (I_{on}/I_{off}) and the number of rows i.e. the complexity of the display. The shape of the transmittance curve is determined by the liquid-crystal material properties, and so these affect the resolution of the image displayed. For a desired contrast ratio of 4, the corresponding on/off ratio for an ECB cell might be 1.82. This gives a maximum number of rows as four, which corresponds to a very-low-resolution display, which would be unusable except for a very basic device. This brief description of a simple electro-optic display operating in the ECB mode illustrates that its performance depends on the dielectric, optical and elastic properties of the liquid crystal material used. Additionally, the properties at the liquid-crystal substrate interface and the geometry of the cell will influence the electro-optic response of the cell.

One performance characteristic that is of very great importance is the speed with which the display information can be changed, since this determines the quality and resolution of moving images. The dynamics of fluids are related to their viscosity, and it has already been shown that the viscous properties of liquid crystals are complicated to describe, and are correspondingly difficult to measure. The complication arises because liquid crystals are elastic fluids, and so there is a coupling between the flow of the fluid and the orientation of the director within the fluid. We have seen that the motion of the director can be described in terms of a rotational viscosity, and the optical properties exploited in displays are related to changes in the director orientation. Thus it is the rotational viscosity that is of primary importance in determining the time response of displays. However, the fact that changes of director orientation cause fluid flow in liquid crystals complicates the process.

The time response of a liquid-crystal display pixel can be illustrated by reference to the ECB display, although other cell configurations modify the behaviour to some extent. In what follows, we shall assume that

Fig. 38.18 Schematic of a directly addressed seven-segment liquid-crystal display

the reorientation of the director within a display pixel does not cause the nematic liquid crystal to flow. For a uniform parallel-aligned nematic-liquid-crystal film, the time for the director to respond depends on the magnitude of the electric field (or voltage) applied to the cell. If the voltage applied is only just greater than the threshold voltage, then the time is very long, while if a large voltage is applied, then the director responds quickly. It is found that the time response can usually be represented as an exponential behaviour, although effects of flow will change this. Neglecting these, a response time τ_{on} can be defined in terms of the change in transmitted light intensity as

$$\frac{I(t) - I(0)}{I_{on} - I(0)} = 1 - \exp\left[-\left(\frac{t}{\tau_{on}}\right)\right]. \qquad (38.22)$$

and the relaxation time for switching on the display is given approximately by [38.38, 39]

$$\tau_{on} = \frac{\gamma_1 \ell^2}{\pi^2 K_{eff}} \left[\left(\frac{V_{on}}{V_{th}}\right)^2 - 1\right]^{-1}. \qquad (38.23)$$

On removing the applied voltage, the display element returns to its off-state, but with a different relaxation time which is independent of the applied voltage, such that

$$\tau_{off} = \frac{\gamma_1 \ell^2}{\pi^2 K_{eff}}; \qquad (38.24)$$

here γ_1 is the rotational viscosity coefficient, and the effective elastic constant, K_{eff} that appears in these expressions is the splay elastic constant, K_1, for the ECB-mode display. These equations can be modified for other display configurations by changing K_{eff}.

Although the ECB-mode display is the simplest that can be envisaged, based on the Fréedericksz effect, there are many disadvantages, in particular with respect to its viewing characteristics, and it has not been used commercially to any significant extent. However, the apparently simple modification of twisting the upper plate by 90° has resulted in the phenomenally successful twisted nematic display, which represents a large part of today's multi-billion-dollar market.

38.3.2 High-Volume Commercial Displays: The Twisted Nematic (TN) and Super-Twisted Nematic (STN) Displays

The simple twisted nematic display is essentially the same as the ECB display depicted in Fig. 38.16, except

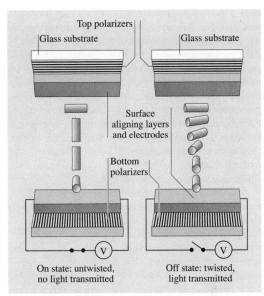

Fig. 38.19 Schematic of a twisted nematic display

that the orientations of the surface director of the containing glass plates are rotated by 90°. However, the TN cell, invented by *Schadt* and *Helfrich* in 1970 [38.40], represented a considerable improvement over earlier devices, and rapidly became the preferred configuration for commercial displays. A schematic representation of the TN cell is given in Fig. 38.19.

This twisted configuration for a liquid crystal film was discovered by Mauguin, who found that instead of producing elliptically polarised light, such a twisted film could rotate the plane of polarisation by an angle equal to the twist angle between the surface directors of the glass plates [38.41]. In fact by working through the optics of twisted films, Mauguin showed that perfect rotation of the plane of polarisation only resulted if the film satisfied the following condition:

$$2\Delta n \ell \gg \lambda. \qquad (38.25)$$

Thus for the cell illustrated in Fig. 38.19, the off-state would be perfectly transmitting. This is known as the *normally white* mode. In reality the cells used for TN displays do not meet the Mauguin condition, and the transmission for a 90° twisted cell between crossed polarisers and incident unpolarised light is given by [38.42]

$$T = \frac{1}{2}\left(1 - \frac{\sin^2\left(\frac{\pi}{2}\sqrt{1+u^2}\right)}{1+u^2}\right), \qquad (38.26)$$

where $u = \frac{2\Delta n \ell}{\lambda}$. Equation (38.26) shows that, for sufficiently large u, the transmission, T is indeed a maximum of 0.5, however it is also a maximum for $u = \sqrt{3}, \sqrt{15}, \sqrt{35}$, etc. These points on the transmission curve correspond to the *Gooch–Tarry minima*; they are labelled as minima, since they were found for a TN cell operating in the *normally black* state. Most commercial cells operate under conditions of the first or second minima so that thin cells can be used, which give faster responses. It is, therefore, important that the birefringence of the liquid-crystal material can be adjusted to match the desired cell thickness, so that the display can have the best optical characteristics in the off-state.

Application of a sufficiently strong electric field across the twisted film of a nematic liquid crystal having a positive dielectric anisotropy causes the director to align along the field direction. Under these circumstances the film no longer rotates the plane of polarised light, and so appears dark. The transmission as a function of voltage for a twisted cell is similar to that shown in Fig. 38.17, except that the transmission varies more strongly with change in voltage above the threshold, and drops to zero much more rapidly. In contrast to the ECB cell discussed above, the threshold voltage for a TN cell depends on all three elastic constants:

$$V_{\text{th}}^{\text{TN}} = \pi \sqrt{\frac{\left(K_1 + \frac{\zeta}{2}(K_3 - 2K_2)\right)}{\varepsilon_0 \Delta \varepsilon}}, \quad (38.27)$$

where ζ is the twist angle (usually $\pi/2$). The relative change of the transmission intensity with voltage of the TN cell is greater than for the ECB cell, and it can be shown that the steepness of the transmission curve increases as the property ratios K_3/K_1 and $\Delta\varepsilon/\varepsilon_\perp$ decrease. The on/off voltage ratio for a TN cell is closer to unity, than for an ECB cell, and so for similar contrast ratios more lines can be addressed: up to about 20 for typical cells and materials. This is significantly larger than for the ECB cell, and so the TN cell allows more complex images to be displayed. There is still a wavelength dependence for the transmission, although this is less marked than for the ECB mode. However, even with the improved multiplexing capabilities of the TN display over the ECB cell, it is still not good enough to use for computer screens. One very successful approach to solve this problem was to modify the TN cell geometry so that instead of a 90° twist, the directors on opposite sides of the cell are rotated by about 270°. This is known as a super-twisted nematic cell.

The concept of the 270° super-twisted nematic (STN) display seems at first sight to be irrational [38.43, 44]. The director is an apolar vector and so there should be no difference between a 270° and a 90° twisted cell. However, it is possible to maintain a director twist greater than 90° if the liquid crystal is chiral. The use of chiral additives in 90° TN cells was already established, since a very small quantity of chiral dopant added to a TN mixture would break the left/right twist degeneracy in the cell and so remove patches of reversed twist, giving a much improved appearance to the display. If the amount of chiral dopant was increased, then the chiral liquid-crystal mixture would develop a significant intrinsic pitch. By adjusting the concentration of the chiral dopant, the pitch of the mixture could be matched to the 270° twist across the cell thickness of ℓ. i.e. $p \approx 4\ell/3$. An STN cell operates in the same way as a TN cell, so that an applied electric field causes the director to rotate towards the field direction, thereby changing the optical retardation through the film and the transmission between external crossed polarisers. However, the additional twist in the STN cell has a significant effect on the optical properties of the nematic film.

The optical characteristics of the STN off-state are usually outside the Mauguin condition, which means that polarised light passing through the cell is not guided, and emerges elliptically polarised. The degree of ellipticity is wavelength dependent, and so in white light the off-state appears coloured, as does the on-state. An ingenious solution to this problem is to have two identical STN cells, one behind the other, but where the second compensating cell has a twist of the opposite sense. In operation, only the first of the cells has a voltage applied to it. The compensation cell acts to subtract the residual birefringence of the liquid crystal layer, and the display now switches between white and black. Despite this additional complexity, the huge advantage of the STN display is the rapid change in optical transmission with increasing voltage, and a full optical analysis shows that under optimum conditions the rate of change of transmission with voltage can become infinite. In modern implementations of the STN display, the residual birefringence can be compensated by an optical film, avoiding the need for double cells. It is easy to operate an STN display with a $V_{\text{on}}/V_{\text{off}}$ ratio of 1.1, which corresponds to an n_{max} of 100. The shapes of the transmission/voltage curve of ECB, TN and STN cells are a direct consequence of the dielectric and elastic properties of the liquid crystal material, but also depend strongly on the configuration of the cells and the surface alignment

In the description given of displays and their performance, some important aspects have been ignored. The surface orientation of the directors has been assumed to be pinned in the surface plane, which is the requirement for a threshold response. However, it has been found that the performance of displays can be greatly improved if this condition is relaxed, and a pre-tilt is introduced to the cell, such that the surface director may make an angle of up to 60° to the plane of the containing glass plates. This pre-tilt can be introduced by different surface treatments, and it depends on the interfacial properties of the liquid crystal. An important performance characteristic of displays is the angle dependence of the contrast ratio, or more simply how the appearance of the displayed image changes with angle of view. This is largely determined by the display design, and can be accurately calculated from the optical properties of the cell. The refractive indices of the liquid crystal will affect the contrast ratio and angle of view, but precise control of these performance parameters is difficult. Improvements to the appearance in terms of the angle of view or brightness of displays have been achieved by the placing precisely manufactured birefringent polymer films behind or in front of liquid-crystal cells.

38.3.3 Complex LC Displays and Other Cell Configurations

The STN configuration described above used passive matrix addressing, and this opened up the possibility of relatively large-area high-resolution displays, which could be used in laptop computers and other hand-held displays. The next step was to introduce colour by dividing each picture element into three sub-pixels with red, green and blue filters. However, the demand in some market sectors for larger displays with improved appearance having higher resolution [extended graphics array (XGA) displays have 1024×768 pixels] overwhelmed the capability of passive matrix addressing, and the alternative method of active matrix addressing is now used for more complex displays. This technique requires each pixel to have its own switch, as in the simple seven-segment display already described. For high-resolution displays hundreds of thousands of switches are provided by thin-film transistors deposited onto the glass substrate, which forms the screen. These are known as thin-film-transistor twisted-nematic (TFT-TN) displays [38.45]. The sophistication of these displays relies on the capabilities of integrated circuit technology, but the properties of the liquid-crystal materials must still be optimised for the device configuration. A representation of a TFT-TN display is given in Fig. 38.20.

Although being described as an active matrix, the TFT-TN display still uses sequential addressing of pixel rows, and so activated pixels must remain switched on while the rest of the display image is created. The TFTs provide a source of voltage to each liquid crystal pixel, which must then hold its charge, as a capacitor, until the image is changed. So another property of liquid crystals becomes important, that of low electrical conductivity, since the charge on a pixel will be lost by conduction through the liquid crystal. This determines the choice of materials for TFT-TN displays. Generally, high dielectric anisotropy is a desirable property for liquid-crystal display mixtures, since it reduces the operating voltage. However, materials with a high dielectric constant tend to have a high electrical conductivity, since charges either from impurities or leached from surfaces will be stabilised in high-dielectric-constant fluids. Thus the selection of suitable materials requires a compromise between its dielectric and conductance properties, and, of course, the all important refractive indices.

Many different cell configurations, which exploit the optical properties of liquid crystals in different ways, have been tried, and some of these have been commercialised to meet particular market requirements. One rather successful approach to the problem of restricted viewing angle has been the development of the in-plane switching mode (IPS) twisted nematic display [38.46, 47]; this is illustrated in Fig. 38.21. The two optical states of the cells are (i) twisted, and (ii) planar (parallel aligned film), and the director is switched between these states by application of an electric field across electrodes on a single plate of the cell. The state

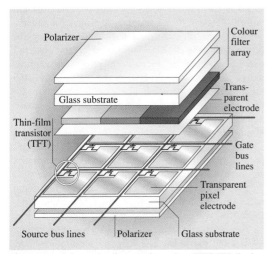

Fig. 38.20 Schematic of a complex colour TFT-TN display

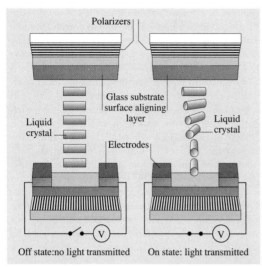

Fig. 38.21 Schematic of an in-plane switching mode display

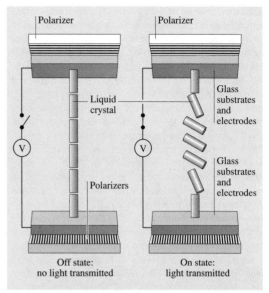

Fig. 38.22 Schematic of a twisted vertically aligned display

that is stabilised by the electric field depends on the dielectric anisotropy of the liquid-crystal material. The preferred configuration uses materials having a negative dielectric anisotropy, so that the off-state is a planar-aligned liquid-crystal film. Application of an electric field to the in-plane electrodes will cause the director at the bottom surface to align perpendicularly to its initial direction, and so induce a twist through the cell (Fig. 38.21).

From an optical point of view, the director is always in the plane of the cell, and this means there is less distortion of an image when viewed at angles other than 90°. Another advantage of the IPS device is that the electric field is confined to the lower plate, and the lines of force do not extend across the cell to the grounded upper plate. This means that a very low electrical conductivity of the liquid-crystal material is less important than for conventional TFT-TN displays. The threshold voltage for the IPS-mode device is given by

$$V_{\text{th}}^{\text{IPS}} = \frac{\pi d}{\ell} \sqrt{\frac{K_2}{\varepsilon_0 \Delta \varepsilon}}, \qquad (38.28)$$

where ℓ is the thickness of the liquid-crystal film, and d is the separation of the in-plane electrodes. Not surprisingly the threshold depends only on the twist elastic constant, which is usually smaller by about a factor of two than the splay and bend elastic constants. While this helps to reduce the operating voltage, the smaller elastic energy associated with the pure twist deformation results in longer switching times. A further disadvantage of the IPS display is that the optical transmission of the cell is reduced by the requirement to have both electrodes deposited on one plate, thereby making smaller the active area available to display the image.

A configuration which shares some of the characteristics of the IPS cell and has been successfully commercialised is the twisted vertically aligned (TVAN) cell. The two optical states for this configuration are uniform vertical (homeotropic) alignment of the director for the off-state, and a twisted geometry for the on-state (Fig. 38.22). The liquid crystal material used has a negative dielectric anisotropy, so application of an electric field between the plates causes the director to align perpendicularly with respect to the field direction. A small quantity of optically active material (chiral dopant) is added to the liquid-crystal mixture to ensure that the switched director adopts a twisted configuration through the cell. The advantages of this cell are good viewing-angle characteristics and high optical contrast. Improvements in display technology continue to be made, often simplifying earlier devices. For example, high-quality displays described as vertically aligned nematic (VAN) devices are now available based on the TVAN configuration, but without the twist. The material requirement here is for a liquid crystal of negative dielectric anisotropy, that will align perpendicularly to an applied electric field.

In the devices described above, one state is defined by the surface conditions of the cell, while the other is defined by the action of the applied electric field. A bistable device is one in which two stable field-free states exist, both of which are accessible by switching with an external field. The first bistable liquid-crystal display to be developed was based on a ferroelectric effect observed in chiral tilted smectic C liquid crystals [38.48]. This ferroelectric smectic display has achieved some limited commercial success in specialist markets, but relies on a surface stabilisation of smectic layers, which is very sensitive to mechanical shock. Recently [38.49], bistable nematic displays have been developed in which two alignment states within a liquid-crystal cell, having different optical transmission, can be stabilised. If one of the substrates of a normal cell is replaced by a surface which has potentially two states of minimum energy corresponding to two surface alignments of the director, then it becomes possible to switch these states selectively using an electric field. A suitable bistable surface is provided by a grooved surface (grating) which has been treated with a surfactant to favour homeotropic alignment of the director at its surface [38.50]. Thus the two surface states correspond to (i) that determined by the grating, and (ii) that determined by the surfactant where the director is perpendicular to the substrate. Combining this intrinsically bistable substrate with a second substrate having a director alignment direction perpendicular to the grating direction gives a cell configuration capable of supporting two optically distinct stable states, which can be switched between using an applied voltage. Various director configurations are possible with this type of cell, and one example is illustrated in Fig. 38.23.

In the absence of any perturbation, the director orientation within the cell will be determined by the homeotropic alignment at one substrate and the alignment at the grating substrate. This hybrid (uniform planar and homeotropic) alignment causes a spatially varying director tilt through the sample. Application of an electric field to a positive-dielectric-anisotropy material will cause the director to align parallel to the applied field, and eventually a fully homeotropic configuration for the director is stabilised. The switch back from the homeotropic state to the hybrid state is thought to be due to a flexoelectric interaction. Other alignment configurations are also possible for the so-called zenithal bistable nematic (ZBD) cell. Displays based on these configurations share the good viewing characteristics of both the IPS and TVAN configurations, but they have the great advantage that the image is retained when the voltage is removed.

All the displays described so far rely on the coupling between an applied electric field and the dielectric properties of the liquid-crystal material, but, as we have shown, other material properties are just as important to the operating characteristics of the display. The appearance of a display depends on the optical properties of the liquid crystal and the cell configuration, but the operating voltage and switching times of a display are crucial in determining the types of application. Changing the nature of the interaction between the switching electric field and the liquid crystal gives rise to another range of possibilities for liquid-crystal devices. Under certain circumstances, a liquid crystal can be made to exhibit permanent ferroelectric (or spontaneous) polarisation, and this couples linearly with an external electric field, in contrast to dielectric properties which couple with the square of the electric field strength. Not surprisingly this makes a big difference to the switching behaviour of liquid crystals.

The final display to be considered is based on flexoelectric coupling between the electric field and the liquid crystal. Flexoelectricity occurs, in principle, with all liquid crystals, chiral or not, and shows itself as a bulk electric polarisation induced by an elastic strain. Conversely application of an electric field can cause an elastic strain. In general, flexoelectricity is rather small and difficult to detect, however it is thought to be responsible for an electro-optic effect observed in chiral nematic liquid crystals, which is being investigated for display applications. The effect, sometimes known as the deformed helix mode [38.51], is similar in some respects to the ferroelectric switching observed in chiral

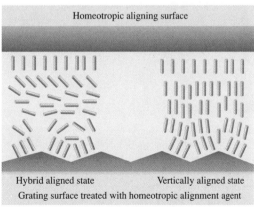

Fig. 38.23 Schematic of a zenithal bistable device

smectic C phases, but there is no longer a requirement for a layered structure. Chiral nematic liquid crystals spontaneously form helical structures in which the director rotates with a pitch determined by the molecular structure. If an electric field is applied perpendicularly to a chiral nematic helix, then there is a tendency for the helix to unwind, depending on the sign of the dielectric anisotropy. Even if the dielectric anisotropy of the material is zero, there is an elastic strain which can generate a polarisation (flexoelectric polarisation), which will interact with an applied electric field. This may be exploited in a device configuration, where a thin film of a chiral nematic liquid crystal, having a small or zero dielectric anisotropy, is aligned such that its helix axis is parallel to the containing glass plates (Fig. 38.24).

Application of an electric field across the plates will cause a distortion of the helix through the splay and bend flexoelectric coefficients, which appears as a rotation of the optic axis in the plane of the film [38.25]. Reversal of the electric field direction will reverse the rotation of the optic axis, with an intrinsic switching time about one hundred times faster than conventional nematic displays. Optically, the effect observed is very similar to that exhibited by smectic ferroelectric displays.

There are many cell configurations that can be used with liquid crystals to produce optical switches, displays or light modulators, and some of the more important have been described. The precise operation and performance of these liquid-crystal devices depends on both the cell design and the material properties of the liquid crystal. To a large extent the configuration of the liquid crystal within the cell is determined by such factors as the surface treatment of the plates enclosing the liquid crystal and the interactions between the surfaces and the liquid crystal. Our understanding of these interactions is very limited at the present time, and much more research is necessary before a quantitative theory can be formulated. However, given the cell configuration, the performance of the liquid-crystal device depends crit-

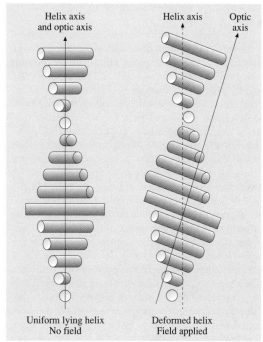

Fig. 38.24 Schematic of a deformed helix mode flexoelectric display

ically on the physical properties of the liquid-crystal material. Thus the electrical switching characteristics will depend on the dielectric properties, while the optical appearance of the device will be determined by the refractive indices. Elastic properties contribute to both the electric field response and the optical appearance, since any deformation of the director will be determined by the elastic properties of the liquid crystal. Finally, the all-important dynamical behaviour will be controlled by the viscous properties of the liquid crystal. All these material properties will be discussed in the next section.

38.4 Materials for Displays

The most important requirement for a liquid-crystal display material is that it should be liquid crystalline over the temperature range of operation of the device. Despite this, some of the first experimental display devices incorporated heaters in order to maintain the material in the liquid-crystal phase: for example, the first liquid-crystal shutter patented by Marconi and early prototype displays developed by the Radio Corporation America (RCA). It was not until the late 1960s that room-temperature nematic liquid crystals suitable for display applications were discovered. The first of these were based on Schiff's bases, which although easy to prepare, were difficult to purify and were susceptible to chemical decomposition in a de-

Table 38.1 Some typical liquid-crystal materials, including selected physical properties (ex – extrapolated from measurements on a nematic solution)

Compound / Transition temperatures (°C) [ref.]	$\Delta\varepsilon$ (T °C)	Δn (T °C)	Rotational Viscosity (γ_1/m Pa s)
CH$_3$O–⟨⟩–CH=N–⟨⟩–C$_4$H$_9$ T_{CrN} 22 °C T_{NI} 47 °C [38.53]	≈ 1	0.2	$\gamma_1 = 109$ [38.52] (37 °C)
C$_5$H$_{11}$–⟨⟩–⟨⟩–CN T_{CrN} 22 °C T_{NI} 35 °C [38.53]	8.5 (29 °C)	0.18 (25 °C)	$\gamma_1 = 102$ [38.52] (25 °C)
C$_5$H$_{11}$–⟨⟩–⟨⟩–F T_{CrI} 34 °C [38.54]	3.2(ex)	0.05(ex)	
C$_5$H$_{11}$–⟨⟩–⟨⟩–CN T_{CrN} 31 °C T_{NI} 55 °C [38.53]	9.9 (48 °C)	0.12 (40 °C)	$\gamma_1 = 128$ [38.52]
C$_5$H$_{11}$–⟨⟩–⟨⟩–CN T_{CrN} 62 °C T_{NI} 85 °C [38.53]	3.5 (78 °C)	0.05	
C$_7$H$_{15}$–⟨⟩–C(=O)O–⟨⟩–CN T_{CrN} 44 °C T_{NI} 57 °C [38.53]	19.9 (40 °C)	0.15	
C$_3$H$_7$–⟨⟩–CH$_2$CH$_2$–⟨⟩–⟨⟩ T_{CrN} 67 °C T_{NI} 82 °C [38.55]	≈ 0	0.15	

vice. One material which attracted particular attention from experimental physicists was MBBA, which has a nematic range of 22–47 °C (Table 38.1). The basic two-ring core linked by an imine group was the structural unit of many compounds having different terminal groups which were prepared for display mixtures in the early 1970s. It was found that mixtures of Schiff's bases often had lower crystal-to-nematic transition temperatures than any of the components, and furthermore would often remain liquid crystalline even below the thermodynamic crystallisation temperature. These two phenomena of eutectic behaviour and super-cooling have been exploited in the development of materials for devices. The early experiments on liquid-crystal displays were primarily focused on nematic or chiral compounds, but we have seen that other displays have been developed which use different liquid-crystal phases, most importantly the chiral smectic C phase.

Early studies established guiding principles for the development of display materials. First, the phase behaviour must be acceptable, i.e. the right phase stable over a suitable temperature range. Secondly, the material must have the correct electrical and optical properties for the particular display application envisaged, and above all must be of sufficient chemical purity to prevent any deterioration in performance over time. Again, guided by the early experiments, suitable display materials require the synthesis of compounds of appropriate chemical structure, and then the formulation of mixtures to optimise the properties. There have been a number of reviews of liquid-crystal materials for displays [38.53, 54, 56–59] and these contain many tables of data on a wide range of compounds. In this Section, we will give a brief account of the basic chemical structures used for materials in modern liquid-crystal displays, and then show how mixtures are devised to give the best possible performance characteristics for different displays. It has to be recognised that many of the details of display materials are matters of commercial confidentiality, and so it is not possible to give precise accounts of materials currently used or under investigation. However, the generic chemical structures and principles used in developing suitable mixtures are generally applicable.

38.4.1 Chemical Structure and Liquid-Crystal Phase Behaviour

There is a huge literature on the relationship between the structure of mesogens and the nature and stability of the liquid-crystal phases they form [38.60]. The studies have embraced empirical correlations of chemical structure and phase behaviour, theoretical calculations for simple particles (hard rods, spherocylinders etc.) representing mesogens, and computer simulations of collections of particles of varying complexities. For the display applications considered in this Chapter, the desired phases are nematic, and occasionally chiral nematic or chiral smectic C. Such phases are formed by molecules having extended structures, which usually require the presence of terminal alkyl chains to reduce the crystallisation temperatures. Components in nematic display mixtures typically have two, three or four carbocyclic rings joined directly or through a variety of linking groups.

38.4.2 The Formulation of Liquid-Crystal Display Mixtures

The two requirements for a liquid crystal to be used in a display are a suitable temperature range of phase stability and appropriate physical properties. These requirements cannot be satisfied for complex displays by a single compound, and commercial display materials may contain up to twenty different components. The formulation of these mixtures is essentially an empirical process, but guided by the results of thermodynamics and experience. The principles behind the preparation of multicomponent mixtures can be illustrated initially by consideration of a binary mixture.

It is well-known that the melting point of a binary mixture of miscible compounds is depressed, sometimes below the melting points of both components. Furthermore, the melting point of the binary mixture may exhibit a minimum at a particular composition, known as the eutectic. This occurs with liquid-crystalline compounds, and provides a method of reducing the lower temperature limit for liquid-crystal phase stability in mixtures. The upper temperature limit of the liquid-crystal range is fixed by the transition to an isotropic liquid. The phase rule of Willard Gibbs predicts that in binary mixtures there will always be a region of two-phase coexistence in the vicinity of a phase transition; that is, the transition from liquid crystal to isotropic occurs over a range of temperatures for which both the isotropic liquid and liquid crystal are stable in the mixture. Because of the weak first-order nature of most liquid crystal to isotropic phase transitions, the two-phase region is small. The character of phase transitions is determined by the corresponding entropy change, and a weak first-order transition has a small ($\approx 2\,\text{J}\,\text{K}^{-1}\text{mol}^{-1}$) associated entropy. If the latter were zero, then the transition would be second order, and there would no longer be a region of two-phase coexistence. The phase diagrams of multicomponent nematic mixtures can be calculated by thermodynamic methods [38.61, 62] and the transition temperatures of the mixtures can vary with composition in a variety of ways. For mixtures of two liquid-crystalline compounds of similar chemical constitution, the variation of the nematic to isotropic transition temperature is approximately linear with composition [38.63].

It is possible to calculate the variation of the melting point with composition using an equation attributed to Schroder and van Laar. For each component i, the mixture composition (mole fraction x_i) and the melting point of the mixture T are related by

$$\ln x_i = -\frac{\Delta H_i}{R}\left(\frac{1}{T} - \frac{1}{T_i}\right), \quad (38.29)$$

where ΔH_i and T_i are, respectively, the latent heat of fusion and melting point of the pure component i. For a binary mixture there are two such equations which can be solved to give the eutectic temperature and composition. In a multicomponent mixture, the set of equations (38.29) can be solved subject to the condition,

$$\sum_i x_i = 1 \tag{38.30}$$

to predict the eutectic of the mixture.

While there is a reasonable thermodynamic basis to the prediction of the phase diagrams of mixtures, the determination of the physical properties of mixtures from the properties of individual components is much more difficult. Given the absence of any better theories, it is common to assume that in mixtures, physical properties such as dielectric anisotropy, birefringence and even viscosity vary linearly with the amount of any component, at least for small concentrations. While this may give an indication of the effect of different components on the properties of a display mixture, it can also be very misleading. One theoretical problem is that, for a mixture at a particular temperature, the orientational order parameters of the different components are not equal. The more anisometric components (e.g. three-ring mesogens) are likely to have a larger orientational order parameter than smaller (two-ring) mesogens. Since the various physical properties of interest in displays depend on the order parameter in different ways, it is difficult to predict the contribution of different components to the overall mixture properties. Despite this, many tables of data for liquid-crystal compounds of interest for display mixtures are prepared [38.60] on the basis of extrapolated measurements on mixtures at low composition, normally < 20% w/w. There is always a problem concerning the temperatures at which to compare the physical properties of liquid crystals and their components. Many measurements are made at room temperature, so that this becomes the temperature for comparison. However, a more useful approach is to compare properties at equal reduced temperatures (or at the same shifted temperature, $T_{NI} - T$), since under these conditions the orientational order parameters are likely to be similar.

38.4.3 Relationships Between Physical Properties and Chemical Structures of Mesogens

Electrical and Optical Properties

These properties include the dielectric permittivity, electrical conductivity and refractive indices. The magnitude of the dielectric anisotropy determines the threshold voltage necessary to switch a display, and influences the transmission/voltage characteristics of the cell. Depending on the particular display configuration, a positive or negative dielectric anisotropy may be required. Refractive indices strongly affect the appearance of a display. Usually the refractive indices or birefringence must be adjusted for a particular cell configuration to give the optimum on/off contrast ratio. Coloration in displays can sometimes occur in materials of high refractive index, and so it is desirable to keep the birefringence as low as possible, compatible with an acceptable contrast ratio. For twisted structures, the magnitude of the birefringence also determines the efficiency of light guiding, and so close control of the values of the principal refractive indices of a display mixture is important. For non-conducting materials, the refractive indices are measures of the dielectric response of a material at very high i.e. optical frequencies, and it is possible to formulate a single theory which relates the dielectric and optical properties of a liquid crystal to its molecular properties. Unfortunately this is not possible for the electrical conductivity. The latter is largely determined by the purity of the liquid crystal, but it is found that the higher the value of the permittivity, the larger the electrical conductivity. Materials of high electrical conductivity tend to leak charge, and so an image may deteriorate during a multiplexing cycle. In general it is desirable to minimise the conductivity of a display mixture, although this was not the case for the first liquid-crystal displays reported [38.64]. These utilised the strong light scattering which results when an electric field is applied to certain nematic materials. The scattering is due to electrohydrodynamic instabilities in liquid-crystal materials which have a significant electrical conductivity. Such materials are not suitable for use in modern, fast-multiplexed displays.

The dielectric anisotropy, $\Delta \varepsilon$, and birefringence, Δn, of a nematic can be related to molecular properties of polarisability and dipole moment using a theory originally developed by *Maier* and *Meier* [38.65]. The birefringence is given by

$$\Delta n \approx \frac{NS}{\varepsilon_0}(\alpha_\ell - \alpha_t), \tag{38.31}$$

where N is the density in molecules per m^3 and $\Delta \alpha = (\alpha_\ell - \alpha_t)$ is the anisotropy of the molecular polarisability. S is the order parameter, defined in Sect. 38.2.1, and small corrections due to the local field anisotropy have been neglected. Such corrections cannot be ignored in the corresponding expression

$$\Delta \varepsilon = \frac{NhFS}{\varepsilon_0}\left[\Delta\alpha + \frac{\mu^2}{2k_B T}(3\cos^2\beta - 1)\right] \tag{38.32}$$

Table 38.2 Materials of high, low and negative birefringence

Compound	Transition temperatures	Δn [ref.] (T °C)
C_3H_7—⌬—C≡C—⌬—CH=CHC$_3$H$_7$	T_{NI} 112 °C	0.31 [38.66] (92 °C)
C_3H_7—(cyclohexyl-cyclohexyl)—CH=CH$_2$	T_{CrSmB} 23 °C T_{SmBN} 35 °C T_{NI} 49 °C	0.052 [38.59]
C_5H_{11}—(bicyclo)—C(=O)O—⌬—OC$_5$H$_{11}$	T_{NI} 93.5 °C	0.074 [38.66] (58.5 °C)
Hexa(phenylethynyl)benzene with six C$_8$H$_{17}$ groups	T_{CrN} 80 °C T_{NI} 96 °C	−0.193 [38.66] (61 °C)

for the dielectric anisotropy, especially for materials of high permittivity. In (38.32) h and F are local-field correction factors, while μ is the molecular dipole moment, and β is the angle between the dipole direction and the long axis of the molecule. For molecules containing a number of dipolar groups, μ is the root mean square of the vector sum of all contributing groups. Both the birefringence and the dielectric anisotropy increase with decreasing temperature, and the detailed variation with temperature is largely determined by the temperature dependence of the order parameter S.

The manipulation of birefringence is achieved by changing the chemical constitution of the mesogen. Thus extending the electronic conjugation along the axis of a mesogen will result in an increase in longitudinal polarisability, and hence an increase in birefringence. Saturated carbocyclic rings, such as cyclohexane, and aliphatic chains generally have small polarisabilities and mesogens containing a predominance of such moieties form low-birefringence liquid crystals. Mixtures for displays require a positive birefringence, which is associated with calamitic or rod-like mesogens. In order to improve the viewing characteristics of displays, optically retarding films are placed in front of the display, and depending on their function, these may be of negative or positive birefringence. The latter can be fabricated by encapsulation or polymerisation of suitable molecules of an extended structure. On the other hand, films of negative birefringence have been made using discotic materials: i. e. mesogenic molecules formed from disc-like structures which have a negative polarisability anisotropy (Sect. 38.1.3). Some examples of liquid crystals having different birefringence are shown in Table 38.2.

Table 38.3 Materials of high, low and negative dielectric anisotropy; the *inset* figure indicates the direction of the total dipole moment with respect to the core of the molecule (1 Debye = 3.33564×10^{-30} Cm)

Compound	Transition temperatures	$\Delta\varepsilon$ [ref.] (T °C)	Total dipole moment μ (Debye)
C_5H_{11}–⬡–⬡–CN	T_{CrN} 22.5 °C T_{NI} 35 °C	11.5 [38.67] (25 °C)	4.8
C_5H_{11}–⬡–⬡–COO–⬡(NC,CN)–OC_4H_9	T_{CrN} 143 °C T_{NI} 150 °C	−10.0 [38.68] (145 °C)	6.4
C_2H_5–⬡–CH_2CH_2–⬡(F)–⬡–C_2H_5	T_{CrN} 13 °C T_{NI} 64 °C	0.0 [38.69] (58 °C)	1.4

The introduction of chirality into liquid crystals has important consequences for their optical properties. The selective reflection of coloured light from the helical structure of a chiral nematic has already been mentioned in Sect. 38.1.2. All chiral materials will rotate the plane of incident polarised light, and the particular optical properties associated with chiral thin films are exploited in many liquid-crystal device applications.

The dipole moment of a molecule is increased if strongly electronegative or electropositive groups are substituted into the structure, with the result that the dielectric permittivity increases. For mesogenic molecules, the locations of the electropositive or electronegative groups are important, since not only the magnitude but also the orientation of the molecular dipole strongly influences the dielectric anisotropy. From (38.32) it can be seen that the dipolar contribution to the dielectric anisotropy may be positive or negative depending on the value of the angle β, since for values of β greater than 54.7° the dipolar contribution to the permittivity anisotropy becomes negative. This is illustrated by the mesogens shown in Table 38.3, where different structures can be designed to give large positive, negative or zero dielectric anisotropy.

The most effective substituent for producing materials of high dielectric anisotropy is the cyano group, and mixtures containing cyano-mesogens were the basis for the rapid development of complex displays in the 1980s and early 1990s. However, these mixtures tended to have relatively high viscosities, which gave rise to slow switching times. Another disadvantage was the high electrical conductivity associated with the high dielectric anisotropies which caused charge leakage during multiplexing, and hence degradation of the image.

As the demands placed on liquid-crystal materials by more sophisticated display technologies have increased, new families of molecules have been synthesised and screened for their physical properties. However, it is no longer the properties of the pure mesogens that are of interest, rather how they behave in mixtures. For this reason, the physical properties of most components of display mixtures are measured in mixtures, and values for the *pure* mesogens are obtained by extrapolation. Data derived in this way are useful in designing display mixtures and for comparison purposes, but cannot be relied upon to give quantitative information about the relationship between molecular structure and physical properties.

The major display technologies using TN, STN or TFT-TN configurations require display mixtures having a positive dielectric anisotropy. Many materials have been developed to improve the electro-optical behaviour of these displays, particularly using fluorine-substituted mesogens to provide the required dielectric and optical properties (for examples see [38.54, 58, 70]). However, within the past seven years, new display configurations have emerged, such as the in-plane switching (IPS) and vertically aligned (VAN and TVAN) nematic modes, which require mixtures with negative dielectric anisotropy. Using the design strategy illustrated above for simple mesogens, it has been possible to prepare a large number of materials with the desired negative dielectric anisotropy. These are again mostly based on fluorine-substituted compounds, and as before their properties have mostly been determined by extrapolation of measurements on mixtures.

Elastic Properties

The property known as elasticity is characteristic of liquid crystals, and distinguishes them from isotropic liquids. It has been shown in Sect. 38.2.3 that the macroscopic orientational disorder of the director in liquid crystals can be represented in terms of three normal modes, designated as splay, twist and bend, and associated with each of these deformations is an elastic constant. Since the elastic properties of display materials contribute to the electro-optic response, their optimisation for particular display configurations is important to maximise the performance of commercial devices. However, despite their importance, the relationships between the magnitudes of elastic constants and the chemical structure of mesogens are poorly understood. There is a good reason for this; the optical and dielectric properties are to a first approximation single particle properties. That is they are roughly proportional to the molecular number density and are also linearly dependent on the order parameter. Because elastic properties are a measure of the change in energy due to displacements of the director, they are related to the orientation-dependent intermolecular forces. Thus, at a molecular level, elastic properties are two-particle properties, and are no longer linearly proportional to the number density. A further consequence is that the elastic properties depend to lowest order on the square of the order parameters. Molecular theories of elasticity in nematic liquid crystals have been developed [38.71] and the simplest results suggest that the different elastic constants can be related to molecular shape

$$K_1 = K_2 \propto \langle x^2 \rangle \text{ and } K_3 \propto \langle z^2 \rangle, \quad (38.33)$$

where $\langle z^2 \rangle$ and $\langle x^2 \rangle$ are the average intermolecular distances parallel and perpendicular to the molecular alignment direction, respectively. Thus theory predicts that for rod-like molecules the splay elastic constant should be smaller than the bend elastic constant, and increasing the molecular length should increase K_3, while increasing the molecular width should increase K_1. This is roughly in accord with experimental results, except that the prediction of equal splay and twist elastic constants is not confirmed (Fig. 38.9). In general, the twist elastic constant is about one half of the splay elastic constant. Hard particle theories [38.72] evaluated for spherocylinders provide further guidance on the relationship of elastic constants to molecular shape. These theoretical results can be presented in a simplified way as follows:

$$\frac{K_1 - K}{K} = \Delta(1 - 3\sigma); \quad \frac{K_2 - K}{K} = -\Delta(2 + \sigma);$$
$$\frac{K_3 - K}{K} = \Delta(1 + 4\sigma), \quad (38.34)$$

where $K = \frac{1}{3}(K_1 + K_2 + K_3)$. The quantities Δ and σ are parameters of the theory, where Δ is approximately equal to the square of the length:width ratio of the spherocylinder, and σ depends on the degree of orientational order. Despite the fact that details of internal chemical structure are ignored, these theoretical results for nematics are in approximate agreement with experimental measurements on simple nematics. If the nematic material has an underlying smectic phase, or if there is a tendency for local smectic-like ordering, this can strongly affect the elastic constants. Both the twist and bend elastic constants are infinite in a smectic phase, and in a nematic phase they diverge as the transition to a smectic phase is approached.

The elastic constants contribute directly to the threshold voltage and the response times of displays. Threshold voltages increase with increasing elastic constants, and the elastic constants responsible depend on the configuration of the display. Thus for the planar-aligned electrically controlled birefringence display (ECB), the threshold voltage depends on K_1, while the switching voltage for TN displays depends on a combination of all three elastic constants (38.27). The IPS display voltage depends only on K_2, and for VAN and TVAN devices, the threshold voltage is determined by K_3. Different combinations of elastic constants determine the transmission/voltage curves, which are important in the multiplexing of complex displays. For example, decreasing the ratio K_3/K_1 increases the steepness of the curve for TN displays, and so increases the number of lines that may be addressed. On the other hand, for the STN display, if the ratio K_3/K_1 is decreased, the number of lines that may be addressed also decreases. Identification of the important elastic constants necessary to optimise the operation of these displays is relatively straightforward; however, manipulation of the components of displays mixtures to give the best results is much more difficult.

Ferroelectric and Flexoelectric Properties

The electro-optic properties considered so far result from interaction of an electric field with the anisotropic permittivity of a material. This might be termed a quadratic response since the dielectric term in the free energy (38.2) is quadratic in the electric field, and as a consequence the electro-optic response does not depend on the sign of the electric field. For materials having a centre of symmetry, such as achiral nematic and smectic liquid crystals, this response is the only one possible. However, if the centro-symmetry of the liquid crystal is broken in some way, then a linear electric polarisation becomes possible, which results in a linear response to an applied electric field. One example of this, in the context of displays, is the chiral smectic C phase, which in the surface-stabilised state exhibits ferroelectricity i.e. a spontaneous electric polarisation. The origin of the symmetry breaking in this case is the chirality of the material, and the polarisation is directed along an axis perpendicular to the tilt plane of the smectic C. Another way in which the symmetry can be broken is through elastic strain. This effect was first described by *Meyer* [38.24], and it can be represented as a polarisation resulting from a splay or bend deformation (Sect. 38.2.4). Since at a molecular level, strain is related to molecular interactions, the flexoelectric response depends on a coupling of intermolecular forces and the molecular charge distribution. Two molecular mechanisms have been identified which contribute to the strain-induced polarisation. If the molecules have a net dipole moment, then the longitudinal component can couple with the molecular shape to give a splay polarisation along the director axis, while the transverse component couples with the shape to give a bend polarisation perpendicular to the director axis. Even in the absence of a net dipole, a quadrupolar charge distribution in a molecule can result in strain-induced polarisation [38.26]. Both contribute to the splay and bend flexoelectric coefficients, but only the dipolar part persists in the sum $e_1 + e_3$. Thus it is common to quote flexoelectric coefficients as a sum and difference rather than as separate coefficients.

The measurement of flexoelectric coefficients has been a challenge to experimentalists, and there is a wide range of values in the literature for standard materials (Sect. 38.2.4). It is, therefore, premature to draw any conclusions about structure/property relationships for flexoelectricity from the limited experimental data available. There have been attempts [38.73, 74] to model flexoelectricity for collections of Gay–Berne particles simulating wedge-shaped molecules. Application of the surface interaction model to flexoelectric behaviour [38.75] has allowed the calculation of flexoelectric coefficients for a number of molecules; these calculations include the quadrupolar contribution. The importance of molecular shape is clearly demonstrated, and in particular changes of shape, either through conformational changes or *cis–trans* isomerisation, have large effects on the magnitude and sign of the flexoelectric coefficients.

Flexoelectric effects contribute to the electro-optic response of nematic displays, especially those with hybrid alignment, i.e. planar on one electrode and homeotropic on the other electrode, but they are not usually considered in the optimisation of mixture properties. However, flexoelectric properties are of direct importance to the operation of displays based on the switching of the direction of the optic axis in chiral nematics: the so-called deformed helix mode [38.25].

Viscous Properties

As explained in Sect. 38.2.5, the flow properties of liquid crystals are complicated. Since the materials are anisotropic, the viscosities in different directions are different, furthermore because of the torsional elasticity, viscous stress can couple with the director orientation to produce complex flow patterns. Thus there are five viscosity coefficients necessary for nematics, in addition to the elastic constants, and as many as 20 viscosities for smectic C liquid crystals [38.76]. To relate all or indeed any of these to molecular structure is a formidable challenge. However, for most liquid-crystal displays, the only viscosity of interest is that which relates to the reorientation of the director: the so-called rotational viscosity. This depends on the temperature and order parameter, and on the forces experienced by the rotating director. The rotational viscosities for all liquid-crystalline materials can be represented by one or other of the following parameterised relations

$$\gamma_1 = aS^x \exp\left(\frac{A(T)}{k_B T}\right)$$

or

$$\gamma_1 = bS^y \exp\left(\frac{B}{T - T_0}\right), \qquad (38.35)$$

Table 38.4 Fluorinated mesogens of positive and negative dielectric anisotropy used in liquid-crystal mixtures for modern displays. All measurements listed have been obtained by extrapolation from measurements on nematic solutions

Compound Transition temperatures (°C) [ref.]	$\Delta\varepsilon$	Δn	Rotational Viscosity (γ_1/m Pa s)
C_5H_{11}—[cyclohexyl]—[difluorophenyl]—OC_2H_5 T_{CrN} 49 °C T_{NI} (13) °C [38.59]	−6.2	0.1	$\gamma_1 = 110$
C_3H_7—[cyclohexyl]—[phenyl]—[difluorophenyl]—OC_2H_5 T_{CrN} 80 °C T_{NI} 173 °C [38.59]	−5.9	0.156	$\gamma_1 = 233$
C_5H_{11}—[cyclohexyl]—[phenyl]—[trifluorophenyl] T_{CrN} 30 °C T_{NI} 58 °C [38.54]	11.3	0.134	$\gamma_1 = 191$ [38.77]
C_5H_{11}—[cyclohexyl]—[phenyl]—[phenyl]—OCF_3 T_{CrSmB} 43 °C T_{SmBN} 128 °C T_{NI} 147 °C [38.54]	9	0.14	$\gamma_1 = 180$ [38.77]
C_3H_7—[cyclohexyl]—[cyclohexyl]—[difluorophenyl]—CH_3 T_{CrN} 67 °C T_{NI} 145 °C [38.59]	−2.7	0.095	$\gamma_1 = 218$
C_5H_{11}—[difluorophenyl]—[phenyl]—[difluorophenyl]—C_5H_{11} T_{CrN} 88 °C T_{NI} 89 °C [38.59]	−4.3	0.2	$\gamma_1 = 210$
C_3H_7—[cyclohexyl]—[cyclohexyl]—COO—[trifluorophenyl] T_{CrN} 56 °C T_{NI} 117 °C [38.59]	11.1	0.067	$\gamma_1 = 175$

where a, b, A and B are material parameters, and S is the order parameter raised to a power of x or y between 0 and 2. The first of these expressions emphasises the diffusional nature of rotational relaxation in a liquid crystal, that is molecules rotating in an external potential. The second expression taken from polymer physics describes rotation in terms of free volume, where T_0 is the temperature at which the free volume becomes zero, and the rotational viscosity infinite i. e. a glass transition.

At the simplest level, the rotational viscosity depends on the molecular shape and size. As the length of the molecule increases, from two rings to three rings etc. γ_1 increases, similarly it increases with the length of the alkyl chain. Varying the nature of the rings in the mesogenic core can have a dramatic effect on the rotational viscosity, which correlates well with free volume, as shown in Fig. 38.25, and the glass temperature of the material.

There is a correlation between the increasing dielectric anisotropy and increasing rotational viscosity which can be attributed to local dipolar intermolecular forces which impede end-over-end rotation of molecules. Thus mesogens having cyano-groups in a terminal position tend to have relatively high viscosities. Other dipolar groups such as fluorine do not have such a deleterious effect on viscosity as the cyano-group, and so fluorine-containing mesogens are increasingly preferred in the formulation of display mixtures.

As with the determination of other properties, the rotational viscosities of mesogenic components are often determined by extrapolation from measurements on mixtures doped with the component under investigation. Such a method only provides approximate values to compare different components, but in the formulation of mixtures for display applications it is only the rotational viscosity of the final mixture that is important. Rotational viscosities of some mesogens of interest for display mixtures are given in the tables accompanying this section. In many instances, the viscosities, as with other properties, have been determined by extrapolation from measurements on mixtures. The measurement of rotational viscosities is experimentally difficult, and some authors prefer to quote the results of bulk-viscosity measurements in terms of a kinematic viscosity. In fact there is a good correlation between kinematic viscosity and rotational viscosity, and where

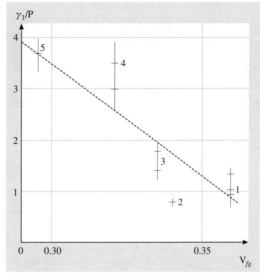

Fig. 38.25 Relationship between the rotational viscosity coefficient γ_1 (Poise = 10^{-1} Pa s) and the geometric free volume (V_{fg}) at 25 °C for bicyclic polar mesogens. Compound numbers represent alkyl series as follows: 1-cyanophenylalkylcyclohexanes; 2-alkylcyanobiphenyls; 3-cyanophenylalkylpyridines; 4-cyanophenylalkylbicyclooctanes; 5-alkoxycyanobiphenyls (After 38.52)

possible both values have been included in the tables. From the various tables, the effect of increasing the molecular length on the viscosity is clearly seen, as is the effect of replacing an F atom with a CN group. Lateral substitution, which produces materials of negative dielectric anisotropy, tends to increase the rotational viscosity. Despite this the fastest nematic displays now use vertical alignment and materials of negative dielectric anisotropy. The operating characteristics of display mixtures depend on the physical properties of individual components in a very complex manner, and optimisation of mixture properties has to be carried out in a concerted way.

References

38.1 P. J. Collings, M. Hird: *Introduction to Liquid Crystals* (Taylor & Francis, London 1997)

38.2 M. A. Bates, G. R. Luckhurst: Computer simulation studies of anisotropic systems. XXX. The phase behaviour and structure of a Gay–Berne mesogen, J. Chem. Phys. **110**, 7087–7108 (1999)

38.3 M. Hird: *The Physical Properties of Liquid Crystals: Nematics*, ed. by D. A. Dunmur, A. Fukuda, G. R. Luckhurst (INSPEC, London 2001) Chap. 1.1

38.4 P. Pasini, C. Zannoni: *Advances in the Simulation of Liquid Crystals* (Kluwer, Dordrecht 1998)

38.5   J. W. Emsley, G. R. Luckhurst, G. H. Shilstone, I. Sage: The preparation and properties of the α,ω-bis(4,4′-cyanobiphenyloxy)alkanes: nematogenic molecules with a flexible core, Mol. Cryst. Liq. Cryst. Lett. **110**, 223–233 (1984)

38.6   G. W. Gray: *The Molecular Physics of Liquid Crystals*, ed. by G. R. Luckhurst, G. W. Gray (Academic, New York 1979) Chap. 12

38.7   T. J. Sluckin, D. A. Dunmur, H. Stegemeyer: *Crystals that Flow* (Taylor & Francis, New York 2004)

38.8   R. B. Meyer, L. Liebert, L. Strzelecki, P. J. Keller: Ferroelectric liquid crystals, J. Phys. (Paris) **L36**, 69–71 (1975)

38.9   S. R. Renn, T. C. Lubensky: Abrikosov dislocation lattice in a model of the cholesteric-to-smectic A transition, Phys. Rev. A **38**, 2132–2147 (1988)

38.10 J. W. Goodby, M. A. Waugh, S. M. Stein, E. Chin, R. Pindak, J. S. Patel: Characterization of a new helical smectic liquid crystal, Nature **337**, 449–451 (1989)

38.11 S. Chandrasekhar, B. K. Sadashiva, K. A. Suresh: Liquid crystals of disc-like molecules, Pramana **9**, 471–480 (1977)

38.12 K. Praefcke: *The Physical Properties of Liquid Crystals: Nematics*, ed. by D. A. Dunmur, A. Fukuda, G. R. Luckhurst (INSPEC, London 2001) Chap. 1.2

38.13 G. G. Nair, D. S. S. Rao, K. S. Prasad, S. Chandrasekhar, S. Kumar: Electrooptic and viewing angle characterisitcs of a display device employing a discotic nematic liquid crystal, Mol. Cryst. Liq. Cryst. **397**, 545–552 (2003)

38.14 N. Boden, R. Bissel, J. Clements, B. Movaghar: Discotic liquid crystals: self-organising molecular wires, Liq. Cryst. Today **6(1)**, 1–4 (1996)

38.15 V. Tsvetkov: Über die Molekülanordnung in der anisotrop-flüssigen Phase, Acta Physicochim **15**, 132–147 (1942)

38.16 G. R. Luckhurst: *Dynamics and Defects in Liquid Crystals*, ed. by P. E. Cladis, P. Palffy-Muhoray (Gordon & Breach, Philadelphia 1998) p. 141

38.17 P. G. de Gennes: *The Physics of Liquid Crystals* (Oxford Univ. Press, Oxford 1974) p. 96

38.18 J. Cognard: Alignment of nematic liquid crystals and their mixtures, Mol. Cryst. Liq. Cryst. **1**, 1–77 (1982), (Suppl.)

38.19 D. W. Berreman: Solid surface shape and the alignment of an adjacent nematic liquid crystal, Phys. Rev. Lett. **28**, 1683–1686 (1972)

38.20 A. Rapini, M. Papoular: Distortion d'une lamelle nématique sous champ magnétique. Conditions d'anchorage aux parois, J. Phys. Colloq. (France) **30**, C-4-54–56 (1969)

38.21 A. Sugimura: *The Physical Properties of Liquid Crystals: Nematics*, ed. by D. A. Dunmur, A. Fukuda, G. R. Luckhurst (INSPEC, London 2001) Chap. 10.2

38.22 F. C. Frank: On the theory of liquid crystals, Trans. Faraday Soc. **25**, 19–28 (1958)

38.23 V. Fréedericksz, V. Zolina: Forces causing the orientation of an anisotropic liquid, Trans. Faraday Soc. **29**, 919–930 (1933)

38.24 R. B. Meyer: Piezoelectric effects in liquid crystals, Phys. Rev. Lett. **22**, 918–921 (1969)

38.25 P. Rudquist, S. T. Lagerwall: On the flexoelectric effect in nematics, Liq. Cryst. **23**, 503–510 (1997)

38.26 A. E. Petrov: *The Physical Properties of Liquid Crystals: Nematics*, ed. by D. A. Dunmur, A. Fukuda, G. R. Luckhurst (INSPEC, London 2001) Chap. 5.5

38.27 J. Prost, J. P. Marcerou: On the microscopic interpretation of flexoelectricity, J. Phys. (Paris) **38**, 315–324 (1977)

38.28 J. K. Moscicki: *The Physical Properties of Liquid Crystals: Nematics*, ed. by D. A. Dunmur, A. Fukuda, G. R. Luckhurst (INSPEC, London 2001) Chap. 8.2

38.29 W. Helfrich: Molecular theory of flow alignment of nematic liquid crystals, J. Chem. Phys. **50**, 100–106 (1969)

38.30 C. Gähwiller: The viscosity coefficients of a room-temperature liquid crystal (MBBA), Mol. Cryst. Liq. Cryst. **20**, 301–318 (1973)

38.31 H. Kneppe, F. Schneider, N. K. Sharma: Rotational viscosity γ_1 of nematic liquid crystals, J. Chem. Phys. **77**, 3203–3208 (1982)

38.32 F. M. Leslie: *The Physical Properties of Liquid Crystals: Nematics*, ed. by D. A. Dunmur, A. Fukuda, G. R. Luckhurst (INSPEC, London 2001) Chap. 8.1

38.33 O. Parodi: Stress tensor for a nematic liquid crystal, J. Phys. (Paris) **31**, 581–584 (1970)

38.34 Y. Björnståhl: Untersuchungen über die anisotropen Flüssigkeiten, Ann. der Phys. **56**, 161–207 (1918)

38.35 B. Levin, N. Levin: Improvements in or relating to light valves, British Patent 441,274 (1936)

38.36 P. Yeh, C. Gu: *Optics of Liquid Crystal Displays* (Wiley, New York 1999)

38.37 P. M. Alt, P. Pleshko: Scanning limitations of liquid crystal displays, IEEE Trans. Electron. Dev. **21**, 146–155 (1974)

38.38 E. Jakeman, E. P. Raynes: Electro-optic response times in liquid crystals, Phys. Lett. **39A**, 69–70 (1972)

38.39 K. Tarumi, U. Finkenzeller, B. Schuler: Dynamic behaviour of twisted nematic cells, Jpn. J. Appl. Phys. **31**, 2829–2836 (1992)

38.40 M. Schadt, W. Helfrich: Voltage-dependent optical activity of a twisted nematic liquid crystal, Appl. Phys. Lett. **18**, 127–128 (1971)

38.41 Ch. Mauguin: Sur les cristaux liquides de Lehmann, Bull. Soc. Fran. Mineral. **34**, 71–117 (1911)

38.42 C. H. Gooch, H. A. Tarry: The optical properties of twisted nematic liquid crystal structures with twist angles $\leq 90°$, J. Phys. D: Appl. Phys. **8**, 1575–1584 (1975)

38.43 T. J. Scheffer, J. Nehring: A new highly multiplexed liquid crystal display, Appl. Phys. Lett. **45**, 1021–1023 (1984)

38.44 C. M. Waters, E. P. Raynes, V. Brimmell: Design of highly multiplexed liquid crystal dye displays, Mol. Cryst. Liq. Cryst. **123**, 303–319 (1985)

38.45 A. J. Snell, K. D. Mackenzie, W. E. Spear, P. G. LeComber, A. J. Hughes: Application of amorphous silicon field effect transistors in addressable liquid crystal display panels, Appl. Phys. **24**, 357–362 (1981)

38.46 G. Baur, R. Kiefer, H. Klausmann, F. Windscheid: In-plane switching: a novel electro-optic effect, Liq. Cryst. Today **5(3)**, 13–14 (1995)

38.47 M. Oh-e, K. Kondo: Electro-optical characteristics and switching behaviour of the in-plane switching mode, Appl. Phys. Lett. **67**, 3895–3897 (1966)

38.48 N. A. Clark, S. T. Lagerwall: Submicrosecond bistable electro-optic switching in liquid crystals, Appl. Phys. Lett. **36**, 899–901 (1980)

38.49 G. P. Bryan-Brown, C. V. Brown, I. C. Sage, V. C. Hui: Voltage-dependent anchoring of a liquid crystal on a grating surface, Nature **392**, 365–367 (1998)

38.50 C. V. Brown, L. Parry-Jones, S. J. Elston, S. J. Wilkins: Comparison of theoretical and experimental switching curves for a zenithally bistable nematic liquid crystal device, Mol. Cryst. Liq. Cryst. **410**, 417–425 (2004)

38.51 J. S. Patel, R. B. Meyer: Flexoelectric electro-optics of a cholesteric liquid crystal, Phys. Rev. Lett. **58**, 1538–1540 (1987)

38.52 V. V. Belyaev: *Physical Properties of Liquid Crystals: Nematics*, ed. by D. A. Dunmur, A. Fukuda, G. R. Luckhurst (INSPEC, London 2001) Chap. 8.4

38.53 G. W. Gray, S. M. Kelly: Liquid crystals for twisted nematic displays, J. Mater. Chem. **9**, 2037–2050 (1999)

38.54 V. F. Petrov: Liquid crystals for AMLCD and TFT-PDLCD applications, Liq. Cryst. **19**, 729–741 (1995)

38.55 H. Takatsu, K. Takeuchi, H. Sato: Mol. Cryst. Lig. Cryst. **100**, 345–355 (1983)

38.56 D. Coates: In: *Liquid Crystals, Applications and Uses*, Vol. 1, ed. by B. Bahadur (World Scientific, Singapore 1990) p. 91

38.57 L. Pohl, U. Finkenzeller: In: *Liquid Crystals, Applications and Uses*, Vol. 1, ed. by B. Bahadur (World Scientific, Singapore 1990) p. 1139

38.58 K. Tarumi, M. Bremer, T. Geelhaar: Recent liquid crystal material development for active matrix displays, Ann. Rev. Mater. Sci. **27**, 423–441 (1997)

38.59 D. Pauluth, K. Tarumi: Advanced liquid crystals for television, J. Mater. Chem. **14**, 1219–1227 (2004)

38.60 D. A. Dunmur, A. Fukuda, G. R. Luckhurst (Eds.): *Physical Properties of Liquid Crystals: Nematics* (INSPEC, London 2001)

38.61 D. S. Hulme, E. P. Raynes, K. J. Harrison: Eutectic mixtures of nematic 4'-substituted 4-cyanobiphenyls, J. Chem. Soc. Chem. Comm., 98–99 (1974)

38.62 D. Demus, Ch. Fietkau, R. Schubert, H. Kehlen: Calculation and experimental verification of eutectic systems with nematic phases, Mol. Cryst. Liq. Cryst. **25**, 215–232 (1974)

38.63 R. L. Humphries, P. G. James, G. R. Luckhurst: A molecular field treatment of liquid crystalline mixtures, Symp. Faraday Trans. **5**, 107–118 (1971)

38.64 G. H. Heilmeier, L. A. Zanoni, L. A. Barton: Dynamic scattering: a new electro-optic effect in certain classes of nematic liquid crystals, Proc. IEEE **56**, 1162–1171 (1968)

38.65 D. A. Dunmur, K. Toriyama: In: *Physical Properties of Liquid Crystals*, ed. by D. Demus, J. Goodby, G. W. Gray, H.-W. Spiess, V. Vill (Wiley-VCH, Weinheim 1999) p. 129

38.66 D. A. Dunmur: *Physical Properties of Liquid Crystals: Nematics*, ed. by D. A. Dunmur, A. Fukuda, G. R. Luckhurst (INSPEC, London 2001) Chap. 7.1

38.67 D. A. Dunmur, M. R. Manterfield, W. H. Miller, J. K. Dunleavy: The dielectric and optical properties of the homologous series of cyano-alkylbiphenyl liquid crystals, Mol. Cryst. Liq. Cryst. **45**, 127–144 (1978)

38.68 K. Toriyama, D. A. Dunmur, S. E. Hunt: Transverse dipole association and negative dielectric anisotropy of nematic liquid crystals, Liq. Cryst. **5**, 1001–1009 (1989)

38.69 D. A. Dunmur, D. A.. Hitchen, X.-J. Hong: The physical and molecular properties of some nematic fluorobiphenylalkanes, Mol. Cryst. Liq. Cryst. **140**, 303–318 (1986)

38.70 S. Naemura: *Physical Properties of Liquid Crystals: Nematics*, ed. by D. A. Dunmur, A. Fukuda, G. R. Luckhurst (INSPEC, London 2001) Chap. 11.2

38.71 H. Gruler: The elastic Constants of a nematic liquid crystal, Z. Naturforsch. **30a**, 230–234 (1975)

38.72 R. G. Priest: Theory of the Frank elastic constants of nematic liquid crystals, Phys. Rev. A **7**, 720–729 (1973)

38.73 J. Stelzer, R. Beradi, C. Zannoni: Flexoelectric effects in liquid crystals formed by pear-shaped molecules. A computer simulation study, Chem. Phys. Lett. **299**, 9–16 (1999)

38.74 J. L. Billeter, R. A. Pelcovits: Molecular shape and flexoelectricity, Liq. Cryst. **27**, 1151–1160 (2000)

38.75 A. Ferrarini: Shape model for the molecular interpretation of the flexoelectric effect, Phys. Rev. E **64**, 021 710–11 (2001)

38.76 I. W. Stewart: *The Static and Dynamic Continuum Theory of Liquid Crystals* (Taylor & Francis, London 2004)

38.77 K. Tarumi, M. Heckmeier: *Physical Properties of Liquid Crystals: Nematics*, ed. by D. A. Dunmur, A. Fukuda, G. R. Luckhurst (INSPEC, London 2001) Chap. 11.4

39. Organic Photoconductors

This Chapter surveys organic photoreceptor devices used in electrophotography. Included in the discussion are the materials (polymers, pigments, charge-transport molecules, etc.), device architecture, fabrication methods, and device electrical characteristics that are critical to the successful functioning of an electrophotographic device (printer).

The Chapter is organized as follows. A brief discussion of the history of xerography and the contributions of Chester Carlson is followed by operational considerations and critical materials properties. The latter includes dark conductivity, photodischarge–charge transport, and photogeneration. Organic photoreceptor characterizations of dark decay, photosensitivity, and electrical-only cycling are discussed in detail. This is followed by discussions of photoreceptor architecture, coating technologies, substrate, conductive layer, and coated layers which carry out specific functions such as smoothing, charge blocking, charge transport, backing, and surface protection.

39.1	Chester Carlson and Xerography	954
39.2	**Operational Considerations and Critical Materials Properties**	956
	39.2.1 Dark Conductivity	956
	39.2.2 Photodischarge–Charge Transport	957
	39.2.3 Photogeneration	963
39.3	**OPC Characterization**	965
	39.3.1 Dark Decay	965
	39.3.2 Photosensitivity	965
	39.3.3 Electrical-Only Cycling	966
39.4	**OPC Architecture and Composition**	967
	39.4.1 OPC Architecture	967
	39.4.2 Coating Technologies	968
	39.4.3 Substrate and Conductive Layer	969
	39.4.4 Smoothing Layer and Charge-Blocking Layer	969
	39.4.5 Charge-Generation Layer (CGL)	970
	39.4.6 Charge-Transport Layer (CTL)	974
	39.4.7 Backing Layer	975
	39.4.8 Overcoat Layer	975
39.5	**Photoreceptor Fabrication**	976
39.6	**Summary**	977
	References	978

Organic photoconductors, devices fabricated from organic photoconductors, and the applications of these devices, are the topics which will be covered in this chapter. The term *organic* is used in the chemical sense to encompass materials with carbon as a major constituent. This includes molecular as well as polymeric materials that are not naturally occurring but are purposefully designed and synthesized in the laboratory for specific physical, chemical, dark-electrical, and photoelectrical characteristics. Organic photoconductors are single, or more commonly, a formulated blend of materials which have photoconductive characteristics. Put simply, these are materials in which the electrical conductivity increases on exposure to light [39.1, 2]. Organic photoreceptors are thin-film devices made from organic photoconductive materials with physical, chemical, dark-electrical, and photoelectrical characteristics designed for optimum performance in specific applications. The most important application is in electrophotography, where organic photoreceptors are utilized in machines for digital printing and copying. In this chapter we will discuss details of organic photoreceptor architecture and their composition and characteristics that have enabled their utilization in electrophotographic printing. In this space it will only be possible to touch briefly on many important aspects of organic photoreceptors but references will be provided for those wishing to delve deeper [39.3–6].

Organic photoconductors are materials and organic photoreceptors are devices made from organic photoconductors. The common acronym for an organic photoreceptor is OPC, which of course stands for organic photoconductor, and simply indicates that acronyms are often inscrutable. The application for which organic photoreceptors have received the most attention and development effort is as the photosensitive image-creating element in electrophotographic-based printers and copiers. In these machines the photoreceptor is a highly uniform, defect-free, multilayer thin film ($\approx 20\text{--}50\,\mu\text{m}$ photoconductor thickness) coated on a metal drum or a flexible belt, with an active area from 160 to over $12\,000\,\text{cm}^2$. Organic photoreceptors for electrophotography were first introduced in the 1960s and, after years of development for improved physical and electrical characteristics, they form the basis of a multi-billion-dollar industry today.

39.1 Chester Carlson and Xerography

Electrophotography has an interesting history [39.7–10] and we will briefly tell this story with emphasis on the enabling aspects of organic photoreceptors in this technology. The story begins with Chester F. Carlson, a patent lawyer who saw the need for a method of reproducing documents. In his kitchen laboratory in Queens, New York, he demonstrated that a charged photoconductor could be used to produce an electrostatic latent image which could subsequently be visualized by contact with oppositely charged particles. The first electrophotographic image was "10.-22.-38 Astoria". The patent application was filed on 4th April 1939 and on 6th October 1942, US patent 2 297 691 was issued [39.11]. Figure 39.1 shows the title page of that patent. In concept, electrophotography is a relatively simple process with six (or seven) steps but it took over 20 years to bring it to fruition and another decade to make it truly practical. The essential steps are all illustrated in the US patent cover page: charging, exposure, developing, transfer and fixing. Not pictured are the final steps, blanket exposure and cleaning, to prepare the photoreceptor for the next electrophotographic cycle.

Carlson had great difficulty finding commercial backing for his invention. He was unable to convince the management at a long list of influential and affluent companies that making copies of documents was an attractive business proposition. In the end the Battelle Development Corporation took on the project and made significant advances towards making it more practical. In Carlson's invention a coating of photoconductive material such as sulfur or anthracene is prepared on a conductive, grounded, metal substrate. The photoconductor is tribocharged by contact with an appropriate material (silk rabbit's fur, etc.). The photoconductor is then image-wise exposed with light of a wavelength that causes the photoconductive response. The areas struck by light become conductive and the surface charge is neutralized. Thus, there is now a difference in surface charge density, and hence surface potential, between the exposed and unexposed areas. This difference in surface potential, the so-called electrostatic latent image, was visualized by contacting the surface with small marking particles, such as spores from the lycopodium club

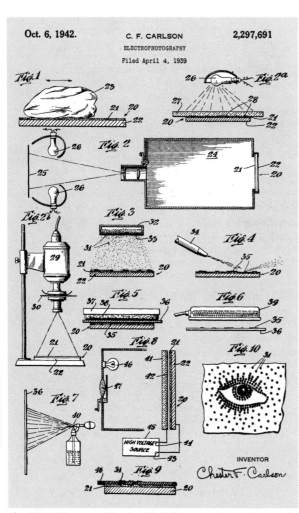

Fig. 39.1 Title page of the first U. S. Patent on electrophotography by C. F. Carlson

moss, with surface charge opposite to the original surface charge. The final copy is produced by transferring the *toned* image on the photoreceptor to the receiver (Carlson accomplished this with a heated sheet of waxed paper). Carlson later dubbed his process *xerography*, meaning dry writing, based on the use of dry powder as the marking particles. Electrophotography is a more general term encompassing, for example, the use of liquid toners in which the charged marking particles are suspended in an insulating fluid.

Two significant early advances were the use of amorphous selenium as the photoconductive material and the development of a practical method (the corotron) for the generation of an air corona for surface charging the photoconductive material. The corotron is simply a fine wire held at high potential (above the air ionization threshold of $\approx 7\,\text{kV}$) and shielded on the back with a grounded metal plate. Xerography was eventually purchased by the Haloid Corporation of Rochester, New York, (later to become the Xerox Corporation) with Carlson as a consultant. The first xerographic product was the XeroX Copier model A introduced in 1949. This initial product had totally manual operation which required several minutes to produce a copy. The photoreceptor was a selenium plate that measured 9×14 inches. From these modest beginnings, advances in engineering the various processes and materials progressed in an evolutionary manner leading to the introduction of the automated copy machine, the Xerox 914 in 1959. This utilized a selenium-coated drum photoreceptor. The evolution continues today and new product introductions have rapidly shifted from copiers (reflection of light from a document to expose the photoreceptor) to digital printers (pixel-wise exposure of the photoreceptor based on data from a previously compiled image file) using drum- or flexible-belt-based organic photoreceptors. Early examples of copiers utilizing organic photoreceptors are the IBM copier 1 (1970) and the Kodak Ektaprint 100 (1975). Initially the photoreceptors were *single-layer* in that the photoconductive materials were all contained in a single layer. For high-speed high-quality applications more complex photoreceptors were introduced. These had multilayer structures with each layer being optimized for a specific function. The details of photoreceptor architecture and composition will be discussed later.

There are some subtle differences in the electrophotographic process between copiers and digital printers. Since the electrostatic image in a copier is produced by reflection of the exposing light from a document (black text for example), areas of the document which are dark (text) reflect no light and the corresponding area of the photoreceptor retains its surface charge. To produce a final print in which the dark areas correspond to dark areas in the original document one must *develop* the image with a marking particle (toner) with charge that is opposite to that on the photoreceptor surface. This process is called *charged area development*. In a digital printer the exposure is controlled by an image file and the image is composed of a microscopic halftone pattern of *pixels*. Pixel size is determined by the desired *resolution* such that 600 dpi corresponds to a pixel size of about $40\,\mu\text{m}$. In a digital process it is possible to expose either those areas which will be developed or undeveloped in the final print. The approach sometimes chosen is to expose only those areas which will be toned since most documents are text-based and text occupies only a small fraction of the total document area. This approach decreases the on-time of the exposing system and lengthens its lifetime. Thus, in this case those areas to be developed have been exposed and have decreased surface potential. This process, *discharged area development*, is carried out by using toner with a charge polarity that is the same as that of the photoreceptor surface. Factors in OPC design relating to their optimization for digital imaging have been investigated but in practice the OPCs used in today's copiers and printers differ little [39.12–14]. Discussion of the chemistry, physics, and engineering involved in toners, developers, and development systems is beyond the scope of this chapter [39.15, 16].

In addition to the photoreceptor some of the major subsystems of a modern digital printer are:

- Exposure: scanning laser or light-emitting diode (LED) arrays.
- Development: magnetic brush or other technology with black and/or colored marking particles plus magnetic carrier.
- Transfer: heat, pressure, etc. to remove toner from the photoreceptor and place it on the receiver. In some systems the toner is transferred first to an intermediate drum or web and then to the final receiver.
- Erase: blanket exposure to return the photoreceptor to a uniform surface-charge state.
- Clean: blade or brush to remove untransferred toner from the photoreceptor surface.
- Fixing: melting the toner particles onto the receiver.
- Computers and software: image capture, rendering, and storage, process control, receiver handling, etc.

- Process control: software and hardware (sensors, voltmeters, etc.) to maintain image quality.
- Sensors and controls: for receiver handling.

A full listing would be much longer and the modern electrophotographic printer is a highly complex system where the hardware, software, and materials have been successfully co-optimized to meet the product aims [39.17–19].

Since the photoreceptor surface is either in physical contact with (development, transfer, cleaning), or exposed to, effluents (fusing, corona charging) from the various subsystems its physical and electrical characteristics must be stable to these interactions throughout its life. Much of the development in OPCs has been to extend the photoreceptor process lifetime. Today the most durable OPCs might be replaced after 100 000 or more imaging cycles.

39.2 Operational Considerations and Critical Materials Properties

The steps involved in the electrophotographic process were detailed previously [39.15, 17]. In this section we discuss some of the underlying physics in the formation of the electrostatic latent image. From the point of view of device physics the OPC is a large-area transducer configured as a belt or drum which converts optical information into a latent charge (primary) image. The photoreceptor material is a high-dielectric-strength insulator that is converted to the electrically conducting state with the application of a field and illumination. Phenomenologically, a high field is applied across the thin-film photoconductor (corona charging) and the film is exposed to radiation absorbed by the material. The photon energy is converted into charge carriers at or near the site of absorption and these drift under the influence of the field. Since the charged photoreceptor is an open-circuit device the surface potential decreases as the charges drift through the material and the photoreceptor is discharged.

In this section we focus on some of the key physics and materials issues governing the field-biased motion of electronic charge through the photoreceptor during light-induced xerographic discharge – the process which leads to latent image formation. Work done in optimizing the design and characterization of photoreceptors in order to service an evolving technology operated in a push–pull relationship with the growth in scientific understanding of photoinduced charge generation, injection, transport and trapping first in amorphous semiconductors and then, in what will concern us here, disordered molecular media.

Because the photoreceptor assembly is a sensitized large-area device, the materials which simultaneously optimize all the required properties have always been glassy coatings. (Polycrystalline media are largely unsuitable for a variety of reasons.) The inorganic materials used especially in light-lens copiers were initially amorphous chalcogenides [39.20, 21] and to a much lesser extent hydrogenated amorphous Si [39.22, 23]. Much of the early work was therefore focused on the physics and chemistry of amorphous semiconductors. In this case photogeneration and the subsequent transport of charge occur in precisely the same medium, and design latitude is clearly restricted by the particular combination of optical and transport characteristics of that single layer. In addition these inorganic films are relatively brittle and therefore unsuitable for applications which require a belt architecture, i.e., one in which the photoreceptor is required to bend around small rollers. On the other hand, polymer-based OPCs are inherently flexible and multi-layer architectures are readily fabricated such that each layer can be optimized for a particular function such as photogeneration and charge transport [39.24]. In fact, all OPCs for high-end applications have been developed according to this principle with the photoconductor divided into a thin light-absorbing charge-generation layer (CGL) adjacent to the electrode and a thicker charge-transport layer (CTL) which transports holes. With this photoreceptor configuration the surface is negatively charged prior to exposure. The details of OPC architecture and the materials chosen for specific layer functions will be discussed later.

39.2.1 Dark Conductivity

The electrophotographic imaging process begins by applying a surface charge to the photoreceptor. A parallel-plate-capacitor model is appropriate and the surface potential and surface charge density are related through the capacitance per unit area as, $Q/A = (C/A)V = (\varepsilon\varepsilon_0/L)V$, where Q/A is the surface charge density (C/cm^2), C/A is the capacitance per unit area (F/cm^2), V is the surface potential (V), ε is the dielectric constant, ε_0 is the permittivity of free space, and L is the photoreceptor thickness. For a typical organic photoreceptor ($L = 25\,\mu\text{m}$, $\varepsilon = 3$, and $V = -500\,\text{V}$) C/A is $3.2 \times 10^{-10}\,\text{F/cm}^2$, corresponding to 10^{12} charges/cm^2. Assuming the surface is composed of molecules with an area of $100\,\text{Å}^2$ ($10^{-14}\,\text{cm}^2$/molecule) only $\approx 1\%$ of the surface molecules are associated with the surface charge. The applied field (V/L) is $2 \times 10^5\,\text{V/cm}$.

Electrophotographic imaging is enabled by a difference in surface potential between exposed and unexposed areas when the imaged area enters the development subsystem. Generally a potential difference of at least 300 V is desired. The materials comprising the bulk of the OPC must be highly insulating. A resistivity of $10^{13}\,\Omega\,\text{cm}$ at a field of $2 \times 10^5\,\text{V/cm}$ will give rise to a dark decay rate of nearly 200 V/s. Since development typically occurs at a fraction of a second to one second after charging, the photoconductor resistivity needs to be $> 10^{13}\,\Omega\,\text{cm}$. Organic polymers such as bisphenol-A polycarbonate meet this requirement and are the major component in OPCs.

There are other sources of dark conductivity such as electrode and/or surface charge injection, bulk charge generation, and/or charge detrapping [39.25]. Electrode injection is prevented with the interposition of a charge-injection blocking layer between the electrode and the photoconductive material. Surface injection is prevented by the chemical composition of the surface layer (which is typically a hole-transporting CTL). Detrapping can be a significant source of dark decay in OPCs and must be controlled by a balance between the process and the materials (the charge-generation material in particular).

39.2.2 Photodischarge–Charge Transport

The photoreceptor must retain charge in the dark and also be photosensitive enough to discharge exposed areas to half their initial charge potential when irradiated with (nominally for a mid-volume laser printer engine) $4-10\,\text{erg/cm}^2$. The xerographic gain or quantum efficiency of supply describes the fractional number of surface charges neutralized per absorbed photon. For a dual-layer OPC it is a complicated convolution of the quantum efficiency of generation in the charge-generation layer, CGL, the efficiency of carrier injection from the CGL to the CTL and the transport parameters of the CTL. For this discussion we will for the moment ignore issues of charge generation and focus on transport. In dual-layer OPCs the majority of photodischarge occurs via charge transport through the CTL with the CGL playing a minor role. In the transport lexicon [39.24, 26] there are essentially two parameters that constitute the figures of merit to characterize charge motion through the polymeric CTL. These are (1) the drift mobility μ, the measure of *how fast* the carrier moves per unit applied field, and (2) the normalized carrier range $\mu\tau$ (τ is the free-carrier lifetime against deep trapping), which is *how far* the injected carrier moves per unit field before becoming immobilized in a deep trap. The time for a photoinjected carrier to traverse the CTL is called the transit time. The transit time t_{tr}, and mobility μ are related to the specimen thickness L according to $t_{\text{tr}} = L/\mu E$. The importance of mobility as a critical parameter in the electrophotographic process can be understood as follows: for an increase in the exposure intensity d(F), the final decrease in surface potential d(V) is proportional to the number of injected carriers and the distance they travel within the CTL. During xerographic discharge, a charge of CV_0 (C is the CTL capacitance and V_0 the initial voltage) traverses the bulk and induces time-dependent variation in the electric field behind the leading edge of the injected carrier front. Thus, as the fastest carriers transit the CTL, the electric field behind them is reduced, and the carriers behind the leading edge transit at a lower field, which in turn makes their velocities lower. Thus during xerographic discharge the transit times of individual photoinjected carriers become dispersed over a wide range, typically about an order of magnitude. For discharge to proceed to completion, even in the complete absence of deep trapping, enough time is required for the slowest carriers in the packet to exit the layer before the photoreceptor reaches the development zone – nominally 0.3–1.0 s after exposure in mid-volume printers. The latter must be allowed for in practice. Thus, carrier mobility in this particular illustration should exceed $10^{-6}\,\text{cm}^2/\text{Vs}$. For example, consider that a dual-layer photoreceptor with a 25-μm CTL, in which there are no deep traps, is subjected to a light flash intense enough to ultimately induce complete discharge (CV_0 of absorbed photons). Consider further that in this CTL the mobility $\mu = 10^{-6}\,\text{cm}^2/\text{Vs}$ at $E = 10^4\,\text{V/cm}$ and that the device is initially charged to 1000 V. It can be calculated that under these conditions the device will have a residual voltage of 20 V after 0.3 s or 7 V after 1 s. Incomplete discharge, unless compensated for, might result in an inadequate toning potential and a toned density less than desired. Note further that, if the mobility is even lower, the results can become totally unacceptable. For example when the mobility is $\mu = 10^{-7}\,\text{cm}^2/\text{Vs}$ under the conditions just described, the residual voltage a full second after exposure is 60 V, even in a completely trap-free CTL. In light of the foregoing illustration for the trap-free case, which sets the mobility benchmark, the effect of traps in the polymeric CTL must be of paramount concern. More precisely, we are concerned with traps whose release time at ambient temperature discernibly exceeds the period of a complete electrophotographic cycle. In the present context we take the latter as the operational definition of a deep trap. With such traps present it is clearly the case that, after repeated charge–expose cycling, some quantity of image degrading charge would remain immobilized in the bulk for times now exceeding the period of a complete xerographic cycle. If ρ is the density of uniformly trapped space charge in a CTL of thickness L and relative dielectric constant ε then there is an associated residual potential V_R, where $V_R = e\rho L^2/2\varepsilon\varepsilon_0$. Here e is the electronic charge and ε_0 is the free-space permittivity. In a nonpolar dielectric medium ($\varepsilon = 3$) with a layer thickness of 25 μm, as few as 10^{13} electronic charges trapped per cm^3 already give rise to a residual of 19 V. Unless process control utilizing electronic feedback correction can be employed, bulk-trapping-induced space-charge buildup during cycling can result in severe image degradation. Residual potential arising from bulk-trapped space charge is a critical electrophotographic process parameter. A residual potential can be related to the normalized carrier range, $\mu\tau$, in the weak-trapping limit from the physically plausible ansatz that the residual potential corresponds to that applied voltage for which the carrier range is about half the specimen thickness L. Thus, $\mu\tau$ is approximately $L^2/2V_R$. On this basis the tolerable trap density is defined by the requirement that the $\mu\tau$ product in practical devices should typically exceed 10^{-6} cm^2/V. Even highly purified polymer will typically contain many chemical impurities in the 1–10 ppm range, which the foregoing calculations show is vastly higher than permissible trap levels. However, chemical impurities, even when present at relatively high concentration, can be rendered trap-inactive by employing molecular design principles to guide the overall choice of active materials. Concepts derived from scientific understanding of photogeneration, injection across interfaces, and electronic transport in disordered organic materials, in combination with the unique compositional flexibility characteristic of the organic solid state, were together responsible for making multilayer OPCs the dominant practical receptor technology for electrophotography.

As discussed above, successful electrophotographic imaging requires that charge transport through the CTL occur with little trapping in the time scale of the electrophotographic cycle. Here we discuss how this is accomplished in an *impure* organic-chemical-based CTL. The CTL is a glassy solid solution of a charge-transport-active moiety dispersed in a polymer binder. The charge-transport moiety can be molecular or a polymer component. The molecular solutions are called molecularly doped polymers (MDP). The transport-active component is typically 40–50% weight fraction of the CTL. Thus, *doping* in the present context is decidedly different from what doping refers to in conventional semiconductor physics [39.27]. In the latter case the dopant is typically introduced at very low concentration to control the relative proportion of electrons and holes in the bands of a semiconductor crystal leaving optical properties largely unaffected. In a semiconductor crystal at ambient temperature the dominant field driven transport mechanism is scattering perturbed motion of charge carriers in the bands. Under these circumstances, mobility decreases algebraically with increasing temperature, while carrier population in the bands is thermally activated. The convolution of these processes manifests in a thermally activated conductivity. On the other hand, in MDPs under equilibrium conditions the transport-active molecule is in the neutral state, and the glassy films are perfect insulators, that is, there are no free carriers

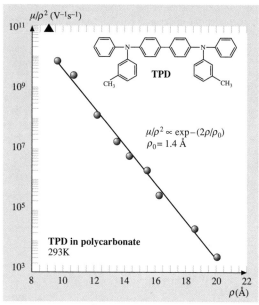

Fig. 39.2 Log (μ/ρ^2) versus average intersite separation, ρ, of TPD molecules in bisphenol-A polycarbonate ($E = 5 \times 10^4$ V/cm). The *triangle symbol* is the hole drift mobility in pure TPD film

present in thermal equilibrium [39.28]. These systems can nevertheless support relatively efficient charge transport under the action of an applied field when in contact with a charge reservoir. At fixed temperature the drift mobility of extrinsic carriers decreases exponentially with increasing average intersite separation of the active molecule (ρ) so the drift mobility of MDPs can be tuned over a broad range by simply adjusting concentration [39.29, 30]. This feature of tunability can readily accommodate a wide range of xerographic process speeds as described above and is therefore advantageous from a technological point of view [39.31]. This is illustrated in Fig. 39.2 for a common hole-transport material, TPD N,N'-diphenyl-N,N'-bis(3-methylphenyl)-(1,1'-biphenyl)-4,4'-diamine (TPD, Fig. 39.4). The data is in conformity with a simple tunneling model where ρ_0 is the wavefunction localization radius and the transport states are sited on the TPD molecules. At 40–50 wt% doping approximately 10^4 *hops* are required for the hole to transit a 20-μm photoreceptor.

Drift mobility of holes in CTL films is most conveniently measured by the canonical small-signal time-of-flight (TOF) technique [39.32, 33]. However, for a given composition of the CTL, and all other conditions analogous, the same drift motilities can be inferred by analyzing transport-limited xerographic discharge in bilayer photoreceptors [39.34, 35]. In TOF, the MDP film is prepared with semitransparent blocking contacts and maintained at a bias which is high enough to insure that the transit time of any excess injected carrier

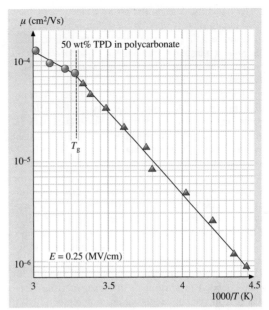

Fig. 39.3 Temperature dependence of the drift mobility in a film of 50 wt% TPD in bisphenol-A polycarbonate ($E = 2.5 \times 10^5$ V/cm). T_g is the glass transition temperature of the film

is shorter than the bulk dielectric relaxation time. The sample is exposed to a very short and weak pulse of strongly absorbed light (typically 337 nm from a ni-

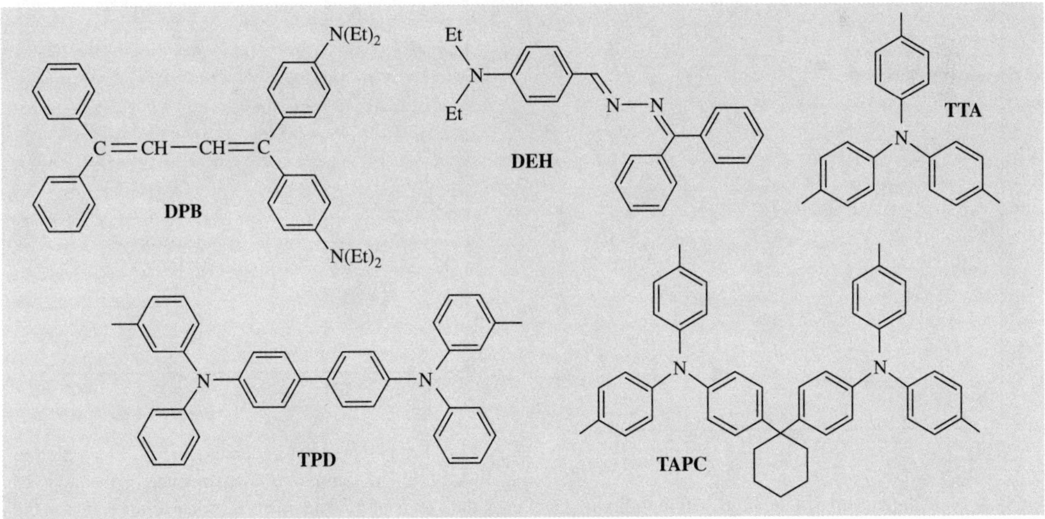

Fig. 39.4 Chemical structures of representative hole transport materials

trogen laser) incident on the positive electrode. Under these circumstances photoexcited transport molecules are oxidized at the positive electrode to the radical cation (the *hole* in transport terminology). This process sets up a chain of redox steps where electrons are progressively transferred from neutral molecules to their neighboring radical cation. The concentration of the advancing pulse of holes is low, such that the applied field remains uniform during their transit. In the ideal case the current from the advancing pulse of holes is constant until the leading edge reaches the counter-electrode, after which it rapidly decreases. The transit time can be extracted from the transient current and from this the mobility. For a given MDP composition drift mobilities are determined for films of known thickness as a function of applied field and temperature. The drift mobility in MDPs always has a thermally activated temperature dependence; thus the log of the drift mobility scales with inverse temperature, as illustrated in Fig. 39.3, where earlier measurements [39.29] have here [39.36] been extended to encompass the characteristic slope change displayed in the glass-transition region [39.37]. However, in a significant number of cases cited in the literature [39.4] the analogous scaling is with the square of the inverse temperature, (i. e. non-simple activation), as predicted by the disorder model [39.38]. The experimentally observed scaling with intersite separation, the temperature dependence, and the high degree of disorder in these systems clearly indicate that electronic transport must involve the field-biased hopping of carriers in an energetically inequivalent manifold of states sited on the transport-active molecules. Apart from their technological importance in electrophotography [39.39] and light-emitting displays [39.40], what has made MDPs a laboratory for the study of hopping transport is that a number of key secondary features in their transport behavior are in fact pervasive in other disordered molecular media [39.41]. There is then the suggestion of a common underlying mechanism susceptible to theoretical treatment. Thus, the field- and temperature-dependent behavior of molecular dispersions is also observed in polymers with transport-active pendant groups such as poly(N-vinylcarbazole) [39.42], in polymers in which small molecule moieties are incorporated in a main

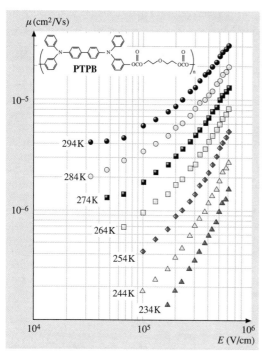

Fig. 39.5 Field dependence of the hole drift mobility in the polymer PTPB parametric in temperature

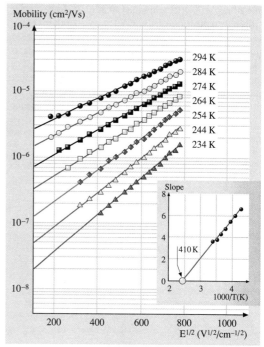

Fig. 39.6 Data set at each temperature in Fig. 39.5 replotted against $E^{1/2}$. The temperature dependence of the slope of each data set is plotted against reciprocal temperature in the *insert*

chain [39.43] in sigma-conjugated polymers [39.44, 45] and most recently in certain poly π-conjugated systems [39.46, 47]. Figure 39.5 is a plot of the field dependence of the hole drift mobility in the polymer polytetraphenylbenzidine (PTPB) (see insert for the molecular structure) versus field and parametric in temperature over a 60 K range [39.48]. PTPB is a glassy polymer in which tetraphenylbenzidine (TPD) moieties are covalently bonded into a main chain [39.43]. The data is replotted in Fig. 39.6 to show the explicit dependence of the log mobility on the square root of the applied field. The field dependence is itself temperature dependent and clearly becomes stronger with decreasing temperature. The two TPD-based hole-transport media are closely related but qualitatively identical results are reproduced in polysilylenes and polygermylenes which have a sigma-conjugated backbone capable of supporting electron delocalization as inferred from analysis of absorption and emission spectra [39.49, 50]. For example, hole transport in PTPB can be represented in an Arrhenius plot over the temperature range illustrated in Fig. 39.7. In this purely phenomenological description the activation decreases with the square root of the applied field, as displayed in the insert. Precisely the same behavior is exhibited when the same phenomenological description is applied to hole drift mobility data in poly(methylphenylsilylene) (PMPS), as shown in Fig. 39.8. Combining drift mobility [39.51] and spectroscopic data [39.52] has in fact suggested that transport in these sigma-conjugated polymers involves the hopping of holes among chromophore-like main-chain segments of varying length whose function for transport is therefore analogous to the TPD sites in PTPB. More recently the convoluted pattern of behavior first identified in MDPs and therefore clearly characteristic of hopping among discrete energetically in equivalent molecular sites has been reported in TOF experiments on π-conjugated systems notably the phenylene vinylenes. The sites, like those in polysilylenes, are interpreted to be domain-like backbone segments of varying conjugation length [39.46, 47]. These combined results and the recurrent pattern of behavior they reveal in systems

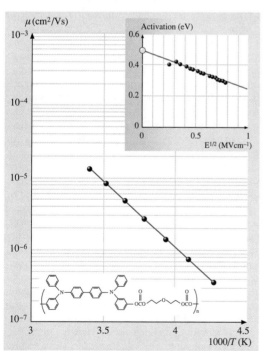

Fig. 39.7 Arrhenius representation of the temperature dependence of hole drift mobility in PTPB. The field dependence of activation is shown in the *insert*

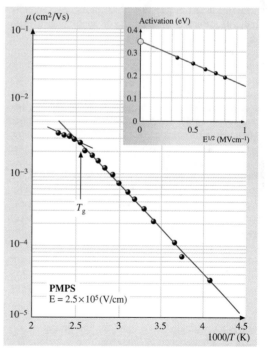

Fig. 39.8 Arrhenius representation of the temperature dependence of hole drift mobility in poly(methylphenylsilylene), PMPS. The field dependence of activation is shown in the insert. T_g is the glass transition temperature of the PMPS film

of widely varying composition and morphology suggest that a transport theory must be founded on universally shared characteristics [39.53]. The theoretical literature stimulated by these observations is in fact extensive and far too detailed to be adequately described here. In the present context it suffices to comment briefly on certain general trends. The activated temperature dependence universal to all these materials systems arises because states in the hopping manifold are energetically inequivalent. Energetic inequivalence can be understood on the basis of site relaxation accompanying polaron formation [39.54–56]. Energetic inequivalence of sites is also a common feature of disordered solids resulting from site-to-site fluctuation of the static and dynamic electrostatic potential [39.38, 57, 58]. The combination of both effects cannot be excluded. In fact, there is a general need to develop the framework to distinguish the relative contributions of these key processes in the analysis of experimental data [39.59, 60]. A particularly challenging issue has been the attempt to account for the special features of the Poole–Frenkel-like field dependence illustrated in Fig. 39.6. Polaron models typically predict that the log mobility should scale linearly with the applied field [39.54, 61]. The disorder model based on analysis of Monte Carlo computer simulation of hopping on a finite lattice as originally proposed by *Bässler* [39.38] was in fact able to self-consistently account for a number of key experimental features in terms of a limited set of disorder parameters. On the other hand, it could only model the observed field dependence over a very narrow range. It was later recognized in the formulation of dipole disorder models [39.62, 63], that slow site-to-site variation, as distinct from fully random fluctuation, in the effective electrostatic potential could more properly account for the field-dependent behavior of mobility commonly observed over several orders of magnitude [39.64–66]. The more general formulation of the disorder model developed by Dunlap and coworkers, treats the disorder potential as fully correlated and therefore slowly varying, and argues that such a supposition is physically plausible on general grounds in glassy solids.

If every chemical impurity remaining in even the most scrupulously processed polymer film could act as a trap, organics as a materials class would be completely excluded as a basis for the design of xerographic photoreceptors. Trap-free transport in CTLs was achieved not by entirely eliminating impurities but by properly designing the transport-active moiety [39.67, 68]. The associated studies first carried out in this technologically critical context on molecularly doped systems [39.69] were instrumental in generally elucidating trap interactions in hopping systems and in first unambiguously illustrating and analyzing the mechanism of trap-controlled hopping transport [39.70]. It was already pointed out that hole transport is supported in MDPs when the neutral transport molecule is donor-like. For electrons the corresponding neutral transport molecule is acceptor-like in character. In each case, the CTLs are typically unipolar when analyzed by TOF or xerographic techniques. As it turns out the principle (simple in hindsight) which applies for example to hole-transport CTLs is to make the transport molecule significantly more donor-like than any of the resident impurity species. The limit is that the material must be stable to air oxidation. From an energetic point of view the impurity levels which then lie above the hole transport states are anti-traps. An important class of small molecules which have these desired characteristics are the aromatic amines and TPD is a prime example of a particularly strong donor in this class of molecules. The trap-free nature of these transport layers is xerographically apparent because there is no buildup of bulk space charge as measured by Kelvin-probe techniques [39.71] even after tens of thousands of CVs of holes are discharged through a photoreceptor with this CTL as a component. The trap-free nature of hole transport can also be independently demonstrated by combining TOF drift mobility and analysis of current–voltage measurements carried out on any transport layer fitted with a semitransparent blocking contact on one surface and an ohmic contact on the opposite face [39.72–74]. Figure 39.9 illustrates the result for the polymer PTPB in which TPD is covalently incorporated as the transport active unit [39.73]. Semitransparent Al on the films exposed surface is used for the TOF measurements of drift mobility. Dark injection occurs under positive bias from the gold-coated mica substrate onto which the PTPB film had initially been deposited. PTPB is highly insulating so that the bulk dielectric relaxation time is always much longer than the transit time of any excess injected carrier at all applied voltages. With the Al contact under positive bias, hole drift mobility is measured as a function of applied voltage V. From the measured drift mobility the trap-free space-charge-limited current density (TFSCLC), J, [39.75] that would be sustained by an ohmic contact under positive bias can be calculated to a good approximation, even when the drift mobility is itself field dependent [39.76, 77] using the expression $J = 9\varepsilon\varepsilon_0\mu V^2/8L^3$. These values are represented by open circles in Fig. 39.9. The electric field in the film is, of course, nonuniform and E in plots of this sort is simply a shorthand for V/L,

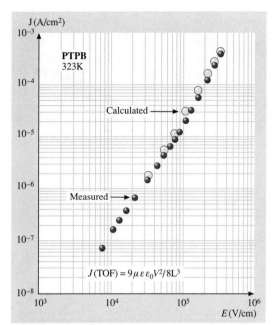

Fig. 39.9 The field dependence of the dark current for a PTPB film at 323 K measured with positive bias applied to an Au coated Mica substrate (*filled circles*). The trap free space charge limited current (*open circles*) is calculated from the hole drift mobilities obtained by the time of flight method

where V is the externally applied voltage. It should be noted that, under the present conditions, the TFS-CLC is in principle the maximum current that the bulk of a transport medium such as PTPB can demand from a contact in the steady state. It should also be noted that the observation that an injecting contact under test is capable of sustaining a space-charge-limited dark current is a prima facie demonstration of ohmicity [39.24]. Filled circles represent the steady-state current measured when positive bias is applied to the substrate. That these calculated and measured currents coincide demonstrates that the injecting Au contact is ohmic and that the PTPB film is trap-free. Similar results have been obtained for other commonly used CTL MDPs such as TPD in bisphenol-A polycarbonate and in films of poly(methylphenylsilylene).

39.2.3 Photogeneration

The CGL material is designed to strongly absorb light from the exposure system (red to near-infrared LEDs and lasers are typical). In a typical dual-layer OPC the CGL material is a thin (submicron) coating of a dispersed pigment. Simply, the pigment material absorbs a photon and mobile charge is created which is *injected* into the CTL and then drifts under the influence of the applied field to discharge the device. Charge generation is a very complex topic [39.78]. In general terms, absorption of a photon produces an excited state (exciton) of a correlated hole–electron pair (valence band and conduction band, respectively) which has some mobility. Measured exciton ranges can be on the order of $0.1\,\mu\text{m}$. The exciton may then undergo recombination, with the release of energy to the environment, or it may be converted into charges. Conversion into charges can occur *spontaneously* or through chemical (impurities or dopants) or physical (crystal defects or the crystal surface) interactions. With the trapping of one of the charge carriers the other is free to drift under the influence of the applied field and geminate recombination is suppressed. The mobile carrier (assume holes) must drift to where it is in contact with the hole transport moieties of the CTL. In fact, in most cases the CTL penetrates and mixes with the CGL so that, even if the CGL is formulated without transport material, we can assume that these molecules are in molecular contact with the pigment particle surface. Now the *hole* is available to oxidize or *inject* into the transport moiety to produce the radical cation. Useful hole-transport materials have relatively low oxidation potentials [$\approx 1\,\text{V}$ versus the saturated calomel electrode (SCE)]. The oxidation potential must be lower than that of the hole at the pigment surface but not so low that the material is unstable to air oxidation. The CTL hole is now transported through the CTL, discharging the photoreceptor, by sequential one-electron redox reactions between adjacent transport molecules as described in the previous section.

The charge-generation mechanism is characterized as *intrinsic* when the hole is generated within the pigment particle without the influence of any externally added materials. In this case the hole is injected into the CTL with concomitant oxidation of a transport moiety at the pigment surface. *Sensitized* generation is said to take place when the photoconductivity is enhanced by an added material (dopant). In this case the dopant is available to react with the pigment exciton at the particle surface. In most cases the dopant is in fact the hole-transport moiety. Dopants may also influence charge generation by forming a (charge-transfer or other) complex which facilitates charge separation. Such complexes may be accessed by direct photoexcitation

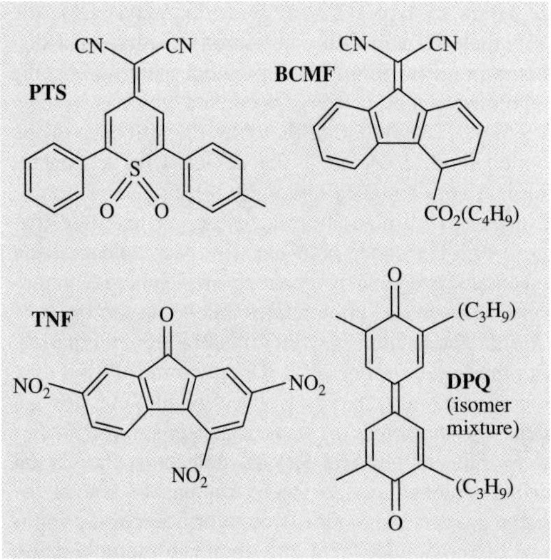

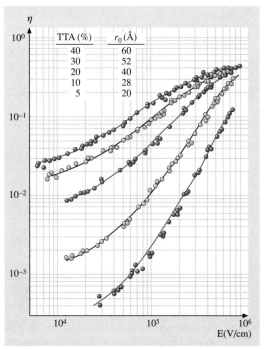

Fig. 39.10 Chemical structures of representative electron transport materials

Fig. 39.11 Quantum efficiency of the field dependence of generation for a dual layer dye-polymer aggregate based OPC. The solid lines were calculated from the Onsager model with $\eta_0 = 0.60$ and r_0 as shown. After [39.88]

or via exciton diffusion. Sensitization of the photoconductivity of poly(N-vinyl carbazole) (PVK) by electron acceptors such as TNF (Fig. 39.10) has been extensively studied [39.79–87]. Charge-generation mechanisms in pigment-based CGLs depend on the pigment molecular structure, crystal structure, morphology, and on the transport molecule used in the formulation.

The fundamental quantity of interest in charge generation is the quantum efficiency as a function of field. A strong field dependence is always observed, as shown in the example of Fig. 39.11. Several models have been proposed to explain charge-generation characteristics. These will be summarized here but more complete discussions can be found in *Pope* and *Swenberg* [39.78], *Popovic* [39.89], *Law* [39.3]; *Borsenberger* and *Weiss* [39.4], and *Weiss* et al. [39.6]. One of the early models to be applied to organic photoconductors was the Onsager model of geminate recombination [39.90]. The 1938 theory gives the probability that a charge pair in thermal equilibrium will separate under the influence of an applied field in competition with geminate recombination. In this model charge separation occurs from an upper vibrational state to produce a *geminate* pair of charges which either recombine due to their Coulombic attraction or are separated under the influence of the applied field. The parameters in this model are the field-dependent quantum efficiency for geminate-pair formation, η_0, and the initial separation distance of the geminate pair (r_0) [39.91]. The data in Fig. 39.11 is for a dual-layer dye–polymer-aggregate-based OPC with varying concentrations of charge-transport material (tri-p-tolylamine, TTA) in the CTL [39.88]. The data has been fit to the 1938 model with $\eta_0 = 0.60$ and r_0 as shown. The model has been refined using a modified basis, also due to *Onsager* [39.92]. Further refinements have involved modifications of the distribution function assumed for the separation distances [39.93, 94]. This model has been successfully used to describe charge generation in amorphous materials [39.95]. However, these models predict a wavelength-dependent quantum efficiency and a photoconduction threshold which is at higher energy than the absorption, neither of which is observed in pigment-based generation materials. A model for charge generation in these materials invokes decay (radiative or radiationless) of the initially formed excited state followed by a competition between field-dependent charge separation and internal conversion [39.96]. Sensitized generation occurs with the reaction of the thermalized excited state with a dopant. Studies of the field dependence of pigment fluorescence quenching and lifetime have been used to sort out the details with specific generation materials [39.89, 97–99].

The quantum efficiency for carrier generation can be defined in several ways, as described above. The efficiency of geminate-pair formation refers to the initially produced charge pair and this is also the efficiency which has been obtained with fluorescence quenching studies. On the other hand, the generation efficiency determined with xerographic discharge always includes some contribution from charge transport because a measured decrease in photoreceptor surface potential is due to charge motion. In addition, very-high-intensity exposures may give rise to (bimolecular) recombination of positive and negative carriers during transport [39.100–104]. Quantum-efficiency determinations from xerographic discharge are carried out with very short, low-light-intensity continuous exposures to minimize the influence of charge transport and recombination (see the following section).

39.3 OPC Characterization

Photoreceptors may be characterized by many techniques. On the one hand one wishes to determine the fundamental characteristics of the materials, while on the other hand one wishes to determine electrophotographic performance-based characteristics of the photoreceptor. The quantum efficiency of charge generation and, depending on the model chosen, the initial separation distance of the charge pair, are common characterizations. Electrophotographic performance characterizations include dark decay, photoinduced discharge sensitivity, and residual potential as well as their stability to electrical-only cycling.

39.3.1 Dark Decay

The quantity of interest is the rate of dark discharge (V/s) as a function of field and environment. However, the dark decay characteristics of a dark-adapted photoreceptor when incrementally corona charged to higher and higher fields will be significantly different from what is observed in the electrophotographic process. This is because OPCs seldom behave as perfect capacitors and there is typically a depletion charging component where previously trapped charges are mobilized and drift in the photoreceptor on the application of the corona charge [39.105]. Thus, in addition to field and environment, the observed dark decay characteristics depend on prior exposure history (trapped charges), corona charger characteristics, and timing. Thus, a given photoreceptor will have a higher dark decay rate if the photoreceptor is charged exactly to the desired initial potential (V_0) and the dark decay rate determined immediately, as opposed to charging well above the desired potential and measuring when the surface potential decreases to the desired potential. Dark decay measurements which are meaningful to the electrophotographic process must be carried out in an apparatus in which the exposures, charging, and timing elements are fixed. Dark decay is an important characteristic of an OPC because, although process control will attempt to maintain V_0, the imaging system in a printer will in general not tolerate a significant V_0 drop.

39.3.2 Photosensitivity

Measurements to determine fundamental characteristics (quantum efficiency of generation materials, mobility, etc.) are often carried on photoreceptors that differ from what might be eventually used in a printer. The layer of interest (CGL or CTL) may be isolated or the device may be modified with the application of a vacuum-deposited electrode (Au is common) on the free surface. Common fundamental characterizations include quantum efficiency as a function of field and wavelength, and mobility. Mobility was discussed in Sect. 39.2.2. The quantum efficiency of carrier generation has several definitions as applied to OPCs. On the one hand it can refer to the number of ion pairs generated per incident or absorbed photon. This fundamental characteristic might be determined experimentally by fluorescence quenching, or calculated based on a model such as the Onsager model discussed above. On the other hand a photodischarge efficiency can be determined based on the decrease in surface potential per photon absorbed or incident. The photoinduced discharge (PIDC) method is commonly used to characterize OPCs as a function of field, wavelength, temperature, humidity, exposure intensity, etc. The photoreceptor is corona charged to apply the desired field and is photodischarged. Using the parallel-plate-capacitor model, $Q = CV$ (terms as defined previously), the change in surface charge density can be calculated from the decrease in the surface potential. The change in surface charge density per unit area per photon incident or absorbed is a measure of photosensitivity. This measure includes the efficiency of charge generation, injection into the CTL, and transport through the CTL. Because of carrier bimolecular recombination and range limitations the quantum efficiency determined using this method may depend on the light intensity and the extent of photodischarge. The photodischarge method is carried out with either continuous or flash exposures. With continuous exposures ($J/cm^2/s$) the surface potential is monitored as a function of time. One measure of OPC photosensitivity which relates to the electrophotographic process is the exposure required to discharge the photoreceptor a fixed amount, often 50%, from an initial potential. This is typically carried out using exposures (wavelength and duration) relevant to a particular electrophotographic process. The photoreceptor sensitivity is usually reported in terms of J/cm^2 (the inverse is often used to report spectral sensitivity). A related method is to determine the initial photodischarge rate (dV/dt). This can be related to the decrease in surface charge density through the capacitance, $dQ/dt = C\,dV/dt$. The light intensity is kept low and the discharge time short to avoid space-charge perturbations during charge transport and to keep the field essentially constant. The ratio of the decrease in surface charge density to the exposure (photons/cm^2) is a measure of photodischarge efficiency. In fact the photodischarge per exposure ($V/J/cm^2$) is another commonly used metric for OPC photosensitivity. When determined as a function of field, data obtained in this manner can be analyzed in terms of the Onsager models. Because photosensitivity depends on field (surface potential per unit OPC thickness) and the exposure produced decrease in surface charge density (related to the change in surface potential through $\Delta Q = C\Delta V$), meaningful comparisons can only be made between OPCs of similar thickness, charged to similar initial potentials, and exposed to similar discharged potentials.

For a process using a flash exposure the photoreceptor discharge is determined at several exposure levels (with the surface potential determined at a fixed time after the exposure) and the data is plotted as surface potential (V) versus log(exposure, J/cm^2). An arbitrary photodischarge point (typically 50%) can be determined from this plot for comparison purposes. The utility of this characterization is that the exposure characteristics and the timing for reading the discharged surface potential directly relate to the printing process. As discussed above, short high-intensity exposures may give rise to Langevan recombination [39.102, 104, 106], resulting in decreased sensitivity (reciprocity failure) relative to low-intensity exposures.

An often overlooked sensitivity factor is that there are more photons per energy unit (J) as the wavelength increases. For example, there are about 30% more photons/J at 820 nm relative to 630 nm. Thus, for a photoreceptor with a given quantum efficiency for carrier generation the sensitivity will increase with increasing wavelength of the exposure source.

39.3.3 Electrical-Only Cycling

Electrophotographic performance characterizations are necessarily carried out on the photoreceptor in an apparatus which has the key elements of the electrophotographic process. The most *relevant* apparatus would of course be the fully configured printer. However, the use of such a complex device is often undesirable, or unnecessarily, for screening purposes. In the laboratory it is common to use *electrical-only* test fixtures in which the process always includes charge and expose, and may include erase (or unique process elements), with the surface potential being determined after the application of each. With an apparatus such as this it is possible to determine process-relevant characteristics such as: chargeability, dark decay, photosensitivity, and residual potential, all as a function of cycling. Of course large changes in any of these characteristics is generally undesirable and long cycling (10–100 kcycles) is often necessary to adequately determine the electrical stability of the photoreceptor to the electrophotographic process. Such stability is a necessary, but not sufficient, condition for eventual commercialization, and extensive testing in the machine for which the photoreceptor was designed is always the final step in the development process. Other factors, often connected to the device physical characteristics (wear rate, corona chemical sensitivity, layer adhesion, seam strength, etc.) may in fact determine the eventual success or failure of a particular OPC in a printer.

39.4 OPC Architecture and Composition

As we have described, OPCs are large-area thin-film devices that are insulating in the dark and become conductive when exposed to light. A high field is applied by exposure to an air corona (or similar charging method) and the photoreceptor is image-wise exposed to produce an electrostatic latent image. The surface potential in the exposed areas is determined by the exposure intensity and the OPC photosensitivity while the surface potential in the unexposed areas is determined by the dark conductivity. The development system deposits charged marking particles (toner) on the photoreceptor surface according to the difference in surface potential between the exposed and unexposed areas. The key photoelectrical processes which enable latent image formation are photoinduced charge generation and the transport of the generated charge through the device. Organic photoreceptors accomplish this in a multilayer architecture with specific materials (monomers and polymers) designed and optimized for specialized functions: binder, charge generation, charge transport, dark conductivity, etc. In this section we will discuss the architecture and composition of organic photoreceptors with details on the function and composition of each layer.

39.4.1 OPC Architecture

The choice of photoreceptor thickness is a balance between capacitance, surface potential, and internal field. Capacitance relates to surface charge density, which determines the achievable developed toner density for toners with a given surface charge per area. Surface potential, in combination with the development system, provides the potential difference to attract toner to the photoreceptor surface. The internal field enables efficient charge generation and transport. In most electrophotographic printing machines the balance is achieved with an OPC thickness of $\approx 20-30\,\mu m$ and a surface potential of 500–1000 V.

A schematic cross section of a fully configured modern electrophotographic photoreceptor is shown in Fig. 39.12. In its most simple form the photoreceptor consists of a conductive substrate and a single layer of photoconductive material to accomplish charge generation and transport. The advantage of this structure is in its simplicity, but the disadvantage is that not all characteristics will be optimum. In the modern embodiment the photoreceptor is constructed with separate layers for charge generation (CGL) and charge transport (CTL). Other layers shown in Fig. 39.12 include: sub-

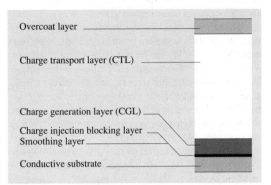

Fig. 39.12 A schematic cross section of a fully configured "modern" electrophotographic photoreceptor

strate, smoothing layer, charge-injection blocking layer, charge-generation layer (CGL), charge-transport layer (CTL), and protective overcoat layer. Additional layers, such as adhesive interlayers, and a back-side anti-curl layer may also be present. OPCs with separate CGL and CTL are termed dual-layer. Each of these layers may be a complex formulation of materials. The advantage in separating layers is that each layer can be optimized for its specialized function. This adds materials expense and manufacturing complexity but has the advantage of superior performance.

In a single-layer photoreceptor when exposed with strongly absorbed radiation directed at the free surface, charge generation will occur near the free surface and the photodischarge will be dominated by the transport of positive or negative charge through the bulk of the layer. The polarity of the mobile charge will be the same as the applied surface potential. In most commercial dual-layer photoreceptors the architecture is substrate–CGL–CTL. Because the CTL is formulated for hole transport, dual-layer OPCs are charged with a negative surface potential. Positive corona charging has some advantages over negative (better uniformity and less ozone production). A single-layer OPC can function with either a positive or negative surface charge depending on the exposure direction and the transport characteristics of the materials used in the formulation. Positive-charging dual-layer photoreceptors with the architecture substrate–CTL–CGL have been commercialized but these require a protective overcoat to prevent the CGL from being damaged by contact with the electrophotographic process elements. The typical substrate–CGL–CTL architecture will function with positive charging if the CTL is formulated for electron transport. Electron-transporting CTLs have been widely studied by not yet commercialized.

In a multilayer photoreceptor the principle is that the functions of each layer are optimized. Thus, with separate CGL and CTL the former is designed for strong absorption of the wavelength of light used in the process and efficient charge generation, and the latter is optimized for the injection of photogenerated charge from the CGL and transport of that charge to the free surface. In general, most charge-generation materials efficiently transport only one polarity of charge (electrons or holes depending on the material) so that it is generally desirable to have the CGL as thin as possible while retaining the required optical density at the exposure wavelength. It is also necessary that the CGL be uniform over the entire area of the photoreceptor to maintain consistent sensitivity for the entire photoreceptor surface. Submicron CGLs are typical of the photoreceptors used in most modern digital printers. The CTL determines the device capacitance and a thickness of $\approx 25\,\mu\text{m}$ is typical. However, the CTL thickness also influences the ultimate latent-image resolution in a digital printer because holes transiting the CTL experience mutual Coulombic repulsion [39.12, 107]. It has been shown that for optimum resolution the photoreceptor thickness should be about one quarter of the pixel size [39.12, 13]. Thus, for 600-dpi imaging ($\approx 40\,\mu\text{m}$ pixel size) the ideal photoreceptor is on the order of $10\,\mu\text{m}$. Thin OPCs have not yet been introduced into digital printers but one might anticipate that this will occur as toners get smaller and the processes of toner transfer and fusing are refined to minimize image disruption [39.108].

39.4.2 Coating Technologies

Hopper coating of continuous webs and dip coating of drums are the two technologies commonly used to fabricate photoreceptors. As with any large-area photoelectrical device absolute cleanliness is required to prevent artifacts in the coated layers. In a multilayer device one must carefully control the interfaces between the layers. For example, a key interface is that between the CGL and CTL. Here it is important that the two layers adhere but do not mix to such a degree that the desired characteristics of either layer are compromised. Attaining the desired degree of mixing is a process of optimization depending on the materials, solvents, and coating method.

In hopper coating the coating solution is continuously pumped through a slot onto the moving web. Film thickness is controlled by the pumping rate and solids content of the coating solution. Precise control of drying temperature, humidity, and air flow enable optimization of the coated layers. Curing of the coated film involves removal of solvent and annealing of the polymer binder(s). In a hopper coating process several layers can be coated in a single pass with sequential hoppers as long as the previous layer is adequately cured. Hopper coatings can be carried out at a very high speed (m/s) to produce a roll of many thousands of feet. To coat subsequent layers the previously coated substrate is re-run through the coating machine. Because the hopper coating process involves contact of the coated surfaces, and winding of the coated web, there is the possibility of film damage. Final inspection for film defects must be carried out and damaged, or imperfectly coated, regions must be removed before construction of the final photoreceptor loops.

In a dip coating process the substrate and coated layers are dipped into a solution or dispersion of the next layer to be coated and slowly withdrawn and cured. The coated layer thickness is determined by the rheological characteristics of the coating solution and the rate of withdrawal. Under these conditions there is the opportunity for extensive interaction between the previously coated layer and the solution into which it is dipped. Thus, formulations which are suitable for hopper coating may be inadequate for dip coating even though the chemical compositions of the final coated layers might be identical after curing. Another consideration with dip coating is that components of the previously coated layers which are soluble in the solvent used in the subsequent dip will contaminate the dip reservoir. Low levels of contamination may be insignificant, but since the amount of fluid removed in each coating operation is a small fraction of the total volume one must make sure that such contamination does not degrade performance.

Other processes such as ring coating (a dipped drum is withdrawn through an annulus which acts as a doctor blade to control the wet coating thickness), spray coating, vacuum coating, etc. may be used for specific applications and materials. In all layers coated from solution it is often necessary to add a small quantity of surface-active agent as a coating aid to enable smooth, defect-free coatings.

In the following sections we will discuss details around the compositions of each of the layers which might be found in an organic photoreceptor for electrophotographic printing. The order of discussion is in the order in which the typical photoreceptor would be constructed: substrate and ground layer, smoothing layer and charge-blocking layer, charge-generation layer (CGL), charge-transport layer (CTL), and protective overcoat layer.

39.4.3 Substrate and Conductive Layer

A common element to all electrophotographic photoreceptors is a substrate on which all the layers are deposited. In a research environment a small square of glass or quartz is convenient, but in commercial applications the substrate is invariably either a polymer film or a metal drum. Some of the material characteristics which might be considered in the choice of a polymer film are: stiffness, toughness, transparency, core set, and surface friction. Polymeric photoreceptor substrates are often 3–7 mil. A common polymer film material is the polyester poly(ethylene terephthalate) (PET). PET film is very durable, optically transparent, and readily available in a wide range of thickness. Since it is highly insulating it must be coated with a conductive material to act as the ground plane. PET or other polymer films are available in large rolls and the coating operations are usually done on the roll rather than cut-to-size sheets. Although a solution-coated conductive layer is possible, the most common ground layers are metals such as Ni, Al, Cr, and Ti, which are applied in a continuous sputtering or similar vacuum technique. The metallized rolls are subsequently coated with the photoconductive layers.

If a metal drum is the chosen substrate it is almost invariably aluminium (1–10 mm wall thickness). The conductive metal serves as the ground layer but typically it receives further treatment before coating the photoconductive layers. The treatment may be chemical such as anodization to form a thin hole-blocking layer, or physical such as turning or sanding. Because these secondary surface treatments add cost, a smoothing layer is often used instead.

The electrical conductivity requirements of the conductive layer are modest. The higher the process speed the greater the conductivity required for adequate surface charge accumulation. Analysis by *Chen* [39.109] demonstrated that the result of inadequate conductivity in corona charging of a photoreceptor is a nonuniform surface potential. The surface potential is decreased depending on the distance from the actual grounding contact. For a typical situation the upper limit of the sheet resistance is $\approx 10^4 \, \Omega$ per square.

Alternative approaches such as *seamless* webs and conductive plastic drums have been developed but have not yet been commercialized.

39.4.4 Smoothing Layer and Charge-Blocking Layer

For xerographic applications it is important that the conductive layer does not inject charge into the photoreceptor in the dark. However, the metals used as grounding layers all have the potential to inject holes into the photoreceptor CGL via direct contact with the charge-generation material or the hole-transport material. The injection and transport of charge in the dark would cause a loss of surface potential (dark decay). Since the formation of a latent image relies on a difference between the rate of dark- and light-induced surface potential decrease, a high dark decay rate means that the process must be amenable to a small imaging potential. Another aspect of dark injection from the conductive layer is that such phenomena often occur in small localized *charge-deficient spots*. In a process where the discharged areas are developed small deposits of toner are found in what should be non-toned background. To counter such processes it is common to interpose a charge- (typically hole-)blocking layer between the ground layer and the CGL. With drum photoreceptors it may be the roughness of the metal surface which initiates the charge injection. In this case it may be necessary to interpose a relatively thick *smoothing layer* between the metal surface and the CGL.

In principle, any good film-forming insulating material will suffice as a charge-blocking layer as long as adhesion to the ground layer and CGL are adequate. Several types of materials have been used for this purpose including polysiloxanes and nylons. Since these materials are insulating, in an electrophotographic charge/expose cycle there will be a residual potential due to the field remaining across the layer. This will build up with cycling depending on the layer conductivity and the process cycle time. Although residual potentials typically decrease when the photoreceptor is rested in the dark, in practice such relaxation is usually incomplete and in subsequent cycling the residual builds up at a faster rate. The physics and chemistry of such hysteresis is not well understood. At some point the residual potential may cause image degradation and necessitate photoreceptor replacement. Thus, the blocking layer must be thick enough to cover the ground layer uniformly but not so thick as to cause an undesirable residual potential. Generally such a layer is coated at around 1 μm or less.

With an aluminium drum a thin anodized layer may prevent charge injection. However, making the drum surface smooth on a submicron scale is expensive and an alternative is to use a relatively thick smoothing/blocking layer to cover the surface irregularities. Smoothing layers are typically metal oxides (titanium for example) in a polymer binder. The oxide loading is high to impart sufficient conductivity to prevent a residual potential. If the metal oxide does not inject holes it also serves as the charge-blocking layer. Such layers are up to around 10 μm thick.

Imaging with monochromatic radiation in digital printers has led to a new imaging artifact caused by interference between the incoming and reflected light. In dual-layer OPCs with the CTL uppermost, if the CGL incompletely absorbs the exposing light it can be reflected from the ground layer. The interference pattern that is produced can be visible in the developed image as a *wood grain* pattern. One way to counter this is to scatter the incoming light before it's reflected. There are many patents describing techniques for carrying this out including roughening of the surface of the conductive layer.

39.4.5 Charge-Generation Layer (CGL)

The CGL is where the imaging light is absorbed and charges are generated. The first major consideration in choice of material is that it strongly absorb the wavelengths of light emitted by the exposure system. In copiers light is reflected from a document onto the charged photoreceptor. Since an original may have colored components the light source must cover the visible range (400–700 nm) and the photoreceptor must have broad sensitivity to visible light. Common light sources in copiers are fluorescent or xenon lamps. Both sources are generally filtered to remove ultraviolet and infrared light. In digital printers the exposing source may be an LED array (680–780 nm) or a scanned laser (633 nm from He–Ne or 740, 780, and 820 nm from laser diodes).

Most modern commercial photoreceptors utilize pigments as charge-generation materials. Pigment-based CGLs comprise submicron crystals of the charge-generation material suspended in a polymer binder. Pigments are milled to the desired size and stabilized with appropriate additives and binders to produce a coating dispersion. Charge-generation materials are generally classified according to their chemical identity. Materials which absorb primarily in the visible region are dye–polymer aggregate, PVK–TNF charge-transfer

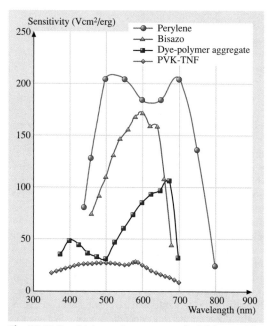

Fig. 39.13 Sensitivity action spectra of visible sensitive OPCs. The chemical structures of the CGL materials are given in Fig. 39.15

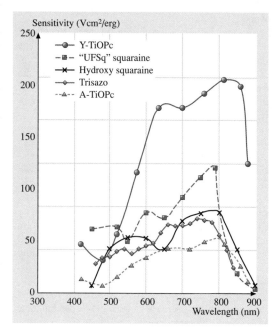

Fig. 39.14 Sensitivity action spectra of near-infrared sensitive OPCs. The chemical structures of the CGL materials are given in Fig. 39.16

complex, perylenes, and bisazos. Phthalocyanines, trisazo, and squarylium pigments are utilized for near-infrared absorption. Figures 39.13 and 39.14 show the sensitivity action spectra and Figs. 39.15 and 39.16 the chemical structures of these charge-generation materials. The sensitivity action spectra in these figures are taken from literature data and the sensitivities converted into consistent units ($V/cm^2/erg$) based on information given in the cited publication. Because the data are from OPCs with different thickness, exposure intensity, exposure time, initial potential, and exposed potential, the sensitivities are not directly comparable. However, the wavelength ranges are accurate. When the exposure is through the CTL (in a negative-charging dual-layer OPC) the photosensitivity will be decreased at wavelengths below ≈ 400 nm because of light absorption by the transport material.

Perylenes, azos, and phthalocyanines are best known for their use as colorants and it is because these materials were readily available that they were investigated as charge-generation materials. A complication with pigments is that some exhibit polymorphism. Over many decades the colorant industry has developed technologies for controlling polymorphism to obtain pigments with specific colorant characteristics and similar technologies have been recently developed to obtain

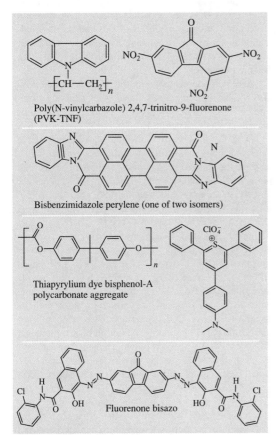

Fig. 39.15 Chemical structures of visible sensitive charge generation materials

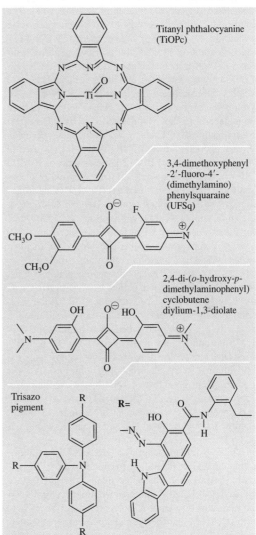

Fig. 39.16 Chemical structures of near-infrared sensitive charge generation materials

polymorphs with the appropriate absorption as well as charge-generation characteristics for electrophotographic applications. Thus, in many cases to prepare a CGL it is first necessary to convert the pigment-generation material into a dispersion of submicron particles of the desired polymorph. Examples of charge-generation materials are discussed below.

The donor–acceptor charge-transfer (CT) complex between poly(N-vinyl carbazole) (PVK) and 2,4,7-trinitrofluoren-9-one (TNF) was the first commercialized organic photoreceptor [39.110, 111]. The photoconductivity characteristics have been extensively studied [39.79–81, 84–86] and reviewed [39.112–115]. With the addition of TNF the carrier-generation efficiency of PVK is increased and extended into the region of CT complex absorption. Based on field quenching of exciplex fluorescence a mechanism for PVK–acceptor systems was proposed involving carrier generation from a nonrelaxed exciplex which is accessed by either direct absorption into the CT state or via an encounter complex between excited PVK and TNF [39.84–87]. As used in the IBM copiers I and II a 1 : 1 molar ratio of PVK to TNF was coated as a single 20-μm layer and electron-dominated transport was used to create the latent image (negative surface charge) with a He–Ne laser exposure (632.8 nm) [39.111]. An exposure of $5.5\,\mu J/cm^2$ was used to decrease the surface potential from -750 to -200 V. The quantum efficiency based on incident photons was ≈ 0.2 while the maximum efficiency was ≈ 0.3 at ≈ 400 nm. The photosensitivity action spectrum for this OPC is shown in Fig. 39.11 [39.110]. Using the photoinduced discharge technique [39.82] the maximum quantum efficiency (charges transported through the device per photon absorbed) was found to be 0.14 at 375 nm and a field of 10^6 V/cm. The PVK–TNF-based OPC had several drawbacks: low photosensitivity, short lifetime in the electrophotographic process, and concerns about the safety of TNF. It was eventually replaced with dual-layer OPCs with CGL materials based on pigments. Acceptor-doped PVK continues to receive attention particularly with C_{60} and related molecules as the acceptor [39.116].

The dye–polymer-aggregate based CGL is unique in that the active material is formed in situ during the coating process. A solution of thiapyrylium dye (2,6-diphenyl-4(4-dimethylaminophenyl)thiapyrylium perchlorate) and bisphenol-A polycarbonate is coated to form a CGL in which the dye and polymer spontaneously aggregate [39.117–119]. The aggregate is a dense filamentary structure which has an absorption spectrum red-shifted from that of the dye. The aggregate filaments transport electrons with high field trapping due to carrier immobilization at filament dead ends. The photosensitivity of the aggregate is several orders of magnitude larger than that of the nonaggregated dye [39.120]. To prepare a CGL the formulation also includes hole-transporting materials. Because the CGL transports both holes and electrons it is possible to use the aggregate as a single layer or as the CGL in a dual-layer photoreceptor. Because the aggregate absorption drops rapidly at wavelengths longer than 700 nm this material is only useful for visible-light exposure systems, Fig. 39.13. In this example the CGL (2 μm) comprised dye and tri-p-tolylamine (TTA) hole-transport material in polycarbonate, and the CTL (13 μm) was a solid solution of 40 wt % tri-p-tolylamine in a polyester binder [39.88]. This was charged to -600 V and exposed with very low-intensity continuous irradiation to obtain the initial rate of photodischarge. At 680 nm the photogeneration efficiency is 0.34 per absorbed photon. The quantum efficiency (Onsager model) was 0.6 and the thermalization distance was dependent on the concentration of hole-transport material ranging from 200 nm (5 wt %) to 600 nm (40 wt %) [39.88], Fig. 39.11.

Azo compounds contain the $-N=N-$ functionality. Bisazo compounds have the general structure $R'-N=N-R-N=N-R''$. Trisazo compounds have the general structure $R(-N=N-R')(-N=N-R'')(-N=N-R''')$. The R substituents generally have extended conjugation to provide the colors which have made bisazo compounds important dyes and pigments for many years. Bisazo CGLs generally absorb in the visible and trisazo into the near-infrared. An early example of a bisazo pigment-based OPC (chlorodiane blue, 4,4′-bis(1″-azo-2″-hydroxy-3″-naphthanilide)-3,3′-dichlorobiphenyl) was used in the IBM 3800 printer. This photoreceptor had over four times the photosensitivity of the PVK–TNF-based OPC [39.121]. Azo charge-generation materials are often synthesized from the reaction of an aromatic amine with an o-hydroxyaromatic carboxylic acid such as naphthol-AS. As CGL materials these materials are used as dispersed pigments. Although bisazo pigments do not generally exhibit polymorphism they can exist as hydroxy-azo or keto-hydrazone tautomers. In some cases, it has been found that the active form is the keto-hydrazone tautomer [39.122]. Because the synthesis of azo compounds is relatively easy there is a considerable literature on the effects of chemical structure on electrophotographic characteristics and carrier-generation efficiency [39.123–127]. In azo pigments charge is generated extrinsically with the pigment exciton oxidizing a hole-transport molecule at the pigment surface [39.127, 128] and it was demonstrated that penetration of the CTL into the CGL occurs during the coating process [39.129]. The photosensitivity action spectrum for a fluorenone bisazo pigment is shown in Fig. 39.13 [39.128]. In this OPC the CGL (0.17 μm) consisted of the pigment in a binder of poly(vinyl butyral) (10 : 4 weight ratio) and the CTL (17.4 μm) a solid solution of hydrazone transport material, p-diethylaminobenzaldehyde diphenylhydrazone (DEH), in polycarbonate (9 : 10 weight ratio). The initial potential was -800 V and the exposure required for a discharge to -400 V determined. The quantum yield of carrier generation from the Onsager model was 0.52 with a thermalization length of 1400 nm. Time-resolved transient absorption spectroscopy has been used to study the formation and decay of the oxidized transport material [39.130]. With a bisazo-based CGL (chlorodiane blue) [39.131] and trisazo-based CGL it was found that the charge-generation efficiency depended on the oxidation potential of the CTL transport materials [39.125]. In the most favorable case the bisazo had a quantum efficiency of 0.25 and the trisazo about 0.5. This was interpreted in terms of the energy gap between the hole at the pigment surface and the transport material. In a study using a series of bisazo pigment CGLs and a common CTL it was found that neither electron donating nor withdrawing substituents favor high photosensitivity [39.126]. Photosensitivity was found to depend inversely on the crystallinity of the pigment, which was interpreted in terms of pigment surface area. The development of high-sensitivity azo-based CGLs has been reviewed by *Murayama* [39.132]. The trisazo pigment-based OPC has an absorption which extends into the near-infrared, making it suitable for laser exposures. Figure 39.14 shows the photosensitivity action spectrum of an OPC with a trisazo pigment CGL [39.133]. The CGL (0.33 μm) consisted of the pigment in poly(vinyl butyral) binder (10 : 4 weight ratio) and the CTL (18.5 μm) was a solid solution of DEH transport material in polycarbonate (9 : 10 weight ratio). The initial surface potential was -700 V and the exposure to discharge to -300 V used to determine the photosensitivity at each wavelength. A photoreceptor with CGL (0.12 μm) and CTL (19.5 μm) had a quantum efficiency for charge generation of 0.46 from 500–800 nm [39.133].

Perylenes used as CGL materials are diimides of perylene-3,4,9,10-tetracarboxylic acid. One example is the N,N'-bis(2-phenethyl)-perylene-3,4:9,10-bis(dicarboximide). In this case the CGL was prepared by vacuum deposition of an amorphous film (0.1 μm). Exposure to dichloromethane (as when overcoating with the charge transport layer) converts the film into a crystalline form with enhanced absorption at 620 nm [39.135]. The mechanism of carrier generation of this pigment in the presence of tri-*p*-tolylamine has been studied by studying the field dependence of quenching of perylene fluorescence [39.136]. The results were interpreted in terms of both intrinsic (directly from the perylene singlet state) and sensitized (interaction of holes on the pigment surface with the amine) charge generation. More recently, a bisbenzimidazole perylene has been developed as a CGL [39.137–139]. Dual-layer OPCs with a CGL of 0.1-μm vacuum-deposited bisbenzimidazole perylene, and a 15-μm CTL of 35 wt % TPD in polycarbonate, exhibit high sensitivity ($E_{1/2}$ 2 erg/cm^2 for -800 V$_0$) out to 700 nm, Fig. 39.13. Recent research has involved the synthesis and study of novel materials such as dimeric perylene pigments [39.140].

Phthalocyanines have received considerable attention as CGL materials for near-infrared exposure systems. Metal-free (H$_2$Pc) [39.141–144], titanyl (TiOPc) [39.134], and hydroxygallium (HOGaPc) [39.145] phthalocyanines have been extensively stud-

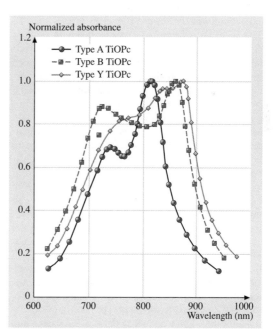

Fig. 39.17 Absorption spectra of polymorphs of TiOPc. After [39.134]

ied as CGL pigments. Although these materials are relatively easy to synthesize their intractability makes them difficult to purify. Except for halogen substituents on the aromatic rings there has been little research into the effect of chemical structure on performance. Instead, much of the development has been to understand and control polymorphism. Compared with solution absorption spectra, the spectra of many CGL pigments are broadened with enhanced absorption in the red and near-infrared. With specific combinations of milling technique, solvent exposure, and thermal treatment many phthalocyanine polymorphs can be produced [39.146–148]. Polymorphs have been characterized by visible and near-infrared absorption spectroscopy, X-ray diffraction, nuclear magnetic resonance (NMR) spectroscopy, infrared spectroscopy, and infra-red (IR) absorption spectroscopy. Examples of TiOPc polymorph absorption spectra are shown in Fig. 39.17 [39.134]. Polymorphs may have very different photogeneration characteristics as CGLs in OPCs. The materials shown in Fig. 39.17 were fabricated into OPCs: CGL, 0.4 μm, pigment and poly(vinyl butyral) 1 : 1 weight ratio; CTL, 20 μm, 43 wt% triarylamine hole-transport material, 4-(4-methylstyryl)-4′,4″-dimethoxytriphenylamine, in polycarbonate. The photosensitivity action spectra for OPCs from the Y- and A-forms of TiOPc are shown in Fig. 39.14. The OPC with the Y-form had a photosensitivity for half photodischarge (600–850 nm) of $0.75\,\text{erg/cm}^2$ (initial potential $-600\,\text{V}$) while the A-form was much less sensitive. In another study, photoreceptors formulated with α-, β-, τ-, and X-forms of H_2Pc had generation efficiencies as follows ($5 \times 10^5\,\text{V/cm}$) $\tau_2 \approx 0.1$, $X_2 \approx 0.09$, $X_1 \approx 0.03$, $\beta_1 \approx 0.01$, $\alpha_2 \approx 0.005$, and $\alpha_1 \approx 0.001$ [39.149]. In the preparation of a generation material a common procedure is to prepare the pigment in the amorphous form by acid pasting or with a specific milling technique, and then to convert the pigment to the desired polymorph with a specific solvent treatment. The desired polymorph must of course be stable to all the subsequent operations. In addition to the dominant influence of crystal form on electrophotographic performance, recent studies have implicated water in carrier generation of Y-form TiOPc [39.134, 150]. A recent paper has reviewed the carrier generation mechanisms in several phthalocyanines [39.151]. In all the pigments studied [$x-H_2Pc$, TiOPc (I) (low sensitivity), TiOPc (IV) (also known as Y, high sensitivity), HOGaPc (V)], fluorescence quenching implicates a neutral excited state (not a charge-transfer state) as the precursor to carrier generation. Studies of the effect of field on fluorescence amplitude and lifetime reveal that with HOGaPc and TiOPc (I and IV) carrier generation comes from a non-relaxed excited state and TiOPc (IV) has an additional generation channel from a trapped state. With $x-H_2Pc$ trapped excitons appear to be the major source of charge carriers.

Squarylium (squaraine) pigments are a class of materials which were first synthesized only recently. Squaryliums [2,4-di-(o-hydroxy-p-dimethylaminophenyl)cyclobutene and 1,4-bis-(4-dimethylamino-2-methylphenyl diylium-1,3-diolate)] were the first pigments to be utilized as a near-infrared-sensitive CGL in an IBM printer [39.131, 152, 153]. The photoreceptor based on the hydroxy squarylium with a CTL (21 μm, DEH in bisphenol-A polycarbonate) had high photosensitivity from 500–800 nm ($\approx 3\,\text{erg/cm}^2$ for a -700 to $-200\,\text{V}$ photodischarge), Fig. 39.14 [39.153]. This class of pigment has been extensively investigated for use as pigment-dispersion CGLs for near-infrared exposure systems [39.137, 154–160]. A nonsymmetrical squaraine (UFSq)-based OPC with CGL, 0.4-μm UFSq in poly(vinyl formal) (80 : 20 weight ratio), and CTL 26-μm 40 wt% TPD in bisphenol-A polycarbonate, exhibited an $E_{1/2}$ photosensitivity of $1.9\,\text{erg/cm}^2$ at 790 nm ($-980\,V_0$) when optimally purified [39.160], Fig. 39.14.

39.4.6 Charge-Transport Layer (CTL)

In all current commercial organic photoreceptors holes are the carriers that dominate photodischarge of the device. Hole transport is carried out by aromatic amines which are either doped into, or are incorporated as part of, the binder polymer. The key molecular characteristic is that aromatic amines have relatively low oxidation potentials. The oxidation potential of the transport moiety must be lower than that of the charge-generation material such that photogenerated holes are injected from the generation material into the charge-transport layer. In a chemical sense the key reaction is oxidation of the charge-transport material by the photogenerated hole on the surface of the generation material. If the oxidation potential of the transport material is too low it will undergo air oxidation, if it is too high its oxidation will require an activation energy and will be inefficient. Molecular orbital studies suggest that molecules with little change in geometry with oxidation will be favorable for hole transport [39.161]. Aromatic amines with oxidation potentials between about 0.8–1.2 eV (relative to SCE) are generally useful as hole-transport materials. The synthesis and study of hole-transport molecules is a very active area. Figure 39.4 shows the chemical structures of some common hole-transport materials: 1,1-bis(*p*-diethylaminophenyl)-4,4-diphenyl-1,3-butadiene (DPB), *p*-diethylaminobenzaldehyde diphenylhydrazone (DEH), tri-*p*-tolylamine (TTA), N,N′-diphenyl-N,N′-bis(3-methylphenyl)-(1,1′-biphenyl)-4,4′-diamine (TPD), and 1,1-bis(di-4-tolylaminophenyl)cyclohexane (TAPC). These are often referred to by acronym and classified according to one of their structural functionalities: arylmethane, diarylamine, enamine, triarylamine, heterocyclic, butadiene, and hydrazone. A more complete listing is available in *Borsenberger* and *Weiss* [39.4].

Charge transport was discussed previously (Sect. 39.2.2). Since transport involves orbital overlap between neighboring charge-transport moieties their concentration needs to be as high as possible. In practice, molecular transport materials are doped, singly or as a mixture, at 40–50% by weight into the binder polymer. The purpose of the binder polymer is to provide an inert, robust, substrate for the transport material. The mobility characteristics of a large number of hole-transport materials have been studied and the results interpreted according to a model where the dissolved transport molecules are disordered in energy and position [39.162–164]. Binder effects on hole transport have been observed. It appears that the presence in the CTL of materials with highly polar substituents decrease hole mobility but in general these effects are not well understood [39.165–170].

Since the CTL is typically the surface that is exposed to the various electrophotographic process elements it must be formulated to resist damage and wear. The glass-transition temperature (T_g) of the CTL is an important characteristic. Too low and the material will be soft and easily damaged, too high and it will be brittle. Generally a T_g around 60 °C is adequate. Polycarbonates and polyesters are two classes of materials commonly used as CTL binder polymers.

Dual-layer devices in which electron transport dominates have been studied extensively but never commercialized. Examples of commercialized electron-transport-dominated photoreceptors are the single-layer dye–polymer aggregate and the PVK–TNF charge-transfer complex. Electron-transport materials are designed to be easily reduced. The reduction potential must be lower than that of the photogenerated electron on the charge-generation material so that it is transferred to the transport material. However, it is believed that oxygen may act as an electron trap so the reduction potential of the transport material must be lower than that required to reduce oxygen dissolved in the polymer film. Figure 39.10 shows the chemical structures of some electron-transport materials. As with hole-transport materials acronyms based on molecular functionality are common: 1,1-dioxo-2-(4-methylphenyl)-6-phenyl-4-(dicyanomethylidene)thiopyran (PTS), *n*-butyl 9-dicyanomethylenefluorene-4-carboxylate (BCMF), 2,4,7-trinitro-9-fluorenone (TNF), 3,3′-dimethyl-5,5′-di-*t*-butyl-4,4′-diphenoquinone and 3,5′-dimethyl-3′,5-di-*t*-butyl-4,4′-diphenoquinone isomer mixture (DPQ); a more complete table can be found in *Borsenberger* and *Weiss* [39.4]. In general the charge mobilities observed with electron-transport materials are much lower than for hole-transport materials. For this reason dual-layer OPCs with electron-transport-based CTLs have not been commercialized. Analysis of the mobility characteristics of electron transport with the disorder model implicates the highly polar nature of all electron-transport molecules [39.171–174].

Both hole- and electron-transport materials have been prepared with the transport moieties as part of a polymer (main chain or pendant) which would serve as a binder with hole-transport functionality. The advantage of a transport-active polymer is that it is a single material. In practice, charge transport is generally not enhanced in these materials and their chemical complexity makes them expensive to synthesize. Other than PVK, molecularly doped polymers are the current choice for CTLs.

39.4.7 Backing Layer

In a web photoreceptor the substrate–photoreceptor may tend to curl due to the residual stress of the coated layers. Such curl can make it difficult to fabricate the web into loops, may degrade loop tracking in the machine, or cause the loop edge to touch other process elements. Any of these attributes would make the photoreceptor unusable. One way to counter film curl is to coat a polymer layer on the back side to compensate for the stresses in the photoreceptor layers. The back-coat polymer must have adequate physical characteristics (friction, wear, etc.) to withstand cycling in the process. The extra coating adds expense but its addition is relatively commonplace. Any suitable polymer, such as a polycarbonate, may be used.

39.4.8 Overcoat Layer

Overcoat layers have received increased attention to lengthen the photoreceptor life by preventing wear, minimizing the effects of surface contacts, the effects of corona effluents (ozone and nitrogen oxides), and liquid attack in liquid-toner-based printers. As with the charge-blocking layer, the simplest approach is to coat a thin insulating polymer onto the photoreceptor surface. The thickness cannot be more than a few µm if one wants to avoid large residual potentials. Several approaches have been developed. One approach is to coat a second CTL with a changed composition to maximize polymer toughness (often by lowering the concentration of transport material). Other approaches are to overcoat the photoreceptor with a silsesquioxane sol-gel several µm thick, or with a submicron refractory layer of diamond-like carbon or silicon nitride.

Silsesquioxane overcoats applied in a sol–gel process have received recent attention [39.175, 176]. One commercial application utilized a sol–gel to overcoat a photoconductor with an uppermost CGL (positive-charging OPC). Because the CGL is sensitive to wear and corona chemicals an overcoat was essential to its implementation in a commercial photoreceptor. The sol is prepared and subsequently hopper-coated and cured at elevated temperature to effect crosslinking. To avoid residual potential some conductivity is necessary but this cannot be so high that the latent image is degraded [39.177].

A polyurethane overcoat in a positive-charging dual-layer OPC has recently been described for use in a liquid-toner-based printer [39.178]. The CGL was based on TiOPc pigment and, to avoid disruption of the CGL during the overcoat coating operation, an aqueous polyurethane dispersion was used.

Diamond-like carbon, silicon nitride, and other [39.179] overcoats are applied in a vacuum process as submicron layers. Although tough, their thinness limits their utility for scratch and wear resistance. Furthermore, depending on their chemical make up they may have sensitivity to corona gasses.

39.5 Photoreceptor Fabrication

As discussed above the two most common photoreceptor configurations are as a loop or a metal drum. After coating the layers of a drum photoreceptor it is ready to use in the process. Generally the drums are inspected for visible defects and samples are taken for process testing and quality control. Since the substrate is conductive the ground contact is easily made by means of a conductive brush contacting the edge or inside of the drum when it is mounted in the machine. On the other hand, after the appropriate layers are coated on a web several more steps are required for fabrication of the photoreceptor. The web is rolled onto a spool at the end of the coating process and subsequently slit to the desired width and rewound. The coated roll is then inspected for visible defects and defective areas are marked for removal. Defective areas are removed when the photoreceptor is cut to the desired length. The cut sheets are then ultrasonically welded to produce a photoconductor loop. Because the conductive layer is buried between the substrate and the coated layers, making a ground contact is not trivial. It is possible to leave one edge of the ground layer uncoated but, because the thin metal would be readily damaged or worn away with use, one typically finds a conductive stripe coated on one edge of the photoreceptor. The only requirement is that the conductive stripe penetrate the coated layers and contact the metal. The ground contact in the machine is often made with a conductive brush contacting the conductive stripe. The printing machine must keep track of the splice for several reasons: cleaning blade, degraded photoconductivity due to the ultrasonic welding, etc. This is sometimes accomplished by applying marks to the back of the photoreceptor, which are detected by the machine.

Fabricating a photoreceptor loop entails several operations not required in the fabrication of drum photoreceptors (slitting, cutting, conductive striping, welding, backside marking, etc.). Web-coating a photoreceptor is an expensive process but many thousands of feet of photoreceptor can be produced in a single coating event. Furthermore all the loop-processing elements are relatively straightforward and inexpensive. The key factor is that web-coating is a continuous process while drum-coating is a batch process. Photoreceptor loops are generally about 10× less costly to produce than drums.

There are advantages and disadvantages to both seamed loop and drum configurations in a printing machine. Loops can be made small for desktop printers or large for high-speed commercial printers. A large web (current commercial webs are as large as $\approx$ 10 feet in circumference) provides more surface area for imaging (several images per loop cycle) and more space for the various electrophotographic process elements. In addition, exposures through a semitransparent conductive layer are possible. Disadvantages are the presence of a seam which must be tracked and makes some processes, such as cleaning with a blade, problematic. Also, the tendency of a web to flutter or move necessitates both tracking and positioning to achieve adequate print registration. Metal drums are seamless and rigid, making it easier to control drum position. However, drums typically have a diameter of 35–300 mm and have limited imaging area and space for process elements. All things being equal a loop which produces several images per cycle will produce more images over its lifetime than a drum. Ultimately, the choice of loop or drum is a printing system issue.

39.6 Summary

Organic photoreceptors have been utilized in electrophotography since 1970. In this chapter we have briefly reviewed the development of organic photoreceptors from the standpoint of the device architecture, materials formulation, and electrical characteristics with emphasis on how the materials used impact on specific device functions.

The use of organic materials has enabled the development of electrophotographic technologies into a multi-billion-dollar business. Today's organic photoreceptor is a large-area photoelectric device capable of very high spatial resolution in the formation of a large-area image, with pixel resolution as small as 20 μm (1200 dpi), to meet demanding image-quality requirements. It is truly an amazing success story. Many of the principles and indeed many of the same materials are being applied to today's highly publicized new technologies including organic light-emitting diodes (OLED) and organic transistors. In this chapter we have described the device requirements and shown how organic photoreceptor development has progressed through a combination of architecture manipulation and materials design.

Organic photoreceptors have to satisfy many diverse requirements to be successfully employed in electrophotography as it is practiced today. These requirements have been met in large part by the principles of: (1) functional separation, and (2) materials design for the enhancement of specific electrical and mechanical characteristics. The concept of functional separation was key to organic photoreceptor development. Rather than try to produce a device with all the desirable characteristics in a single layer it is more productive to have a multi-layer architecture with each layer being optimized for a specific characteristic. Not surprisingly these layers are commonly named after those characteristics which are paramount: the charge-generation layer (CGL), the charge-transport layer (CTL), the charge-blocking layer, the protective overcoat layer, etc. In this chapter we have described the characteristics of each of these layers and demonstrated how materials have been chosen and new materials synthesized to optimize their characteristics.

It is worthwhile to take a minute to review the electrophotographic process and consider the characteristics which have been built into the device.

- Spatial uniformity. The thickness of each of the coated layers (each varies from submicron to a few tens of μm) and the total thickness (controlled mainly by the CTL) must be very uniform so that the device capacitance (a major factor in determining surface charging characteristics) is uniform (on the order of a few percent variability). The CGL generation characteristics must be uniform, which means that the optical density of the CGL at the exposure wavelength must be uniform, to produce a uniform surface potential after exposure (on the order of a few percent variability). Thickness must be uniform for the smallest pixel (20 μm for 1200-dpi imaging) over the entire area of the device ($> 1000\,\mathrm{cm}^2$). The ability to coat large areas of such thin layers to this degree of uniformity is a great technological achievement.
- Chemical resistance. Photoreceptors must resist chemical attack. Because corona charging is the most common method of applying a surface potential, the photoreceptor must be stable to the acid and oxidizing species produced in the corona. These chemicals (ozone, NO_x, and HNO_x) are highly reactive and the layer that is exposed (usually the CTL) must not be degraded. This is achieved by judicious choice of materials, the addition of stabilizers (antioxidants) to the CTL, or a protective overcoat.
- Electrical uniformity and stability. The photoreceptor dark electrical and photoelectrical characteristics (surface potentials with no exposure, imaging exposure, and erase exposure) must be very uniform and stable to prolonged electrophotographic cycling. The initial (dark) and exposed potentials must be the same for each pixel over the entire device area and

must remain stable with cycling. If these characteristics drift uniformly with cycling, machine process control must be implemented to maintain image quality. If these characteristics drift nonuniformly the photoreceptor must be discarded when image artifacts (ghosting) appear.

- Photosensitivity. The sensitivity of the photoreceptor to light, often expressed as the energy needed to produce a defined decrease in surface potential, must be mated with the exposure used in the process in terms of exposure energy and wavelength. This has been achieved by the synthesis and pigment form manipulation of charge-generation materials. Today's high-sensitivity photoreceptors can approach a quantum efficiency of unity for charge generation (at a typical applied field of ≈ 20 V/cm). But, most importantly photoreceptors with lower sensitivity can be fabricated depending on the needs of the imaging system.
- Photodischarge rate (switching time). The time it takes for photogenerated carriers to transit the device limits the printing process speed. With the synthesis of highly stable hole-transport materials and the principle of molecularly doped polymers it has been possible to prepare devices with hole transit times that permit printing speeds approaching 200 pages/min.
- Mechanical strength. The surface of the photoreceptor that is exposed to the abrasive process elements (often the development and cleaning systems), must not wear too rapidly or unevenly and must be stable to crazing or other forms of damage. Typically the CTL is the surface layer, and considering that it is typically a polymer with as much as 50% by weight of charge-transport molecules, it is remarkable that materials can be chosen such that significant wear occurs only after tens of thousands of cycles. Where warranted, very tough protective overcoats have been developed which have little effect on the overall OPC device performance.

Today multilayer organic photoreceptors that meet all of the requirements listed above are made commercially on a huge scale. This success is due in large measure to the skill of the scientists and engineers who design the materials and processes for photoreceptor fabrication, and who then integrate it into the electrophotographic printing system such that acceptable and consistent image quality is obtained for a photoreceptor life, which is routinely 200 000 imaging cycles or longer if protective overcoat layers are used. This is an amazing story, an organic photoreceptor device with *adjustable* sensitivity at any desired wavelength from the visible to the near-infrared, with stable dark and photoelectrical characteristics over many thousands of imaging cycles, such that the characteristics of a 40-μm pixel area are identical over an area which might be greater than 1000 cm^2.

References

39.1 J. Mort, D. M. Pai: *Photoconductivity and Related Phenomena* (Elsevier, New York 1976)

39.2 N. V. Joshi: *Photoconductivity Art, Science, and Technology* (Marcel Dekker, New York 1990)

39.3 K.-Y. Law: Chem. Rev. **93**, 449 (1993)

39.4 P. M. Borsenberger, D. S. Weiss: *Organic Photoreceptors for Xerography* (Marcel Dekker, New York 1998)

39.5 P. M. Borsenberger, D. S. Weiss: Photoreceptors: Organic Photoreceptors. In: *Handbook of Imaging Materials*, ed. by A. S. Diamond, D. S. Weiss (Marcel Dekker, New York 2002) p. 369

39.6 D. S. Weiss, J. R. Cowdery, R. H. Young: Electrophotography. In: *Molecular-Level Electronics, Imaging and Information, Energy and Environment*, 2, Vol. 5, ed. by V. Balzani (Wiley-VCH, Weinheim 2001) Chap. 2

39.7 R. M. Schaffert, C. D. Oughton: J. Opt. Soc. Am. **38**, 991 (1948)

39.8 J. H. Dessauer, G. R. Mott, H. Bogdonoff: Photogr. Eng. **6**, 250 (1955)

39.9 J. Mort, I. Chen: Appl. Solid State Sci. **5**, 69 (1975)

39.10 J. Mort: *The Anatomy of Xerography* (McFarland, Jefferson 1989)

39.11 C. F. Carlson: Electrophotography, US Patent 2 297 691 (1942)

39.12 I. Chen: J. Imaging Sci. **34**, 15 (1990)

39.13 I. Chen: Nature of Latent Images Formed on Single Layer Organic Photoreceptors. In: *Proc. IS&T NIP18: 2002 Int. Conf. Digital Printing Technol.* (Society Imaging Science and Technology, Springfield 2002) p. 404

39.14 S. Jeyadev, D. M. Pai: J. Imaging Sci. Technol. **40**, 327 (1996)

39.15 D. M. Pai, B. E. Springett: Rev. Mod. Phys. **65**, 163 (1993)

39.16 L. B. Schein: *Electrophotography and Development Physics* (Laplacian, Morgan Hill 1996)

39.17 R. M. Schaffert: *Electrophotography* (Focal, New York 1975)

39.18 M. E. Scharfe: *Electrophotography Principles and Optimization* (Wiley, New York 1984)

39.19 E. M. Williams: *The Physics and Technology of Xerographic Processes* (Wiley-Interscience, New York 1984)

39.20 R. G. Enck, G. Pfister: Amorphous Chalcogenides. In: *Photoconductivity and Related Phenomena*, ed. by J. Mort, D. M. Pai (Elsevier, New York 1976) p. 215

39.21 S. O. Kasap: Photoreceptors: The Chalcogenides. In: *Handbook of Imaging Materials*, ed. by A. S. Diamond, D. S. Weiss (Marcel Dekker, New York 2002) p. 329

39.22 J. Mort: Applications of Amorphous Silicon and Related Materials in Electronic Imaging. In: *Handbook of Imaging Materials*, ed. by A. S. Diamond, D. S. Weiss (Marcel Dekker, New York 2002) p. 629

39.23 R. Joslyn: Photoreceptors: Recent Imaging Applications for Amorphous Silicon. In: *Handbook of Imaging Materials*, ed. by A. S. Diamond, D. S. Weiss (Marcel Dekker, New York 2002) p. 425

39.24 A. Rose: *Photoconductivity and Related Processes* (Interscience, New York 1963)

39.25 J. C. Scott, G. S. Lo: Dark Decay in Organic Photoconductors. In: *Proc. 6th Int. Symp. Adv. Non-Impact Printing Technol.*, ed. by R. J. Nash (Society Imaging Science and Technology, Springfield 1991) p. 403

39.26 R. H. Bube: *Photoelectronic Properties of Semiconductors* (Cambridge Univ. Press, Cambridge 1992)

39.27 S. M. Sze: *Physics of Semiconductor Devices*, 2 edn. (Wiley, New York 1981)

39.28 M. A. Abkowitz, M. Stolka: Electronic Transport in Polymeric Photoreceptors: PVK in Polysilylenes. In: *Proc. Int. Symp. Polym. Adv. Technol.*, ed. by M. Lewin (VCH, Weinheim 1988) p. 225

39.29 D. M. Pai, J. F. Yanus, M. Stolka: J. Phys. Chem. **88**, 4714 (1984)

39.30 M. Stolka, J. F. Yanus, D. M. Pai: J. Phys. Chem. **88**, 4707 (1984)

39.31 M. Stolka, M. A. Abkowitz: Mater. Res. Soc. Symp. **277**, 3 (1992)

39.32 F. K. Dolezalek: Experimental Techniques. In: *Photoconductivity and Related Phenomena*, ed. by J. Mort, D. M. Pai (Elsevier, New York 1976) p. 27

39.33 A. R. Melnyk, D. M. Pai: Photoconductivity Measurements. In: *Physical Methods in Chemistry*, ed. by B. W. Rossiter, J. F. Hamilton, R. C. Baetzold (Wiley, New York 1992) p. 321

39.34 I. Chen: J. Appl. Phys. **43**, 1137 (1972)

39.35 I. Chen, J. Mort: J. Appl. Phys. **43**, 1164 (1972)

39.36 M. Abkowitz: J. Reinforced Plastics Composites **16**, 1303 (1997)

39.37 M. A. Abkowitz, M. Stolka, M. Morgan: J. Appl. Phys. **52**, 3453 (1981)

39.38 H. Bässler: Phys. Status Solidi (b) **107**, 9 (1984)

39.39 M. Stolka, M. A. Abkowitz: TBD a Contribution to the Success of Organic Materials. In: *Practical Applications of Organic Electronic Materials*, ed. by M. Iwamoto, I. S. Pu, S. Taniguchi (Science Forum, Tokyo 1994) p. 265

39.40 C. W. Tang, S. A. VanSlyke: Appl. Phys. Lett. **51**, 913 (1987)

39.41 M. A. Abkowitz, H. Bässler, M. Stolka: Philos. Mag. B **63**, 201 (1991)

39.42 W. D. Gill: J. Appl. Phys. **43**, 5033 (1972)

39.43 M. A. Abkowitz, J. S. Facci, W. W. Limburg, J. Janus: Phys. Rev. B: Cond. Matter **46**, 6705 (1992)

39.44 R. G. Kepler, J. M. Zeigler, L. A. Harrah, S. R. Kurtz: Phys. Rev. B **35**, 2818 (1987)

39.45 M. A. Abkowitz, M. Stolka: Synth. Met. **50**, 395 (1992)

39.46 D. Hertel, H. Baessler, U. Scherf, H. H. Horhold: J. Chem. Phys. **110**, 9214 (1999)

39.47 C. Im, H. Baessler, H. Rost, H. H. Horhold: J. Chem. Phys. **113**, 3802 (2000)

39.48 J. S. Facci, M. A. Abkowitz, W. W. Limburg, D. Renfer, J. Yanus: Mol. Cryst. Liq. Cryst. **194**, 55 (1991)

39.49 J. Michl, J. W. Downing, T. Karatsu, A. J. McKinley, G. Poggi, G. M. Wallraff, R. Sooriyakumaran: Pure Appl. Chem. **60**, 959 (1988)

39.50 R. D. Miller, J. Michl: Chem. Rev. **89**, 1359 (1989)

39.51 M. A. Abkowitz, M. Stolka: Polym. Preprints (ACS Div. Polym. Chem.) **31**, 254 (1990)

39.52 R. D. Miller, J. R. Rabolt, R. Sooriyakumaran, G. N. Fickes, B. L. Farmer, H. Kuzmany: ACS Polym. Preprints **28**, 422 (1987)

39.53 M. A. Abkowitz, M. Stolka: Philos. Mag. Lett. **58**, 239 (1988)

39.54 D. Emin: Adv. Phys. **24**, 305 (1975)

39.55 L. B. Schein, A. Rosenberg, S. L. Rice: J. Appl. Phys. **60**, 4287 (1986)

39.56 L. B. Schein: Molec. Cryst. Liquid Cryst. **183**, 41 (1990)

39.57 H. Bässler: Phys. Status Solidi **175**, 15 (1993)

39.58 H. Baessler, G. Schoenherr, M. A. Abkowitz, D. M. Pai: Phys. Rev. B: Cond. Matter **26**, 3105 (1982)

39.59 P. E. Parris, V. M. Kenkre, D. H. Dunlap: Phys. Rev. Lett. **87**, 126601 (2001)

39.60 D. E. Dunlap, P. E. Parris, V. M. Kenkre: Proc. SPIE **3799**, 88 (1999)

39.61 L. B. Schein: Philos. Mag. B **65**, 795 (1992)

39.62 S. V. Novikov, A. V. Vannikov: J. Phys. Chem. **99**, 14573 (1995)

39.63 S. V. Novikov, D. H. Dunlap, V. M. Kenkre: Proc. SPIE **3471**, 181 (1998)

39.64 D. H. Dunlap: Phys. Rev. B **52**, 939 (1995)

39.65 D. H. Dunlap, P. E. Parris, V. M. Kenkre: Phys. Rev. Lett. **77**, 542 (1996)

39.66 S. V. Novikov, D. H. Dunlap, V. M. Kenkre, P. E. Parris, A. V. Vannikov: Phys. Rev. Lett. **81**, 4472 (1998)

39.67 M. A. Abkowitz: Philos. Mag. B **65**, 817 (1992)

39.68 M. Stolka, M. A. Abkowitz: Synth. Met. **54**, 417 (1993)

39.69 H. J. Yuh, D. Abramsohn, M. Stolka: Philos. Mag. Lett. **55**, 277 (1987)

39.70 F. W. Schmidlin: Phys. Rev. B **16**, 2362 (1977)

39.71 M. A. Abkowitz, R. C. Enck: J. Phys. Colloq. C **4**(1), 443 (1981)

39.72 M. A. Abkowitz: Characterization of Metal Interfaces to Molecular Media from Analysis of Transient and Steady State Electrical Measurements. In: *Conjugated Polymer and Molecular Interfaces*, ed. by W. R. Salaneck et al. (Marcel Dekker, New York 2002) p. 545

39.73 M. A. Abkowitz, J. S. Facci, M. Stolka: Chem. Phys. **177**, 783 (1993)

39.74 M. A. Abkowitz, J. S. Facci, M. Stolka: Appl. Phys. Lett. **63**, 1892 (1993)

39.75 M. A. Lampert, P. Mark: *Current Injection in Solids* (Academic, New York 1970) p. 17

39.76 P. N. Murgatroyd: J. Phys. D **3**, 151 (1970)

39.77 R. H. Young: Philos. Mag. Lett. **70**, 331 (1994)

39.78 M. Pope, C. E. Swenberg: *Electronic Processes in Organic Crystals* (Clarendon, Oxford 1982)

39.79 Y. Hayashi, M. Kuroda, A. Inami: Bull. Chem. Soc. Jpn. **39**, 1660 (1966)

39.80 H. Hoegl: J. Phys. Chem. **69**, 755 (1965)

39.81 P. J. Regensburger: Photochem. Photobiol. **8**, 429 (1968)

39.82 H. Hoegl, H. G. Barchietto, D. Tar: Photochem. Photobiol. **16**, 335 (1972)

39.83 P. J. Melz: J. Chem. Phys. **57**, 1694 (1972)

39.84 M. Yokoyama, Y. Endo, H. Mikawa: Bull. Chem. Soc. Jpn. **49**, 1538 (1976)

39.85 M. Yokoyama, Y. Endo, A. Matsubara, H. Mikawa: J. Chem. Phys. **75**, 3006 (1981)

39.86 M. Yokoyama, S. Shimokihara, A. Matsubara, H. Mikawa: J. Chem. Phys. **76**, 724 (1982)

39.87 M. Yokoyama, H. Mikawa: Photogr. Sci. Eng. **26**, 143 (1982)

39.88 M. B. O'Regan, P. M. Borsenberger, E. H. Magin, T. Zubil: J. Imaging Sci. Technol. **40**, 1 (1996)

39.89 Z. D. Popovic: Carrier Generation Mechanisms in Organic Photoreceptors. In: *Proc. 9th Int. Cong. Adv. Non-Impact Printing Technol.* (Society Imaging Science and Technology, Springfield 1993) p. 591

39.90 L. Onsager: Phys. Rev. **54**, 554 (1938)

39.91 P. M. Borsenberger, A. I. Ateya: J. Appl. Phys. **50**, 909 (1979)

39.92 L. Onsager: J. Chem. Phys. **2**, 599 (1934)

39.93 C. Braun: J. Chem. Phys. **80**, 4157 (1984)

39.94 S. N. Smirnov, C. Braun: J. Imaging Sci. Technol. **43**, 425 (1999)

39.95 D. M. Pai: Geminate Recombination in some Amorphous Materials. In: *Physics of Disordered Materials*, ed. by A. Dler, H. Fritzsche, S. R. Ovshinsky (Plenum, New York 1985) p. 579

39.96 J. Noolandi, K. M. Hong: J. Chem. Phys. **70**, 3230 (1979)

39.97 T. Niimi, M. Umeda: J. Appl. Phys. **74**, 465 (1993)

39.98 Z. D. Popovic: J. Chem. Phys. **76**, 2714 (1982)

39.99 Z. D. Popovic: Chem. Phys. **86**, 311 (1984)

39.100 R. C. Hughes: J. Chem. Phys. **58**, 2212 (1973)

39.101 J. W. Kerr, G. H. S. Rokos: J. Phys. D: Appl. Phys. **10**, 1151 (1977)

39.102 I. Chen: J. Appl. Phys. **49**, 1162 (1978)

39.103 W. Mey, E. I. P. Walker, D. C. Hoesterey: J. Appl. Phys. **50**, 8090 (1979)

39.104 R. H. Young: J. Appl. Phys. **60**, 272 (1986)

39.105 J.-Y. Moisan, B. André, R. Lever: Chem. Phys. **153**, 305 (1991)

39.106 A. V. Buettner, W. Mey: Photogr. Sci. Eng. **26**, 80 (1982)

39.107 K. Aizawa, M. Takeshima, H. Kawakami: A Study of 1-dot Latent Image Potential. In: *Proc. NIP17: Int. Conf. Digital Printing Technol.* (Society Imaging Science and Technology, Springfield 2001) p. 572

39.108 T. Iwamatsu, T. Toyoshima, A. Nobuyuki, Y. Mutou, Y. Nakajima: A Study of High Resolution Latent Image Forming and Development. In: *Proc. IS&T NIP 15: 1999 Internat. Conf. on Digital Printing Technol.* (Society Imaging Science and Technology, Springfield 1999) p. 732

39.109 I. Chen: J. Imaging Sci. Technol. **37**, 396 (1993)

39.110 R. M. Schaffert: IBM J. Res. Devel. **15**, 75 (1971)

39.111 U. Vahtra, R. F. Wolter: IBM J. Res. Devel. **22**, 34 (1978)

39.112 J. M. Pearson: Pure Appl. Chem. **49**, 463 (1977)

39.113 R. C. Penwell, B. N. Ganguly, T. W. Smith: J. Polym. Sci., Makromol. Rev. **13**, 63 (1978)

39.114 W. D. Gill: Polymeric Photoconductors. In: *Photoconductivity and Related Phenomena*, ed. by J. Mort, D. M. Pai (Elsevier, New York 1976) p. 303

39.115 M. Hatano, K. Tanikawa: Prog. Organic Coat. **6**, 65 (1978)

39.116 F. Li, Y. Li, Z. Guo, Y. Mo, L. Fan, F. Bai, D. Zhu: Solid State Commun. **107**, 189 (1998)

39.117 W. J. Dulmage, W. A. Light, S. J. Marino, C. D. Salzberg, D. L. Smith, W. J. Staudenmayer: J. Appl. Phys. **49**, 5543 (1978)

39.118 P. M. Borsenberger, A. Chowdry, D. C. Hoesterey, W. Mey: J. Appl. Phys. **49**, 5555 (1978)

39.119 J. M. Perlstein: Structure and Charge Generation in Low-Dimensional Organic Molecular Self-Assemblies. In: *Electrical Properties of Polymers*, ed. by D. A. Seanor (Academic, New York 1982) p. 59

39.120 P. M. Borsenberger, D. C. Hoesterey: J. Appl. Phys. **51**, 4248 (1980)

39.121 D. McMurtry, M. Tinghitella, R. Svendsen: IBM J. Res. Devel. **28**, 257 (1984)

39.122 J. Pacansky, R. J. Waltman: J. Am. Chem. Soc. **114**, 5813 (1992)

39.123 G. DiPaola-Baranyi, C. K. Hsiao, A. M. Hor: J. Imaging Sci. **34**, 224 (1990)

39.124 K.-Y. Law, I. W. Tarnawskyj: J. Imaging Sci. Technol. **37**, 22 (1993)

39.125 M. Umeda, T. Shimada, T. Aruga, T. Niimi, M. Sasaki: J. Phys. Chem. **97**, 8531 (1993)

39.126 K.-Y. Law, I. W. Tarnawskyj, Z. D. Popovic: J. Imaging Sci. Technol. **38**, 118 (1994)

39.127 M. Umeda: J. Imaging Sci. Technol. **43**, 254 (1999)

39.128 M. Umeda, T. Niimi, M. Hashimoto: Jpn. J. Appl. Phys. **29**, 2746 (1990)

39.129 T. Niimi, U. Umeda: J. Appl. Phys. **76**, 1269 (1994)

39.130 K. Takeshita, Y. Sasaki, T. Shoda, T. Murayama: Time-Resolved Absorption Study on the Photocarrier Generation Process in Layered Organic Photoreceptors: A role of Delocalized Holes in Photocarrier Generation. In: *Proc. IS&T NIP19: 2003 Int. Conf. Digital Printing Technol.* (Society Imaging Science and Technology, Springfield 2003) p. 683

39.131 P. J. Melz, R. B. Champ, L. S. Chang, C. Chiou, G. S. Keller, L. C. Liclican, R. R. Neiman, M. D. Shattuck, W. J. Weiche: Photogr. Sci. Eng. **21**, 73 (1977)

39.132 T. Murayama: The Design of High Performance Organic Photoconductors. In: *Proc. IS&T NIP17: Int. Conf. Digital Printing Technol.* (Society Imaging Science and Technology, Springfield 2001) p. 557

39.133 M. Umeda, M. Hashimoto: J. Appl. Phys. **72**, 117 (1992)

39.134 Y. Fujimaki, H. Tadokoro, Y. Oda, H. Yoshioka, T. Homma, H. Moriguchi, K. Watanabe, A. Konishita, N. Hirose, A. Itami, S. Ikeuchi: J. Imaging Technol. **17**, 202 (1991)

39.135 E. H. Magin, P. M. Borsenberger: J. Appl. Phys. **73**, 787 (1993)

39.136 Z. D. Popovic, R. Cowdery, I. M. Khan, A.-M. Hor, J. Goodman: J. Imaging Sci. Technol. **43**, 266 (1999)

39.137 R. O. Loutfy, A. M. Hor, P. Kazmaier: Properties and Application of Organic Photoconductive Materials: Molecular Design of Organic Photoconductive Polycyclic Aromatic Diimides Compounds. In: *Proc. 32nd Symp. Macromolecules*, ed. by T. Saegusa, T. Higashimura, A. Abe (Blackwell Scientific, Oxford 1988) p. 437

39.138 R. O. Loutfy, A. M. Hor, P. Kazmaier, M. Tam: J. Imaging Sci. **33**, 151 (1989)

39.139 Z. D. Popovic, A.-M. Hor, R. O. Loutfy: Chem. Phys. **127**, 451 (1988)

39.140 J. M. Duff, G. Allen, A.-M. Hor, S. Gardner: Synthesis, Spectroscopy and Photoconductivity of Dimetric Perylene Bisimide Pigments. In: *Proc. IS&T NIP15: 1999 Int. Conf. Digital Printing Technol.* (Society Imaging Science and Technology, Springfield 1999) p. 655

39.141 A. Kakuta, Y. Mori, S. Takano, M. Sawada, I. Shibuya: J. Imaging Sci. Technol. **11**, 7 (1985)

39.142 A. Shimada, M. Anzai, A. Kakuta, T. Kawanishi: IEEE Trans. Ind. Appl. **1A23**, 804 (1987)

39.143 Y. Kanemitsu, S. Imamura: J. Appl. Phys. **67**, 3728 (1990)

39.144 T. Enokida, R. Hirohashi, S. Mizukami: J. Imaging Sci. **35**, 235 (1991)

39.145 K. Daimon, K. Nukada, Y. Sakaguchi, R. Igarashi: J. Imaging Sci. Technol. **40**, 249 (1996)

39.146 T. Enokida, R. Hirohashi, T. Nakamura: J. Imaging Sci. **34**, 234 (1990)

39.147 S. Takano, Y. Mimura, N. Matsui, K. Utsugi, T. Gotoh, C. Tani, K. Tateishi, N. Ohde: J. Imaging Sci. Technol. **17**, 46 (1991)

39.148 T. I. Martin, J. D. Mayo, C. A. Jennings, S. Gardner, C. K. Hsiao: Solvent Induced Transformations of the Polymorphs of Oxytitanium Phthalocyanine (TiOPc). In: *Proc. IS&T Eleventh Int. Cong. Adv. Non-Impact Printing Technol.* (Society Imaging Science and Technology, Springfield 1995) p. 30

39.149 Y. Kanemitsu, A. Yamamoto, H. Funada, Y. Masumoto: J. Appl. Phys. **69**, 7333 (1991)

39.150 Z. D. Popovic, M. I. Khan, S. J. Atherton, A.-M. Hor, J. L. Goodman: Time-Resolved Fluorescence Quenching and Carrier Generation in Titanyl Phthalocyanine (TiOPc)-Humidity Effects. In: *Electrical and Related Properties of Organic Solids*, ed. by R. W. Munn, A. Miniewicz, B. Kuchta (Kluwer Academic, Dordrecht 1997) p. 207

39.151 Z. D. Popovic, M. I. Khan, A.-M. Hor, J. L. Goodman, J. F. Graham: Study of the Photoconductivity Mechanism in Phthalocyanine Pigment Particles by Electric Field Modulated Time Resolved Fluorescence. In: *Proc. IS&Ts NIP19: 2003 Int. Conf. Digital Printing Technol.* (Society Imaging Science and Technology, Springfield 2003) p. 687

39.152 R. E. Wingard: IEEE Ind. Appl. **37**, 1251 (1982)

39.153 R. B. Champ: SPIE Proc. **759**, 40 (1987)

39.154 R. O. Loutfy, C. K. Hsiao, P. M. Kazmaier: Photogr. Sci. Eng. **27**, 5 (1983)

39.155 K.-Y. Law: J. Imaging Sci. **31**, 83 (1987)

39.156 K.-Y. Law, F. C. Bailey: J. Imaging Sci. **31**, 172 (1987)

39.157 R. O. Loutfy, A.-M. Hor, C.-K. Hsiao, G. Baranyi, P. Kazmaier: Pure Appl. Chem. **60**, 1047 (1988)

39.158 P. M. Kazmaier, R. Burt, G. Di-Paola-Baranyi and C.-K. Hsiao, R. O. Loutfy, T. I. Martin, G. K. Hamer, T. L. Bluhm, M. G. Taylor: J. Imaging Sci. **32**, 1 (1988)

39.159 G. DiPaola-Baranyi, C. K. Hsiao, P. M. Kazmaier, R. Burt, R. O. Loutfy, T. I. Martin: J. Imaging Sci. **32**, 60 (1988)

39.160 K.-Y. Law: Chem. Mater. **4**, 605 (1992)

39.161 K. Sakanoue, M. Motoda, M. Sugimoto, S. Sakaki: J. Phys. Chem. **103**, 5551 (1999)

39.162 P. M. Borsenberger, E. H. Magin, M. Van der Auweraer, F. C. De Schryver: Phys. Stat. Sol. (a) **140**, 9 (1993)

39.163 P. M. Borsenberger, R. Richert, H. Bässler: Phys. Rev. B **47**, 4289 (1993)

39.164 M. Van der Auweraer, F. C. De Schryver, P. M. Borsenberger, H. Bässler: Adv. Mater. **6**, 199 (1994)

39.165 P. M. Borsenberger, H. Bässler: Phys. Status Solidi (b) **170**, 291 (1992)

39.166 A. Dieckmann, H. Bässler, P. M. Borsenberger: J. Chem. Phys. **99**, 8136 (1993)

39.167 P. M. Borsenberger, J. J. Fitzgerald: J. Phys. Chem. **97**, 4815 (1993)

39.168 R. H. Young: Philos. Mag. B **72**, 435 (1995)

39.169 A. Fujii, T. Shoda, S. Aramaki, T. Murayama: J. Imaging Sci. Technol. **43**, 430 (1999)

39.170 T. Shoda, T. Murayama, A. Fujii: Relationship Between Molecular Properties of Hole Transport Molecules and Field Dependence of Hole Mobility. In: *Proc. NIP17: Int. Conf. Digital Printing Technol.* (Society Imaging Science and Technology, Springfield 2001) p. 550

39.171 P. M. Borsenberger, W. T. Gruenbaum: J. Polym. Sci.: Polym. Phys. **34**, 575 (1996)

39.172 P. M. Borsenberger, E. H. Magin, M. R. Detty: J. Imaging Sci. Technol. **39**, 12 (1994)

39.173 P. M. Borsenberger, H.-C. Kan, E. H. Magin, W. B. Vreeland: J. Imaging Sci. Technol. **39**, 6 (1995)

39.174 P. M. Borsenberger, W. T. Gruenbaum, M. B. O'Regan, L. J. Rossi: J. Polym. Sci.: Polym. Phys. **33**, 2143 (1995)

39.175 W. T. Ferrar, D. S. Weiss, J. R. Cowdery-Corvan, L. G. Parton: J. Imaging Sci. Technol. **44**, 429 (1999)

39.176 X. Jin, D. S. Weiss, L. J. Sorriero, W. T. Ferrar: J. Imaging Sci. Technol. **47**, 361 (2003)

39.177 D. S. Weiss, J. R. Cowdery, W. T. Ferrar, R. H. Young: J. Imaging Sci. Technol. **40**, 322 (1996)

39.178 N.-J. Lee, H.-R. Joo, K.-Y. Yon, Y. No: Development of Positive Charging Multi-Layered Organic Photoconductor for Liquid Electrophotographic Process. In: *Proc. IS&T NIP19: 2003 Int. Conf. Digital Printing Technol.* (Society Imaging Science and Technology, Springfield 2003) p. 670

39.179 Y. C. Chan, X. S. Miao, E. Y. B. Pun: J. Mater. Res. **13**, 2042 (1998)

40. Luminescent Materials

This chapter surveys the field of solid-state luminescent materials, beginning with a discussion of the different ways in which luminescence can be excited. The internal energy-level structures of luminescent ions and centres, particularly rare-earth ions, are then discussed before the effects of the vibrating host lattice are included. Having set the theoretical framework in place, the chapter then proceeds to discuss the specific excitation process for photo-stimulated luminescence and thermally stimulated luminescence before concluding by surveying current applications, including plasma television screens, long-term persistent phosphors, X-ray storage phosphors, scintillators, and phosphors for white LEDs.

40.1 Luminescent Centres............................ 985
    40.1.1 Rare-Earth Ions 985
    40.1.2 Transition-Metal Ions 986
    40.1.3 s^2 Ions 987
    40.1.4 Semiconductors 987

40.2 Interaction with the Lattice.................. 987

40.3 Thermally Stimulated Luminescence...... 989

40.4 Optically (Photo-)Stimulated Luminescence 990

40.5 Experimental Techniques – Photoluminescence 991

40.6 Applications... 992
    40.6.1 White Light-Emitting Diodes (LEDs) 992
    40.6.2 Long-Persistence Phosphors....... 992
    40.6.3 X-Ray Storage Phosphors 993
    40.6.4 Phosphors for Optical Displays 994
    40.6.5 Scintillators 994

40.7 Representative Phosphors 995

References 995

Luminescent materials are substances which convert an incident energy input into the emission of electromagnetic waves in the ultraviolet (UV), visible or infrared regions of the spectrum, over and above that due to black-body emission. A wide range of energy sources can stimulate luminescence, and their diversity provides a convenient classification scheme for luminescence phenomena, which is summarised in Table 40.1. Pho-

Table 40.1 Types of luminescence

Designation	Excitation	Trigger	Acronym
Photoluminescence	UV, visible photons	–	PL
Radioluminescence	X-ray, gamma rays, charged particles	–	RL
Cathodoluminescence	Energetic electrons	–	CL
Electroluminescence	Electric field	–	EL
Thermoluminescence	Photons, charged particles	Heat	TSL
Optically/photo-stimulated luminescence	Photons, charged particles	Visible/IR photons	OSL, PSL

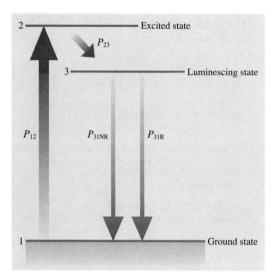

Fig. 40.1 Optical pumping cycle for a generic photoluminescent system

The terms phosphorescence and fluorescence are often used in connection with luminescent materials. This classification is based on the time-domain response of the luminescent system. Figure 40.1 shows a generic photoluminescent system where incident UV radiation excites a system from a ground state 1 with probability per unit time P_{12} into an excited state 2. The system decays with probability P_{23} to a luminescing level 3, from which there is a probability P_{31R} and P_{31NR} of radiative and nonradiative decay, respectively, to the ground state. Nonradiative decay generally involves phonon emission.

First suppose that the transition probabilities are such that $P_{12}, P_{21} \gg P_{31}(= P_{31R} + P_{31NR})$. If the optical pumping (excitation) is abruptly stopped, the population of the luminescing state 3 decays as,

$$N_3 = N_3(0) \exp(-P_{31}t) \qquad (40.1)$$

and the rate of luminescent energy emission is $-(h\nu P_{31R}/P_{31}) \mathrm{d}N_3/\mathrm{d}t$ for an energy difference $E_{31} = h\nu$ between states 1 and 3. Hence the luminescence intensity at distance r from the sample is,

$$I = (h\nu) N_3(0) P_{31R} \exp(-P_{31}t) /(4\pi r^2) \qquad (40.2)$$

and the characteristic luminescence lifetime is $\tau = (P_{31})^{-1}$. Thus the lifetime is governed by both radiative and nonradiative processes, whilst the intensity of the luminescence depends on the relative magnitude of P_{31R}.

This discussion provides the basis for understanding the terms fluorescence and phosphorescence applied to luminescent materials. A material is often classified as one or the other according to the relative magnitude of $\tau = (P_{31})^{-1}$, with 10 ns being set in a relatively arbitrary way as the boundary between a *fast* fluorescent system and a *slow* phosphorescent one. For comparison, the theoretical lifetime for spontaneous emission for a strongly allowed hydrogen atom 2p → 1s transition is about 0.2 ns.

However, by this definition phosphorescence can also arise from luminescent states with short lifetimes which are populated through ones with long lifetimes. In Fig. 40.1, if state 2 is long-lived in the sense that $P_{23} \ll P_{31}$ then the measured lifetime for emission from the luminescing state will be $\tau = (P_{23})^{-1}$, and the system will be labelled phosphorescent, even though the luminescing level itself has a very short lifetime. In consequence, a second classification [40.2] is based on whether or not the luminescing level is fed by a metastable state which sets the lifetime. Sometimes the metastable state is a long-lived intermediate form

toluminescence, where the luminescence is stimulated by UV or visible light, is a widely used materials science technique for characterising dopants and impurities, and finds applications in lighting technologies such as fluorescent lamps. Radioluminescence involves excitation by ionising radiation, and is used in scintillators for nuclear particle detection; the special case of stimulation by energetic electrons is called cathodoluminescence, the name arising from early atomic physics experiments involving gas discharges. The major application of cathodoluminescence is in cathode ray tubes for television sets and computer monitors. Electroluminescence involves collisional excitation by internal electrons accelerated by an applied electric field, and with a much lower energy than in the case of cathodoluminescence. Electroluminescence finds applications in panel lighting used in liquid-crystal display (LCD) back-planes, and in light-emitting diodes.

There are other forms of luminescence which we mention for completeness but will not discuss further: bioluminescence and chemiluminescence where the energy input is from chemical or biochemical reactions, sonoluminescence (sound wave excitation), and triboluminescence (strain or fracture excitation).

There are several books which describe the luminescence of materials in more depth than is possible in a short article, for example that edited by *Vij* [40.1] and the monograph by *Blasse* and *Grabmaier* [40.2].

of energy storage which can be triggered by an external stimulus to undergo a transition to a fluorescent level. Thus in Table 40.1, thermally stimulated and optically stimulated luminescence involve a metastable state consisting of trapped electrons and holes, which can be triggered to recombine by heating or by optical stimulation; the recombination energy is transferred to a fluorescing centre.

Overall, the fluorescence/phosphorescence classification is somewhat nebulous, and it is debateable whether the classification is necessary or desirable.

40.1 Luminescent Centres

A wide variety of centres give rise to luminescence in semiconductors and insulating materials, including rare-earth ions, transition-metal ions, excitons, donor–acceptor pairs, and ions with a d^{10} or s^2 electronic configuration ground state. Some luminescence spectra consist of broad emission bands arising from the interaction between the electronic system of the luminescent centre and the vibrations of the atoms or ions, which surround it; the broad bands arise from simultaneous transitions of both electronic and vibrational systems. For others, such as the rare earths, the spectra comprise sharp lines arising from purely electronic transitions, and the effect of the environment is felt mainly through their effects on the lifetimes of the states. Thus in discussing the physical background to luminescence, it is simplest to start with a discussion of rare-earth luminescence, where the effect of vibrations can be initially ignored.

40.1.1 Rare-Earth Ions

The trivalent rare-earth ions have n electrons ($n = 1-14$) in the 4f shell. In a free ion, the eigenstates resulting from the various atomic interactions are labelled by the total spin S and orbital angular momenta L. Spin–orbit coupling breaks up each L, S multiplet of degeneracy $(2S+1)(2L+1)$ into sub-multiplets labelled by the total angular momentum $J = L + S$, where J can range from $L - S$ to $L + S$. The $4f^n$ orbitals lie within the outer $5s^2$ and $5p^6$ filled shells, which partly shield them from the effects of a crystalline environment. The effects of the latter are quantitatively described by the *crystal field* [40.3], and this term in the Hamiltonian splits the J multiplets into $2J + 1$ sublevels, the so-called crystal-field splitting. Some of these crystal-field levels may still be two- or threefold degenerate, depending upon the symmetry of the environment. Odd-electron systems always have at least twofold (Kramers) degeneracy. The resulting energy-level structures are complicated, and are summarised in the classic *Dieke diagram* [40.3]. The original has been updated by *Carnall* et al. [40.4] and is reproduced in many books and papers on rare-earth ion spectroscopy [40.2]. In Fig. 40.2, we show a schematic version (not accurately to scale) of the diagram appropriate for Pr^{3+} with $n = 2$, which serves for our discussion. The crystal-field splitting is usually smaller than the spin–orbit splitting, and is illustrated schematically by the vertical extent of the bands in Fig. 40.2. The multiplet labels follow the usual $^{2S+1}L_J$ system.

In the figure we also show a generic highest excited band which does not belong to the $4f^2$ configuration. In the specific case of the rare earths this can be the excited-state configuration resulting from promo-

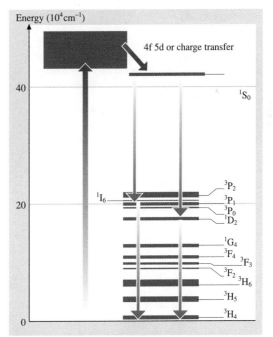

Fig. 40.2 Schematic energy-level diagram for Pr^{3+}

tion of one electron to the 5d state, giving an overall $4f^{n-1}5d^1$ configuration; more generally it could be a so-called charge transfer band, which corresponds to the transfer of one electron from the ligands to the luminescent ion. The relative location and importance of these bands varies with the luminescent ion and the crystalline environment, but they play an important role in the excitation of luminescence.

Excitation and luminescence transitions within the various levels in Fig. 40.2 are governed by the golden rule of quantum mechanics [40.5] for interactions with the electromagnetic field; in summary the probability of a transition between two states i and j is proportional to square of the matrix element $<i|H|j>$, where H is the time-dependent perturbation Hamiltonian representing the interaction of the electrons with the electromagnetic field. The proportionality constant contains the light intensity in the case of excitation. The perturbation can be expanded in a power series involving electric and magnetic multipoles of the electronic system, but of these the electric dipole term is dominant, with the magnetic dipole term being much smaller by a factor of more than five orders of magnitude. Since the electric dipole operator er has odd parity, the matrix element for transitions $r_{ij} = <i|er|j>$ is necessarily zero unless i and j have opposite parity. This is the most important selection rule governing luminescence: transitions between states of the same parity have zero transition probability and so are *forbidden* (Laporte's rule). In the case of the rare earths, all states of a single $4f^n$ configuration have the same parity, and so all optical transitions within the configuration are strictly forbidden. But this rule is relaxed by several considerations. First, if the crystalline environment lacks inversion symmetry, the crystal field admixes a small fraction of the excited configurations (eg $4f^1 5d^1$ for Pr^{3+}) of opposite parity into the ground configuration, which makes such transitions weakly allowed. Secondly, the selection rule for magnetic dipole transitions is that they are allowed between states of the same parity, although they are typically about five orders of magnitude weaker than for electric dipole ones. Finally, odd-parity vibrations and an electron–phonon interaction produce a similar configuration admixture effect to static lattice odd-parity mixing although this effect is more important for 3d ions than for 4f ones.

With only weak transitions possible within the 4f configuration, one might wonder how it would be possible to optically excite any significant luminescence. The answer lies in the 5d or charge-transfer bands which either lie at higher energies, or overlap with the upper levels of the 4f configuration. These give rise to strong absorption and efficient pumping. Relaxation can occur via the parity-allowed transitions to the upper levels of the 4f configuration, and from there via single or multiple radiative emissions back to the ground state. Because these intra-configurational transitions are only weakly allowed, the lifetimes are generally quite long, of the order of μs–ms. Figure 40.2 shows some of the observed transitions in the case of the Pr^{3+} ion. There are further constraints on possible transitions which arise from an analysis of the angular momenta of the initial and final states. For example, a transition between two states both of which have $J = 0$ is forbidden since there is no angular momentum change as required for a photon; similarly for dipole transitions we require $\Delta J = 0, \pm 1$.

The transition probability per second for spontaneous emission [40.6] is given by

$$P_{ij} = \frac{64\pi^4 \nu^3}{3hc^3} |r_{ij}|^2 ,\qquad(40.3)$$

where ν is the frequency of the transition, h is Planck's constant, c is the velocity of light, and $|r_{ij}|$ is the matrix element of the electric dipole operator er_{ij} between the two states i and j, and e is the electronic charge. For absorption, this must be multiplied by N, the mean number of photons with energy $h\nu$, which thus incorporates the effect of the incident beam intensity. Experimentally, one measures an absorption coefficient k as a function of energy $k(E)$ [40.6] which is linked to P_{ij} through,

$$\int k(E) \mathrm{d}E = N_i \left(\frac{\pi e^2 h}{nmc} \right) \left(\frac{n^2+2}{3} \right)^2 f_{ij} ,\qquad(40.4)$$

where n is the refractive index of the crystal environment, m is the electronic mass, N_i is the concentration of the luminescent centres, and f is the *oscillator strength* for the transition. For both absorption and emission, the dimensionless oscillator strength f_{ij} defined as [40.6]

$$f_{ij} = \frac{8\pi^2 m\nu}{3he^2} |r_{ij}|^2 \qquad(40.5)$$

is often quoted to compare the relative transition probabilities. For an electron harmonic oscillator, $\sum f_{ij} = 1$, and so oscillator strengths of the order of $0.1-1$ are strongly allowed transitions.

40.1.2 Transition-Metal Ions

Transition-metal ions from the 3d series are characterised by a much stronger interaction with the crystalline environment than the 4f ions since there is no equivalent of screening by the 5s, 5p outer shells. In addition, the spin–orbit coupling is weaker, and so the order of perturbation is reversed: the atomic L, S multiplets are split by the crystal field, with spin–orbit coupling being a smaller interaction. Intra-configurational transitions are again strictly forbidden, but become weakly allowed by inter-configurational mixing through odd-parity crystal fields, and by odd-parity vibrations. The result of this is that the strongest selection rule after parity is that transitions should have $\Delta S = 0$, since the electric dipole operator does not involve spin. The other major difference, again due to the strength of the crystal-field interaction, is that transitions which are purely electronic, the so-called zero-phonon lines, are rarely observed. Rather what are seen are broad bands which correspond to the simultaneous excitation of an electronic transition and vibrational transitions, which overlap to give the broad observed bands. In particular, transitions involving odd-parity vibrations have a high transition probability through the effect of configuration admixing. This will be considered in a following section. The most commonly observed luminescent ions are those from the d^3 configuration (Cr^{3+}, Mn^{4+}) and from the d^5 configuration (Mn^{2+}).

40.1.3 s^2 Ions

The $5s^2$ (e.g. Sn^{2+} and Sb^{3+}) ions and $6s^2$ (e.g. Tl^+, Pb^{2+}, Bi^{3+}) ions are of considerable importance because transitions to and from the excited s^1p^1 states are Laporte-allowed. The interaction of the p state with the crystalline environment can be very strong, and so broad spectra are often observed.

40.1.4 Semiconductors

Luminescence in semiconductors is dominated by near-band-gap luminescence arising from recombination of electrons and holes. This process is most efficient in direct-band-gap materials such as ZnS and GaP rather than indirect-gap materials such as Si and Ge because the transition probability requires conservation of wavevector, but the photon wavevector is ≈ 0 on the scale of the Brillouin zone. Hence creation or destruction of a phonon is required for band-to-band luminescence in indirect-gap materials, which is less probable. The near-edge emission may correspond to luminescence from a variety of shallow energy-level structures such as free or trapped excitons, or from donor–acceptor recombination. These are both example of electronic systems with spatially extensive wavefunctions, in contrast to the atomically localised 3d and 4f wavefunctions considered earlier. However, it is also possible to observe *deep-level* luminescence from transition-metal ions and rare earths in semiconductors provided that the electron affinities and band-gap energies are such that the pertinent energy levels fall in the band gap. Since semiconductors are discussed elsewhere in this volume, we shall not consider them further here.

40.2 Interaction with the Lattice

For rare-earth ions, the interaction with the vibrations of the crystal lattice can be ignored for most purposes; the observed luminescence spectrum consists of sets of sharp electronic transitions. But for other luminescent ions which interact strongly with the vibrating ions of the surrounding crystal, the incorporation of the latter is critical to explaining the observed spectra. The simplest model of ion–lattice interactions is to consider only the N nearest neighbour ions and their atomic displacements X_n, Y_n, Z_n, ($n = 1, N$) in Cartesian coordinates. The vibrational Hamiltonian involves cross terms in these coordinates, but may be transformed to harmonic form if symmetry-adapted forms of these coordinates (*normal modes*) are used instead of the actual displacements. For example, the so-called breathing mode Q_b, for an octahedrally coordinated ion takes the form

$$Q_b = (Z_1 - Z_2 + X_3 - X_4 + Y_5 - Y_6)/6 \,. \quad (40.6)$$

If all the other modes have zero amplitude, the ions move radially towards or away from the central luminescent ion. The key point in considering the influence of the crystal lattice is that the vibrational potential energy is just the variation of the electronic energy with ionic displacement, or equivalently with the normal modes (within the spirit of the Born–Oppenheimer approximation). Put another way, the crystal field depends on the ion positions so that the electron and lattice quantum-mechanical systems are linked through this electron–lattice coupling. We can therefore expect a difference in the harmonic vibrational potential from one electronic state to another, so that it will in general have the form $(1/2)m\omega_i^2(Q - Q_{0i})^2$; i.e. both the magnitude of the potential and the position of the minima Q_{i0} will depend on the electronic state i. The vibrational states in the harmonic approximation are just the usual simple harmonic oscillator states with energies $(n + 1/2)h\nu_i$,

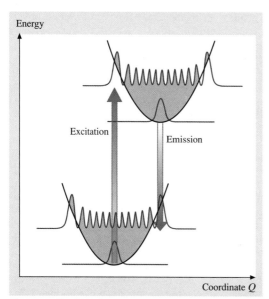

Fig. 40.3 Configuration coordinate diagram for excitation/emission cycle

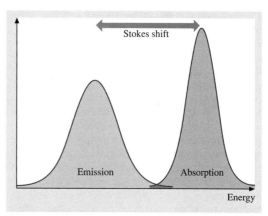

Fig. 40.4 Absorption and emission line shapes for strong electron–lattice coupling

with n an integer. Thus we arrive at a *configuration coordinate* diagram such as that shown in Fig. 40.3, with the potential energies for a ground state g and excited state e being offset parabolas of different curvature. The square of the vibrational wavefunction is shown for the ground state and that with $n = 11$. We note that for a classical oscillator there would be a peak in the probability function at the extreme lengths of travel, corresponding to the maxima for $n = 11$ at the outer limits of the wavefunction in Fig. 40.3.

We consider first luminescence from the excited electronic state/lowest vibrational state. The transition probability may be calculated using the Franck–Condon principle, that is we assume that the duration of the electronic transition is much shorter than a vibrational period, so that the Q remains constant during a transition. The transition is therefore taken to be *vertical* on the configuration coordinate diagram. The transition probability is just $\langle \phi_e | \phi_g \rangle^2$, the square of the vibrational wavefunction overlap, multiplied by the electronic transition probability considered earlier. From Fig. 40.3, this overlap will be a maximum for some vibrational state other than the ground state, unless the positions of the minima of the two potential energy curves coincide accidentally. (In Fig. 40.3, this maximum would be for the $n = 11$ ground vibrational state). Put another way, the maximum transition probability is not for the zero-phonon transition (no change in vibrational state), but corresponds to the creation of a finite number of phonons. A range of transitions is allowed, and the result in a semiclassical analysis, allowing for finite line widths, is a Gaussian-shaped band. The analysis has to be extended to include finite temperatures and other modes, but the overall result is that the emission line shape is approximately a Gaussian centred at an energy lower than that of the difference between the minima of the two potential curves. The same argument can be applied to the excitation process, and again an approximately Gaussian line shape results but this time centred on an energy above that of the difference in potential-energy minima. Figure 40.4 shows the overall result; the difference between the maxima of the two curves, known as the Stokes shift, is an indicator of the degree of electron–lattice coupling.

It is clear from the diagram that the, in cases where there is strong electron–lattice coupling of this type, that: (a) there will be very little intensity in the zero-phonon line, and (b) there will be a large Stokes shift between the energies for maximum absorption and maximum emission. Thus the luminescence of transition-metal ions, colour centres, and closed-shell ground-state ions (s^2, d^0), which have strong interactions with the lattice in the excited states, are typically broad bands with only occasionally weak, sharp zero-phonon lines being observed. For luminescence from within the 4f states of the rare earths, the reverse is true; we are in the weak-coupling regime, and zero-phonon lines are the predominant features of the spectrum.

40.3 Thermally Stimulated Luminescence

Thermally stimulated luminescence (TSL), or simply thermoluminescence (TL), refers to luminescence induced by thermally stimulated recombination of trapped electrons and holes in materials which have been subject to prior irradiation. The irradiation, which may be in the form of UV light, X-rays, gamma rays, or energetic electrons, creates free electrons and holes, most of which promptly recombine, but some of which are locally trapped at defect centres such as impurities and vacancies. If the trap binding energies are sufficiently large, thermal promotion of the electron or hole to the conduction band or valence band, respectively, is improbable at the irradiation temperature, and so these charge carriers remain trapped after irradiation. However, if the sample is then heated, thermally assisted recombination becomes increasingly probable, and the result is an initially increasing light output with increasing temperature until the traps are depleted, whereupon the light intensity drops. The curve of light intensity versus temperature $I(T)$ is known as a glow curve, and may be analysed to extract the trap depths and concentrations. A comprehensive review of the field has been given by *McKeever* [40.7], and a shorter discussion is given by *Vij* [40.1].

The process is shown schematically for a simple system comprising a single electron trapping level T and a single recombination centre R in Fig. 40.5.

Irradiation results in a trapped electron at trap T and a trapped hole at R. The trapped hole binding energy is larger than that of the trapped electron, so the latter is depopulated first, with a probability P which has the form,

$$P = s \exp(-E/kT), \qquad (40.7)$$

where s and E are the attempt frequency and the activation energy respectively. We have shown the hole trap and recombination centre as being one and the same in Fig. 40.5, but it is also possible that the recombination energy is transferred to a separate luminescent centre.

In the measurement process, the sample is heated at a fast and linear rate, typically 1–10 K/s, whilst the light emission is monitored by a sensitive filter/photomultiplier combination, or by a monochromator system. The resulting glow curve may be fitted to a theoretical curve to extract the trap parameters. Generally the light output is quite weak, and TSL systems have sensitivity as one of their prime design factors, so cooled detectors and photon counting are commonplace. Above about 400 °C, thermal radiation from the sample/heater is a problem and must be eliminated by filtering or by subtraction of a glow curve recorded using a thermally bleached sample.

The main uses of TSL are in determining trap depths and irradiation doses. In archaeological and geological applications, a comparison is made between accumulated natural dose and a dose from a radioisotope; by combining this with a measurement of the activity of the surroundings, a date since last thermal or optical erasure

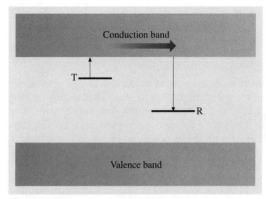

Fig. 40.5 Thermally stimulated luminescence process

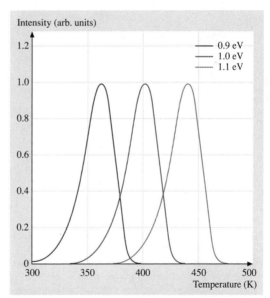

Fig. 40.6 Computed glow curves for first-order kinetics for fixed escape frequency and various trap depths

of the object can be determined. The nature of the traps is often poorly understood for these chemically complex samples. For medical dosimetry, room-temperature stable traps with high thermoluminescent output are required. At present, the material of choice is LiF doped with a few hundred ppm of Mg and Ti.

The mathematical form of the TSL glow curve depends on the physical model used for the TL process. In the simplest case, assuming first-order kinetics [40.1, 7], the light intensity I at temperature T is given by,

$$I(T) = n_0 s \exp(-E/k_B T)$$
$$\times \exp\left[-(s/\beta)\int_{T_0}^{T} \exp(-E/k_B T)\, dT\right],$$
(40.8)

where n_0 is the number of occupied traps at time $t = t_0$ when the temperature is T_0, β is the heating rate in K/s, and k_B is Boltzmann's constant. The result is a glow curve whose peak position varies approximately linearly with trap depth E, as shown in Fig. 40.6 (for fixed escaped frequency $s = 3.3 \times 10^{11}$ s^{-1} and heating rate 1 K/s).

Of course, the peak position also depends on the escape frequency s, but is less sensitive to s than to E. For second-order kinetics [40.1, 7], the glow curve equation becomes

$$I(T) = \left(n_0^2 s / N\right) \exp(-E/k_B T)$$
$$\bigg/ \left[1 + (n_0 s / N\beta)\int_{T_0}^{T} \exp(-E/k_B T)\, dT\right]^2,$$
(40.9)

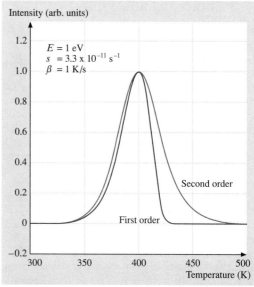

Fig. 40.7 Comparison of first- and second-order glow curves

where N is the number of available traps. The two forms both give glow curves of similar shape, but the first-order form shows an asymmetric form with a sharp fall off on the high-temperature side of the peak, whilst the second-order form is more symmetric as shown in Fig. 40.7. The parameters are extracted by least-squares fitting of these expressions to the experimental glow curves, which is cumbersome due to the integral, but *Kitis* et al. [40.8] have given analytical approximations to (40.8) and (40.9). In practice, many glow curves do not follow first- or second-order kinetics precisely.

40.4 Optically (Photo-)Stimulated Luminescence

Thermally stimulated luminescence is sometimes also accompanied by optically stimulated luminescence (OSL), in which one of the trapped carriers is excited by optical stimulation to a level from which it can recombine with the conjugate carrier by tunnelling, or completely to one of the bands so that recombination is achieved through what is essentially a photoconductivity effect. For OSL to be significant, there must be an appreciable optical transition probability, so not all TSL centres are OSL active. The stimulation energy measured in OSL generally differs from that determined from TSL because of the Franck–Condon principle. The OSL effect finds practical application in dosimetry, e.g. [40.9], and in X-ray storage phosphors used for medical imaging, as described later.

40.5 Experimental Techniques – Photoluminescence

A typical traditional photoluminescence measuring system involves a broad-spectrum source, either a combined tungsten filament for the visible spectrum and deuterium lamp for the UV, or a xenon flash lamp. The lamp emission is passed through a grating monochromator and so selectively excites the luminescence. Band-pass or band-edge filters are generally required to eliminate unwanted second- and higher-order diffraction maxima from the grating. The luminescence is efficiently gathered by low-f/number optics and fed to a second grating monochromator, also equipped with filters, to monitor and analyse the luminescence. The final detector may be a photomultiplier, or preferably a charge-coupled device or photodiode array for improved data collection efficiency at multiple wavelengths.

This arrangement relies on good monochromation/filtering to remove what is sometimes a relatively strong component of scattered light from the beam analysed by the emission monochromator. An alternative method of removing scattered light is to use time discrimination, by replacing the source by a xenon flash lamp (for example, as in the common Perkin Elmer LS55B luminescence spectrometer). The flash has a duration of about $10\,\mu s$, so any scattered light has decayed away to an insignificant level when the emitted beam is sampled some time ($0.1-10\,ms$) after the flash. A timed electronic gate is used to sample the emission immediately after a flash and just before the next flash; the difference between these two sampled signals gives the short-term luminescence whilst the second sample alone gives the long-term luminescence with *long* and *short* being relative to the flash repetition period. Of course, luminescence with lifetimes shorter than the pulse width ($\approx 10\,\mu s$) cannot be readily measured with this system. The luminescence intensity is normalised with respect to the excitation intensity by steering a sample of the excitation beam to a rhodamine dye cell which has a quantum efficiency of essentially unity for wavelengths below about 630 nm. The fluorescence from the dye is measured with a second photomultiplier.

To minimise the effect of scattered light, a conventional laser with its intrinsically narrow linewidth and high intensity is a very convenient replacement for a broad-spectrum lamp, but suffers from the disadvantage of a fixed wavelength. Typical lasers of interest are nitrogen (pulsed), argon (UV lines, or frequency-doubled visible lines), krypton, and the new generations of GaN/GaInN blue/violet/UV laser diodes. For rare-earth spectroscopy, or other systems which are characterised by narrow absorption lines, it is very useful to have a scanning dye laser as the excitation source. This *selective excitation* facility enables *tagging* of particular luminescent levels with excited states belonging to the same centre, so that a picture of the energy-level structure of each luminescent centre can be built up in cases where several such centres contribute to the overall luminescence.

For decay kinetics on faster time scales, fluorimeters such as those developed initially by the Spex company (now Horiba) use a fast modulator and phase-sensitive detection to measure the phase shift between fluorescence and excitation; it is claimed that fluorescence decays can be measured with a resolution of 25 ps this way. An alternative is the time-correlated single-photon-counting technique which can measure decay constants in the ps–ns range. In this method, the excitation comes from a fast laser pulse, and the light level reaching the photomultiplier or micro-channel plate detector is reduced to such a low level that less than one photon per excitation pulse is detected. The time delay between the photon detection and the time of the pulse is measured, and a histogram produced of numbers of detected photons versus arrival time taken over a large number of excitation pulses. For efficient data collection, a high repetition rate and fast-pulse laser are required, often a Ti–sapphire laser. The wavelength for Ti–sapphire is too large for stimulating many materials directly with single-photon excitation, but stimulation is nonetheless possible by a two-photon excitation process, or by the use of a nonlinear crystal acting as a frequency doubler to produce laser output at one half the wavelength of the basic laser.

There are a number of specialist techniques, such as hole-burning, fluorescence line-narrowing, and photon echo methods associated with the use of lasers with either very narrow line widths or short pulse duration which have developed in a parallel way to techniques first introduced in nuclear magnetic resonance, and which are mainly used to investigate the dynamics and quantum mechanics of the luminescent species rather than the material in which they are contained. *Meijerink* gives a review of experimental luminescence techniques [40.1] which includes a short discussion of these specialist techniques.

40.6 Applications

The largest market for luminescent materials has traditionally been in the areas of lighting, through fluorescent tubes, and in cathode-ray-tube screen phosphors for image display. Both of these areas can now be regarded as mature in terms of materials development. However, new discoveries in the past decade, and the advent of new technologies, have rekindled interest in phosphor materials. Some of the current areas of activity in applications are outlined below.

40.6.1 White Light-Emitting Diodes (LEDs)

The development of blue, violet and UV LEDs based on GaN, InGaN and other semiconductors and alloys has stimulated great interest in the possibility of producing a *white-light LED* for use in lighting applications. LEDs are now available with emissions which peak as low as 365 nm in the ultraviolet. The concept is to use the blue emission to stimulate luminescence from yellow, or red and green phosphors, so that when mixed with residual blue light from the LED the result is simulated white light. For UV LEDs the LED output is used to stimulate blue, red and green phosphors.

The first generation of white LEDs from companies such as Nichia relied on a YAG:Ce phosphor to convert some of the emission from a 465-nm GaInN LED into an orange/yellow emission centred at 550 nm; the combination of blue and yellow simulates white light. More recent versions of this scheme include $Sr_3SiO_5:Eu^{2+}$, which emits at 570 nm, and is claimed [40.10] to be more efficient than YAG:Ce, and a CaSiAlON:Eu ceramic phosphor which offers improved thermal stability [40.11]. Improved colour rendition is obtained by using more phosphors and better balance between the various emissions, for example by replacing the YAG:Ce with $SrGa_2S_4:Eu$ (green) and $SrY_2S_4:Eu$ (red). Most recent developments based on 375 nm UV LEDs have used multiple emissions to achieve even better colour balance; for example *Kim* et al. [40.12] use $Sr_3MgSi_2O_8:Eu^{2+}$, Mn^{2+}, which has blue (Eu^{2+}), yellow(Eu^{2+}) and red emissions (Mn^{2+}).

A second development in this area is the substitution of another semiconductor for the phosphor – the so-called photon recycling technique. In this method, a layer of AlGaInP is used to absorb some of the blue incident radiation and down-convert it to the complementary colour. The advantage is that the fabrication/integration process is simpler compared to combining phosphor and semiconductor technology.

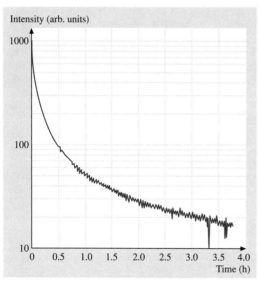

Fig. 40.8 Dark decay of persistent luminescence in a commercial lighting strip based on Nemoto $SrAl_2O_4:Eu/Dy$ material

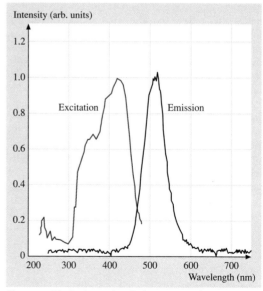

Fig. 40.9 Excitation and emission spectra of a commercial lighting strip based on Nemoto $SrAl_2O_4:Eu/Dy$ material

40.6.2 Long-Persistence Phosphors

It has been discovered in the past decade that some materials show a very long-lived but intense afterglow, arising from thermal emission of charge carriers from deep traps, followed by electron–hole recombination at or near a luminescent ion, i.e. room-temperature thermoluminescence. The persistence is for periods of several hours, much longer than the well-known ZnS:Cu persistent phosphor. This new class of materials have widespread applications in areas such as signage and passive emergency lighting for public buildings, aircraft cabins etc. The material is activated or *charged* by the blue/UV content of solar radiation or fluorescent indoor lighting during normal conditions; in subsequent dark conditions the energy is released as an afterglow. No power supply is required, the operation is entirely passive, and so a high degree of reliability is assured. Generally the phosphor powder ($\approx 10\,\mu$ grain size) is mixed with a resin binder and applied as a thick (≈ 1 mm) coating. The materials which have been used are rare-earth-doped strontium or calcium aluminate [$CaAl_2O_4$:Eu/Nd (blue), $SrAl_2O_4$:Eu/Dy (green), $Sr_4Al_{14}O_{25}$:Eu/Dy (blue–green), and Y_2O_2S:Eu/Mg/Ti (orange)]. Figures 40.8 and 40.9 show the decay and emission/excitation characteristics of a commercial phosphor strip of this type, based on Luminova® $SrAl_2O_4$:Eu/Dy powder sourced from the Nemoto company. The decay is clearly non–exponential, but after the initial rapid decay the intensity may be fitted with an exponential with a decay constant of 3.1 h.

The traps which are responsible for the long-lived decays in these systems have not been clearly identified, and it is likely that there are several involved. *Jia* et al. [40.13, 14] have studied persistent violet/UV and blue (≈ 400 nm, 450 nm) luminescence from rare-earth ions in $CaAl_2O_4$ and $BaAl_2O_4$ and report that the thermally stimulated luminescence from these materials contains multiple glow peaks. They also describe how co-doping the calcium aluminate with Tb^{3+} results in a green persistent phosphor as a result of energy transfer from the cerium to the terbium ion; this mechanism could be the basis for convenient colour control, including white persistent phosphors. *Aitasalo* et al. [40.15] report that the nature of the traps is affected by the particular rare-earth ion or couple.

40.6.3 X-Ray Storage Phosphors

Optically stimulated or photo-stimulated luminescence is the basis for an X-ray imaging technology known commercially as computed radiography (CR). CR was the first of a number of imaging techniques which are steadily replacing traditional photographic film methods. Several other techniques such as those based on amorphous selenium photoconductor/flat panels and a-Si arrays, and scintillator CCD/complementary metal–oxide–semiconductor (CMOS) arrays have also emerged in recent years, but CR is still the dominant of the new technologies on the basis of numbers of installed units. The many advantages of CR over photographic film, and the principles of the method are detailed in [40.16,17], but the basic mechanism is that incident X-rays create electron–hole pairs in the material. Most of these promptly recombine, but some are trapped at defects and impurities, and remain trapped for periods of hours after the X-ray source is turned off. In an X-ray storage phosphor (XRSP), one of the trapped carriers is optically stimulable, and can be excited to the conduction band or valence band, or to a level from which it can recombine by tunnelling with the conjugate trapped carrier. The resulting recombination energy is transferred to a luminescent ion, and the intensity of the photo-stimulated luminescence (PSL) is in direct proportion to the incident X-ray intensity. The dominant material used in current XRSP systems is $BaFBr_{1-x}I_x$:Eu^{2+}, where the electron traps are F-centres, the hole traps are unidentified, and the luminescence is the 5d–4f transition of the Eu^{2+} ion. One disadvantage of systems based on this (powder) phosphor is that when the image is extracted with a raster-scanned He–Ne laser beam, light scattering from the powder grains means that material outside the focussed laser spot is also stimulated, limiting the spatial resolution to around 200 μm, which is inadequate for applications such as mammography. Several ways to overcome this are currently under development. RbBr:Eu is also an X-ray storage phosphor [40.18] and can be grown by vapour deposition in a columnar form.

Fig. 40.10 PSL image of a BC549 transistor recorded on a glass-ceramic imaging plate

The columnar structure has a light-guiding property, restricting the scattering effect, and improving the resolution. A second development is that of glass-ceramic storage phosphors, where PSL active crystals are embedded in a glass [40.19–21]; the combination of particle size, separation and refractive-index mismatch means that these composite materials are semitransparent and the problem of scattering of read-out light is reduced. Figure 40.10 shows an image of a BC549 transistor recorded in a glass-ceramic X-ray storage phosphor.

40.6.4 Phosphors for Optical Displays

There are several new technologies being developed to replace cathode ray tubes for domestic televisions, including plasma display panels (PDPs). In these units, each pixel is a sealed cell containing a mixture of Xe and Ne in a dielectric-shielded electrode structure (for a review, see *Boeuf* [40.22]). An alternating current (AC) voltage applied between the electrodes results in a glow discharge being set up in the gas, and a Xe dimer vacuum UV (VUV) emission predominantly between 147 and 190 nm occurs. (In comparison, the mercury discharge in a conventional fluorescent tube emits primarily at 254 nm.) The UV discharge excites red, blue, or green phosphors coated on one of the cells; each colour is activated by an adjacent electrode. The requirements for efficient output from these phosphors differ from conventional tubes since the latter were chosen on the basis of their luminescence efficiency at a 254-nm pump wavelength, and their resistance to degradation by the UV light and chemical attack by Hg^+ ions. These requirements are evidently different for the PDP technology; in addition the phosphors must have a significant reflection coefficient in the visible to optimise the light output [40.23], and the surface quality is of greater significance due to the short penetration depth of the VUV. The phosphors which have been used so far include $BaMgAl_{10}O_{17}:Eu^{2+}$ (blue), $Zn_2SiO_4:Mn^{2+}$ (green), and $(Y, Gd)BO_3:Eu^{3+}$ and $Y_2O_3:Eu^{3+}$ (red). The blue phosphor is prone to degradation.

The widespread introduction of Xe excimer excitation in PDPs can be expected to stimulate applications in other lighting technologies. In this regard, the possibility of so-called *quantum cutting* is of much interest. This recognises that the energy of a VUV photon is equivalent to two or more visible photons, so that quantum efficiencies in excess of 100% can in principle be achieved. The difficulty lies in finding a luminescent ion system whose energy-level system provides for both efficient pumping and two-photon luminescence in the visible. One example which has been reported to have a quantum efficiency of up to $\approx 145\%$ is Pr^{3+} in YF_3 and other hosts [40.24, 25], where the excitation is through the allowed $4f^2 \rightarrow 4f^1 5d^1$ or host transitions. The system then decays to the $^1S_0$ excited state of the $4f^2$ configuration from which two-photon decay is possible through successive $^1S_0 \rightarrow {}^1I_6$ and $^3P_0 \rightarrow {}^3H_J, {}^3F_J$ transitions, as shown in Fig. 40.2. (The intermediate step from $^1I_6$ to $^3P_0$ is provided by a nonradiative transition.) A difficulty is that the photon for the first transition is in the UV region of the spectrum, and so it is necessary to incorporate a second luminescent ion pumped by this transition to convert the UV to visible output, and the visible quantum efficiency is necessarily reduced.

40.6.5 Scintillators

Although semiconductor detectors of ionising radiation are making increasing inroads into the particle detection market, traditional scintillators are still widely used and are indispensable for some applications. The operating principle is that an incident gamma ray creates a large number of electron–hole pairs in the scintillating material directly or indirectly through the photoelectric effect, Compton scattering, or pair production, and that some of the energy of recombination appears as photon emission from luminescent ions. Charged particles such as protons produce electron–hole pairs through the Coulomb interaction with the band electrons. The *scintillation* or multiphoton bursts which signal the event is detected by a photomultiplier, and the height of the output pulse from the photomultiplier is proportional to the energy of the particle. A pulse-height analyser sorts the pulses according to energy, and so an energy spectrum can be obtained. Key figures of merit for a scintillator material are the numbers of photons per MeV of particle energy, the radiative lifetime of the luminescent ion (since possible pulse overlap limits the maximum count rate which can be measured), and the weighted density ρZ_{eff}^4, which reflects the gamma sensitivity. (Here Z_{eff} is the effective atomic number.) A recent review of scintillators has been given by *van Eijk* [40.26]. The most widely used scintillator for many years has been NaI:Tl, but many different scintillators are being investigated, driven by the need for improved performance and lower cost for medical applications such as positron emission tomography (PET) and single-photon emission computed tomography (SPECT), and for large-scale elementary-particle facilities such as those at the Centre Européen pour la Recherche Nucléaire (CERN). In the latter regard, the Crystal Clear collaborative project and other programs have resulted in several new materials such as $LaBr_3$, $LaCl_3$, Lu_2SiO_5, Gd_2SiO_5, and $LuAlO_3$, all Ce-doped, and undoped $Bi_4Ge_3O_{12}$ and $PbWO_4$, with typical performance figures of 10 000–50 000 photons/MeV and lifetimes of 10–50 ns. For gamma spectroscopy the pulse-height resolution is critical and $LaBr_3$:Ce has twice the resolution of NaI:Tl. The fastest scintillators are based on core-valence luminescence where a hole created in a core level recombines with an electron in the valence band. For example, BaF_2 shows this *cross-luminescence* effect with a lifetime as short as 600 ps. The effect is only shown by materials with a core-valence (CV) band energy gap less than the usual band gap, otherwise CV luminescence is absorbed.

40.7 Representative Phosphors

To conclude, we present in Table 40.2 a list of several luminescent materials of practical significance. The table is intended to be representative rather than comprehensive. It is noticeable from the table that just a few ions are responsible for a large number of applications, and primarily as oxides.

Table 40.2 Some luminescent materials of practical significance

Host	Dopants	Colour	Excitation	Application
$Bi_4Ge_3O_{12}$	–	Blue	Ionising radiation	Scintillator
ZnS	Ag^+	Blue	Electrons	Colour TV screens
$Zn_{0.68}Cd_{0.32}S$	Ag^+	Green	Electrons	Colour TV screens
$Y_3Al_5O_{12}$	Ce^{3+}	Yellow	Blue, violet	White LED
Gd_2SiO_5	Ce^{3+}	UV	Ionising radiation	Scintillator
ZnS	Cu^+	Green	Electrons	Colour TV screens
BaFBr	Eu^{2+}	UV/blue	X-rays	X-ray imaging
$BaMgAl_{10}O_{17}$	Eu^{2+}	Blue	UV	fluorescent lamps, plasma displays
Sr_3SiO_5	Eu^{2+}	Blue	UV	White LED
$SrGa_2S_4$	Eu^{2+}	Green	UV	White LED
$SrAl_2O_4$	Eu^{2+}, Dy^{3+}	Green	UV, violet	Persistent phosphor
$CaAl_2O_4$	Eu^{2+}, Nd^{3+}	Blue	UV, violet	Persistent phosphor
Y_2O_3	Eu^{3+}	Red	Electrons, UV	Colour TV screens, fluorescent lamps
Sr_2SiO_4	Eu^{3+}	Yellow	UV	White LED
$(Y, Gd)BO_3$	Eu^{3+}	Red	UV	Plasma displays
Y_2O_3	Eu^{3+}	Red	UV	Plasma displays
SrY_2S_4	Eu^{3+}	Red	UV	White LED
LiF	Mg^{2+} and Ti^{4+}	UV//blue	Ionizing radiation	TL dosimetry
ZnS	Mn^{2+}	Yellow	Electric field	Panel displays
Zn_2SiO_4	Mn^{2+}	Green	UV	Plasma displays
$CeMgAl_{11}O_{19}$	Tb^{3+}	Green	UV	Fluorescent lamps

References

40.1 D. J. Vij: *Luminescence of Solids* (Plenum, New York 1998)
40.2 G. Blasse, B. C. Grabmeier: *Luminescent Materials* (Springer, Berlin, Heidelberg 1994)
40.3 G. H. Dieke: *Spectra and Energy Levels of Rare Earth Ions in Crystals* (Interscience, New York 1968)
40.4 W. T. Carnall, G. L. Goodman, K. Rajnak, R. S. Rana: J. Chem. Phys. **90**, 343 (1989)

40.5 E. Merzbacher: *Quantum Mechanics* (Wiley, New York 1970)

40.6 D. Curie: *Luminescence in Crystals* (Methuen, London 1962)

40.7 S. W. S. McKeever: *Thermoluminescence of Solids* (Cambridge Univ. Press, Cambridge 1985)

40.8 G. Kitis, J. M. Gomex-Ros, J. W. N. Tuyn: J. Phys. D. **31**, 2636–2641 (1998)

40.9 L. Botter-Jensen, S. W. S. McKeever, A. G. Wintle: *Optically Stimulated Luminescence Dosimetry* (Elsevier, Amsterdam 2003)

40.10 J. K. Park, C. H. Kim, H. D. Park, S. Y. Choi: Appl. Phys. Lett. **84**, 1647–1649 (2004)

40.11 S. Ken, O. Koji, K. Naoki, O. Masakazu, T. Daiichiro, H. Naoto, Y. Yominobu, X. Rong-Jun, S. Takayuki: Opt. Lett. **29**, 2001–2003 (2004)

40.12 P. L. Kim, P. E. Jeon, Y. H. Park, J. C. Choi, L. P. Park: Appl. Phys. Lett. **85**, 3696–3698 (2004)

40.13 D. Jia, R. S. Meltzer, W. M. Yen: Appl. Phys. Lett. **80**, 1535–1537 (2002)

40.14 D. Jia, X. J. Wang, E. van der Kolk, W. M. Yen: Opt. Commun. **204**, 247–251 (2002)

40.15 T. Aitasalo, A. Durygin, J. Holsa, J. Niittykoski, A. Suchocki: J. Alloys Comp. **380**, 4–8 (2004)

40.16 S. Schweizer: Phys. Status Solidi **187**, 335–393 (2001)

40.17 J. A. Rowlands: Phys. Med. Biol. **47**, R123–R166 (2002)

40.18 P. Hackenschmied, G. Schierning, A. Batentschuk, A. Winnacker: J. Appl. Phys. **93**, 5109–5113 (2003)

40.19 S. Schweizer, L. Hobbs, M. Secu, J.-M. Spaeth, A. Edgar, G. V. M. Williams: Appl. Phys. Lett. **83**, 449–451 (2003)

40.20 M. Secu, S. Schweizer, A. Edgar, G. V. M. Williams, U. Rieser: J. Phys. C: Condens. Matter **15**, 1–12 (2003)

40.21 A. Edgar, G. V. M. Williams, S. Schweizer, M. Secu, J.-M. Spaeth: Curr. Appl. Phys. **4**, 193–196 (2004)

40.22 J. P. Boeuf: J. Phys. D **36**, R53–R79 (2003)

40.23 H. Bechtel, T. Juestel, H. Glaeser, D. U. Wiechert: J. Soc. Inform. Display **10**, 63–67 (2002)

40.24 S. Kuck, I. Sokolska, M. Henke, M. Doring, T. Scheffler: J. Lumin. **102–103**, 176–181 (2003)

40.25 A. B. Vink, P. Dorenbos, C. W. E. Van Eijk: J. Solid State Chem. **171**, 308–312 (2003)

40.26 C. W. E. Van Eijk: Nuclear Instruments and Methods in Physics Research **A 460**, 1–14 (2001)

41. Nano-Engineered Tunable Photonic Crystals in the Near-IR and Visible Electromagnetic Spectrum

Photonic crystals offer a well-recognized ability to control the propagation of modes of light in an analogous fashion to the way in which nanostructures have been harnessed to control electron-based phenomena. This has led to proposals and indeed demonstrations of a wide variety of photonic-crystal-based photonic devices with applications in areas including communications, computing and sensing, for example. In such applications, photonic crystals can offer both a unique performance advantage, as well as the potential for substantial miniaturization of photonic systems. However, as this review outlines, two-dimensional (2-D) and three-dimensional (3-D) structures for the spectral region covering frequencies from the ultraviolet to the near-infrared ($\approx 2\,\mu m$) are challenging to fabricate with appropriate precision, and in a cost-effective and also flexible way, using traditional methods. Naturally, a key concern is how amenable a given approach is to the intentional incorporation of selected defects into a particular structure. Beyond passive structures, attention turns to so-called active photonic crystals, in which the response of the photonic crystal to light can be externally changed or tuned. This capability has widespread potential in planar lightwave circuits for telecommunications, where it offers mechanisms for selective switching, for example. This review discusses alternative proposals for tuning of such photonic crystals.

41.1	PC Overview	998
	41.1.1 Introduction to PCs	998
	41.1.2 Nano-Engineering of PC Architectures	999
	41.1.3 Materials Selection for PCs	1000
41.2	Traditional Fabrication Methodologies for Static PCs	1001
	41.2.1 2-D PC Structures	1001
	41.2.2 3-D PC Structures	1007
41.3	Tunable PCs	1011
	41.3.1 Tuning the PC Response by Changing the Refractive Index of the Constituent Materials	1011
	41.3.2 Tuning PC Response by Altering the Physical Structure of the PC	1012
41.4	Summary and Conclusions	1014
References		1015

Photonic crystals (PCs) are periodic, dielectric, composite structures in which the interfaces between the dielectric media behave as light-scattering centers. PCs consist of at least two component materials having different refractive indices, and which scatter light due to their refractive-index contrast. The one, two, or three-dimensional (1-D, 2-D, or 3-D) periodic arrangement of the scattering interfaces may, under certain conditions, prevent light with wavelengths comparable to the periodicity dimension of the PC from propagating through the structure. The band of forbidden wavelengths is commonly referred to as a *photonic band gap* (PBG). Thus, PCs are also commonly referred to as photonic-band-gap (PBG) structures.

PCs have great potential for providing new types of photonic devices. The continuing demand for photonic devices in the areas of communications, computing, and signal processing, using photons as information carriers, has made research into PCs an emerging field with considerable resources allocated to their technological development. PCs have been proposed to offer a means for controlling light propagation in submicron-scale volumes – the photon-based equivalent of a semiconductor chip – consisting of optical devices integrated together onto a single compact circuit. Proposed applications of PCs for the telecommunication sector include optical cavities, high-Q filters, mirrors, channel add/drop filters, superprisms and compact waveguides for use in so-called planar lightwave circuits (PLCs).

Practical applications of PCs generally require manmade structures, as photonic devices are designed primarily for light frequencies ranging from the ultraviolet to the near-IR regime (i.e., ≈ 100 nm to $\approx 2\,\mu$m, respectively) and PCs having these corresponding periodicities are not readily available in nature. 1-D PCs in this wavelength range may be easily fabricated using standard thin-film deposition processes. However, 2-D and 3-D PC structures are significantly more difficult to fabricate and remain among the more challenging nanometer-scale architectures to realize with cost-effective and flexible patterning using traditional fabrication methodologies. Recently, there has been considerable interest in PC-based devices that has driven advanced fabrication technologies to the point where techniques are now available to fabricate such complex structures reliably on the laboratory scale. In addition to traditional semiconductor nanostructure patterning methods based on advanced patterning/etching techniques developed by the semiconductor industry, novel synthesis methods have been identified for 2-D and 3-D periodic nanostructured PC arrays. There are several excellent reports reviewing these fabrication techniques in the literature [41.1–4] and this growing field has already been the subject of numerous recent reviews, special issues, and books in the area of theoretical calculations (both band-structure and application simulations), 2-D PC structures, 3-D PC structures, and opal-based structures [41.5].

Recently, there has been great interest in exploring the use of PCs for active applications in the field of telecommunications, such as in the area of PLCs (e.g., for optical switching). In such applications, the PC properties should be adjustable to create *tunable* photonic band gaps. This development increases the functionality of all present applications of PCs by allowing the devices in such applications to be adjustable, or tunable. We review recent developments in the engineering of tunable nanometer-scale architectures in 2-D and 3-D. This review aims to organize this ever-changing volume of information such that interested theorists can design structures that may be easily fabricated with certain materials, and such that technologists can try to meet existing fabrication gaps and issues with current systems.

The chapter begins with a brief introduction to PCs, followed by general criteria used to determine appropriate methodologies for the nano-engineering of tunable PCs. Finally, overviews of the most common fabrication methods for tunable 2-D and 3-D PCs will be given.

41.1 PC Overview

41.1.1 Introduction to PCs

The simplest PC structure is a multilayer film, periodic in 1-D, consisting of alternating layers of material with different refractive indices (Fig. 41.1). Theoretically, this 1-D PC can act as a perfect mirror for light with wavelengths within its photonic band gap, and for light incident normal to the multilayer surface. 1-D PCs are found in nature, as seen for example in the iridescent colors of abalone shells, butterfly wings and some crystalline minerals [41.6], and in manmade 1-D PCs (i.e., also known as Bragg gratings). The latter are widely used in a variety of optical devices, including dielectric mirrors, optical filters and in optical fiber technology.

The center frequency and size (i.e., frequency band) or so-called stop band of the PBG depends on the refractive index contrast (i.e., n_1/n_2, where n_1 and n_2 represent the refractive index of the first and second material, respectively) of the component materials in the system. Figure 41.1 is an example of a 1-D PC, with a periodic arrangement of low-loss dielectric materials. This multilayer film is periodic in the z-direction and extends to infinity in the x- and y-directions. In 1-D, a photonic band gap occurs between every set of bands, at either the edge or at the center of the Brillouin zone – photonic band gaps will appear whenever n_1/n_2 is not equal to unity [41.7]. For such multilayer structures, corresponding photonic band gap diagrams show that, the smaller the contrast, the smaller the band gaps [41.7]. In 1-D PCs, if light is not incident normal to the film surface, no photonic band gaps will exist. It is also important to note tha,t at long wavelengths (i.e., at wavelengths much larger than the periodicity of the PC), the electromagnetic wave does not probe the fine structure of the crystal lattice and effectively sees the structure as a homogeneous dielectric medium.

The phenomena of light waves traveling in 1-D periodic media was generalized for light propagating in any direction in a crystal, periodic in all three dimensions, in 1987 when two independent researchers suggested that

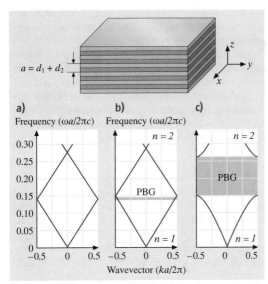

Fig. 41.1a–c Example of a 1-D photonic crystal. For a structure with $d = 0.5a$, the corresponding band gap diagrams are shown for: (**a**) GaAs bulk ($\varepsilon = 13$); (**b**) GaAs/GaAlAs multilayer ($\varepsilon_1 = 13$, $\varepsilon_2 = 12$); and (**c**) GaAs/air multilayer ($\varepsilon_1 = 13$, $\varepsilon_2 = 1$) (from [41.7])

light propagation in 3-D could be controlled using 3-D PCs [41.8,9]. By extending the periodicity of the 1-D PC to 2-D and 3-D, light within a defined frequency range may be reflected from any angle in a plane (in 2-D PBG structures) or at any angle (in 3-D PBG structures).

Since the periodicity of PCs prevents light of specific wavelengths (i. e., those within the photonic band gap) from propagating through them in a given direction, the intentional introduction of *defects* in these structures allows PCs to control and confine light. Propagation of light with wavelengths that were previously forbidden can now occur through such defect states located within the photonic band gap. Defects in such PCs are defined as regions having a different geometry (i. e., spacing and/or symmetry) and/or refractive-index contrast from that of the periodic structure. For example, in a 2-D PC consisting of a periodic array of dielectric columns separated by air spaces, a possible defect would include the removal of a series of columns in a line. Specific wavelengths of light forbidden from propagating through defect-free regions would then be able to propagate through the line defect, but not elsewhere. Indeed, by appropriately eliminating further columns, light may be directed to form optical devices, including, for example, a low-loss 90° bend in a 2-D

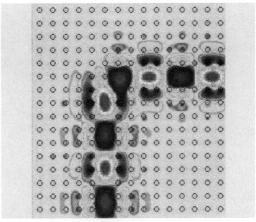

Fig. 41.2 Theoretical simulation of a low-loss 90° bend in a 2-D waveguide [41.7]

waveguide, as shown by theoretical simulation [41.7] (Fig. 41.2).

Clearly, photons controlled and confined in small structures, with size on the order of the wavelength of light, using the extremely tight bend radii offered by PCs would facilitate miniaturization and the fabrication of PLCs [41.7]. In addition, since the periodicity of the PC gives rise to the existence of band gaps which change the dispersion characteristics of light at given frequencies, defect-free PCs give rise to other interesting phenomena, including highly dispersive elements, through the so-called superprism effect [41.10]. Possible designs of PC-based optical devices using such properties have been extensively explored [41.1, 11], generating much excitement in the field of optical telecommunications [41.6].

41.1.2 Nano-Engineering of PC Architectures

Most of the promising applications of 2-D and 3-D PCs depend on the center frequency and frequency range of the photonic band gaps. A so-called *complete*, *full*, or *true* PBG is defined as one that extends throughout the entire Brillouin zone in the photonic band structure – that is, for all directions of light propagation for photons of an appropriate frequency. An incomplete band gap is commonly termed a *pseudo-gap* or a *stop band*, because it only occurs for reflection/transmission along a particular propagation direction. A complete gap occurs when stop-band frequencies overlap in all directions in 3-D. The center frequencies and stop-band

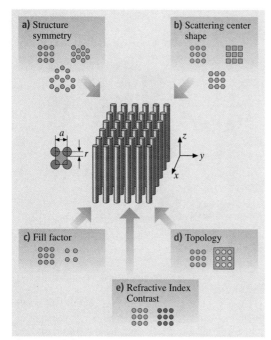

Fig. 41.3 PBG parameters that affect the center frequency and associated stop band (shown in *top view*, in *boxes*) relative to the 2-D periodic arrangement of one material (i.e., the rods, in *grey*) embedded in a second material (i.e., air), periodic in the *x–y* plane (shown in the *central figure*)

locations of the PBGs critically depend on the unit-cell structure [41.7–9, 12, 13]. In particular, the PC properties depend on the symmetry of the structure (i.e., the unit-cell arrangements), the scattering element shape within the unit cell, the fill factor (i.e., the relative volume occupied by each material), the topology, and the refractive-index contrast (Fig. 41.3).

It has been shown that a triangular lattice with circular-cross-section scattering elements in 2-D, or a face-centered-cubic/diamond lattice with spherical scattering elements in 3-D, tend to produce larger PBGs [41.7]. Also, as discussed above, the dielectric contrast is an important determining factor for 2-D and 3-D PC structures. The lower the dimension of the structure, the more readily are PBG manifested, since overlap of the PBGs in different directions is more likely (something which is a certainty in 1-D structures). In 3-D structures, calculations have determined that the minimum dielectric contrast (n_1/n_2) required to obtain a full PBG is about 2.0 [41.2, 12, 14]. For full PBGs, the ideal structure typically consists of a dielectric–air combination, to obtain both the greatest dielectric contrast as well as to reduce losses associated with light propagation in optical materials other than air [41.7].

The relationship between the fill factor, structure symmetry, and scattering element shape on the size and location of the PBG is complex [41.7, 15–18] and will not be discussed in detail here. However, it is clear that the ability to adjust, or tune, one or more of these parameters and thus tune the PC properties, is a very exciting development for future PC device applications.

41.1.3 Materials Selection for PCs

The physical architecture of a given PC is just one of the design considerations – another important one is the optical properties of the PC materials. In particular, the refractive index of a given material and its electronic band gap determine the performance and appropriate range of frequencies of PC devices fabricated from such a material.

For large PBGs, component materials need to satisfy two criteria; firstly, having a high refractive-index contrast, and secondly, having a high transparency in the frequency range of interest. This may be a challenge to satisfy at optical wavelengths. Suitable classes of PC materials include conventional semiconductors and ceramics, since at wavelengths longer than their absorption edge (or electronic band gap), they can have both high refractive indices and low absorption coefficients. In addition, these materials have other very useful electronic and optical properties that may complement the functions served by the presence of the PBG.

Another, often overlooked, consideration for PC material selection is the ability to translate the desired bulk single-crystal properties of well-known materials into nanometer-scale PC properties. Such structures typically have extremely large surface areas, and the microstructure of the constituent PC elements must be controlled during fabrication. Consequently, the final properties of the PC elements often vary from those of the bulk properties. This is extremely relevant when considering the functionality of a final PC device, such as those relying on electronic properties (e.g., lasing).

Finally, it is interesting to note that there is a strong correlation between the fabrication methods selected and the component materials. For example, top-down dry etching of semiconductors (e.g., Si, GaAs, InP, etc.) into nanostructured 2-D arrays is well characterized and relatively commonplace, whereas the same cannot be said for ceramic materials. The opposite is generally true for chemical or bottom-up synthesis using sol–gel technology, in which ceramic PCs are relatively easy to fabricate when sol–gel-based infiltration techniques are used (e.g., in 3-D inverse opal fabrication). It is also clear that the fabrication technique primarily determines the PC structure that can be fabricated, which in turn, determines the PC properties.

41.2 Traditional Fabrication Methodologies for Static PCs

Unlike for electrons, where there are natural length scales, for photons there is no such length scale [41.19]. This lack of an absolute length scale ensures that the physics of PCs is scalable: resizing the system rescales the energy in such a way that the spectrum, in units of c/a (where c is the speed of light and a is the lattice parameter of the PC), is independent of the system size [41.19]. Since $\omega/(2\pi c/a) = a/\lambda$, it is customary to use such dimensionless units to measure the photon energy [41.19]. Besides spatial scaling, there is also dielectric scaling – that is, there is no fundamental scale for ε. Two systems whose dielectric functions scale by a factor, $\varepsilon'(r) = \varepsilon(r)/s^2$ have spectra that are scaled by the same factor: $\omega' = s\omega$, which means that increasing the dielectric function by a factor of four decreases the photon energy by a factor of two [41.19].

The apparent scalability of PCs is misleading because it suggests that PCs are truly size-independent, which in practice is not the case. As the size of the PC elements decrease, the effect of structural inhomogeneities become increasingly important, leading to inherent size limitations for particular architectures and for particular materials systems. This is most evident in changes in effective refractive index (e.g., due to the formation of inter-granular phases, surface oxides, etc.) and functionality.

2-D systems are increasingly being used for microphotonic applications and consist of periodic repetitions of objects in a 2-D arrangement [41.19]. 3-D systems present a modulation in the third direction, such as a stack of spheres [41.19]. To date, 2-D and 3-D PCs have been fabricated using various approaches, including those based on electron-beam lithography, self-assembly and templating. The stringency of the requirements for a fabrication methodology, are dictated by the application of a given PC, in conjunction with associated requirements for material and architecture. Currently, there remains a great need for fabrication methods for mass production of nanometer-scale PCs, offering flexibility in material composition and design, while overcoming the low throughput of serial lithographic patterning, and the low reproducibility and material restrictions of self-assembly methods. Many excellent reports reviewing these fabrication techniques are available in the literature [41.1–4]. For readers unfamiliar with traditional PC fabrication techniques, the following sections review typical 2-D and 3-D PC fabrication methods, based on lithography, self-assembly and hybrid techniques.

41.2.1 2-D PC Structures

Introduction to 2-D PBG Structures

The advantage of the 2-D PC structure is that, due to its planar form, it is easier to fabricate using existing lithographic methods, it retains its original single-crystal properties, and it can be simply integrated into desired PLC designs. However, there are specific design limitations that must be considered for 2-D PC structures. The main problem that has been identified both theoretically and experimentally is diffraction losses in the third dimension (i.e., for light directed out of the plane of the 2-D PC) [41.1, 2].

Calculations often assume a perfect and infinite 2-D material. However, the aspect ratio in experimental structures fabricated using typical micromachining techniques (e.g., reactive-ion etching) tend to be very limited [41.2]. Thus, in order to confine light in the third dimension (i.e., the out-of-plane dimension), the structure must either be large compared to the beam size (i.e., many wavelengths deep), or must be sufficiently confined within a waveguide to experience the full interaction with the periodic lattice [41.1]. As a consequence, 2-D PC structures are now typically fabricated in three principal forms: (1) a 2-D array of air holes in a dielectric (or dielectric pillars of very high aspect ratio in air), (2) a slab waveguide consisting of a 2-D array of holes perforating a thin membrane (with the membrane thickness approximately equal to the hole diameter) and (3) a heterostructure waveguide (i.e., similar to those used for confining light and carriers in semiconductor laser heterostructures) with holes drilled through the heterostructure [41.20]. New fabrication methodologies need to be able to address the fabrication issues for each of these three types of low-loss 2-D PC structures in order to implement these structures in practical device applications.

2-D Lithographic PC Fabrication Methods

The majority of semiconductor PCs are synthesized using the advanced lithographic and etching capabilities developed for the semiconductor microelectronic industry [41.1, 2, 4, 21]. One major benefit of being able to fabricate PCs from single crystals using top-down lithography is that the properties (e.g., nonlinear electro-optic properties) are typically superior to those for polycrystalline structures produced using bottom-up (e.g., template-based) processes [41.22]. Utilizing the former materials enables significant modulation of the optical properties suitable for developing tunable PC devices. Also, in top-down fabrication, the optical and dielectric properties of the bulk material may be characterized before PC fabrication, allowing them to be more effectively modeled after PC fabrication. Indeed, semiconductor-based PBG structures made using lithographic processes tend to be more reproducible than self-assembled PC structures. Most importantly, there is much greater flexibility in the design and implementation of lithographically fabricated structures, and hence to implementing functional devices. Finally, PC-based PLCs are ideal for optical system integration.

For high refractive-index contrast (i.e., larger photonic stop bands), the preferred PBG structure is a topologically patterned substrate (i.e., an air–solid PC), which may be fabricated using multistep submicron machining based on lithography. In this process, a radiation-sensitive resist is patterned to transfer structures to the bulk material. The patterned resist is developed (i.e., the properties of the exposed material are changed), followed by etching and/or metal thin-film deposition to define features in the bulk material. The majority of submicron machining processes involve exposure of the resist with either electromagnetic radiation (e.g., optical, UV or X-ray photons) or charged particles (electrons, low-energy heavy ions, high-energy light ions) [41.23]. Although resist exposure is essentially a surface micromachining technique and therefore planar, various wet and dry etching techniques have been successfully used to produce topologically varying microstructures in the axial direction [41.23].

Similar to all resist-based processes, the resist itself limits the ultimate minimum size and maximum density of nanostructured patterns [41.24, 25], and organic residues (from processes resist) are often sources of contamination [41.26]. Organic resists are chosen as a compromise between sufficient resilience for selected postprocessing procedures, and having sufficiently high sensitivity and contrast (i.e., a low energy threshold for resist exposure and total resist exposure occurring only above a certain energy threshold with no resist exposure occurring below, respectively) to the incident exposure [41.27] – this tends to be a difficult compromise to satisfy in practice for traditional resist materials.

Lithographic methods provide a simple and straightforward approach for the intentional introduction of defects into these 2-D systems (due to the simple surface patterning), and this top-down method may be the only way to retain the bulk-like semiconducting properties that are required for certain functionalities (e.g., such as lasing).

Electromagnetic Radiation Patterning. Generally, using electromagnetic radiation to micromachine structures requires masks to expose a resist material in a spatially selective manner, with subsequent development of the exposed resist to produce microstructures. Masks typically incorporate multiple repeat patterns, enabling multiple microstructures to be fabricated in a single exposure, which significantly reduces costs [41.23]. Masks tend to be less useful for patterning with charged particles, since the high energy deposited into the mask during exposure can produce undesirable mask instabilities due to heat expansion, stress and damage [41.23]. Charged-particle micromachining is therefore normally limited to direct-write processes, where a focused charged-particle beam is scanned over a material in a specific pattern to produce microstructures. Such a direct-write process has the advantage of not requiring a mask, but suffers from relatively low throughput as a serial write process, and hence losses its attraction for multiple component production [41.23].

A main consideration of patterning approaches for PC fabrication is the minimum required feature size. Diffraction effects occurring for wavelengths of light below 100 nm impose a fundamental limitation on achievable structure sizes using optical lithography. Micromachined features smaller than about 250 nm are essentially beyond current readily available optical lithography systems [41.23], although recent advances (e.g., in UV lithography) will probably make this size regime achievable in the near future using a parallel patterning processes. Other techniques such as X-ray lithography that are capable of high-volume production, by allowing parallel exposure for larger-area nanometer-scale patterning, are hampered by challenges including making suitable high-resolution masks, as well as the significant capital cost required to achieve the high photon flux suitable for mass production of nanometer-scale components [41.23].

Charged-Particle Beams. Limitations on the minimum feature size using traditional photolithography have motivated a number of studies using serial patterning with charged-particle beams (e.g., electron-beam lithography (EBL) or, less commonly, focused ion-beam (FIB) lithography) to obtain the small feature sizes required in PC structures.

The advantages of EBL and FIB for defining sub-100-nm feature sizes in a mask-less fashion are offset by the fact that high-volume production is commercially impractical using such serial writing techniques [41.23, 28, 29]. Conventional ion implantation, though widely available and robust, has thus far been incapable of patterning densely organized nanostructures due to ion scattering [41.27, 30] and mask limitations [41.31–35].

Another consideration for choosing lithographic techniques for sub-100-nm-scale feature fabrication is that all the steps in this multistep process have their own limitations. Indeed, the finite size of the molecular compound comprising the resist sets the ultimate limit on pattern resolution [41.24, 36]. Since all traditional submicron micromachining techniques except FIB require resist exposure, this is a major factor for determining resolution – for example, the ultimate resolution of EBL is not determined by the electron optical system, which is ≈ 0.1 nm, but by the resolution of the resist and by the subsequent fabrication process [41.37]. Also, electrons in the beam scatter very easily from electrons in the resist material, and both scattered and secondary electrons cause a lateral spread in the patterned material due to diffusion effects (i.e., proximity effects) [41.23]. Thus, the very high spatial resolution achievable at the surface using EBL deteriorates as the beam penetrates into the resist, making EBL unsuitable for high-resolution topographically patterned structures using resist alone [41.23]. The advantage of mask-less processing for high-definition surface structures in resist and the inherently slower processing mean that EBL has been limited to mask-making and direct-writing on wafers for specialized applications such as fabrication of 2-D PCs for research purposes only [41.23].

Currently, the most commonly used method to fabricate 2-D PCs in the optical regime is EBL-based mask processing, combined with etching [41.1]. This approach has been used to pattern different semiconductor materials, including Si [41.4] and GaAs [41.1, 4, 7] (Fig. 41.4). Some detrimental effects arising from dry etching have been reported, including damage and ion

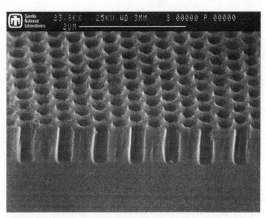

Fig. 41.4 Example of a 2-D PC structure in GaAs [41.7]. The air hole radius is 122.5 nm, the lattice parameter is 295 nm, and the holes are 600 nm deep

channeling that increase nonradiative recombination in active microstructures [41.1]. Other post-mask procedures used to fabricate 2-D PC structures include vertical selective oxidation and lift-off techniques [41.1]. In vertical selective oxidation, an *all solid* 2-D PC structure may be fabricated, which maintains the physical integrity of the crystal and allows deposition of electrical contacts afterwards. However, the oxidation process usually also occurs laterally, while typical desired configurations for 2-D PC crystals are structured in the vertical direction. Even with improved vertically selective oxidation techniques, the depth-to-diameter aspect ratios of pillars are around 1 : 1, which is significantly less than the 10 : 1 ratios achieved using dry etching techniques [41.1]. Also, volume shrinkage of up to 13% [41.1] can lead to strain, the formation of micro-cracks and other structural problems. 2-D PC membranes have been fabricated using epitaxial lift-off/substrate removal, but this method is limited to materials with sufficient wet-etch selectivity [41.1].

Despite its extremely low throughput, focused ion beams are presently used for direct, resistless, nanometer-scale surface modification, micromachining, and ion implantation. This is primarily because such patterning cannot be achieved with any other method. However, FIB patterning is inherently slow and expensive, and requires sophisticated optics for ion focusing. Also, the patterning resolution and proximity of nanometer-scale features is greatly limited by the significant FIB tail distribution [41.26, 36] and the highly dose-dependent sputtering yield of FIBs [41.38]. The use of FIB patterning has not been widely used to fabricate PCs.

Table 41.1 Summary of current submicron resist-based machining techniques [41.23]

Technique	Potential sub–100nm capability	Requires mask	Direct write	Facility availability	High-volume production capability	3-D capability
Optical lithography	No	Yes	No	Widespread industrial use	Yes	No[a]
X-ray lithography	Yes: depends on mask	Yes	No	Scarce and expensive	Yes	Yes
UV lithography	Yes: depends on mask	Yes	No	Scarce	Yes	Yes
Electron-beam writing	Yes: currently achieved	No	Yes	Medium availability	No	No
Low-energy ion beam writing	Yes: depends on focusing	No	Yes	Scarce	No	No[b]
High-energy ion beam writing	Yes: depends on focusing	No	Yes	Very scarce	No	Yes

[a] Optical lithography is a surface technique. However, new supplementary dry etching techniques such as plasma etching are being used to produce submicron 3-D structures from 2-D surface structures in silicon

[b] Low-energy ion-beam micro-machining (FIB) relies on sputtering atoms from the surface. 3-D structures can be produced by continuous erosion at a given location although, in practice, this is far too slow for practical 3-D microcomponent production.

Light ions collimated by stencil masks can be used to expose resists in a parallel fashion in both ion-projection lithography and masked ion-beam lithography [41.27]. The resist is exposed in a similar manner to X-ray and EBL. Light ions are generally preferred over heavier ions in ion lithography because they have the greatest range in resists (thick resists can be used to give low defect densities [41.31]), but protons also scatter more than heavier ions at the end of the implant range, limiting resolution and nanometer-scale feature proximity. However, the use of heavier ions to pattern, following the traditional approach used for semiconductor doping, can lead to mask erosion. This effect is negligible when light ions (e.g., hydrogen) are used, but can become pronounced for higher-dose implantation of heavier ions [41.27]. A summary of current lithography-based submicron micromachining techniques is given in Table 41.1.

2-D Self-Assembled PC Fabrication Methods

For 2-D PC structures, electrochemical methods (i.e., anodic etching and growth) and fiber-pulling have used to prepare extremely high-aspect-ratio ($\gg 50:1$) large-area nanometer-scale hole arrays (Fig. 41.5). These types of structures represent near-ideal infinite 2-D PC structures [41.20]. However, these methods are limited to specific materials (e.g., silicon [41.41] and alumina [41.39] for electrochemical methods, or glass [41.40] by fiber-pulling), and processes such as electrochemical etching are often limited in their capability to control fill factor, structure symmetry and scattering element shape. 2-D PC structures with long-range periodicity obtained by electrochemical etching often require surface pre-patterning by serial lithography to obtain lattice dimensions in the visible to near-IR range.

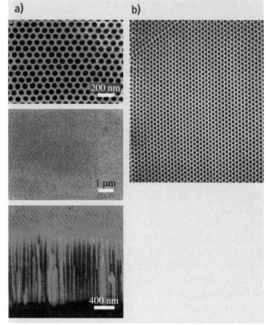

Fig. 41.5a,b Examples of 2-D self-organized PC structures of: (a) NCA [41.39], (b) NCG (nano channel glass) [41.40]

Table 41.2 Parameters of current high-aspect-ratio 2-D periodic structures [41.42–44, 46]

Nanotemplate type	Alumina	Silicon	Glass
Pore size (nm)	$1-10^2$	$400-10^4$	> 10
Pore density (cm^{-2})	10^9-10^{12}	10^6-10^9	10^{10}
Pore arrangement	Pseudo-ordered[a]	Disordered[a]	Ordered
Thickness (nm)	10^3	10^3	10^3
Temperature tolerance	$< 450\,°C$	$< 900\,°C$	$< 600\,°C$

[a] May be ordered using pre-patterning processes [41.42, 45]

Some examples of easily available, high-aspect-ratio nanoporous arrays include electrochemically grown porous alumina (nanochannel alumina, or NCA) [41.39], electrochemically grown silicon [41.42] and fiber-pulled nanochannel array glass (Table 41.2) [41.43, 44].

Nanochannel Alumina. NCA may be fabricated economically with well-characterized properties. Pseudo-ordered regions can be easily fabricated using electrochemical etching, compared with porous silicon, which requires back-side illumination in addition to an applied bias. The pore arrangement is completely disordered in silicon if no pre-patterning with optical or EBL is used [41.46].

Aluminium is electrochemically oxidized to alumina (Al_2O_3) under positive polarization. For over a century, the growth of disordered pore arrangements have been observed and studied for selected electrolytes that weakly dissolve alumina. However, in 1995 it was first discovered that, after an extended period of anodization, self-ordered porous alumina films with pores arranged in a hexagonal pattern are obtained at the growth front [41.39]. Although knowledge of how to fabricate such NCA templates is now widespread, the formation mechanism of hexagonal NCA has still not been completely elucidated. Mechanical stress between neighboring pores due to the volume expansion of alumina with respect to the aluminium substrate has been proposed as a mechanism for the self-ordering [41.46].

Controlled single-domain porous arrays may be obtained by lithographically pre-patterning the aluminium substrate, and applying the optimum potential for the corresponding inter-pore distance [41.47]. Recently, in addition to the hexagonal lattice, square and honeycomb lattices with square or triangular pore shapes have also been obtained by appropriate pre-patterning [41.48].

Extremely high-aspect-ratio NCA templates can be fabricated with periodicities ranging from ≈ 80 to ≈ 600 nm. The domain sizes (i.e., areas with the same orientations) are controlled by the anodization conditions and increase with time to micrometer-scale sizes. The thickness of the NCA template can be controlled by the electrochemical etching time, from ≈ 50 nm to $> 10\,\mu m$ thick.

Current Alternative 2-D PC Fabrication Methods.
Templating of 2-D Structures. 2-D PCs have been fabricated by templating a secondary material into a PC backbone, that can either remain in the final structure as a composite material or be removed to form a dielectric/air PC structure. In many cases, a dielectric/air PC structure is preferred due to the higher dielectric contrast, but depending on the original PC backbone material, it may not be possible for the secondary materials to be removed after templating.

Over the past decade NCA templates have been commonly used to fabricate various nanostructured arrays by either dry etching or by growth through templates [41.45, 49, 50]. This method of using self-assembled templates for nanostructure patterning has been used to pattern physical topology, resulting in either nanostructured hole arrays (e.g., by dry etching through the templates), or growth of nanostructured *wire* arrays [e.g., by metalorganic chemical vapor deposition (MOCVD), MBE, and chemical-solution deposition]. The effectiveness of NCA as a templating tool results from the fact that the alumina may be easily selectively chemically etched compared with semiconductor materials, and forms a robust mask for dry etching (compared to, for example, patterned resist materials).

Combining Self-Organized 2-D Structures with Ion Implantation. Ion-beam techniques may be used to pattern, for example, sub-100-nm-resolution features on thin membranes, but they are unsuitable for deep sub-surface patterning, particularly at the end of the implantation range. One reason for this is that stencil-type masks need to be used, due to the relatively short range of energetic ions compared to, for example, X-rays, which do not require stencil-type masks [41.24, 27]. In the past, high-aspect-ratio stencil masks for ion implantation were fabricated using techniques such as EBL, but such stencil masks were limited by the low pattern density achievable using such fabrication methods, requiring the use of self-complementary masks with multiple offset exposures to create printed patterns [41.34]. Also, the minimum resolution using nanoscale ion implantation through masks is so sensitive to the mask's wall angle [41.32, 33] that advanced photoresists and lithographic techniques need to be used to achieve such ideal masks.

Investigations have been reported on new ways to combine self-organized 2-D PC structures with conventional, broad-beam ion implantation to achieve nanoscale pattern transfer [41.51, 52]. Using self-organized templates as masks significantly simplifies the process, as no resolution-limiting resist or no time-intensive serial patterning is necessary, and these ideal, stencil-type masks allow direct nanostructure array transfer with significantly less scattering than solid masks. The use of self-organized masks allows this process to be economically feasible for large-area nanostructure array patterning. This approach transfers the desired pattern from the self-organized template into a desired material that cannot be patterned on the same scale using self-organization.

Despite concerns about mask edge defects [41.53–55], extended radiation damage accumulation extending below the mask edges [41.30, 35] and possible mask distortion from stress and temperature [41.31], it has been shown that nanoscale features could be retained using conventional ion implantation combined with high-aspect-ratio masks to make PC-type structures [41.51, 52] (Fig. 41.6). Energetic, charged particles were accelerated into a substrate using broad-area parallel implantation through NCA templates to pattern 2-D nanostructured arrays of ions in both semiconductor (i.e., InP) and ceramic (i.e., $SrTiO_3$) single crystals [41.51, 52]. The ion-implanted volume was selectively etched away after implantation to form 2-D nanostructured air–hole arrays in the substrate material. The array pattern was successfully fabricated over the entire implant area ($\approx 1\,cm^2$), with resulting minimum feature sizes of ≈ 40 nm. Such a direct and parallel method, combining ion implantation with a nanoscale mask, can be immediately beneficial, particularly in applications where using FIB is impractically slow (e.g., for substrate nanostructure array pre-patterning for selective or electrochemical etching). In addition, complex profiles (e.g., membrane formation) were shown to be possible using this technique. This method allows ions to be implanted into nanoscale areas, which may result in additional substrate modification, including, but not limited to, micromachining of implanted areas to fabricate topologically patterned nanostructured arrays.

The ability to process structures of different form and shape using ion-implantation techniques offers opportunities for nano-machining, and enables material processing on a nanometer scale – this is applicable to materials which cannot be structured using standard lithography (e.g., samples with pronounced topography [41.38]). This nanoscale patterning method can also be used to directly pattern substrates to create electrochemical initiation sites for fabrication of ordered porous Si and porous GaAs structures for light emission [41.56] and for 3-D PC fabrication in Si [41.57]. In addition, such array pattern transfer is applicable to pre-patterning of self-assembled templates themselves,

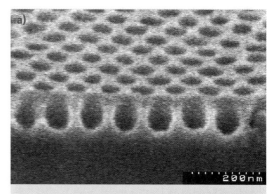

Fig. 41.6a,b Example of nanoscale features formed in a $SrTiO_3$ (100) substrate using conventional ion implantation through high-aspect-ratio masks combined with selective wet etching [41.51, 52]. Scanning electron microscope images for: (**a**) tilt view of single implant and (**b**) multiple energy implants to create membrane structure

as a substitute for lithographic pre-patterning for the fabrication of ordered nanochannels in materials such as Si and Al$_2$O$_3$ [41.46].

41.2.2 3-D PC Structures

Introduction to 3-D PBG Structures
A main advantage of 3-D PBG structures over 2-D PBG structures is that the 3-D PBG structures can overcome the problem of diffraction losses in the third (out-of-plane) dimension [41.1, 2]. However, this advantage is countered by the significant challenges associated with fabricating 3-D PBG structures. Fabricated 3-D structures reported to date lack the reproducibility and design versatility required for functional devices [41.21].

3-D Lithographic Fabrication Methods
Fabrication of 3-D PCs using top-down micromachining techniques has been more difficult than for 2-D PCs.

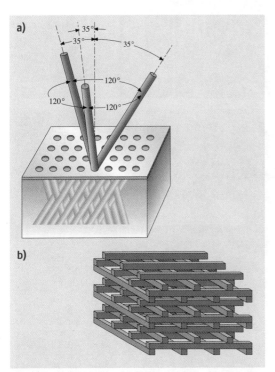

Fig. 41.7a,b Examples of 3-D photonic crystal structures: (**a**) *Yablonovite*, obtained by drilling holes in a homogeneous dielectric material (15 GHz); and (**b**) *wood-pile* structure, obtained by stacking micromachined Si rods (500 GHz) [41.2]

The first experimental demonstration of a full 3-D photonic band gap was performed in the microwave region (i.e., at 15 GHz, λ ≈ 2 cm), due to the relative ease of fabrication of appropriate structures on the centimeter scale [41.58] (Fig. 41.7). Another case of a full PBG was demonstrated at 500 GHz (λ ≈ 600 μm) with a 3-D structure fabricated by assembling micromachined silicon dielectric rods [41.59] (Fig. 41.7) into log-pile structures [41.60, 61].

In a 3-D lattice, theoretical calculations have determined that the band gap position is generally about one half to one third of the periodicity scale [41.62]. This translates into patterning the dielectric materials with features of 100–300 nm in all three dimensions, with required alignment accuracies among the different layers exceeding tens of nanometers [41.62]. In addition, with air as the second dielectric, the empty volume required for large PBGs is so large (≈ 70–80%) that the mechanical stability of the structure becomes very poor, adding another challenge for the etching procedure [41.2]. Often 3-D PC structures are so full of defects after such etching that the systems no longer exhibit significant PC behavior [41.2].

Due to difficulties in fabricating structures in the optical or near-infrared frequencies [41.2, 62, 63], only a few experimental systems have recently demonstrated the signature of a complete PBG. Lithographically derived 3-D structures require many complex fabrication steps, and it takes many months to make a single structure. Some examples of prototype 3-D PC include log-pile structures fabricated using layer-by-layer patterning (i.e., repeated deposition and etching) of Si [41.64] (Fig. 41.8), and structures fabricated using wafer fusion combined with etching in GaAs and InP [41.65] (Fig. 41.8). In the former case, processing difficulties have limited the number of layers in such 3-D periodic structures (i.e., to below 10) and in the latter cases, to materials other than those commonly used in microelectronics [41.62]. Sequential deposition of dielectric films on a patterned substrate may be used to produce 3-D periodic structures in the *autocloning* process [41.66], which is considered to be a relatively cheap and flexible process. However, the range of structures that can be produced with this technique is limited. Also, the technique does not allow the component dielectrics to be continuously connected in 3-D [41.67].

Hole-drilling has also been implemented using X-ray lithography [41.68] and can be scaled to optical frequencies by directional etching through a mask on a semiconductor surface [41.69]. Other

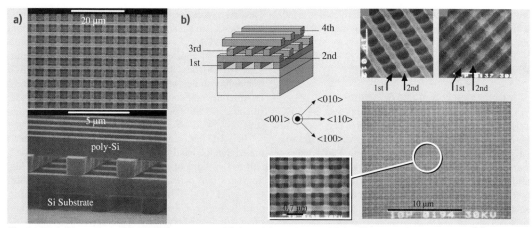

Fig. 41.8a,b Examples of 3-D PBG structures fabricated by micromachining/lithographic techniques: (**a**) advanced lithography [41.64], (**b**) wafer fusion [41.65]

fabrication methods include direct laser writing, but its resolution is currently only suitable for infrared structures.

3-D Self-Assembly Methods: Inverse Opals

The technological challenges involved in extending microlithographic techniques to 3-D patterning have made self-assembly methods attractive for 3-D PC fabrication. The benefits of using this type of bottom-up assembly for 3-D fabrication include its low capital cost, and its parallel and rapid nature, which allows the ordered fabrication of 3-D PBG-type structures over macroscopic length scales [41.3, 62, 73, 74]. However, self-assembly methods for 3-D PC fabrication are often material-specific, and templated bottom-up PCs tend to have diluted properties compared to bulk single crystals, typically containing unintentional defects and lacking reproducibility.

Self-organized 3-D colloidal crystal structures in a face-centered-cubic lattice (i.e., artificial opals) can be fabricated through sedimentation of colloidal suspensions (i.e., typically spheres, but other structures, such as micelles formed from rod–coil block copolymers have been used). Previously, colloidal crystals were not considered appropriate for PCs with band gaps in the visible and IR ranges because the colloids [typically SiO_2 or polymer materials such as polystyrene (PS) [41.75, 76]

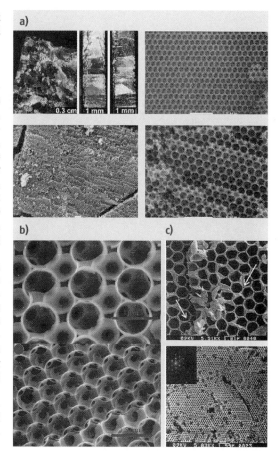

Fig. 41.9a–c Examples of 3-D inverse opal PBG structures: (**a**) graphite (using CVD infiltration) [41.70], (**b**) Si (using CVD infiltration) [41.71], (**c**) titania (using liquid-phase chemical reaction) [41.72]

or poly (methyl-methacrylate) (PMMA)] have refractive indices ≈ 1.5, which are too low to obtain a full PBG [41.77–79]. Such structures have weak refractive-index contrast, but theoretical calculations have shown that the final porosity of the artificial opal is too small to ever obtain a full PBG, even with very high refractive-index contrast [41.1, 3, 73].

These colloidal structures may also be infiltrated with a second material of higher refractive index to potentially improve the dielectric contrast [41.77, 80], while providing the required dimensional control and low solid fraction [41.74]. So called *inverse* structures can be formed when the original colloidal structure is removed to leave a macroporous structure. Calculations have shown that these inverse structures give rise to larger PBGs for the same refractive-index contrast [41.81]. The porous materials obtained using this approach have also been referred to as *inverse opals* or *inverted opals* because they have an open, periodic 3-D framework complementary to that of the parent opal. Modeling suggests that a porous material consisting of an opaline lattice of interconnected air balls (embedded in an interconnected matrix with a higher refractive index) should produce a full PBG. Optimum photonic effects require that the volume fraction of the matrix material should be 20–30%. Template-based synthesis based on colloids is attractive because the periodicity of this system can be conveniently tuned and a wide variety of materials with relatively high refractive index can be used [41.73]. There have been many examples of inverse opals using materials such as carbon, titania, zirconia, Si and Ge reported in the literature [41.3, 70–74, 82–89] (Fig. 41.9).

The inverse structure is made by infiltrating the interstitial void spaces (≈ 26% by volume) between the colloidal spheres in the opaline array with a fluid precursor. The fluid may be a pure liquid (e.g., a liquid metal alkoxide or a molten metal), a solution (e.g., a salt solution), a vapor (e.g., using CVD), or a colloidal dispersion of nanocrystals (e.g., colloidal gold particles) [41.73, 74]. Alternately, suspensions containing nanoparticle precursors (with sizes of ≈ 1–50 nm) and monodisperse spheres can be co-precipitated into an ordered structure [41.73, 74]. The void spaces between the colloidal spheres has also been filled using a variety of materials by electrochemical deposition [41.73]. The fidelity of this procedure is mainly determined by van der Waals interactions, the wetting of the template surface, kinetic factors such as the filling of the void spaces in the template, and the volume shrinkage of precursors during solidification [41.73].

The choice of wall material depends on the desired optical functionality of the PC, as its properties depend on the transparency region of the dielectric material. The availability of a precursor that can infiltrate the voids between the colloidal spheres without significantly swelling or dissolving the template is critical. For example, high-temperature infiltration techniques exclude the use of polymer templates. The precursor must be carefully chosen to allow for shrinkage occurring during solidification and the final grain size, which affects the smoothness, density, effective refractive index, PC wall structure and the mechanical properties of the final 3-D PC [41.74]. The approach and conditions used for infiltration can affect the filling efficiencies (i.e., the effective wall density) which, in turn, affect the mechanical strength and reduces the average refractive index of the walls [41.74, 82]. Choice of template also depends on the method of solidification of the precursor and whether the template can be selectively removed from the wall material, either chemically or by heating during the final fabrication step.

Porous structures are produced by the removal of the colloidal templates, either by burning, extracting with a solvent, or, in the case of silica colloids, by dissolving with dilute hydrofluoric (HF) acid. The method of template removal depends on the colloid material and on the properties of the infiltrated precursor. Silica sphere template arrays are removed by dissolution in aqueous HF solutions [41.74, 88]. Most metal-oxide inverse opals have typically been synthesized using polymer spheres as templates [41.74]. Polymer sphere template arrays are often removed thermally, with the added benefit of chemically converting the sol–gel precursors in the infiltrated phase to the desired oxide product [41.74, 88]. If precursor solidification is feasible at low temperatures, or if the precursor has components that would be destroyed by thermal template removal, polymer spheres can also be removed chemically using organic solvents, such as toluene, or tetrahydrofuran (THF) and acetone mixtures, or by photodegradation [41.74, 88].

Fabrication of such structures is not trivial: some of the difficulties include incomplete infiltration, structural shrinkage after infiltration, and microstructural variations (affecting the refractive index). A major problem with chemical assembly is that residual disorder drastically reduces the PBG width and the intensity of the reflectance peaks [41.74, 85]. Such disorder can arise from several different sources. Firstly, there are defects in the initial opal template, such as stacking faults or small deviations in sphere size or shape. It is not clear whether the imperfections present in current inverse opals are small enough to observe a complete PBG [41.3]. While progress has been made in reducing such defects, numerical simulations predict that even a 2% deviation in sphere size (or lattice constant) can close a PBG in even high-refractive-index inverted opals [41.3, 74]. Secondly, significant disorder can arise from the infiltration process itself. Even the best samples to date have variations in the amount of guest material filling the lattice [41.3]. Since these inhomogeneities must be much smaller than the optical wavelength, the elimination of residual order remains a challenge with such methods [41.3].

Although 3-D inverse opals are fascinating research objects, they are not amenable for practical photonic applications because of difficulties in integrating them into PLCs. Thus, there is strong interest in fabricating 3-D inverse opals in pseudo-2-D form, or supported on a substrate as an inverse opal film, to be compatible with traditional optoelectronic elements for system integration [41.1, 86, 91–93]. Recently, ordered colloidal crystal films have been formed by convective assembly [41.86] by particle confinement in thin slits between two solid plate boundaries [41.73, 86] and by compacting particles in the dry state [41.86, 91, 93–96]. Recently, patterned substrates were successfully used to control slow sedimentation of colloidal particles to improve the long-range order and to induce growth along a specific crystal plane [41.74, 86]. However, growth of defect-free crystalline templates with desired symmetry and orientation remains one of the challenges of the field.

Liquid sol–gel precursors can be used to form a variety of oxide materials and are very convenient for infiltration into artificial opal templates [41.88, 97]. However, the typical high percentage of volume shrinkage (typically 15–30%) involved in the solidification of such sol–gel precursors during the precursor-to-oxide conversion process has often resulted in random cracks in the final inverse opal films [41.73, 74, 88], often preventing the fabrication of large, ordered, inverse opal film structures. Colloidal template films have been infiltrated by soaking the template into the precursor sol–gel solution [41.74, 85, 98], but achieving a dense, complete filling of the void spaces within the 3-D template film has proven to be challenging [41.97]. Another approach used for large-area inverse opal film fabrication involves the crystallization of colloids in a liquid medium containing nanoparticles of the desired material, which are subsequently packed into a dense, 3-D lattice [41.74, 97]. The main limitation of this method is that nanometer-sized particles are not always available for the desired wall ma-

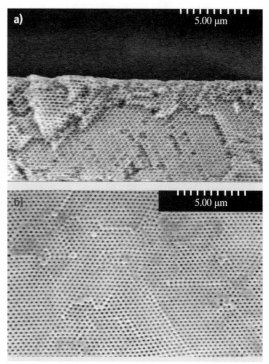

Fig. 41.10a,b Scanning electron microscope images [41.90] of Pb-doped BST inverse opal thin film: (**a**) cross section; and (**b**) top view

terial, and so only very specific materials may be made into inverse opals [41.88]. Another method of fabricating high-quality inverse opal films derived from liquid precursors is by infiltrating a sol–gel liquid precursor into a colloidal crystal template film by spin-coating, which results in very uniform filling of the template [41.90] (Fig. 41.10). By comparison, solvent-extracted samples shrink about 10% less [41.88]. Thus, for organic polymers, it has been possible to fabricate such 3-D porous structures as free-standing thin films that are several square centimeters in area.

Alternative 3-D PBG Fabrication Methods

Alternative approaches to lithography or self-assembly for 3-D fabrication exist. Some notable examples include laser-guided stereolithography, in which structures can also be written directly layer-by-layer by using a tightly focused laser beam to cause chemical vapor deposition [41.99], and two-photon-induced deposition or polymerization [41.100, 101]. Currently, feature sizes that can be routinely achieved using these methods are on the order of a few micrometers due to focusing and diffraction effects.

A 3-D holographic patterning process using multiple laser beams has recently been investigated; a 3-D laser interference pattern exposes a photosensitive polymer precursor (i. e., a photoresist) rendering the exposed areas insoluble and unexposed areas are dissolved away to leave a 3-D PC formed of cross-linked polymer with air-filled voids [41.102]. This process allows flexible design of the structure of a unit cell and thus of the optical properties of the microstructured material. Of concern is that such structures have relatively weak refractive-index contrast, and the processing approach is quite demanding [41.62].

41.3 Tunable PCs

Although conventional PCs offer an ability to control light propagation or confinement through the introduction of defects, once such defects are introduced, the propagation or confinement of light in these structures is not controllable. Thus, discretionary switching of light, for example, or rerouting of optical signals, is not available with fixed defects in PCs. There are two approaches that have been pursued to tune the properties of PCs: these are by tuning the refractive index of the constituent materials, or by altering the physical structure of the PC. In the latter case, the emphasis has been principally on changing the lattice constant, although other approaches are also relevant (i. e., including for example, the fill factor, structure symmetry and scattering element shape). In the text below, we discuss the state of the art in both of these areas.

41.3.1 Tuning the PC Response by Changing the Refractive Index of the Constituent Materials

We discuss recent progress in using four approaches within this category – namely, tuning the PC response by using: (i) light, (ii) applied electric fields, (iii) temperature or electrical field in infiltrated PCs, and (iv) by changing the concentration of free carriers (using electric field or temperature) in semiconductor-based PCs.

PC Refractive-Index Tuning Using Light
One approach to modifying the behavior of PCs is to use intense illumination of the PC by one beam of light to change the optical properties of the crystal in a nonlinear fashion; this, in turn, can thus control the properties of the PC for another beam of light. An example of a nonlinear effect includes the flattening of the photon dispersion relation near a PBG, which relaxes the constraints on phase matching for second- and third-harmonic generation [41.103]. In addition, light location near defects can enhance a variety of third-order nonlinear processes in 1-D PCs [41.104].

A new approach for PBG tuning has been proposed based on photo-reversible control over molecular aggregation, based on using the photochromic effect in dyes [41.105]. This can cause a reversible change over the photonic stop band. Structures that were studied included opal films formed from 275-nm-diameter silica spheres, infiltrated with photochromic dye: two dyes were considered – namely, 1,3-dihydro-1,3,3-trimethyl-spiro-[2H-indol-2,3′-[3H]-naphth[2,1-b] [1,4]oxazine] (SP) and cis-1,2-dicyano-1,2-bis(2,4,5-trimethyl-3-trienyl)ethane (CMTE). For the SP dye opal, a reversible 15-nm shift in the reflectance spectrum was observed following UV irradiation, which was ascribed to changes in the reflective index due to resonant absorption near the stop band. Smaller shifts (of about 3 nm) were observed for the CMTE dye opal. The recovery process on cessation of UV illumination was quite slow, taking about 38 s for the SP dye.

Nonlinear changes in refractive index have also been studied in PCs consisting of 220-nm-diameter SA (self assembly) polystyrene spheres infiltrated with water [41.106]. In these studies, the optical Kerr effect was used to shift the PBG. $40\,\text{GW/cm}^2$ [41.2] of peak pump power at $1.06\,\mu\text{m}$ (35-ps pulses at 10-Hz repetition rate) was used to shift the PBG 13 nm. The large optical nonlinearity originates from the delocalization of conjugated π-electrons along the polymer chains, leading to a large third-order nonlinear optical susceptibility. The time response was measured as a function of the delay time between pump and probe, and confirmed a response time of several picoseconds.

PC Refractive-Index Tuning Using an Applied Electric Field
There have been a number of studies focusing on using ferroelectric materials to form PCs [41.90, 98, 107, 108]. The application of an electric field to such materials can be used to change the refractive index and tune the PC optical response. For example, lead lanthanum zirconate titanate (PLZT) inverse opal structures have been fabri-

cated by infiltration using 350-nm-diameter polystyrene sphere templates and annealing at 750 °C [41.107]. The films were formed on indium tin oxide (ITO) coated glass to enable electric-field-induced changes in reflectivity to be measured due to the electro-optic effect. Applying voltages of up to about 700 V across films of thickness of about 50 mm only achieved a few nm of peak shift, attributed to the very modest changes in refractive index from the applied field (i. e., from 2.405 to 2.435, as a result of the bias). It should be noted, however, that the intrinsic response of the electro-optic effect in these materials is known to be in the GHz range and hence highly suitable for rapid tuning. Other reports include the formation of inverse opal barium strontium titanate (BST) PCs using infiltration of polystyrene opals [41.90, 98]. BST is most interesting in that it provides a high-refractive-index material and a factor of at least two times higher breakdown field strength than PLZT, and hence offers a much wider range of applied fields for tuning. Reports of using other ferroelectric materials include a high-temperature infiltration process for the ferroelectric copolymer, poly(vinylidene fluoride-trifluoroethane), infiltrated into 3-D silica opals with sphere diameters of 180, 225 and 300 nm [41.108].

Refractive-Index Tuning of Infiltrated PCs

In this case the approach has been to consider modulation of the refractive index of a PC infiltrated with a tunable medium – in particular, the most popular approach has been to use liquid crystals to infiltrate porous 2-D and 3-D PC structures. Such liquid crystals can behave as ferroelectrics whose refractive index may be tuned using either an applied electric field, or by thermal tuning. Reported results for this approach, using 3-D inverse opal structures, have been restricted to infiltration of ferroelectric liquid-crystal material into a silicon/air PC [41.109]. Reported changes in the refractive index of the liquid crystal are 1.4–1.6 under an applied field [41.109]. However, since the ferroelectric liquid has a higher refractive index than air, the infiltration results in a significant decrease in the refractive index contrast. This means that the original full photonic band gap of the inverse opal silicon PC no longer exists, and the practical utility of the silicon structure as a PC is effectively lost. Theoretical simulations have shown that partial surface wetting of the internal inverse opal surface can retain the full photonic band in silicon [41.109], but it is questionable whether such a complex structure can ever be practically fabricated. Finally, the presence of ferroelectric liquid crystal surrenders one of the main advantages of the original concept for PC structures: that

is, permitting light propagation in air [41.110]. Temperature tuning of the liquid-crystal material was shown to result in very small changes in refractive index (changes in $n < 0.01$ over a 70 °C change) and thus only provide minimal shifts in the transmittance through the PC over a large temperature range [41.111].

PC Refractive-Index Tuning by Altering the Concentration of Free Carriers (Using Electric Field or Temperature) in Semiconductor-Based PCs

An elegant way to rapidly tune the PBG of semiconductor-based PCs is to adjust the refractive index by modulating the free-carrier concentration using an ultra-fast optical pulse [41.112]. Using this approach, reflectivity of a two-dimensional silicon-based honeycomb PCs with 412-nm air holes (100 μm in length) in a 500-nm periodic array, was studied with a pump–probe approach [41.112]. By varying the delay between the pump the speed of PBG tuning was measured to be about 0.5 ps. The reflectance relaxation (corresponding to the return of the PBG to its original position) occurred on a timescale of 10–100 ns, characteristic of recombination of excess electrons and holes. Although these results are very encouraging, this approach cannot suppress or correct for light-scattering losses caused by structural imperfections, which remain an important consideration for currently fabricated PCs.

41.3.2 Tuning PC Response by Altering the Physical Structure of the PC

The second approach that we discuss for tuning the response of a PC is based on changes to the physical structure of the PC. We discuss the following approaches for tuning using this approach: tuning using (i) temperature, (ii) an applied magnetic field, (iii) strain/deformation, (iv) piezoelectric effects, and (v) using micro-electro-mechanical systems (MEMS) [i. e., actuation].

Tuning PC Response Using Temperature

An example of this approach is a study of temperature tuning of PCs fabricated from self-assembled polystyrene beads [41.113]. The PBG in these structures was fine-tuned by annealing samples at temperatures of 20–100 °C, resulting in a continuous blue shift of the stop band wavelength from 576 nm to 548 nm. New stop bands appeared in the UV transmission spectra when the sample was annealed above about 93 °C – the glass-transition temperature of the polystyrene beads.

Tuning PC Response Using Magnetism

An example of this approach is the use of an external applied magnetic field to adjust the spatial orientation of a PC [41.97]. This can find application in fabricating photonic devices such as tunable mirrors and diffractive display devices. These authors fabricated magnetic PCs by using monodisperse polystyrene beads self-assembled into a ferro-fluid consisting of magnetite particles, with particle sizes below 15 nm [41.97]. On evaporation of the solvent, a cubic PC lattice was formed with the nanoparticles precipitating out into the interstices between the spherical polystyrene colloids. These authors then showed how the template could be selectively removed by calcination or wet etching to reveal an inverse opal of magnetite – such structures being proposed as suitable for developing magnetically tunable PCs.

Tuning PC Response Using Strain

The concept in this case is quite straightforward – deforming or straining the PC changes the lattice constant or arrangement of dielectric elements in the PC with concomitant change in the photonic band structure. Polymeric materials would appear to be most suited to this methodology, owing to their ability to sustain considerable strains. However, concerns of reversibility and speed of tuning would clearly need to be addressed. Theoretical predictions for the influence of deformation on such systems include a report on a new class of PC based on self-assembling cholesteric elastomers [41.114]. These elastomers are highly deformable when subjected to external stress. The high sensitivity of the photonic band structure to strain, and the opening of new PBGs have been discussed [41.114]. Charged colloidal crystals were also fixed in a poly(acrylamide) hydrogel matrix to fabricate PCs whose diffraction peaks were tuned by applying mechanical stress [41.115]. The PBG shifted linearly and reversibly over almost the entire visible spectral region (from 460 nm to 810 nm).

Modeling of the photonic band structure of 2-D silicon-based triangular PCs under mechanical deformation was also reported [41.116]. The structures considered consisted of a silicon matrix with air columns. The authors showed that while a 3% applied shear strain provides only minor modifications to the PBG, uniaxial tension can produce a considerable shift. Other modeling includes a study of how strain can be used to tune the anisotropic optical response of 2-D PCs in the long-wavelength limit [41.117]. These calculations showed that the decrease in dielectric constant per unit strain is larger in the direction of the strain than normal to it. Indeed, the calculated birefringence is larger than that of quartz. They suggest that strain-tuning of this birefringence has attractive application in polarization-based optical devices.

To appreciate the sensitivity of such structures to mechanical tuning, it is instructive to refer to some recent work on PMMA inverse opal PC structures that were fabricated using silica opal templates [41.118]. Under the application of uniaxial deformation of these PCs, the authors found a blue shift of the stop band in the transmission spectrum – the peak wavelength of the stop band shifted from about 545 nm in the undeformed material to about 470 nm under a stretch ratio of about 1.6.

Another practical approach that has been applied to physically tuning PC structures is that of thermal annealing. One such study showed how the optical properties of colloidal PCs consisting of silica spheres can be tuned through thermal treatment [41.119]. This was attributed to both structural and physio-chemical modification of the material on annealing. A shift in the minimum transmission from about 1000 nm (un-annealed) to about 850 nm (for annealing at about 1000 °C) was demonstrated, or about a maximum shift in Bragg wavelength of $\approx 11\%$.

A quite novel application of strain-tuning was recently reported [41.120]. The authors studied 2-D PCs consisting of arrays of coupled optical microcavities fabricated from vertical-cavity surface-emitting laser structures. The influence of strain, as manifested by shifts in the positions of neighboring rows of microcavities with respect to each other, corresponded to alternating square or quasi-hexagonal shear-strain patterns. For strains below a critical threshold value, the lasing photon mode-locked to the corresponding mode in the unstrained PC. At the critical strain, switching occurred between the square and hexagonal lattice modes.

Finally, there has been a proposal for using strain in a PC to tune the splitting of a degenerate photon state within the PBG, suitable for implementing tunable PC circuits [41.121]. The principle applied is analogous to the static Jahn–Teller effect in solids. These authors showed that this effect is tunable by using the symmetry and magnitude of the lattice distortion. Using this effect the design of an optical valve that controls the resonant coupling of photon modes at the corner of a T-junction waveguide structure has been discussed.

Tuning PC Response Using Piezoelectric Effects

In this section we discuss using piezoelectric effects to physically change PC structures and hence tune them.

A proposal was made for using the piezoelectric effect to distort the original symmetry of a two-dimensional PC from a regular hexagonal lattice to a quasi-hexagonal lattice under an applied electric field [41.122]. The original bands decomposed into several strained bands, dependent on the magnitude and direction of the applied field. In the proposed structures, the application of $\approx 3\%$ shear strain is shown to be suitable for shifting 73% of the original PBG, which they refer to as the tunable-bandgap regime. An advantage of such an approach is that such structures are suggested to be capable of operation at speeds approaching MHz. Another report [41.123] discusses the design and implementation of a tunable silicon-based PBG microcavity in an optical waveguide, where tuning is accomplished using the piezoelectric effect to strain the PC; this was carried out using integrated piezoelectric microactuators. These authors report on a 1.54-nm shift in the cavity resonance at 1.56 μm for an applied strain of 0.04%.

There have also been reports of coupling piezoelectric-based actuators to PCs. One such report [41.124] discusses a poly(2-methoxyethyl acrylate)-based PC composite directly coupled to a piezoelectric actuator to study static and dynamic stop-band tuning characteristics; the stop band of this device could be tuned through a 172-nm tuning range, and could be modulated at up to 200 Hz.

Tuning PC Response Using Micro-Electro-Mechanical Systems (MEMS) Actuation

There have been a number of interesting developments in this field including PC-based air-bridge devices consisting of suspended 1-D PC mirrors separated by a Fabry–Perot cavity (gap) [41.125]. When such structures are mechanically perturbed, there can be a substantial shift in the PBG due to strain. The authors discuss how a suite of spectrally tunable devices can be envisioned based on such structures – these include modulators, optical filters, optical switches, WDM (wavelength division multiplexing), optical logical circuits, variable attenuators, power splitters and isolators. Indeed, the generalization of these concepts beyond 1-D was discussed in a recent patent [41.126] that covers tunable PC structures. This report [41.126], as well as others [41.127], consider the extension of these ideas to form families of micromachined devices. These authors [41.127] modeled and implemented a set of micromachined vertical resonator structures for 1.55-μm filters consisting of two PC (distributed Bragg reflector [DBR]) mirrors separated by either an air gap or semiconductor heterostructure. Electromechanical tuning was used to adjust the separation between the mirrors and hence fine-tune the transmission spectrum. The mirror structures were implemented using strong-index-contrast InP/air DBRs giving an index contrast of 2.17, and weak-index-contrast (0.5) silicon nitride/silicon dioxide DBRs. In the former case, a tuning range of over 8% of the absolute wavelength was achieved; varying the inter-membrane voltage up to 5 V gave a tuning range of over 110 nm. Similar planar structures are discussed in other papers [41.128] where the mirrors are formed from two slabs of PC separated by an adjustable air gap [41.126]. These structures have been shown to be able to perform as either flat-top reflection or all-pass transmission filters, by varying the distance between the slabs, for normally incident light. Unlike all previously reported all-pass reflection filters, based on Gires–Tournois interferometers using multiple dielectric stacks, their structure generates an all-pass transmission spectrum, significantly simplifying signal extraction and optical alignment – also the spectral response is polarization-independent owing to the 90° rotational symmetry of their structure.

41.4 Summary and Conclusions

This chapter discussed the fabrication and properties of PCs, with emphasis on developing tunable structures. New fabrication methodologies and also tuning schemes offers a glimpse of the far-reaching prospects for developing photonic devices which can, in a discretionary fashion, control the propagation of modes of light in an analogous fashion to the way in which nanostructures have been harnessed to control electron-based phenomena. Analogous to the evolution of electronic systems, one can anticipate a path toward development of compact active integrated photonic systems, as envisioned based on the technology outlined in this review.

References

41.1 T. F. Krauss, R. M. de la Rue: Photonic crystals in the optical regime – past, present, and future, Prog. Quantum Electron. **23**, 51–96 (1999)

41.2 V. Berger: From photonic band gaps to refractive index engineering, Opt. Mater. **11**, 131–142 (1999)

41.3 D. J. Norris, Y. A. Vlasov: Chemical approaches to three-dimensional semiconductor photonic crystals, Adv. Mater. **13**(6), 371–376 (2001)

41.4 V. Mizeikis, S. Juodkazis, A. Marcinkevicius, S. Matsuo, H. Misawa: Tailoring and characterization of photonic crystals, J. Photochem. Photobiol. C **2**, 35–69 (2001)

41.5 Photonics Nanostruct. **1**(1), 1–78 (2003): whole edition is dedicated to fundamentals and applications of photonic crystals

41.6 S. G. Johnson, J. D. Joannopoulos: *Photonic Crystals: Road from Theory to Practice* (Kluwer Academic, Boston 2002)

41.7 J. D. Joannopoulos, R. D Meade, J. N. Winn: *PCs: Moulding the Flow of Light* (Princeton Univ. Press, Princeton 1995)

41.8 S. John: Strong localization in certain disordered dielectric super-lattices, Phys. Rev. Lett. **58**(23), 2486–2489 (1987)

41.9 E. Yablonovitch: Inhibited spontaneous emission in solid state physics and electronics, Phys. Rev. Lett. **58**(20), 2059–2062 (1987)

41.10 H. Kosaka, T. Kawashima, A. Tomita, M. Notomi, T. Tamamura, T. Sato, S. Kwakami: Superprism phenomena in photonic crystals, Phys. Rev. B **58**(16), R10096–R10099 (1998)

41.11 V. Berger: Photonic crystals and photonic structures, Cur. Opin. Solid State Mater. Sci. **4**, 209–216 (1999)

41.12 K. M. Ho, C. T. Chan, C. M. Soukoulis: Existence of a photonic gap in periodic dielectric structures, Phys. Rev. Lett. **65**(25), 3152–3155 (1990)

41.13 S. Satpathy, Z. Zhang, M. R. Salehpour: Theory of photon bands in three-dimensional periodic dielectric structures, Phys. Rev. Lett. **64**(11), 1239–1242 (1990)

41.14 E. Yablonovitch: Photonic band-gap structures, J. Opt. Soc. Am. B **10**(2), 283–295 (1993)

41.15 Y. Xia: Photonic Crystals, Adv. Mater. **13**(6), 369 (2001)

41.16 C. Anderson, K. Giapis: Larger two-dimensional photonic band gaps, Phys. Rev. Lett. **77**, 2949–2952 (1996)

41.17 R. D. Meade, A. M. Rappe, K. D. Brommer, J. D. Joannopoulos: Nature of the photonic band gap: Some insights from a field analysis, J. Opt. Soc. Am. B **10**(2), 328–332 (1993)

41.18 R. D. Meade, A. M. Rappe, K. D. Brommer, J. D. Joannopoulos, O. L. Alerhand: Accurate theoretical analysis of photonic band-gap materials, Phys. Rev. B **48**(11), 8434–8437 (1993)

41.19 C. Lopez: Materials aspects of PCs, Adv. Mater. **15**(20), 1679–1704 (2003)

41.20 C. Weisbuch, C. H. Benisty, S. Olivier, M. Rattier, C. J. M. Smith, T. F. Krauss: Advances in photonic crystals, Phys. Status Solidi B **221**(93), 93–99 (2000)

41.21 C. Weisbuch, H. Benisty, M. Rattier, C. J. M. Smith, T. F. Krauss: Advances in 2D semiconductor PCs, Synth. Mater. **116**, 449–452 (2001)

41.22 S. L. Swartz: Topics in electronic ceramics, IEEE Trans. Electr. Ins. **25**(5), 935–987 (1990)

41.23 F. Watt: Focused high energy proton beam micromachining: A perspective view, Nucl. Instrum. Methods Phys. Res. **158**, 165–172 (1999)

41.24 D. K. Ferry, R. O. Grondin: *Physics of Submicron Devices* (Plenum, New York 1991)

41.25 J. Gierak, D. Mailly, G. Faini, J. L. Pelouard, P. Denk, F. Pardo, J. Y. Marzin, A. Septier, G. Schmmid, J. Ferre, R. Hydman, C. Chappert, J. Flicstein, B. Gayral, J. M. Gerard: Nano-fabrication with focused ion beams, Microelectron. Eng. **57-58**, 865–875 (2001)

41.26 K. Gamo: Nanofabrication by FIB, Microelectron. Eng. **32**, 159–171 (1996)

41.27 J. Melngailis, A. A. Mondelli, I. L. Berry III, R. Mohondro: A review of ion projection lithography, J. Vac. Sci. Technol. B **16**(3), 927–957 (1998)

41.28 P. Peercy: The drive to miniaturization, Nature **406**, 1023–1026 (2000)

41.29 T. Ito, S. Okazaki: Pushing the limits of lithography, Nature **406**, 1027–1031 (2000)

41.30 N. Peng, C. Jeynes, R. P. Webb, I. R. Chakarov, M. G. Blamire: Monte Carlo simulations of masked ion beam irradiation damage profiles in $YBa_2Cu_3O_{7-\delta}$ thin films, Nucl. Instrum. Methods Phys. Res. B **178**, 242–246 (2001)

41.31 J. L. Bartelt: Masked ion beam lithography: An emerging technology, Sol. State Technol. **29**(5), 215–220 (1986)

41.32 D. P. Stumbo, J. C. Wolfe: Contrast of ion beam proximity printing with non-ideal masks, J. Vac. Sci. Technol. B **12**(6), 3539–3542 (1994)

41.33 T. Devolder, C. Chappert, Y. Chen, E. Cambril, H. Launois, H. Bernas, J. Ferre, J. P. Jamet: Patterning of planar magnetic nanostructures by ion irradiation, J. Vac. Sci. Technol. B **17**(6), 3177–3181 (1999)

41.34 P. Ruchhoeft, J. C. Wolfe, R. Bass: Ion beam aperture-array lithography, J. Vac. Sci. Technol. B **19**(6), 2529–2532 (2001)

41.35 Y. Hsieh, Y. Hwang, J. Fu, Y. Tsou, Y. Peng, L. Chen: Dislocation multiplication inside contact holes, Microelectron. Reliab. **39**, 15–22 (1999)

41.36 J. Gierak, A. Septier, C. Vieu: Design and realization of a very high-resolution FIB nanofabrication instrument, Nucl. Instrum. Methods Phys. Res. A **427**, 91–98 (1999)

41.37 A. N. Broers, A. C. F. Hoole, J. M. Ryan: Electron beam lithography – resolution limits, Microelectron. Eng. **32**, 131–142 (1996)

41.38 C. Lehrer, L. Frey, S. Petersen, H. Ryssel: Limitations of focused ion beam nano-machining, J. Vac. Sci. Technol. B **19**(6), 2533–2538 (2001)

41.39 H. Masuda, K. Fukuda: Ordered metal nanohole arrays made by a two-step replication of honeycomb structures of anodic alumina, Science **268**, 1466–1468 (1995)

41.40 R. Tonucci, B. Justus, A. Campillo, C. Ford: Nanochannel array glass, Science **258**, 783–785 (1992)

41.41 V. Lehmann, H. Foll: Formation mechanism and properties of electrochemically etched trenches in n-type silicon, J. Electrochem. Soc. **137**(6), 653–659 (1990)

41.42 A. Birner, R. B. Wehrspohn, U. M. Gosele, K. Busch: Silicon-based photonic crystals, Adv. Mater. **13**(6), 377–388 (2001)

41.43 J. Martin: Nanomaterials: A membrane-based synthetic approach, Science **266**, 1961–1966 (1994)

41.44 J. I. Martin, J. Nogues, K. Liu, J. L. Vicent, I. K. Schuller: Ordered magnetic nanostructures: Fabrication and properties, J. Magn. Mater. **256**(1-3), 449–501 (2003)

41.45 H. Masuda, M. Ohya, H. Asoh, M. Nakao, M. Nohtomi, T. Tamamura: Photonic crystals using anodic porous alumina, Jpn. J. Appl. Phys. Pt. 2 **38**(12A), L1403–1405 (1999)

41.46 R. Wehrspohn, J. Schilling: Electrochemically prepared pore arrays for photonic-crystal applications, MRS Bull., 623–626 (2001)

41.47 A. P. Li, F. Muller, A. B. K. Nielsch, U. Gosele: Hexagonal pore arrays with a 50–420 nm interpore distance formed by self-organization in anodic alumina, J. Appl. Phys. **84**(11), 6023–6026 (1998)

41.48 H. Masuda, H. Asoh, M. Watanabe, K. Nishio, M. Nakao, T. Tamamura: Square and triangular nanohole array architectures in anodic alumina, Adv. Mater. **13**, 189–192 (2001)

41.49 M. Nakao, S. Oku, T. Tamamura, K. Yasui, H. Masuda: GaAs and InP nanohole arrays fabricated by reactive beam etching using highly ordered alumina membranes, Jpn. J. Appl. Phys. Pt. 1 **38**(2B), 1052–1055 (1999)

41.50 J. Liang, H. Chik, A. Yin, J. Xu: Two-dimensional lateral superlattices of nanostructures: Nonlithographic formation by anodic membrane template, J. Appl. Phys. **91**(4), 2544–2546 (2002)

41.51 N. Matsuura, T. W. Simpson, C. P. McNorgan, I. V. Mitchell, X. Mei, P. Morales, H. E. Ruda: Nanometer-scale pattern transfer using ion implantation. In: *Three-Dimensional Nano-engineered Assemblies*, MRS Proc., Vol. 739, ed. by T. M. Orlando, L. Merhari, D. P. Taylor, K. Ikuta (Mater. Res. Soc., Boston 2002)

41.52 N. Matsuura, T. W. Simpson, I. V. Mitchell, X. Mei, P. Morales, H. E. Ruda: Ultra-high density, non-lithographic, sub-100 nm pattern transfer by ion implantation an selective chemical etching, Appl. Phys. Lett. **81**(25), 4826–4828 (2002)

41.53 E. Rimini: *Ion Implantation: Basics to Device Fabrication* (Kluwer Academic, Norwellt 1995)

41.54 G. Hobler: Monte Carlo simulation of two-dimensional implanted dopant distributions at mask edges, Nucl. Instrum. Methods Phys. Res. B **96**, 155–162 (1995)

41.55 M. M. Faye, C. Vieu, G. B. Assayag, P. Salles, A. Claverie: Lateral damage extension during masked ion implantation into GaAs, J. Appl. Phys. **80**(8), 4303–4307 (1996)

41.56 P. Schmuki, L. Erickson: Direct micro-patterning of Si and GaAs using electrochemical development of focused ion beam implants, Appl. Phys. Lett. **73**, 2600–2602 (1998)

41.57 K. Wang, A. Chelnokov, S. Rowson, P. Garouche, J.-M. Lourtioz: Three-dimensional Yablonovite-like photonic crystals by focused ion beam etching of macroporous silicon, Mater. Res. Soc. Symp. Proc. **637**, E1.4.1–E1.4.5 (2001)

41.58 E. Yablonovitch, T. Gmitter, K. Leung: Photonic band structure: The face-centered-cubic case employing non-spherical atoms, Phys. Rev. Lett. **67**, 2295–2298 (1991)

41.59 E. Ozbay, A. Abeyta, G. Tuttle, M. Tringides, R. Biswas, C. Chan, C. Soukoulis, K. Ho: Measurement of a three-dimensional photonic band gap in a crystal structure made of dielectric rods, Phys. Rev. B **50**, 1945–1948 (1994)

41.60 K. M. Ho, C. T. Chan, C. M. Soukoulis, R. Biswas, M. Sigalas: Photonic band gaps in three dimensions: New layer-by-layer periodic structures, Sol. State Commun. **89**(5), 413–481 (1994)

41.61 H. S. Sözüer, J. P. Dowling: Photonic band calculations for woodpile structures, J. Mod. Opt. **41**(2), 231–239 (1994)

41.62 Y. Xia, B. Gates, Z-Y. Li: Self-assembly approaches to three-dimensional photonic crystals, Adv. Mater. **13**(6), 409–413 (2001)

41.63 A. Moroz: Three-Dimensional complete photonic-band-gap structures in the visible range, Phys. Rev. Lett. **83**(25), 5274–5277 (1999)

41.64 S. Lin, J. Fleming, D. Hetherington, B. Smith, R. Biswas, K. Ho, M. Sigalas, W. Zubrzycki, S. Kurtz, J. Bur: A three-dimensional photonic crystals operating at infrared wavelengths, Nature **394**, 251–253 (1998)

41.65 S. Noda, K. Tomoda, N. Yamamoto, A. Chutinan: Full three-dimensional photonic bandgap crystals at near-infrared wavelengths, Science **289**, 604–606 (2000)

41.66 S. Kawakami: Fabrication of sub-micrometre 3D periodic structures composed of Si/SiO$_2$, Electron. Lett. **33**(4), 1260–1261 (1997)

41.67 C. T. Chan, S. Datta, K. M. Ho, C. M. Soukoulis: A7 structure: A family of photonic crystals, Phys. Rev. B **50**(3), 1988–1991 (1994)

41.68 G. Feiertag, W. Ehrfeld, H. Freimuth, H. Kolle, H. Lehr, M. Schmidt, M. M. Sigalas, C. M. Soukoulis, G. Kiriakidis, T. Pedersen, J. Kuhl, W. Koenig: Fabrication of photonic crystals by deep x-ray lithography, Appl. Phys. Lett. **71**(11), 1441–1443 (1997)

41.69 C. Cheng, A. Scherer: Fabrication of photonic bandgap crystals, J. Vac. Sci. Technol. B **13**(6), 2153–3113 (1995)

41.70 A. A. Zakhidov, R. H. Baughman, Z. Iqbal, C. Cui, I. Khayrullin, S. O. Dantas, J. Marti, V. G. Ralchenko: Carbon structures with three-dimensional periodicity at optical wavelengths, Science **282**, 897–901 (1998)

41.71 A. Blanco, E. Chomski, S. Grabtchak, M. Ibisate, S. John, S. W. Leonard, C. Lopez, F. Meseguer, H. Miguez, J. P. Mondia, G. A. Ozin, O. Toader, H. M. van Driel: Large-scale synthesis of a silicon photonic crystals with a complete three-dimensional bandgap near 1.5 micrometres, Nature **405**, 437–440 (2000)

41.72 J. E. G. J. Wijnhoven, W. L. Vos: Preparation of photonic crystals made of air spheres in titania, Science **281**, 802–804 (1998)

41.73 Y. Xia, B. Gates, Y. Yin, Y. Lu: Mono-dispersed colloidal spheres: Old materials with new applications, Adv. Mater. **12**(10), 693–713 (2000)

41.74 A. Stein: Sphere templating methods for periodic porous solids, Micropor. Mesopor. Mater. **44–45**, 227–239 (2001)

41.75 J. Martorell, N. M. Lawandy: Observation of inhibited spontaneous emission in a periodic dielectric structure, Phys. Rev. Lett. **65**(15), 1877–1880 (1990)

41.76 I. I. Tarhan, G. H. Watson: Photonic band structure of FCC colloidal crystals, Phys. Rev. Lett. **76**(2–8), 315–318 (1996)

41.77 Y. A. Vlasov, M. Deutsch, D. J. Norris: Single-domain spectroscopy of self-assembled photonic crystals, Appl. Phys. Lett. **76**(12), 1627–1629 (2000)

41.78 D. C. Reynolds, F. Lopez-Tejeira, D. Cassagne, F. Garcia-Vidal, C. Jouanin, J. Sanchez-Dehesa: Spectral properties of opal-based photonic crystals having a SiO$_2$ matrix, Phys. Rev. B **60**(16), 11422–11426 (1999)

41.79 V. N. Bogomolov, S. V. Gaponenko, I. N. Germanenko, A. M. Kapitonov, E. P. Petrov, N. V. Gaponenko, A. V. Prokofiev, A. N. Ponyavina, N. I. Silvanovich, S. M. Samoilovich: Photonic band gap phenomenon and optical properties of artificial opals, Phys. Rev. E **55**(6), 7619–7625 (1997)

41.80 S. G. Romanov, A. V. Fokin, R. M. De La Rue: Stop-band structure in complementary three-dimensional opal-based photonic crystals, J. Phys. Condens. Matter **11**, 3593–3600 (1999)

41.81 R. Biswas, M. M. Sigalas, G. Subramania, K. M. Ho: Photonic band gaps in colloidal systems, Phys. Rev. B **57**(9), 3701–3705 (1998)

41.82 A. Richel, N. P. Johnson, D. W. McComb: Observation of Bragg reflection in photonic crystals synthesized from air spheres in a titania matrix, Appl. Phys. Lett. **76**(14), 1816–1818 (2000)

41.83 B. T. Holland, C. F. Blanford, A. Stein: Synthesis of macroporous minerals with highly ordered three-dimensional arrays of spheroidal voids, Science **281**, 538–540 (1998)

41.84 M. S. Thijssen, R. Sprik, J. E. G. J. Wijnhoven, M. Megens, T. Narayanan, A. Lagendijk, W. L. Vos: Inhibited light propagation and broadband reflection in photonic air-sphere crystals, Phys. Rev. Lett. **83**(14), 2730–2733 (1999)

41.85 F. Meseguer, A. Blanco, H. Miguez, F. Garcia-Santamaria, M. Ibisate, C. Lopez: Synthesis of inverse opals, Coll. Surf. A **202**, 281–290 (2002)

41.86 O. D. Velev, E. W. Kaler: Research news: Structured porous materials via colloidal crystal templating: From inorganic oxides to metals, Adv. Mater. **12**(7), 531–534 (2000)

41.87 O. D. Velev, A. M. Lenhoff: Colloidal crystals as templates for porous materials, Current Opin. Coll. Interf. Sci. **5**, 56–63 (2000)

41.88 F. Blanford, H. Yan, R. C. Schroden, M. Al-Daous, A. Stein: Gems of chemistry and physics: Macroporous metal oxides with 3D order, Adv. Mater. **13**(6), 401–407 (2001)

41.89 A. M. Kapitonov, N. V. Gaponenko, V. N. Bogomolov, A. V. Prokofiev, S. M. Samoilovich, S. V. Gaponenko: Photonic stop band in a three-dimensional SiO$_2$/TiO$_2$ lattice, Phys. Stat. Sol. (a) **165**(1), 119–123 (1998)

41.90 N. Matsuura, S. Yang, P. Sun, H. E. Ruda: Development of highly-ordered, ferroelectric inverse opal films using sol-gel infiltration, Appl. Phys. A **81**, 379–384 (2005)

41.91 S. M. Yang, H. Miguez, G. A. Ozin: Opal circuits of light – planarized micro photonic crystals chips, Adv. Funct. Mater. **12**(6–7), 425431 (2002)

41.92 A. Polman, P. Wiltzius: Materials science aspects of PCs, MRS Bull., 608–610 (2001)

41.93 V. L. Colvin: From opals to optics: Colloidal photonic crystals, MRS Bull., 637–641 (2001)

41.94 S. H. Park, D. Qin, Y. Xia: Crystallization of mesoscale particles over large areas, Adv. Mater. **10**(3), 1028–1032 (1998)

41.95 P. Jiang, J. Bertone, K. Hwang, V. Colvin: Single-crystal colloidal multi-layers of controlled thickness, Chem. Mat. **11**, 2132–2140 (1999)

41.96 Y. A. Vlasov, X.-Z. Bo, J. C. Sturm, D. J. Norris: On-chip natural assembly of silicon photonic bandgap crystals, Nature **414**, 289–293 (2001)

41.97 B. Gates, Y. Xia: Photonic crystals that can be addressed with an external magnetic field, Adv. Mater. **13**(21), 1605–1608 (2001)

41.98 I. Soten, H. Miguez, S. M. Yang, S. Petrov, N. Coombs, N. Tetreault, N. Matsuura, H. E. Ruda, G. A. Ozin: Barium titanate inverted opals – synthesis, characterization, and optical properties, Adv. Funct. Mater. **12**(1), 71–77 (2002)

41.99 M. C. Wanke, O. Lehmann, K. Muller, Q. Wen, M. Stuke: Laser rapid prototyping of photonic band-gap microstructures, Science **275**, 1284–1286 (1997)

41.100 B. H. Cumpston, S. P. Ananthavel, S. Barlow, D. L. Dyer, J. E. Ehrlich, L. L. Erskine, A. A. Heikal, S. M. Kuebler, I.-Y. S. Lee, D. McCord-Maughon, J. Qin, H. Rockel, M. Rumi, X.-L. Wu, S. R. Marder, J. W. Perry: Two-photon polymerization initiators for three-dimensional optical data storage and microfabrication, Nature **398**, 51–54 (1999)

41.101 H-B. Sun, S. Matsuo, H. Misawa: Three-dimensional photonic crystals structures achieved with two-photon-absorption photo-polymerization of resin, Appl. Phys. Lett. **74**(6), 786–788 (1999)

41.102 M. Campbell, D. Sharp, M. Harrison, R. Denning, A. Turberfield: Fabrication of photonic crystals for the visible spectrum by holographic lithography, Nature **404**, 53–56 (2000)

41.103 J. Martorell, R. Vilaseca, R. Corbalan: Second harmonic generation in a photonic crystal, Appl. Phys. Lett. **70**(6), 702–704 (1997)

41.104 H. Inouye, Y. Kanemitsu: Direct observation of non-linear effects in a one dimensional photonic crystal, Appl. Phys. Lett. **82**(8), 1155–1157 (2003)

41.105 Z.-Z. Gu, T. Iyoda, A. Fujishima, O. Sato: Photo reversible regulation of optical stop bands, Adv. Mater. **13**(7), 1295–1298 (2001)

41.106 X. Hu, Q. Zhang, Y. Liu, B. Cheng, D. Zhang: Ultrafast three-dimensional tunable photonic crystal, Appl. Phys. Lett. **83**(13), 2518–2520 (2003)

41.107 B. Li, J. Zou, X. J. Wang, X. H. Liu, J. Zi: Ferroelectric inverse opals with electrically tunable photonic band gap, Appl. Phys. Lett. **83**(23), 4704–4706 (2003)

41.108 T. B. Xu, Z. Y. Cheng, Q. M. Zhang, R. H. Baughman, C. Cui, A. A. Zakhidov, J. Su: Fabrication and characterization of three dimensional periodic ferroelectric polymer-silica opal composites and inverse opals, J. Appl. Phys. **88**(1), 405–409 (2000)

41.109 S. John, K. Busch: Photonic bandgap formation and tunability in certain self-organizing systems, J. Lightwave Technol. **17**(11), 1931–1943 (1999)

41.110 J. D. Joannopoulos: The almost-magical world of photonic crystals, Braz. J. Phys. **26**(1), 53–67 (1996)

41.111 X. Yoshino, Y. Kawagishi, M. Ozaki, A. Kose: Mechanical tuning of the optical properties of plastic opal as a photonic crystals, Jpn. J. Appl. Phys. Pt. 2 **38**(7A), L786–78 (1999)

41.112 S. W. Leonard, H. M. van Driel, J. Schilling, R. B. Wehrspohn: Ultrafast band edge tuning of a two dimensional silicon photonic crystal via free carrier injection, Phys. Rev. B **66**, 161102-1–111102-4 (2002)

41.113 B. Gates, S. H. Park, Y. Xia: Tuning the photonic bandgap properties of crystalline arrays of polystyrene beads by annealing at elevated temperatures, Adv. Mater. **12**(9), 653–656 (2000)

41.114 P. A. Bermel, M. Warner: Photonic bandgap structure of highly deformable self-assembling systems, Phys. Rev. E **65**(1), 010702(R)-1–010702(R)-4 (2001)

41.115 Y. Iwayama, J. Yamanaka, Y. Takiguchi, M. Takasaka, K. Ito, T. Shinohara, T. Sawada, M. Yonese: Optically tunable gelled photonic crystal covering almost the entire visible light wavelength region, Langmuir **19**(4), 977–980 (2003)

41.116 S. Jun, Y-S. Cho: Deformation-induced bandgap tuning of 2D silicon-based photonic crystals, Opt. Express **11**(21), 2769–2774 (2003)

41.117 C-S. Kee, K. Kim, H. Lim: Tuning of anisotropic optical properties of 2D dielectric photonic crystals, Physica B **338**, 153–158 (2003)

41.118 K. Sumioka, H. Kayashima, T. Tsutsui: Tuning the optical properties of inverse opal photonic crystals by deformation, Adv. Mater. **14**(18), 1284–1286 (2002)

41.119 H. Miguez, F. Meseguer, C. Lopez, A. Blanco, J. S. Moya, J. Requena, A. Mifsud, V. Fornes: Control of the photonic crystal properties of fcc-packed sub-micrometer SiO_2 spheres by sintering, Adv. Mater. **10**(6), 480–483 (1998)

41.120 H. Pier, E. Kapon, M. Moser: Strain effects and phase transitions in photonic crystal resonator crystals, Nature **407**, 880–883 (2000)

41.121 N. Malkova, V. Gopalan: Strain tunable optical valves at T-junction waveguides in photonic crystals, Phys. Rev. B **68**, 245115-1–245115-6 (2003)

41.122 S. Kim, V. Gopalan: Strain tunable photonic band gap crystals, Appl. Phys. Lett. **78**(20), 3015–3017 (2001)

41.123 C. W. Wong, P. T. Rakich, S. G. Johnson, M. Qi, H. I. Smith, E. P. Ippen, L. C. Kimmerling, Y. Jeon, G. Barbastathis, S-G. Kim: Strain-tunable silicon photonic bandgap microcavities in optical waveguides, Appl. Phys. Lett. **84**(8), 1242–1244 (2004)

41.124 S. H. Foulger, P. Jiang, A. Lattam, D. W. Smith, J. Ballato, D. E. Dausch, S. Grego, B. R. Stoner: Photonic crystal composites with reversible high-frequency stop band shifts, Adv. Mater. **15**(9), 685–689 (2003)

41.125 S. Rajic, J. L. Corbeil, P. G. Datskos: Feasibility of tunable MEMS photonic crystal devices, Ultramicroscopy **97**, 473–479 (2003)

41.126 N. Matsuura, H. E. Ruda, B. G. Yacobi: Configurable photonic device, US Patent 09/918398 [pending]

41.127 H. Hiller, J. Daleiden, C. Prott, F. Römer, S. Irmer, V. Rangelov, A. Tarraf, S. Schüler, M. Strassner: Potential for a micromachined actuation of ultra-wide continuously tunable optoelectronic devices, Appl. Phys. B **75**, 3–13 (2002)

41.128 W. Suh, S. Fan: Mechanically switchable photonic crystal filter with either all-pass transmission or flat-top reflection characteristics, Opt. Lett. **28**(19), 1763–1765 (2003)

42. Quantum Wells, Superlattices, and Band-Gap Engineering

This chapter reviews the principles of band-gap engineering and quantum confinement in semiconductors, with a particular emphasis on their optoelectronic properties. The chapter begins with a review of the fundamental principles of band-gap engineering and quantum confinement. It then describes the optical and electronic properties of semiconductor quantum wells and superlattices at a tutorial level, before describing the principal optoelectronic devices. The topics covered include edge-emitting lasers and light-emitting diodes (LEDs), resonant cavity LEDs and vertical-cavity surface-emitting lasers (VCSELs), quantum cascade lasers, quantum-well solar cells, superlattice avalanche photodiodes, inter-sub-band detectors, and quantum-well light modulators. The chapter concludes with a brief review of current research topics, including a discussion of quantum-dot structures.

42.1 Principles of Band-Gap Engineering and Quantum Confinement 1022
    42.1.1 Lattice Matching 1022
    42.1.2 Quantum-Confined Structures 1023

42.2 Optoelectronic Properties of Quantum-Confined Structures........... 1024
    42.2.1 Electronic States in Quantum Wells and Superlattices.............. 1024
    42.2.2 Interband Optical Transitions...... 1026
    42.2.3 The Quantum-Confined Stark Effect................................... 1028
    42.2.4 Inter-Sub-Band Transitions........ 1029
    42.2.5 Vertical Transport 1030
    42.2.6 Carrier Capture and Relaxation ... 1031

42.3 Emitters.. 1032
    42.3.1 Interband Light-Emitting Diodes and Lasers............................... 1032
    42.3.2 Quantum Cascade Lasers 1033

42.4 Detectors 1034
    42.4.1 Solar Cells................................ 1034
    42.4.2 Avalanche Photodiodes.............. 1034
    42.4.3 Inter-Sub-Band Detectors 1035
    42.4.4 Unipolar Avalanche Photodiodes. 1035

42.5 Modulators 1036

42.6 Future Directions 1037

42.7 Conclusions....................................... 1038

References 1038

The need for efficient light-emitting diodes and lasers operating over the whole of the visible spectrum and also the fibre-optic windows at 1.3 μm and 1.55 μm drives research into new direct-gap semiconductors to act as the active materials. Since the emission wavelength of a semiconductor corresponds to its band-gap energy, research focuses on engineering new materials which have their band gaps at custom-designed energies. This science is called *band-gap engineering*.

In the early years of semiconductor optoelectronics, the band gaps that could be achieved were largely determined by the physical properties of key III–V materials such as GaAs and its alloys such as AlGaAs and InGaAs. Then in 1970 a major breakthrough occurred when *Esaki* and *Tsu* invented the semiconductor quantum well and superlattice [42.1]. They realised that developments in epitaxial crystal growth could open the door to new structures that exploit the principles of quantum confinement to engineer electronic states with custom-designed properties. They foresaw that these quantum-confined structures would be of interest both to research scientists, who would be able to explore uncharted areas of fundamental physics, and also to engineers, who would learn to use their unique properties for device applications. Their insight paved the way for a whole new breed of devices that are now routinely found in a host of everyday applications ranging from compact-disc players to traffic lights.

The emphasis of the chapter is on the optoelectronic properties of quantum-well and superlattice structures. We begin by outlining the basic principles of band-gap engineering and quantum confinement. We will then discuss the electronic states in quantum-confined structures and the optical properties that follow from them. In Sects. 42.3–42.5 we will explain the principles of the main optoelectronic devices that employ quantum wells and superlattices, namely emitters, detectors and modulators. Finally we will indicate a few interesting recent developments that offer exciting prospects for future devices before drawing the chapter to its conclusion. A number of texts cover these topics in more detail (e.g. [42.2–5]), and the interested reader is referred to these sources for a more comprehensive treatment. A description of the purely electronic properties of low-dimensional structures may be found in [42.6].

42.1 Principles of Band-Gap Engineering and Quantum Confinement

42.1.1 Lattice Matching

The art of band-gap engineering relies heavily on developments in the science of crystal growth. Bulk crystals grown from the melt usually contain a large number of impurities and defects, and optoelectronic devices are therefore grown by epitaxial methods such as liquid-phase epitaxy (LPE), molecular-beam epitaxy (MBE) and metalorganic vapour-phase epitaxy (MOVPE), which is also called metalorganic chemical vapour deposition (MOCVD) (Chapt. 14). The basic principle of epitaxy is to grow thin layers of very high purity on top of a bulk crystal called the substrate. The system is said to be *lattice-matched* when the lattice constants of the epitaxial layer and the substrate are identical. Lattice-matching reduces the number of dislocations in the epitaxial layer, but it also imposes tight restrictions on the band gaps that can be engineered easily, because there are only a relatively small number of convenient substrate materials available.

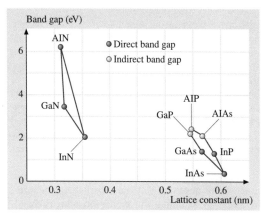

Fig. 42.1 Room-temperature band gap of a number of important optoelectronic III–V materials versus their lattice constant

We can understand this point more clearly by reference to Fig. 42.1. This diagram plots the band-gap energy E_g of a number of important III–V semiconductors as a function of their lattice constant. The majority of optoelectronic devices for the red/near-infrared spectral regions are either grown on GaAs or InP substrates. The simplest case to consider is an epitaxial layer of GaAs grown on a GaAs substrate, which gives an emission wavelength of 873 nm (1.42 eV). This wavelength is perfectly acceptable for applications involving short-range transmission down optical fibres. However, for long distances we require emission at 1.3 μm or 1.55 μm, while for many other applications we require emission in the visible spectral region.

Let us first consider the preferred fibre-optic wavelengths of 1.3 μm and 1.55 μm. There are no binary semiconductors with band gaps at these wavelengths, and so we have to use alloys to tune the band gap by varying the composition (Chapt. 31). A typical example is the ternary alloy $Ga_xIn_{1-x}As$, which is lattice-matched to InP when $x = 47\%$, giving a band gap of 0.75 eV (1.65 μm). $Ga_{0.47}In_{0.53}As$ photodiodes grown on InP substrates make excellent detectors for 1.55-μm radiation, but to make an emitter at this wavelength, we have to increase the band gap while maintaining the lattice-matching condition. This is achieved by incorporating a fourth element into the alloy – typically Al or P, which gives an extra design parameter that permits band-gap tuning while maintaining lattice-matching. Thus the quaternary alloys $Ga_{0.27}In_{0.73}As_{0.58}P_{0.42}$ and $Ga_{0.40}In_{0.60}As_{0.85}P_{0.18}$ give emission at 1.3 μm and 1.55 μm, respectively, and are both lattice-matched to InP substrates.

Turning now to the visible spectral region, it is a convenient coincidence that the lattice constants of GaAs and AlAs are almost identical. This means that we can grow relatively thick layers of $Al_xGa_{1-x}As$ on GaAs substrates without introducing dislocations and other defects. The band gap of $Al_xGa_{1-x}As$ varies quadratically with x according to:

$$E_g(x) = (1.42 + 1.087x + 0.438x^2)\,\text{eV},\quad (42.1)$$

but unfortunately the gap becomes indirect for $x > 43\%$. We can therefore engineer direct band gaps of 1.42–1.97 eV, giving emission from 873 nm in the near infrared to 630 nm in the red spectral range. Much work has been done on quaternary alloys such as AlGaInP, (Chapt. 31) but it has not been possible to make blue- and green-emitting devices based on GaAs substrates to date, due to the tendency for arsenic and phosphorous compounds to become indirect as the band gap increases.

The approach for the blue end of the spectrum preferred at present is to use nitride-based compounds. (Chapt. 32) Early work on nitrides established that their large direct gaps made them highly promising candidates for use as blue/green emitters [42.7]. However, it was not until the 1990s that this potential was fully realised. The rapid progress followed two key developments, namely the activation of p-type dopants and the successful growth of strained $In_xGa_{1-x}N$ quantum wells which did not satisfy the lattice-matching condition [42.8]. The second point goes against the conventional wisdom of band-gap engineering and highlights the extra degrees of freedom afforded by quantum-confined structures, as will now be discussed.

42.1.2 Quantum-Confined Structures

A quantum-confined structure is one in which the motion of the electrons (and/or holes) are confined in one or more directions by potential barriers. The general scheme for classifying quantum-confined structures is given in Table 42.1. In this chapter we will be concerned primarily with *quantum wells*, although we will briefly refer to *quantum wires* and *quantum dots* in Sect. 42.6. Quantum size effects become important when the thickness of the layer becomes comparable with the de Broglie wavelength of the electrons or holes.

Table 42.1 Classification of quantum-confined structures. In the case of quantum wells, the confinement direction is usually taken as the z-axis

Structure	Confined directions	Free directions (dimensionality)
Quantum well	1 (z)	2 (x, y)
Quantum wire	2	1
Quantum dot (or box)	3	none

If we consider the free thermal motion of a particle of mass m in the z-direction, the de Broglie wavelength at a temperature T is given by

$$\lambda_{\text{deB}} = \frac{h}{\sqrt{mk_BT}}. \quad (42.2)$$

For an electron in GaAs with an effective mass of $0.067m_0$, we find $\lambda_{\text{deB}} = 42\,\text{nm}$ at 300 K. This implies that we need structures of thickness $\approx 10\,\text{nm}$ in order to observe quantum-confinement effects at room temperature. Layers of this thickness are routinely grown by the MBE or MOVPE techniques described in Chapt. 14.

Figure 42.2 shows a schematic diagram of a GaAs/AlGaAs quantum well. The quantum confinement is provided by the discontinuity in the band gap at the interfaces, which leads to a spatial variation of the conduction and valence bands, as shown in the lower half of the figure. The Al concentration is typically chosen to be around 30%, which gives a band-gap discontinuity of 0.36 eV according to (42.1). This splits roughly 2 : 1 between the conduction and valence bands, so that electrons see a confining barrier of 0.24 eV and the holes see 0.12 eV.

If the GaAs layers are thin enough, according to the criterion given above, the motion of the electrons and holes will be quantised in the growth (z) direction, giving

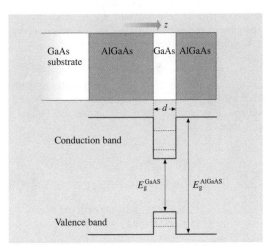

Fig. 42.2 Schematic diagram of the growth layers and resulting band diagram for a GaAs/AlGaAs quantum well of thickness d. The quantised levels in the quantum well are indicated by the *dashed lines*. Note that in real structures a GaAs buffer layer is usually grown immediately above the GaAs substrate

rise to a series of discrete energy levels, as indicated by the dashed lines inside the quantum well in Fig. 42.2. The motion is in the other two directions (i.e. the x–y plane) is still free, and so we have quasi two-dimensional (2-D) behaviour.

The quantisation of the motion in the z-direction has three main consequences. Firstly, the quantisation energy shifts the effective band edge to higher energy, which provides an extra degree of freedom in the art of band-gap engineering. Secondly, the confinement keeps the electrons and holes closer together and hence increases the radiative recombination probability. Finally, the density of states becomes independent of energy, in contrast to three-dimensional (3-D) materials, where the density of states is proportional to $E^{1/2}$. Many of the useful properties of the quantum wells follow from these three properties.

The arrangement of the bands shown in Fig. 42.2 in which both the electrons and holes are confined in the quantum well is called *type I band alignment*. Other types of band alignments are possible in which only one of the carrier types are confined (*type II band alignment*). Furthermore, the flexibility of the MBE and MOVPE growth techniques easily allows the growth of *superlattice* (SL) structures containing many repeated quantum wells with thin barriers separating them, as shown in Fig. 42.3. Superlattices behave like artificial one-dimensional periodic crystals, in which the periodicity is designed into the structure by the repetition of the quantum wells. The electronic states of SLs form delocalised *minibands* as the wave functions in neighbouring wells couple together through the thin barrier that separates them. Structures containing a smaller number of repeated wells or with thick barriers that prevent coupling between adjacent wells are simply called *multiple quantum well* (MQW) structures.

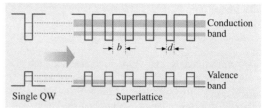

Fig. 42.3 Schematic diagram of a superlattice, showing the formation of minbands from the energy levels of the corresponding single quantum well (QW). The structure forms an artificial one-dimensional crystal with period $(d+b)$, where d and b represent the thickness of the QW and barrier regions respectively. The width of the minibands depends on the strength of the coupling through the barriers. It is frequently the case that the lowest hole states do not couple strongly, and hence remain localised within their respective wells, as shown in the figure

In the next section we will describe in more detail the electronic properties of quantum wells and superlattices. Before doing so, it is worth highlighting two practical considerations that are important additional factors that contribute to their usefulness. The first is that band-gap tunability can be achieved without using alloys as the active material, which is desirable because alloys inevitably contain more defects than simple compounds such as GaAs. The second factor is that the quantum wells can be grown as strained layers on top of a lattice with a different cell constant. A typical example is the $In_xGa_{1-x}N/GaN$ quantum wells mentioned above. These layers do not satisfy the lattice-matching condition, but as long as the total $In_xGa_{1-x}N$ thickness is less than the critical value, there is an energy barrier to the formation of dislocations. In practice this allows considerable extra flexibility in the band-gap engineering that can be achieved.

42.2 Optoelectronic Properties of Quantum-Confined Structures

42.2.1 Electronic States in Quantum Wells and Superlattices

Quantum Wells

The electronic states of quantum wells can be understood by solving the Schrödinger equation for the electrons and holes in the potential wells created by the band discontinuities. The simplest approach is the infinite-well model shown in Fig. 42.4a. The Schrödinger equation in the well is

$$-\frac{\hbar^2}{2m_w^*}\frac{d^2\psi(z)}{dz^2} = E\psi(z), \quad (42.3)$$

where m_w^* is the effective mass in the well and z is the growth direction. Since the potential barriers are infinite, there can be no penetration into the barriers, and we must therefore have $\psi(z) = 0$ at the interfaces. If we choose our origin such that the quantum well runs from $z=0$

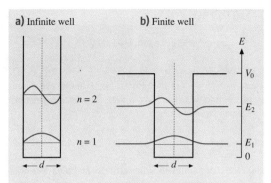

Fig. 42.4a,b Confined states in a quantum well of width d. (a) A perfect quantum well with infinite barriers. (b) A finite well with barriers of height V_0. The wave functions for the $n = 1$ and $n = 2$ levels are sketched for both types of well

to $z = d$, d being the width of the well, the normalised wave functions take the form (see e.g. [42.9]):

$$\psi_n(z) = \sqrt{2/d} \sin k_n z , \qquad (42.4)$$

where $k_n = (n\pi/d)$ and the quantum number n is an integer (≥ 1). The energy E_n is given by

$$E_n = \frac{\hbar^2 k_n^2}{2m_w^*} = \frac{\hbar^2}{2m_w^*}\left(\frac{n\pi}{d}\right)^2 . \qquad (42.5)$$

The wave functions of the $n = 1$ and $n = 2$ levels are sketched in Fig. 42.4a.

Although the infinite-well model is very simplified, it nonetheless provides a good starting point for understanding the general effects of quantum confinement. Equation (42.5) shows us that the energy is inversely proportional to d^2, implying that narrow wells have larger confinement energies. Furthermore, the confinement energy is inversely proportional to the effective mass, which means that lighter particles experience larger effects. This also means that the heavy- and light-hole states have different energies, in contrast to bulk semiconductors in which the two types of hole states are degenerate at the top of the valence band.

Now let us consider the more realistic finite-well model shown in Fig. 42.4b. The Schrödinger equation in the well is unchanged, but in the barrier regions we now have:

$$-\frac{\hbar^2}{2m_b^*}\frac{d^2\psi(z)}{dz^2} + V_0\psi(z) = E\psi(z), \qquad (42.6)$$

where V_0 is the potential barrier and m_b^* is the effective mass in the barrier. The boundary conditions require that the wave function and particle flux $(1/m^*) d\psi/dz$ must be continuous at the interface. This gives a series of even and odd parity solutions which satisfy

$$\tan(kd/2) = \frac{m_w^* \kappa}{m_b^* k} , \qquad (42.7)$$

and

$$\tan(kd/2) = -\frac{m_b^* k}{m_w^* \kappa} , \qquad (42.8)$$

respectively. k is the wave vector in the well, given by

$$\frac{\hbar^2 k^2}{2m_w^*} = E_n , \qquad (42.9)$$

while κ is the exponential decay constant in the barrier, given by

$$\frac{\hbar^2 \kappa^2}{2m_b^*} = V_0 - E_n . \qquad (42.10)$$

Solutions to (42.7) and (42.8) are easily found by simple numerical techniques [42.9]. As with the infinite well, the eigenstates are labelled by the quantum number n and have parities of $(-1)^{n+1}$ with respect to the axis of symmetry about the centre of the well. The wave functions are approximately sinusoidal inside the well, but decay exponentially in the barriers, as illustrated in Fig. 42.4b. The eigen-energies are smaller than those of the infinite well due to the penetration of the barriers, which means that the wave functions are less well confined. There is only a limited number of solutions, but there is always at least one, no matter how small V_0 might be.

As an example we consider a typical GaAs/$Al_{0.3}Ga_{0.7}As$ quantum well with $d = 10$ nm. The confinement energy is 245 meV for the electrons and 125 meV for the holes. The infinite well model predicts $E_1 = 56$ meV and $E_2 = 224$ meV for the electrons, whereas (42.7), (42.8) give $E_1 = 30$ meV and $E_2 = 113$ meV. For the heavy (light) holes the infinite-well model predicts 11 meV (40 meV) and 44 meV (160 meV) for the first two bound states, instead of the more accurate values of 7 meV (21 meV) and 29 meV (78 meV) calculated from the finite-well model. Note that the separation of the electron levels is greater than $k_B T$ at 300 K, so that that the quantisation effects will be readily observable at room temperature.

Strained Quantum Wells

Even more degrees of freedom for tailoring the electronic states can be achieved by epitaxially stacking semiconductor layers with different lattice constants to form *strained quantum wells*. Examples include

In$_x$Ga$_{1-x}$As on GaAs, and Si$_{1-x}$Ge$_x$ on Si. Large biaxial strain develops within the x–y plane of a quantum well grown on a substrate with a different lattice constant. In order to avoid the buildup of misfit dislocations at the interfaces, the strained layers need to be thinner than a certain critical dimension. For example, a defect-free strained In$_x$Ga$_{1-x}$As layer on GaAs requires a thickness less than around 10 nm when $x = 0.2$. Since the band gap is related to the lattice constant, the strain induces a shift of the band edges which, in turn, affects many other properties. It is due to some of these effects that strained QW structures have become widely exploited in optoelectronic devices. (Chapt. 37)

The most significant effect of the strain is to alter the band gap and remove the valence-band degeneracy near the Γ valley. The splitting of the valence band is a consequence of the lattice distortion, which reduces the crystal symmetry from cubic to tetragonal [42.10]. There are essentially two types of strain. Compressive strain occurs when the active layer has a larger lattice constant than the substrate, for example in In$_x$Ga$_{1-x}$As on GaAs. In this case, the band gap increases and the effective mass of the highest hole band decreases, while that of next valence band increases. The opposite case is that of tensile strain, which occurs when the active layer has a smaller lattice constant than the substrate, such as Si$_{1-x}$Ge$_x$ on Si. The ordering of the valence bands is opposite to the case of compressive strain, and the overall band gap is reduced.

Superlattices

The analytical derivation of the allowed energy values in a superlattice (SL) is similar to that for a single QW, with the appropriate change of the boundary conditions imposed by the SL periodicity. The mathematical description of a superlattice is similar to a one-dimensional crystal lattice, which allows us to borrow the formalism of the band theory of solids, including the well-known Kronig–Penney model [42.9]. Within this model, the electron envelope wave function $\psi(z)$ can be expressed as a superposition of Bloch waves propagating along the z-axis. For a SL with a barrier height V_0, the allowed energy is calculated numerically as a solution of the transcendental equation involving the Bloch wave vector:

$$\cos(ka) = \cos(kd)\cos(\kappa' b)$$
$$-\frac{k^2 + \kappa'^2}{2k\kappa'}\sin(kd)\sin(\kappa' b),$$
$$E > V_0, \quad (42.11)$$

$$\cos(ka) = \cos(kd)\cos(\kappa b)$$
$$-\frac{k^2 - \kappa^2}{2k\kappa}\sin(kd)\sin(\kappa b),$$
$$E < V_0, \quad (42.12)$$

where $a \equiv (b+d)$ is the period, and k and κ are given by (42.9) and (42.10), respectively. The decay constant κ' is given by:

$$E - V_0 = \frac{\hbar^2 \kappa'^2}{2m_b^*}. \quad (42.13)$$

The electronic states in superlattices can be understood in a more qualitative way by reference to Fig. 42.3 and making use of the analogy with the tight-binding model of band formation in solids. Isolated atoms have discrete energy levels which are localised on the individual atom sites. When the atoms are brought close together, the energy levels broaden into bands, and the overlapping wave functions develop into extended states. In the same way, repeated quantum-well structures with large values of the barrier thickness b (i. e. MQWs) have discrete levels with wave functions localised within the wells. As the barrier thickness is reduced, the wave functions of adjacent wells begin to overlap and the discrete levels broaden into *minibands*. The wave functions in the minibands are delocalised throughout the whole superlattice. The width of the miniband depends on the cross-well coupling, which is determined by the barrier thickness and the decay constant κ (42.10). In general, the higher-lying states give rise to broader minibands because κ decreases with E_n. Also, the heavy-hole minibands are narrower than the electron minibands, because the cross-well coupling decreases with increasing effective mass.

42.2.2 Interband Optical Transitions

Absorption

The optical transitions in quantum wells take place between electronic states that are confined in the z-direction but free in the x–y plane. The transition rate can be calculated from Fermi's golden rule, which states that the probability for optical transitions from the initial state $|i\rangle$ at energy E_i to the final state $|f\rangle$ at energy E_f is given by:

$$W(i \to f) = \frac{2\pi}{\hbar}|\langle f|e\mathbf{r}\cdot\boldsymbol{\mathcal{E}}|i\rangle|^2 g(\hbar\omega), \quad (42.14)$$

where $e\mathbf{r}$ is the electric dipole of the electron, $\boldsymbol{\mathcal{E}}$ is the electric field of the light wave, and $g(\hbar\omega)$ is the joint density of states at photon energy $\hbar\omega$. Conservation of energy requires that $E_f = (E_i + \hbar\omega)$ for absorption, and $E_f = (E_i - \hbar\omega)$ for emission.

Let us consider a transition from a confined hole state in the valence band with quantum number n to a confined electron state in the conduction band with quantum number n'. We apply Bloch's theorem to write the wave functions in the following form:

$$|i\rangle = u_v(r) \exp(i\mathbf{k}_{xy} \cdot \mathbf{r}_{xy}) \psi_{hn}(z)$$
$$|f\rangle = u_c(r) \exp(i\mathbf{k}_{xy} \cdot \mathbf{r}_{xy}) \psi_{en'}(z), \quad (42.15)$$

where $u_v(r)$ and $u_c(r)$ are the envelope function for the valence and conduction bands, respectively, $\mathbf{k}_{xy}$ is the in-plane wave vector for the free motion in the x–y plane, $\mathbf{r}_{xy}$ being the xy component of the position vector, and $\psi_{hn}(z)$ and $\psi_{en'}(z)$ are the wave functions for the confined hole and electron states in the z-direction. We have applied conservation of momentum here so that the in-plane wave vectors of the electron and hole are the same.

On inserting these wave functions into (42.14) we find that the transition rate is proportional to both the square of the overlap of the wave functions and the joint density of states [42.11]:

$$W \propto |\langle \psi_{en'}(z) | \psi_{hn}(z) \rangle|^2 \, g(\hbar\omega). \quad (42.16)$$

The wave functions of infinite wells are orthogonal unless $n = n'$, which gives a selection rule of $\Delta n = 0$. For finite wells, the $\Delta n = 0$ selection rule is only approximately obeyed, although transitions between states of different parity (i.e. Δn odd) are strictly forbidden. The joint density of states is independent of energy due to the quasi-2-D nature of the quantum well.

Figure 42.5a illustrates the first two strong transitions in a typical quantum well. These are the $\Delta n = 0$ transitions between the first and second hole and electron levels. The threshold energy for these transitions is equal to

$$\hbar\omega = E_g + E_{hn} + E_{en}. \quad (42.17)$$

The lowest value is thus equal to $(E_g + E_{h1} + E_{e1})$, which shows that the optical band gap is shifted by the sum of the electron and hole confinement energies. Once the photon energy exceeds the threshold set by (42.17), a continuous band of absorption occurs with the absorption coefficient independent of energy due to the constant 2-D density of states of the quantum well. The difference between the absorption of an ideal quantum well with infinite barriers and the equivalent bulk semiconductor is illustrated in Fig. 42.5b. In the quantum well we find a series of steps with constant absorption coefficients, whereas in the bulk the absorption varies as

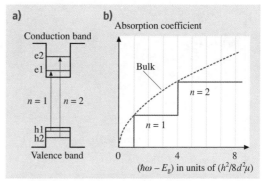

Fig. 42.5a,b Interband optical transitions between confined states in a quantum well. (**a**) Schematic diagram showing the $\Delta n = 0$ transitions between the $n = 1$ and $n = 2$ sub-bands. (**b**) Absorption spectrum for an infinite quantum well of thickness d with a reduced electron–hole mass μ in the absence of excitonic effects. The absorption spectrum of the equivalent bulk semiconductor in shown by the *dashed line* for comparison

$(\hbar\omega - E_g)^{1/2}$ for $\hbar\omega > E_g$. Thus the transition from 3-D to 2-D alters the shape of the absorption curve, and also causes an effective shift in the band gap by $(E_{h1} + E_{e1})$.

Up to this point, we have neglected the Coulomb interaction between the electrons and holes which are involved in the transition. This attraction leads to the formation of bound electron–hole pairs called excitons. The exciton states of a quantum well can be modelled as 2-D hydrogen atoms in a material with relative dielectric constant ε_r. In this case, the binding energy E^X is given by [42.12]:

$$E^X(\nu) = \frac{\mu}{m_0} \frac{1}{\varepsilon_r^2} \frac{1}{(\nu - 1/2)^2} R_H, \quad (42.18)$$

where ν is an integer ≥ 1, m_0 is the electron mass, μ is the reduced mass of the electron–hole pair, and R_H is the Rydberg constant for hydrogen (13.6 eV). This contrasts with the standard formula for 3-D semiconductors in which E^X varies as $1/\nu^2$ rather than $1/(\nu - 1/2)^2$, and implies that the binding energy of the ground-state exciton is four times larger in 2-D than 3-D. This allows excitonic effects to be observed at room temperature in quantum wells, whereas they are only usually observed at low temperatures in bulk semiconductors.

Figure 42.6 compares the band-edge absorption of a GaAs MQW sample with that of bulk GaAs at room temperature [42.13]. The MQW sample contained 77 GaAs quantum wells of thickness 10 nm with thick $Al_{0.28}Ga_{0.72}As$ barriers separating them. The shift of

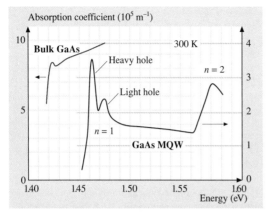

Fig. 42.6 Absorption spectrum of a 77-period GaAs/Al$_{0.28}$Ga$_{0.72}$As MQW structure with 10-nm quantum wells at room temperature. The absorption spectrum of bulk GaAs is included for comparison. (After [42.13], © 1982 AIP)

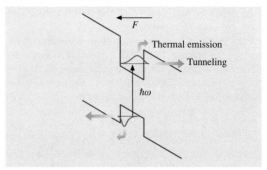

Fig. 42.7 Schematic band diagram and wave functions for a quantum well in a DC electric field F

the band edge of the MQW to higher energy is clearly observed, together with the series of steps due to each $\Delta n = 0$ transition. The sharp lines are due to excitons, which occur at energies given by

$$\hbar\omega = E_\text{g} + E_{\text{h}n} + E_{\text{e}n} - E^\text{x}. \qquad (42.19)$$

Equation (42.18) predicts that E^X should be around 17 meV for the ground-state exciton of an ideal GaAs quantum well, compared to 4.2 meV for the bulk. The actual QW exciton binding energies are somewhat smaller due to the tunnelling of the electrons and holes into the barriers, with typical values of around 10 meV. However, this is still substantially larger than the bulk value and explains why the exciton lines are so much better resolved for the QW than the bulk. The absorption spectrum of the QW above the exciton lines is approximately flat due to the constant density of states in 2-D, which contrasts with the rising absorption of the bulk due to the parabolic 3-D density of states. Separate transitions are observed for the heavy and light holes. This follows from their different effective masses, and can also be viewed as a consequence of the lower symmetry of the QW compared to the bulk.

Emission

Emissive transitions occur when electrons excited in the conduction band drop down to the valence band and recombine with holes. The optical intensity $I(\hbar\omega)$ is proportional to the transition rate given by (42.14) multiplied by the probability that the initial state is occupied and the final state is empty:

$$I(\hbar\omega) \propto W(\text{c} \rightarrow \text{v}) f_\text{c}(1 - f_\text{v}), \qquad (42.20)$$

where f_c and f_v are the Fermi–Dirac distribution functions in the conduction and valence bands, respectively. In thermal equilibrium, the occupancy of the states is largest at the bottom of the bands and decays exponentially at higher energies. Hence the luminescence spectrum of a typical GaAs QW at room temperature usually consists of a peak of width $\sim k_\text{B}T$ at the effective band gap of $(E_\text{g} + E_{\text{hh}1} + E_{\text{e}1})$. At lower temperatures the spectral width is affected by inhomogeneous broadening due to unavoidable fluctuations in the well thickness. Furthermore, in quantum wells employing alloy semiconductors, the microscopic fluctuations in the composition can lead to additional inhomogeneous broadening. This is particularly true of InGaN/GaN quantum wells, where indium compositional fluctuations produce substantial inhomogeneous broadening even at room temperature.

The intensity of the luminescence peak in quantum wells is usually much larger than that of bulk materials because the electron–hole overlap is increased by the confinement. This leads to faster radiative recombination, which then wins out over competing nonradiative decay mechanisms and leads to stronger emission. This enhanced emission intensity is one of the main reasons why quantum wells are now so widely used in diode lasers and light-emitting diodes.

42.2.3 The Quantum-Confined Stark Effect

The *quantum-confined Stark effect* (QCSE) describes the response of the confined electron and hole states in quantum wells to a strong direct-current (DC) electric field applied in the growth (z) direction. The field is usually applied by growing the quantum wells inside a p–n junction, and then applying reverse bias to the diode. The magnitude of the electric field F is given by:

$$F = \frac{V^{\text{built-in}} - V^{\text{bias}}}{L_i}, \qquad (42.21)$$

where $V^{\text{built-in}}$ is the built-in voltage of the diode, V^{bias} is the bias voltage, and L_i is the total thickness of the intrinsic region. $V^{\text{built-in}}$ is approximately equal to the band-gap voltage of the doped regions (≈ 1.5 V for a GaAs diode).

Figure 42.7 gives a schematic band diagram of a quantum well with a strong DC electric field applied. The field tilts the potential and distorts the wave functions as the electrons tend to move towards the anode and the holes towards the cathode. This has two important consequences for the optical properties. Firstly, the lowest transition shifts to lower energies due to the electrostatic interaction between the electric dipole induced by the field and the field itself. At low fields the dipole is proportional to F, and the red shift is thus proportional to F^2 (the quadratic Stark effect). At higher fields, the dipole saturates at a value limited by ed, where e is the electron charge and d the well width, and the Stark shift is linear in F. Secondly, the parity selection rule no longer applies due to the breaking of the inversion symmetry about the centre of the well. This means that *forbidden* transitions with Δn equal to an odd number become allowed. At the same time, the $\Delta n = 0$ transitions gradually weaken with increasing field as the distortion to the wave functions reduces the electron–hole overlap.

Figure 42.8 shows the normalised room-temperature photocurrent spectra of a GaAs/Al$_{0.3}$Ga$_{0.7}$As MQW p–i–n diode containing 9.0-nm quantum wells at 0 V and -10 V applied bias. These two bias values correspond to field strengths of around 15 kV/cm and 115 kV/cm respectively. The photocurrent spectrum closely resembles the absorption spectrum, due to the field-induced escape of the photoexcited carriers in the QWs into the external circuit (Sect. 42.2.5). The figure clearly shows the Stark shift of the absorption edge at the higher field strength, with a red shift of around 20 meV (≈ 12 nm) at -10 V bias for the hh1 $\to$ e1 transition. The intensity of the line weakens somewhat due to the reduction in the electron–hole overlap, and there is lifetime broadening caused by the field-assisted tunnelling. Several parity-forbidden transitions are clearly observed. The two most obvious ones are identified with arrows, and correspond to the hh2 $\to$ e1 and hh1 $\to$ e2 transitions, respectively.

A striking feature in Fig. 42.8 is that the exciton lines are still resolved even at very high field strengths. In bulk GaAs the excitons ionise at around 5 kV/cm [42.11], but in QWs the barriers inhibit the field ionisation, and excitonic features can be preserved even up to ≈ 300 kV/cm [42.14]. The ability to control the absorption spectrum by the QCSE is the principle behind a number of important modulator devices, which will be discussed in Sect. 42.5.

In the case of a superlattice, such as that illustrated in Fig. 42.3, a strong perpendicular electric field can break the minibands into discrete energy levels local to each QW, due to the band-gap tilting effect represented in Fig. 42.7. The possibility of using an electric field to modify the minibands of a superlattice is yet another remarkable ability of band-gap engineering to achieve control over the electronic properties by directly using fundamental principles of quantum mechanics.

42.2.4 Inter-Sub-Band Transitions

The engineered band structure of quantum wells leads to the possibility of *inter-sub-band (ISB) transitions*, which take place between confined states within the conduction or valence bands, as illustrated schematically in Fig. 42.9. The transitions typically occur in the infrared spectral region. For example, the e1 $\leftrightarrow$ e2 ISB transition in a 10-nm GaAs/AlGaAs quantum well occurs at around 15 μm. For ISB absorption transitions we must first dope the conduction band so that there is a large population of electrons in the e1 level, as shown

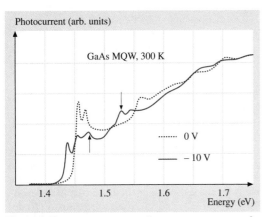

Fig. 42.8 Room-temperature photocurrent spectra for a GaAs/Al$_{0.3}$Ga$_{0.7}$As MQW p–i–n diode with a 1-μm-thick i-region at zero bias and -10 V. The quantum well thickness was 9.0 nm. The *arrows* identify transitions that are forbidden at zero field. (After [42.14], © 1991 IEEE)

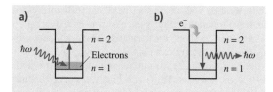

Fig. 42.9 (a) Inter-sub-band absorption in an n-type quantum well. (b) Inter-sub-band emission following electron injection to a confined upper level in the conduction band

in Fig. 42.9a. This is typically achieved by n-type doping of the barriers, which produces a large electron density as the extrinsic electrons drop from the barriers to the confined states in the QW. Undoped wells are used for ISB emission transitions, and electrons must first be injected into excited QW states, as shown in Fig. 42.9b.

The basic properties of ISB transitions can be understood by extension of the principles outlined in Sect. 42.2.2. The main difference is that the envelope functions for the initial and final states are the same, since both states lie in the same band. The transition rate for a conduction-band ISB transition then turns out to be given by:

$$W^{ISB} \propto |\langle \psi_{en'}(z) |z| \psi_{en}(z) \rangle|^2 \, g(\hbar\omega) \,, \quad (42.22)$$

where n and n' are the quantum numbers for the initial and final confined levels. The z operator within the Dirac bracket arises from the electric-dipole interaction and indicates that the electric field of the light wave must be parallel to the growth direction. Furthermore, the odd parity of the z operator implies that the wave functions must have different parities, and hence that Δn must be an odd number. The use of ISB transitions in infrared emitters and detectors will be discussed in Sects. 42.3.2, 42.4.3 and 42.4.4.

42.2.5 Vertical Transport

Quantum Wells

Vertical transport refers to the processes by which electrons and holes move in the growth direction. Issues relating to vertical transport are important for the efficiency and frequency response of most QW optoelectronic devices. The transport is generally classified as either *bipolar*, when both electrons and holes are involved, or *unipolar*, when only one type of carrier (usually electrons) is involved. In this section we will concentrate primarily on bipolar transport in QW detectors and QCSE modulators. Bipolar transport in light-emitting devices is discussed in Sects. 42.2.6

and 42.3.1, while unipolar transport in quantum cascade lasers is discussed in Sect. 42.3.2.

In QW detectors and QCSE modulators the diodes are operated in reverse bias. This produces a strong DC electric field and tilts the bands as shown in Fig. 42.7. Electrons and holes generated in the quantum wells by absorption of photons can escape into the external circuit by tunnelling and/or thermal emission, as illustrated schematically in Fig. 42.7.

The physics of tunnelling in quantum wells is essentially the same as that of α-decay in nuclear physics. The confined particle oscillates within the well and attempts to escape every time it hits the barrier. The escape rate is proportional to the attempt frequency ν_0 and the transmission probability of the barrier. For the simplest case of a rectangular barrier of thickness b, the escape time τ_T is given by:

$$\frac{1}{\tau_T} = \nu_0 \exp(-2\kappa b) \,, \quad (42.23)$$

where κ is the tunnelling decay constant given by (42.10). The factor of 2 in the exponential arises due to the dependence of the transmission probability on $|\psi(z)|^2$. The situation in a biased quantum well is more complicated due to the non-rectangular shape of the barriers. However, (42.23) allows the basic trends to be understood. To obtain fast tunnelling we need thin barriers and small κ. The second requirement is achieved by keeping m_b^* as small as possible and by working with a small confining potential V_0. The tunnelling rate increases with increasing field, because the average barrier height decreases.

The thermal emission of electrons over a confining potential is an old problem which was originally applied to the heated cathodes in vacuum tubes. It has been shown that the thermal current fits well to the classical Richardson formula:

$$J_E \propto T^{1/2} \exp\left(-\frac{e\Phi}{k_B T}\right) \,, \quad (42.24)$$

with the work function Φ replaced by $[V(F) - E_n]$, $V(F)$ being the height of the barrier that must be overcome at the field strength F [42.15]. The emission rate is dominated by the Boltzmann factor, which represents the probability that the carriers have enough thermal kinetic energy to escape over the top of the barrier. At low fields $V(F) \approx (V_0 - E_n)$, but as the field increases, $V(F)$ decreases as the barriers tilt over. Hence the emission rate (like the tunnelling rate) increases with increasing field. The only material-dependent parameter that enters the Boltzmann factor is the barrier height. Since this is insensitive both to the effective masses and to the barrier

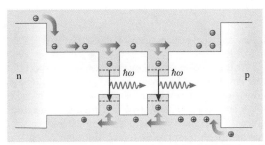

Fig. 42.10 Schematic representation of the drift of injected carriers and their subsequent capture by quantum wells. Light emission occurs when electrons and holes are captured in the same quantum well and then recombine with each other

thickness, the thermal emission rate can dominate over the tunnelling rate in some conditions, especially at room temperature in samples with thick barriers. For example, the fastest escape mechanism in GaAs/Al$_{0.3}$Ga$_{0.7}$As QWs at room temperature can be the thermal emission of holes, which have a much smaller barrier to overcome than the electrons [42.16].

Superlattices

The artificial periodicity of superlattice structures gives rise to additional vertical transport effects related to the phenomenon of Bloch oscillations. It is well known that an electron in a periodic structure is expect to oscillate when a DC electric field is applied. This effect has never been observed in natural crystals because the oscillation period – equal to h/eFa, where a is the unit cell size – is much longer than the scattering times of the electrons. In a superlattice, by contrast, the unit cell size is equal to $(d+b)$ (See Fig. 42.3) and the oscillation period can be made much shorter. This allows the electron to perform several oscillations before being scattered. The oscillatory motion of the electrons in a superlattice was first observed by two groups in 1992 [42.17, 18]. The following year, another group directly detected the radiation emitted by the oscillating electron wave packet [42.19]. The subject has since developed greatly, and THz-frequency emission has now been achieved from GaAs/AlGaAs superlattices even at room temperature [42.20].

42.2.6 Carrier Capture and Relaxation

In a QW light-emitting device, the emission occurs after carriers injected from the contacts are transported to the active region and then captured by the QWs. The capture and subsequent relaxation of the carriers is thus of crucial importance. Let us consider the band-edge profile of a typical QW diode-laser active region, as illustrated in Fig. 42.10. The active region is embedded between larger band-gap cladding layers designed to prevent thermally assisted carrier leakage outside the active region. Electrons and holes are injected from the n- and p-doped cladding layers under forward bias and light emission follows after four distinct process have taken place: (1) relaxation of carriers from the cladding layers to the confinement barriers (CB); (2) carrier transport across the CB layers, by diffusion and drift; (3) carrier capture into the quantum wells; and (4) carrier relaxation to the fundamental confined levels.

The carrier relaxation to the CB layers occurs mainly by longitudinal optical (LO) phonon emission. The CB layer transport is governed by a classical electron fluid model. The holes are heavier and less mobile than the electrons, and hence the ambipolar transport is dominated by the holes. Carrier nonuniformities, such as carrier pile up at the p-side CB region due to the lower mobility of the holes, are taken into account in the design of the barrier layers. The carrier capture in the QWs is governed by the phonon-scattering-limited carrier mean free path. It is observed experimentally that the capture time oscillates with the QW width. Detailed modelling reveals that this is related to a resonance between the LO phonon energy and the energy difference between the barrier states and the confined states within the well [42.21]. As another design guideline, the QW widths must be larger or at least equal to the phonon-scattering-limited carrier mean free path at the operating temperature in order to speed up the carrier capture. Finally, the relaxation of carriers to the lowest sub-band occurs on a sub-picosecond time scale if the inter-sub-band energy separation is larger than the LO phonon energy. Carrier–carrier scattering can also contribute to an ultrafast thermalisation of carriers, on a femtosecond time scale at the high carrier densities present inside laser diodes. Many of these processes have been studied in detail by ultrafast laser spectroscopy [42.22].

Carrier capture and escape are complementary vertical transport mechanisms in MQW structures. In the design of vertical transport-based MQW devices, one process must often be sped up at the expense of making the other as slow as possible. For example, in order to enhance the performance of QW laser diodes, the carrier confinement capability of the MQW active region must be optimised in terms of minimising the ratio between the carrier capture and escape times [42.23].

42.3 Emitters

42.3.1 Interband Light-Emitting Diodes and Lasers

Quantum wells have found widespread use in light-emitting diode (LED) and laser diode applications for a number of years now. As discussed in Sects. 42.1.2 and 42.2.2, there are three main reasons for this. Firstly, the ability to control the quantum-confinement energy provides an extra degree of freedom to engineer the emission wavelength. Secondly, the change of the density of states and the enhancement of the electron–hole overlap leads to superior performance. Finally, the ability to grow strained layers of high optical quality greatly increases the variety of material combinations that can be employed, thus providing much greater flexibility in the design of the active regions.

Much of the early work concentrated on lattice-matched combinations such as GaAs/AlGaAs on GaAs substrates. GaAs/AlGaAs QW lasers operating around 800 nm has now become industry-standard for applications in laser printers and compact discs. Furthermore, the development of high-power arrays has opened up new applications for pumping solid-state lasers such as Nd : YAG. Other types of lattice-matched combinations can be used to shift the wavelength into the visible spectral region and also further into the infrared. QWs based on the quaternary alloy $(Al_yGa_{1-y})_xIn_{1-x}P$, are used for red-emitting laser pointers [42.24], while $Ga_{0.47}In_{0.53}As$ QWs and its variants incorporating Al are used for the important telecommunication wavelengths of 1300 nm and 1550 nm.

The development of strained-layer QW lasers has greatly expanded the range of material combinations that are available. The initial work tended to focus on $In_xGa_{1-x}As$/GaAs QWs grown on GaAs substrates. The incorporation of indium into the quantum well shifts the band edge to lower energy, thereby giving emission in the wavelength range 900–1100 nm. An important technological driving force has been the need for powerful sources at 980 nm to pump erbium-doped fibre amplifiers [42.25]. Furthermore, as mentioned in Sect. 42.2.1, the strain alters the band structure and this can have other beneficial effects on the device performance. For example, the compressive strain in the $In_xGa_{1-x}As$/GaAs QW system has been exploited in greatly reducing the threshold current density. This property is related to the reduced effective mass of the holes and hence the reduced density of states. An extensive account of the effects of strain on semiconductor layers and the performance of diode lasers is given in [42.26].

At the other end of the spectral range, a spectacular development has been the $In_xGa_{1-x}N$/GaN QWs grown on sapphire substrates. These highly strained QWs are now routinely used in ultrabright blue and green LEDs, and there is a growing interest in developing high-power LED sources for applications in solid-state lighting [42.27]. Commercial laser diodes operating around 400 nm have been available for several years [42.8], and high-power lasers suited to applications in large-capacity optical disk video recording systems have been reported [42.28]. Lasers operating out to 460 nm have been demonstrated [42.29], and also high-efficiency ultraviolet light-emitting diodes [42.30]. At the same time, much progress has been made in the application of AlGaN/GaN quantum-well materials in high-power microwave devices [42.31, 32].

A major application of quantum wells is in *vertical-cavity surface-emitting lasers* (VCSELs). These lasers emit from the top of the semiconductor surface, and have several advantages over the more-conventional edge-emitters: arrays are readily fabricated, which facilitates their manufacture; no facets are required, which avoids complicated processing procedures; the beam is circular, which enhances the coupling efficiency into

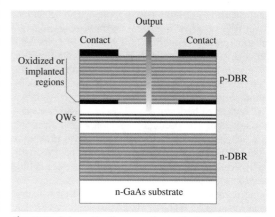

Fig. 42.11 Schematic diagram of a vertical-cavity surface-emitting laser (VCSEL). The quantum wells (QWs) that comprise the gain medium are placed at the centre of the cavity formed between two distributed Bragg reflector (DBR) mirrors. Oxidised or proton-implanted regions provide the lateral confinement for both the current and the optical mode

optical fibres; and their small size leads to very low threshold currents. For these reasons the development of VCSELs has been very rapid, and many local-area fibre networks operating around 850 nm currently employ VCSEL devices. This would not have been possible without the high gain coefficients that are inherent to the QW structures.

Figure 42.11 gives a schematic diagram of a typical GaAs-based VCSEL. The device contains an active QW region inserted between two distributed Bragg reflector (DBR) mirrors consisting of AlGaAs quarter-wave stacks made of alternating high- and low-refractive-index layers. The structure is grown on an n-type GaAs substrate, and the mirrors are appropriately doped n- or p-type to form a p–n junction. Electrons and holes are injected into the active region under forward bias, where they are captured by the QWs and produce gain at the lasing wavelength λ. The quantum wells are contained within a transparent layer of thickness $\lambda/2n_0$, where n_0 is the average refractive index of the active region. The light at the design wavelength is reflected back and forth through the gain medium and adds up constructively, forming a laser resonator. Oxidised or proton-implanted regions provide lateral confinement of both the current and the optical mode. Reviews of the design and properties of VCSELs may be found in [42.34] and [42.35].

The conventional VCSEL structures grown on GaAs substrates operate in the wavelength range 700–1100 nm [42.36]. Some of these structures are lattice-matched, but others – notably the longer-wavelength devices which incorporate strained InGaAs quantum wells – are not. Much work is currently focussed on extending the range of operation to the telecommunication wavelengths of 1300 nm and 1550 nm. Unfortunately, it is hard to grow DBR mirrors with sufficient reflectivity on InP substrates due to the low refractive-index contrast of the materials, and thus progress has been slow. Recent alternative approaches based on GaAs substrates will be mentioned in Sect. 42.6.

Resonator structures such as the VCSEL shown in Fig. 42.11 can be operated below threshold as resonant-cavity LEDs (RCLEDs). The presence of the cavity reduces the emission line width and hence increases the intensity at the peak wavelength [42.37]. Furthermore, the narrower line width leads to an increase in the bandwidth of the fibre communication system due to the reduced chromatic dispersion [42.38]. A review of the progress in RCLEDs is given in [42.39].

42.3.2 Quantum Cascade Lasers

The principles of infrared emission by ISB transitions were described in Sect. 42.2.4. Electrons must first be injected into an upper confined electron level as shown in Fig. 42.9b. Radiative transitions to lower confined states with different parities can then occur. ISB emission is usually very weak, as the radiative transitions have to compete with very rapid nonradiative decay by phonon emission, (Sect. 42.2.6). However, when the electron density in the upper level is large enough, population inversion can occur, giving rapid stimulated emission and subsequent laser operation. This is the operating concept of the *quantum cascade (QC) laser* first demonstrated in 1994 [42.40]. The laser operated at 4.2 µm at temperatures up to 90 K. Although the threshold current for the original device was high, progress in the field had been very rapid. A comprehensive review of the present state of the art is given in [42.33], while a more introductory overview may be found in [42.41].

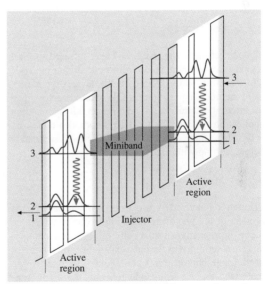

Fig. 42.12 Conduction band diagram for two active regions of an InGaAs/AlInAs quantum cascade laser, together with the intermediate miniband injector region. The levels in each active region are labelled according to their quantum number n, and the corresponding wave function probability densities are indicated. Laser transitions are indicated by the wavy arrows, while electron tunnelling processes are indicated by the straight arrows. (After [42.33], © 2001 IOP)

The quantum-well structures used in QC lasers are very complicated, and often contain hundreds of different layers. Figure 42.12 illustrates a relatively simple design based on lattice-matched $In_{0.47}Ga_{0.53}As/Al_{0.48}In_{0.52}As$ quantum wells grown on an InP substrate. The diagram shows two active regions and the miniband injector region that separates them. A typical operational laser might contain 20–30 such repeat units. The population inversion is achieved by resonant tunnelling between the $n = 1$ ground state of one active region and the $n = 3$ upper laser level of the next one. The basic principles of this process were enunciated as early as 1971 [42.42], but it took more than 20 years to demonstrate the ideas in the laboratory. The active regions contain asymmetric coupled quantum wells, and the laser transition takes place between the $n = 3$ and $n = 2$ states of the coupled system. The separation of the $n = 2$ and $n = 1$ levels is carefully designed to coincide with the LO-phonon energy, so that very rapid relaxation to the ground state occurs and the system behaves as a four-level laser. This latter point is crucial, since the lifetime of the upper laser level is very short (typically ≈ 1 ps), and population inversion is only possible when the lifetime of the lower laser level is shorter than that of the upper one. The lasing wavelength can be varied by detailed design of the coupled QW active region. The transition energy for the design shown in Fig. 42.12 is 0.207 eV, giving emission at 6.0 μm. Further details may be found in [42.33].

A very interesting recent development has been the demonstration of a QC laser operating in the far-infrared spectral region at 67 μm [42.43]. Previous work in this spectral region had been hampered by high losses due to free-carrier absorption and the difficulties involved in designing the optical waveguides. The device operated up to 50 K and delivered 2 mW. These long-wavelength devices are required for applications in the THz frequency range that bridges between long-wavelength optics and high-frequency electronics.

42.4 Detectors

Photodetectors for the visible and near-infrared spectral regions are generally made from bulk silicon or III–V alloys such as GaInAs. Since these devices work very well, the main application for QW photodetectors is in the infrared spectral region and for especially demanding applications such as avalanche photodiodes and solar cells. These three applications are discussed separately below, starting with solar cells.

42.4.1 Solar Cells

The power generated by a solar cell is determined by the product of the photocurrent and the voltage across the diode. In conventional solar cells, both of these parameters are determined by the band gap of the semiconductor used. Large photocurrents are favoured by narrow-gap materials, because semiconductors only absorb photons with energies greater than the band gap, and narrow-gap materials therefore absorb a larger fraction of the solar spectrum. However, the largest open-circuit voltage that can be generated in a p–n device is the built-in voltage which increases with the band gap of the semiconductor. Quantum-well devices can give better performance than their bulk counterparts because they permit separate optimisation of the current- and voltage-generating factors [42.44]. This is because the built-in voltage is primarily determined by the band gap of the barrier regions, whereas the absorption edge is determined by the band gap of the quantum wells. The drawback in using quantum wells is that it is difficult to maintain high photocurrent quantum efficiency in the low-field forward-bias operating conditions in solar cells.

Recent work in this field has explored the added benefits of the versatility of the design of the QW active region [42.45] and also the possibility of using strained QWs. In the latter case, a tradeoff arises between the increase in both the absorption and the number of interface dislocations (which act as carrier traps) with the number of QWs. A way round this compromise is to use strain balance. An example is the case of $In_xGa_{1-x}As/GaAs_{0.94}P_{0.06}$ QW solar cells grown on GaAs substrates, in which the compressive strain of the InGaAs QWs is compensated with the tensile-strained $GaAs_{0.94}P_{0.06}$ barriers, such that the overall active region could be successfully lattice-matched to the substrate [42.46].

42.4.2 Avalanche Photodiodes

Avalanche photodiodes (APDs) are the detectors of choice for many applications in telecommunications and single-photon counting. The avalanche multiplica-

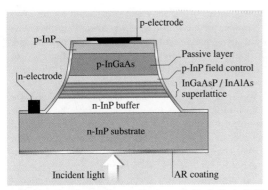

Fig. 42.13 Schematic representation of an InGaAs/InP/InGaAsP/InAlAs superlattice avalanche photodiode (SL-APD). Light is absorbed in the bulk InGaAs layer and the resulting photocurrent is multiplied by the avalanche process in the InGaAsP/InAlAs superlattice region. The spatial periodicity of the superlattice reduces the dark current

tion mechanism plays a critical role in determining the photodetection gain, the noise and the gain–bandwidth product. Commercially available III–V semiconductor APDs are typically engineered with different band-gap materials in the absorption and multiplication regions [42.47]. The absorption layer has a relatively narrow band gap (such as $In_{0.53}Ga_{0.47}As$) to allow for larger absorption, whereas the multiplication region has a wider band gap (such as InP or $In_{0.52}Al_{0.48}As$) to reduce the dark current at the high electric fields required. It has been demonstrated that the dark current can be further reduced by incorporating adequately designed superlattices into the multiplication layer to form a *superlattice avalanche photodiode* (SL-APD) as shown in Fig. 42.13 [42.48]. Essentially, the miniband formation (Fig. 42.3) in the SL corresponds to a larger effective band gap and thus reduces the probability of band-to-band tunnelling at the high electric fields required for carrier impact ionisation and subsequent carrier multiplication.

It has been proposed that the excess noise factor can be reduced further by designing multiplication layers with SL [42.49] or *staircase* [42.50] structures to enhance the ionisation rate of one carrier type relative to the other due to their different band-edge discontinuities [42.51]. However, this proposal is still under active theoretical and experimental scrutiny, because most of the electrons populate the higher energy satellite valleys (X and L) at the high electric fields required for avalanche gain [42.52]. Recent Monte Carlo simulations [42.53], backed up by experimental evidence [42.54], have shown that the ratio of the electron and hole ionisation rates, commonly used as a figure of merit in bulk multiplication layers [42.55], is not critically affected by the band-gap discontinuities. In fact, these recent models emphasise that it is the spatial modulation of the impact ionisation probability, associated with the periodic band-gap discontinuity, which leads to a reduction in the multiplication noise.

42.4.3 Inter-Sub-Band Detectors

The principles of infrared absorption by ISB transitions were described in Sect. 42.2.4. Infrared detectors are required for applications in defence, night vision, astronomy, thermal mapping, gas-sensing, and pollution monitoring. *Quantum-well inter-sub-band photodetectors* (QWIPs) are designed so that the energy separation of the confined levels is matched to the chosen wavelength. A major advantage of QWIPs over the conventional approach employing narrow-gap semiconductors is the use of mature GaAs-based technologies. Furthermore, the detection efficiency should in principle be high due to the large oscillator strength that follows from the parallel curvature of the in-plane dispersions for states within the same bands [42.56]. However, technical challenges arise from the requirement that the electric field of the light must be polarised along the growth (z) direction. This means that QWIPs do not work at normal incidence, unless special steps are taken to introduce a component of the light field along the z-direction. Various approaches have been taken for optimum light coupling, such as using bevelled edges, gratings or random reflectors [42.57].

Despite their promising characteristics, QWIPs have yet to be commercialised. The main issue is the high dark current at higher operating temperatures. The dark current is governed by the thermionic emission of ground-state electrons directly out of the QW above 45 K [42.58]. Overcoming such technical difficulties has made possible the demonstration of long-wavelength large-format focal-plane array cameras based on ISB transitions [42.59].

42.4.4 Unipolar Avalanche Photodiodes

The combination of resonant inter-sub-band (ISB) photodetection (Sect. 42.4.3) and avalanche multiplication has been studied for exploiting the possibility of designing a unipolar avalanche photodiode (UAPD) [42.60]. UAPDs rely on impact ionisation involving only one

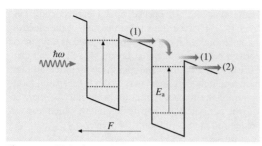

Fig. 42.14 Unipolar carrier multiplication in a multiple QW structure at field strength F: (1) is the primary electron resulting from photodetection, while (2) is the secondary electron resulting from the impact ionisation of the QW by the primary electron

type of carrier in order to achieve gain in photoconductive detectors for mid- and far-infrared light. The unipolar impact ionisation occurs when the kinetic energy of the primary carrier exceeds an activation energy E_a defined as the transition energy between the QW ground state and the QW top state (Fig. 42.14). A single QW then releases an extra electron each time it is subject to an impact with an incoming electron.

The impact ionisation probability for this process is given by the product between the carrier capture probability and the carrier escape probability under a mechanism of carrier–carrier scattering in the QW. The subsequent electron transport towards further multiplication events occurs through a sequence of escape, drift and kinetic energy gain under the applied electric field, relaxation, capture and QW ionisation. Ultimately, the multiplication gain in an UAPD is governed by the QW capture probability, the number of QWs and the field uniformity over the QW sequence. The unipolar nature of the multiplication process must be preserved in order to avoid field nonuniformities stemming from spatial-charge variation caused by bipolar carrier transport across the multiplication region.

Interest in a purely unipolar multiplication mechanism was originally motivated by the possibility of reduced noise in comparison to bipolar APDs (Sect. 42.4.2), where band-to-band transitions lead to gain fluctuations manifested as excess noise [42.55]. For this purpose, the QWs in an UAPD structure are typically tailored such that the inter-sub-band activation energy E_a is smaller than the inter-band impact ionisation activation energy that would be responsible for bipolar avalanche multiplication. However, recent studies have shown that unipolar avalanche multiplication is also accompanied by an excess noise factor, such that the noise gain exceeds the photoconductive gain [42.61], thus limiting the practical applications of UAPDs.

42.5 Modulators

In Sect. 42.2.3 we noted that the optical properties of quantum-well diodes are strongly modified by the application of voltages through the quantum-confined Stark effect (QCSE). Referring to Fig. 42.8, we see that at wavelengths below the heavy-hole exciton at 0 V, the absorption increases as the voltage is applied, which provides a mechanism for the modulation of light. For example, the amount of light transmitted at wavelengths close to the band edge would change with the voltage applied. Moreover, since changes of absorption are accompanied by changes of the refractive index through the Kramers–Kronig relationship, it is possible to make QCSE phase modulators as well [42.62]. In addition to the standard GaAs/AlGaAs devices operating around 800 nm, QCSE modulators have been demonstrated in several other material systems, such as GaInAs-based structures for the important telecommunications wavelength at 1.5 μm [42.63].

The operation of GaAs-based QCSE transmission modulators at normal incidence is hampered by the fact that the substrates are opaque at their operating wavelength. One way round this problem

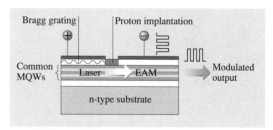

Fig. 42.15 Schematic diagram of an integrated quantum-well waveguide electroabsorption modulator (EAM) and distributed feedback (DFB) laser. The laser is forward-biased while the EAM is reverse-biased. The p-contacts of the two electrically independent devices are separated by proton implantation. The light emitted by the laser is guided through the EAM region, resulting in a modulated output beam when data pulses are applied to the EAM

is to etch the substrate away, but this is a difficult process, and a much better solution is to include a mirror underneath the quantum wells so that the modulated light does not have to pass through the substrate [42.64]. In many practical applications, however, mostly involving the integration of QCSE modulators with MQW light emitters on a common substrate, the waveguide geometry is the configuration of choice [42.65]. In this architecture, the light beam propagates along the waveguide from the emitter to the *electroabsorption modulator* (EAM) region, as shown in Fig. 42.15. The QCSE modulator transmits the incoming laser light when no voltage is applied and absorbs the beam when the MQW stack is suitably biased.

The most successful commercial impact of QCSE modulators has been in the integration of EAMs with distributed feedback (DFB) or distributed Bragg reflector (DBR) MQW diode lasers in waveguide configurations, as shown in Fig. 42.15. These devices have been used for optical coding in the C-band (1525–1565 nm), at 10 Gb/s or higher data transmission speeds [42.66]. The combination of a continuous laser and a high-speed modulator offers better control of the phase chirp of the pulses than direct modulation of the laser output itself [42.67]. In particular, the chirp factor is expected to become negative if the photogenerated carriers can be swept out fast enough in the EAM, which is desirable for long-distance data transmission through optical fibers [42.68].

A promising step toward the merger between band-gap-engineered semiconductors and mature very-large-scale integration (VLSI) silicon architectures has been achieved when III–V semiconductor QCSE modulator structures have been integrated with state-of-the-art silicon complementary metal–oxide–semiconductor (CMOS) circuitry [42.69]. Through this hybrid technology, thousands of optical inputs and outputs could be provided to circuitry capable of very complex information processing. The idea of using light beams to replace wires in telecommunications and digital computer systems has thus become an attractive technological avenue in spite of various challenges implied [42.70].

42.6 Future Directions

The subject of quantum-confined semiconductor structures moves very rapidly and it is difficult to see far into the future. Some ideas have moved very quickly from the research labs into the commercial sector (e.g. VCSELs), while others (e.g. quantum cascade lasers) have taken many years to come to fruition. We thus restrict ourselves here to a few comments based on active research fields at the time of writing.

One idea that is being explored in detail is the effects of lower dimensionality in quantum-wire and quantum-dot structures (Table 42.1). Laser operation from one-dimensional (1-D) GaAs quantum wires was first demonstrated in 1989 [42.72], but subsequent progress has been relatively slow due to the difficulty in making the structures. By contrast, there has been an explosion of interest in zero-dimensional (0-D) structures following the discovery that quantum dots can form spontaneously during MBE growth in the Stranski–Krastanow regime. A comprehensive review of this subject is given in [42.73].

Figure 42.16 shows an electron microscope image of an InAs quantum dot grown on a GaAs crystal by the Stranski–Krastanow technique. The dots are formed because of the very large mismatch between the lattice constants of the InAs and the underlying GaAs. The strain that would be produced in a uniform layer is so large that it is energetically favourable to form small clusters. This then leads to the formation of islands of InAs with nanoscale dimensions, which can then be encapsulated within an optoelectronic structure by overgrowth of further GaAs layers.

The ability to grow quantum-dot structures directly by MBE has led to very rapid progress in the deployment of quantum dots in a variety of applications. It remains unclear at present whether quantum dots really lead to superior laser performance [42.74]. The intrin-

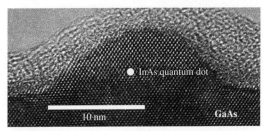

Fig. 42.16 Transmission electron microscope image of an uncapped InAs quantum dot grown on GaAs by the Stranski–Krastanow technique. (After [42.71], © 2000 APS)

sic gain of the dots is higher than that of a quantum well [42.75], and the threshold current is less sensitive to temperature [42.76]. However, the volume of the gain medium is necessarily rather small, and the benefits of the lower dimensionality cannot be exploited to the full. At present, one of the most promising applications for quantum dots is in long-wavelength lasers [42.77]. As mentioned in Sect. 42.3.1, the production of VCSELs at 1300 nm and 1550 nm has proven to be difficult using conventional InP-based QW structures due to the low refractive-index contrast of the materials that form the DBR mirrors. The use of InAs/GaAs quantum dots as the active region circumvents this problem and allows the benefits of mature GaAs-based VCSEL technology.

Another very exciting potential application for quantum dots is in quantum information processing. High-efficiency single-photon sources are required for quantum cryptography and also quantum computation using linear optics. Several groups have demonstrated single photon emission after excitation of individual InAs quantum dots (see e.g. [42.78, 79]), and one group has demonstrated an electrically driven single-photon LED [42.80]. After these proofs of principle, the challenge now lies ahead to establish the quantum-dot sources in working quantum information-processing systems.

At the same time as exploring the effects of lower dimensionality, many other groups are working on new QW materials. One of the most promising recent developments is the dilute nitride system for applications in long-wavelength VCSELs and solar cells [42.81]. It has been found that the inclusion of a small fraction of nitrogen into GaAs leads to a sharp decrease in the band gap due to very strong band-bowing effects. This then allows the growth of InGaAsN structures that emit at 1300 nm on GaAs substrates [42.77, 82]. The field is developing very rapidly, with 1300-nm VCSELs and 1500-nm edge emitters already demonstrated [42.83, 84].

42.7 Conclusions

Semiconductor quantum wells are excellent examples of quantum mechanics in action. The reduced dimensionality has led to major advances in both the understanding of 2-D physics and the applied science of optoelectronics. In some cases, QWs have enhanced the performance of conventional devices (e.g. LEDs and edge-emitting lasers), and in others, they have led to radically new devices (e.g. VCSELs, quantum cascade lasers, QCSE modulators). At present, the main commercial use for QW optoelectronic devices is in LEDs, laser diodes and QCSE modulators. It remains to be seen whether some of the other devices described here (QW solar cells, SL-APDs, QWIPs) will come to commercial fruition, and whether systems of lower dimensionality will eventually replace QWs in the same way that QWs have replaced bulk devices.

References

42.1   L. Esaki, R. Tsu: IBM J. Res. Develop. **14**, 61–5 (1970)
42.2   G. Bastard: *Wave Mechanics Applied to Semiconductor Heterostructures* (Wiley, New York 1988)
42.3   M. Jaros: *Physics and Applications of Semiconductor Microstructures* (Clarendon, Oxford 1989)
42.4   C. Weisbuch, B. Vinter: *Quantum Semiconductor Structures* (Academic, San Diego 1991)
42.5   S. O. Kasap: *Optoelectronics and Photonics: Principles and Practices* (Prentice Hall, Upper Saddle River 2001)
42.6   M. J. Kelly: *Low-Dimensional Semiconductors* (Clarendon, Oxford 1995)
42.7   Paul J. Dean: III–V Compound Semiconductors. In: *Electroluminescence*, ed. by J. I. Pankove (Springer, Berlin, Heidelberg 1977) pp. 63–132
42.8   S. Nakamura, S. Pearton, G. Fasol: *The Blue Laser Diode*, 2nd edn. (Springer, Berlin, Heidelberg 2000)
42.9   S. Gasiorowicz: *Quantum Physics*, 2nd edn. (Wiley, New York 1996)
42.10   E. P. O'Reilly: Semicond. Sci. Technol. **4**, 121–137 (1989)
42.11   M. Fox: *Optical Properties of Solids* (Clarendon, Oxford 2001)
42.12   M. Shinada, S. Sugano: J. Phys. Soc. Jpn. **21**, 1936–46 (1966)
42.13   D. A. B. Miller, D. S. Chemla, D. J. Eilenberger, P. W. Smith, A. C. Gossard, W. T. Tsang: Appl. Phys. Lett. **41**, 679–81 (1982)
42.14   A. M. Fox, D. A. B. Miller, G. Livescu, J. E. Cunningham, W. Y. Jan: IEEE J. Quantum Electron. **27**, 2281–95 (1991)

42.15 H. Schneider, K. von Klitzing: Phys. Rev. B **38**, 6160–5 (1988)

42.16 A. M. Fox, R. G. Ispasoiu, C. T. Foxon, J. E. Cunningham, W. Y. Jan: Appl. Phys. Lett. **63**, 2917–9 (1993)

42.17 J. Feldmann, K. Leo, J. Shah, D. A. B. Miller, J. E. Cunningham, T. Meier, G. von Plessen, A. Schulze, P. Thomas, S. Schmitt-Rink: Phys. Rev. B **46**, 7252–5 (1992)

42.18 K. Leo, P. Haring Bolivar, F. Brüuggemann, R. Schwedler, K. Köhler: Solid State Commun. **84**, 943–6 (1992)

42.19 C. Waschke, H. G. Roskos, R. Schwedler, K. Leo, H. Kurz, K. Köhler: Phys. Rev. Lett. **70**, 3319–22 (1993)

42.20 Y. Shimada, K. Hirakawa, S.-W. Lee: Appl. Phys. Lett. **81**, 1642–4 (2002)

42.21 P. W. M. Blom, C. Smit, J. E. M. Haverkort, J. H. Wolter: Phys. Rev. B **47**, 2072–2081 (1993)

42.22 J. Shah: *Ultrafast Spectroscopy of Semiconductors and Semiconductor Nanostructures*, 2nd edn. (Springer, Berlin, Heidelberg 1999)

42.23 R. G. Ispasoiu, A. M. Fox, D. Botez: IEEE J. Quantum Electron. **36**, 858–63 (2000)

42.24 P. Blood: Visible-emitting quantum well lasers. In: *Semiconductor Quantum Optoelectronics*, ed. by A. Miller, M. Ebrahimzadeh, D. M. Finlayson (Institute of Physics, Bristol 1999) pp. 193–211

42.25 N. Chand, S. N. G. Chu, N. K. Dutta, J. Lopata, M. Geva, A. V. Syrbu, A. Z. Mereutza, V. P. Yakovlev: IEEE J. Quantum Electron. **30**, 424–40 (1994)

42.26 E. P. O'Reilly, A. R. Adams: IEEE J. Quantum Electron. **30**, 366–79 (1994)

42.27 M. R. Krames, J. Bhat, D. Collins, N. F. Gargner, W. Götz, C. H. Lowery, M. Ludowise, P. S. Martin, G. Mueller, R. Mueller-Mach, S. Rudaz, D. A. Steigerwald, S. A. Stockman, J. J. Wierer: Phys. Stat. Sol. A **192**, 237–245 (2002)

42.28 M. Ikeda, S. Uchida: Phys. Stat. Sol. A **194**, 407–13 (2002)

42.29 S. Nagahama, T. Yanamoto, M. Sano, T. Mukai: Phys. Stat. Sol. A **194**, 423–7 (2002)

42.30 S. Kamiyama, M. Iwaya, H. Amano, I. Akasaki: Phys. Stat. Sol. A **194**, 393–8 (2002)

42.31 L. F. Eastman, V. Tilak, V. Kaper, J. Smart, R. Thompson, B. Green, J. R. Shealy, T. Prunty: Phys. Stat. Sol. A **194**, 433–8 (2002)

42.32 W. S. Tan, P. A. Houston, P. J. Parbrook, G. Hill, R. J. Airey: J. Phys. D: Appl. Phys. **35**, 595–8 (2002)

42.33 C. Gmachl, F. Capasso, D. L. Sivco, A. Y. Cho: Rep. Prog. Phys. **64**, 1533–1601 (2001)

42.34 K. J. Eberling: Analysis of vertical cavity surface emitting laser diodes (VCSEL). In: *Semiconductor Quantum Optoelectronics*, ed. by A. Miller, M. Ebrahimzadeh, D. M. Finlayson (Institute of Physics, Bristol 1999) pp. 295–338

42.35 M. SanMiguel: Polarisation properties of vertical cavity surface emitting lasers. In: *Semiconductor Quantum Optoelectronics*, ed. by A. Miller, M. Ebrahimzadeh, D. M. Finlayson (Institute of Physics, Bristol 1999) pp. 339–366

42.36 O. Blum Spahn: Materials issues for vertical cavity surface emitting lasers (VCSEL) and edge emitting lasers (EEL). In: *Semiconductor Quantum Optoelectronics*, ed. by A. Miller, M. Ebrahimzadeh, D. M. Finlayson (Institute of Physics, Bristol 1999) pp. 265–94

42.37 E. F. Schubert, Y.-H. Wang, A. Y. Cho, L.-W. Tu, G. J. Zydzik: Appl. Phys. Lett. **60**, 921–3 (1992)

42.38 N. E. Hunt, E. F. Schubert, R. F. Kopf, D. L. Sivco, A. Y. Cho, G. J. Zydzik: Appl. Phys. Lett. **63**, 2600–2 (1993)

42.39 R. Baets: Micro-cavity light emitting diodes. In: *Semiconductor Quantum Optoelectronics*, ed. by A. Miller, M. Ebrahimzadeh, D. M. Finlayson (Institute of Physics, Bristol 1999) pp. 213–64

42.40 J. Faist, F. Capasso, D. L. Sivco, C. Sirtori, A. L. Hutchinson, A. Y. Cho: Science **264**, 553–6 (1994)

42.41 F. Capasso, C. Gmachl, D. L. Sivco, A. Y. Cho: Phys. Today **55**(5), 34–40 (2002)

42.42 R. A. Kazarinov: Sov. Phys. Semicond. **5**, 707–9 (1971)

42.43 R. Köhler, A. Tredicucci, F. Beltram, H. E. Beere, E. H. Linfield, A. G. Davies, D. A. Ritchie, R. C. Iotti, F. Rossi: Nature **417**, 156–9 (2002)

42.44 K. Barnham, I. Ballard, J. Barnes, J. Connolly, P. Griffin, B. Kluftinger, J. Nelson, E. Tsui, A. Zachariou: Appl. Surf. Sci. **113/114**, 722–733 (1997)

42.45 R. H. Morf: Physica E **14**, 78–83 (2002)

42.46 N. J. Ekins-Daukes, K. W. J. Barnham, J. P. Connolly, J. S. Roberts, J. C. Clark, G. Hill, M. Mazzer: Appl. Phys. Lett. **75**, 4195–5197 (1999)

42.47 J. Wei, J. C. Dries, H. Wang, M. L. Lange, G. H. Olsen, S. R. Forrest: IEEE Photon. Technol. Lett. **14**, 977–9 (2002)

42.48 A. Suzuki, A. Yamada, T. Yokotsuka, K. Idota, Y. Ohiki: Jpn. J. Appl. Phys. **41**, 1182–5 (2002)

42.49 F. Capasso, W. T. Tsang, A. L. Hutchinson, G. F. Williams: Appl. Phys. Lett. **40**, 38–40 (1982)

42.50 G. Ripamonti, F. Capasso, A. L. Hutchinson, D. J. Muehlner, J. F. Walker, R. J. Malek: Nucl. Instrum. Meth. Phys. Res. A **288**, 99–103 (1990)

42.51 R. Chin, N. Holonyak, G. E. Stillman, J. Y. Tang, K. Hess: Electron. Lett. **16**, 467–9 (1980)

42.52 C. K. Chia, J. P. R. David, G. J. Rees, P. N. Robson, S. A. Plimmer, R. Grey: Appl. Phys. Lett. **71**, 3877–9 (1997)

42.53 F. Ma, X. Li, S. Wang, K. A. Anselm, X. G. Zheng, A. L. Holmes, J. C. Campbell: J. Appl. Phys. **92**, 4791–5 (2002)

42.54 P. Yuan, S. Wang, X. Sun, X. G. Zheng, A. L. Holmes, J. C. Campbell: IEEE Photon. Technol. Lett. **12**, 1370–2 (2000)

42.55 M. A. Saleh, M. M. Hayat, P. P. Sotirelis, A. L. Holmes, J. C. Campbell, B. E. A. Saleh, M. C. Teich: IEEE Trans. Electron. Dev. **48**, 2722–31 (2001)

42.56 L. C. West, S. J. Eglash: Appl. Phys. Lett. **46**, 1156–8 (1985)

42.57 S. D. Gunapala, G. Sarusi, J. S. Park, T. Lin, B. F. Levine: Phys. World **7**(10), 35–40 (1994)

42.58 S. D. Gunapala, S. V. Bandara: Quantum well infrared photodetector (QWIP) focal plane arrays. In: *Semiconductors and Semimetals*, Vol. 62, ed. by M. C. Liu, F. Capasso (Academic, New York 1999) pp. 197–282

42.59 S. D. Gunapala, S. V. Bandara, J. K. Liu, E. M. Luong, N. Stetson, C. A. Shott, J. J. Block, S. B. Rafol, J. M. Mumolo, M. J. McKelvey: IEEE Trans. Electron. Dev. **47**, 326–332 (2000)

42.60 B. F. Levine, K. K. Choi, C. G. Bethea, J. Walker, R. J. Malik: Appl. Phys. Lett. **51**, 934–6 (1987)

42.61 H. Schneider: Appl. Phys. Lett. **82**, 4376–8 (2003)

42.62 J. S. Weiner, D. A. B. Miller, D. S. Chemla: Appl. Phys. Lett. **50**, 842–4 (1987)

42.63 R. W. Martin, S. L. Wong, R. J. Nicholas, K. Satzke, M. Gibbons, E. J. Thrush: Semicond. Sci. Technol. **8**, 1173–8 (1993)

42.64 G. D. Boyd, D. A. B. Miller, D. S. Chemla, S. L. McCall, A. C. Gossard, J. H. English: Appl. Phys. Lett. **50**, 1119–21 (1987)

42.65 A. Ramdane, F. Devaux, N. Souli, D. Delprat, A. Ougazzaden: IEEE J. Quantum Electron. **2**, 326–35 (1996)

42.66 T. Ido, S. Tanaka, M. Suzuki, M. Koizumi, H. Sano, H. Inoue: J. Lightwave Technol. **14**, 2026–33 (1996)

42.67 Y. Miyazaki, H. Tada, T. Aoyagi, T. Nishimura, Y. Mitsui: IEEE J. Quantum Electron. **38**, 1075–80 (2002)

42.68 G. Agrawal: *Fiber Optic Communication Systems* (Wiley, New York 1993)

42.69 K. W. Goosen, J. A. Walker, L. A. D'Asaro, S. P. Hui, B. Tseng, R. Leibenguth, D. Kossives, D. D. Bacon, D. Dahringer, L. M. F. Chirovsky, A. L. Lentine, D. A. B. Miller: IEEE Photon. Technol. Lett. **7**, 360–2 (1995)

42.70 D. A. B. Miller: IEEE J. Sel. Top. Quantum Electron. **6**, 1312–7 (2000)

42.71 P. W. Fry, I. E. Itskevich, D. J. Mowbray, M. S. Skolnick, J. J. Finley, J. A. Barker, E. P. O'Reilly, L. R. Wilson, I. A. Larkin, P. A. Maksym, M. Hopkinson, M. Al-Khafaji, J. P. R. David, A. G. Cullis, G. Hill, J. C. Clark: Phys. Rev. Lett. **84**, 733–6 (2000)

42.72 E. Kapon, D. M. Hwang, R. Bhat: Phys. Rev. Lett. **63**, 430–3 (1989)

42.73 D. Bimberg, M. Grundmann, Nikolai N. Ledentsov: *Quantum Dot Heterostructures* (Wiley, Chichester 1998)

42.74 M. Grundmann: Physica E **5**, 167–84 (2000)

42.75 M. Asada, Y. Miyamoto, Y. Suematsu: IEEE J. Quantum Electron. **22**, 1915–21 (1986)

42.76 Y. Arakawa, H. Sakaki: Appl. Phys. Lett. **40**, 939–41 (1982)

42.77 V. M. Ustinov, A. E. Zhukov: Semicond. Sci. Technol. **15**, R41–R54 (2000)

42.78 P. Michler, A. Kiraz, C. Becher, W. V. Schoenfeld, P. M. Petroff, L. Zhang, E. Hu, A. Imamoglu: Science **290**, 2282–5 (2000)

42.79 C. Santori, M. Pelton, G. Solomon, Y. Dale, Y. Yamamoto: Phys. Rev. Lett. **86**, 1502–5 (2001)

42.80 Z. Yuan, B. E. Kardynal, R. M. Stevenson, A. J. Shields, C. J. Lobo, K. Cooper, N. S. Beattie, D. A. Ritchie, M. Pepper: Science **295**, 102–5 (2002)

42.81 A. Mascarenhas, Y. Zhang: Current Opinion Solid State Mater. Sci. **5**, 253–9 (2001)

42.82 A. Yu. Egorov, D. Bernklau, B. Borchert, S. Illek, D. Livshits, A. Rucki, M. Schuster, A. Kaschner, A. Hoffmann, Gh. Dumitras, M. C. Amann, H. Riechert: J. Cryst. Growth **227–8**, 545–552 (2001)

42.83 G. Steinle, H. Riechert, A. Yu. Egorov: Electron. Lett. **37**, 93–5 (2001)

42.84 D. Gollub, M. Fischer, A. Forchel: Electron. Lett. **38**, 1183–4 (2002)

43. Glasses for Photonic Integration

Inorganic glasses are the workhorse materials of optics and photonics. In addition to offering a range of transparency windows, glasses provide flexibility of processing for the realization of fibers, films, and shaped optical elements. Traditionally, the main role of glass has been as a passive material. However, a significant attribute of glasses is their ability to incorporate dopants such as nanoparticles or active ions. Hence, glasses promise to play an increasingly important role in active photonics, as laser, amplification, switching, and nonlinear media.

For photonic integration, many of the attributes of glasses are particularly compelling. Glasses allow numerous options for thin film deposition and integration on arbitrary platforms. The possibility of controlling the viscosity of a glass during processing can be exploited in the realization of extremely low loss microphotonic waveguides, photonic crystals, and microcavities. The metastable nature of glass can enable the direct patterning of photonic elements by energetic beams.

This chapter provides an overview of these unique properties of glasses, from the perspectives of the technology options they afford and the

43.1	**Main Attributes of Glasses as Photonic Materials**............................ 1042	
	43.1.1 The Glass Transition as Enabler ... 1043	
	43.1.2 Metastability 1046	
	43.1.3 Glass as Host Material................ 1049	
43.2	**Glasses for Integrated Optics** 1050	
	43.2.1 Low Index Glassy Films 1050	
	43.2.2 Medium Index Glassy Films 1051	
	43.2.3 High Index Glassy Films 1051	
43.3	**Laser Glasses for Integrated Light Sources** 1053	
	43.3.1 Advantages of Glass-based Light Sources 1053	
	43.3.2 Alternative Glass Hosts................ 1054	
	43.3.3 Progress Towards Integrated Light Sources in Glass 1056	
43.4	**Summary** 1057	
	References 1059	

practical limitations they present. Further, an overview is provided of the main families of glassy inorganic films studied for integrated optics. Finally, the main features of rare earth doped glasses are reviewed, with an emphasis on their potential for implemention of compact integrated light sources and amplifiers.

Inorganic glasses have played a central role in optical science and technology, and more generally within the electrical and electronic engineering disciplines [43.1]. Amongst optical materials, the unique advantages of glasses are well known [43.2,3]. Glasses can be worked relatively easily into various forms, such as bulk lenses, fibers, and thin films. This is dramatically illustrated by the modern technology for the manufacture of low-loss silica fibers, in which a hair-thin glass fiber (with tightly controlled geometrical and material properties) is drawn from a heated glass preform at rates as high as 20 m/s or greater [43.4]. Glass can be manufactured with excellent homogeneity and without grain boundaries, so that scattering of light is acceptably low. While perfect crystals typically exhibit even lower levels of scattering, they are difficult to realize in large sizes and with arbitrary shapes. Optical losses in polycrystalline materials, on the other hand, are generally excessive due to scattering from grain boundaries. Finally, the standard glass compositions (oxide, halide, and chalcogenide glasses and hybrids of those) provide transparency windows from the UV to the mid-infrared (see Fig. 43.1). It has been estimated [43.2] that more than 90% of optical components are based on glasses. Given the massive worldwide installation of telecommunications fibers since the 1980s [43.5], and depending on the definition assigned to the term 'component', this statistic might actually be closer to 100%.

Befitting their central role in optical technology, there are numerous excellent reviews covering the optical properties and photonic functionalities of glasses [43.3, 6]. Further, one of the distinguishing attributes of glasses is their flexibility of composition. Hundreds of commercial glasses are available, and many more have been studied in research laboratories and reported in the academic literature. Thus, it is not possible to capture the full range of glassy materials (even restricted to inorganic glasses) within a short review paper. The recent monograph by *Yamane* and *Asahara* [43.6] provides a comprehensive and highly recommended treatment of glass technology and its application to photonics. The present contribution has relatively modest aims:

1. With a focus specifically on *integrated* photonics, the main advantages of glasses are discussed and the most studied glass systems are reviewed in brief. Integrated photonics (integrated optics) refers to guided wave photonic devices fabricated on planar platforms. Glass substrates (with waveguides defined by ion diffusion) have played an important role in integrated optics, but the advantages of glasses (compared to crystalline materials such as III-V semiconductors or lithium niobate) are less compelling in that case. For this reason, emphasis is placed on glassy thin films that can enable photonic functionality on other substrates. In practice, substrates of interest might be printed circuit boards or semiconductor wafers.
2. For the most part, the discussion is restricted to inorganic glasses. Some reference is made to organic glasses, since their properties and advantages parallel those of inorganic glasses in many respects, and because certain practical aspects of organic glasses have received greater attention in the literature. Good reviews covering polymer materials for integrated photonics are available [43.8].
3. To date, the most important inorganic glass system for integrated photonics is the silica on silicon planar lightwave circuit (PLC) platform. Several recent papers [43.9, 10] provide an overview of that technology, so only brief mention is made here. Rather, the goal is to provide some insight on advantages that

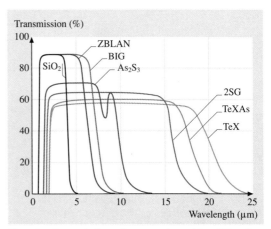

Fig. 43.1 Transmission plots are shown for SiO_2 glass and some representative fluoride and chalcogenide glasses. ZBLAN and BIG are heavy-metal fluoride glasses, 2SG is a selenide glass, and TeXAs and TeX are tellurium-based chalcogenide glasses. (After [43.7])

are globally associated with glassy materials. Further, an emphasis is placed on the unique advantages (lower processing temperatures, higher refractive index contrast, active functionality, etc.) enabled by some glass systems.

With these goals in mind, the chapter is organized as follows. In Sect. 43.1, we highlight the main advantages of glasses for photonic integration. These attributes include the unique processing options afforded by the existence of the glass transition and by the metastability of glasses. Section 43.2 provides an overview of the most-studied materials for glass-based integrated optics. This includes the standard glass formers (oxide, halide, chalcogenide glasses) as well as several materials that are generally classified as ceramics (Si_3N_4, Al_2O_3, etc.) but which can be realized as amorphous thin films. In Sect. 43.3, some unique advantages of glasses for realization of integrated light sources are reviewed. Section 43.4 provides a summary and some thoughts on future prospects for glasses in photonic integration.

43.1 Main Attributes of Glasses as Photonic Materials

The essential characteristic of a glass is the existence of a glass transition temperature (T_g). In short, if a viscous liquid is cooled fast enough, the liquid passes through the freezing temperature of the material without making a phase transition to the solid crystalline state (see Fig. 43.2). Rather, it becomes a supercooled liquid, whose rate of change of specific volume (or enthalpy) with decreasing temperature remains constant

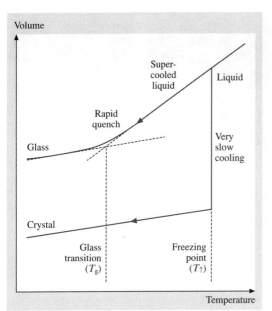

Fig. 43.2 Schematic plot illustrating the change in specific volume of a glass with temperature (after [43.11])

transition between states of very low viscosity (molten liquid), low viscosity (supercooled liquid), and high viscosity (frozen liquid; i.e. glass). This remarkable variation in mechanical properties with temperature, spanning several orders of magnitude in viscosity, underpins both the ancient and modern manufacture of glass products (by molding, extrusion, float processes, etc.). It is interesting to note that T_g can be defined as the approximate temperature at which the shear viscosity attains 10^{12} Pa s. Further, viscosity varies rapidly in the vicinity of T_g. For so-called strong glass formers (such as SiO_2), the viscosity η varies according to an Arrhenius equation of the form $\eta = A \exp(B/k_B T)$, where A and B are temperature-independent constants and k_B is the Boltzmann constant [43.12]. To form glasses into desired shapes, the temperature is typically controlled such that the viscosity is in the 10^2 to 10^7 Pa s range, depending on the process and the stability of the particular glass composition [43.6, 15].

It was once held that only certain materials could form glasses, but it is now believed that the glassy state is attainable for all types of materials [43.11]. Accordingly, the literature contains references to various families of glasses, including inorganic (ceramic), organic (polymeric), and metallic glasses.

and equivalent to that of the molten liquid. At T_g, the slope of these curves (for example, specific volume versus temperature) changes abruptly (but in a continuous way) to a value that is approximately that of the corresponding crystal. Below T_g, the material is an amorphous solid or glass.

While glass forming has been a staple of human technological innovation for thousands of years, the physical processes underlying the glass transition are still a matter of some debate [43.11–13]. Several empirical facts are well known, however. The glass transition temperature of a particular material is not a constant, but is somewhat dependent on processing details such as the rate of cooling from the melt, or more generally on the technique used to form the glass (melt quenching, thin film deposition, etc.) and even the age of the glass [43.14]. T_g is slightly higher for a faster rate of cooling from a melt, for example, but in general $T_g \approx 2T_m/3$, where T_m is the melting/freezing temperature of the material [43.12]. For glasses that cannot be formed easily by melt quenching, such as amorphous Si [43.13], the direct experimental determination of T_g is not always possible, and indirect methods are sometimes used to estimate an effective T_g.

Central to the technological application of glasses is the ability to control their viscosity by way of temperature [43.5]. For stable glass formers, it is possible to

43.1.1 The Glass Transition as Enabler

The surface tension of liquids is a well-known enabler of optical quality surfaces [43.16]. Glass devices can be formed in the liquid state, and can thus accommodate manufacturing processes that are assisted by surface tension mediated self-assembly. The manufacture of modern single mode fibers is a good example of the enabling nature of the glass transition. In a typical process [43.4, 6, 17], highly purified glass layers are deposited either on the inside walls of a fused silica tube by chemical vapor deposition (CVD) or on the outside surface of a ceramic rod by flame hydrolysis deposition (FHD). The hollow glass cylinder (after removal of the ceramic rod in the latter case) is then heated (sintered) to $\approx 1600\,°C$, such that it collapses to form a solid glass preform (typically 1 m in length and 2 cm in diameter). $1600\,°C$ is the approximate softening point of silica glass, typically defined as corresponding to a viscosity of $\approx 10^7$ Pa s [43.6]. The fiber drawing process involves local heating of the preform to a temperature of approximately $2000\,°C$ (corresponding to a viscosity of $\approx 10^5$ Pa s). Thus, viscosity control is critical at nearly all stages of the process. Further, the drawing process allows tremendous control over the optical and geometrical properties of the fiber, in part due to the self-regulating properties of surface tension effects in the supercooled liquid. As an added benefit, inhomogeneities in the original preform tend to be averaged out in the process [43.18]. The amazing properties of silica fibers (tensile strength comparable to that of steel, low loss, and extremely tight tolerances for material and geometrical parameters) are well known, and attest to the technological versatility of (silica) glasses. It is also interesting to note that the standard technique for splicing sections of silica fibers together, which is achieved with less than 0.1 dB loss [43.5], is based on a reflow process wherein surface tension contributes to their alignment.

Within integrated optics, the manufacture and optimization of PLC devices relies directly on the glass transition in many cases [43.9, 10]. PLC devices are manufactured in doped SiO_2 glasses deposited (by CVD, FHD, or sol-gel techniques) on a silicon wafer. The undercladding is often pure SiO_2, which can be formed by thermal oxidation of the silicon wafer. Typically, dopants in the core and uppercladding layers act to modify both the refractive index and the thermo-mechanical properties of the glass. In particular, controlled reflow (by heating the glasses near T_g) of these layers often plays an important part in the process. Reflow is used to consolidate the interface between layers, to smoothen the rough sidewalls of the waveguide core (formed by dry etching processes), and to assist in gap filling and planarization of the cladding layer over the patterned core layer [43.9], [43.18–20]. Typically, the layers are doped such that the glass transition decreases in the upward direction. In that way, a layer can be reflowed without affecting the underlying layers. As an example, a 'pinned based reflow' is shown in Fig. 43.3 [43.19]. The circle-section shaped core arises from selective reflow of the core material while its interface with the undercladding glass remains fixed. The upper boundary of the core is shaped by surface tension.

Surface tension driven manufacture of more exotic optical components has been widely studied. Examples include silica-glass based optical microcavities [43.21–23], nm-scale silica wires [43.24], photonic crystal fibers [43.25], microball lenses [43.26], and microlenses [43.27, 28]. In these cases, the reduction of scattering losses by surface tension mediated smoothening of interfaces is particularly important.

As discussed further in Sect. 43.2, high index contrast waveguides, photonic crystals, and microcavities are major themes within integrated optics. High index contrast (between the core and cladding of a waveguide, for example) is desirable because it enables strong con-

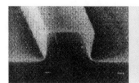

Fig. 43.3 A glass waveguide core is shown before (*left*) and after (*right*) a selective reflow process (From [43.19]

finement of light to small waveguides or small resonant cavities. This is essential to the realization of high-density integrated optics [43.9]. The downside of high index contrast is that scattering at interfaces increases with index contrast. This is clarified by the so-called Tien model for waveguide loss due to interfacial scattering in a symmetric slab waveguide [43.29]:

$$\alpha_S = \frac{2\sigma^2 k_0^2 h}{\beta} \Delta n^2 E_S^2 \qquad (43.1)$$

where σ is the standard deviation (characteristic amplitude) of the roughness, k_0 is the free-space wavenumber, h is the inverse of the penetration depth of the mode into the cladding, β is the propagation constant of the mode, and E_S is the normalized electric field amplitude at the core-cladding interface. Further, $\Delta n^2 = n_{core}^2 - n_{clad}^2$, where n_{core} and n_{clad} are the core and cladding refractive indices, respectively. This equation explicitly shows the strong dependence of scattering loss on index contrast. A more accurate expression, derived by considering the statistics of roughness (not just the amplitude), is the Payne-Lacey model [43.30]:

$$\alpha_S = 4.34 \frac{\sigma^2}{k_0 \sqrt{2} d^4 n_{core}} \cdot g \cdot f \qquad (43.2)$$

where d is the core thickness, g is a function that depends mainly on the core-cladding index offset, and f is a function that depends on the statistics of the roughness. If $r(z)$ is the random function describing the line-edge roughness (i.e. the deviation of the core-cladding interface from a straight line), the autocorrelation of $r(z)$ is often well described by an exponential function [43.31]:

$$R(u) = \langle r(z) r(z+u) \rangle \approx \sigma^2 \exp\left(\frac{-|u|}{L_c}\right) \qquad (43.3)$$

where L_c is the so-called correlation length. It can be shown that for waveguide geometries of typical interest, low scattering loss requires both σ and L_c to be low. As a second illustration of the impact of interface roughness, it can be shown that the surface scattering limited quality (Q) factor for whispering gallery modes in a spherical microsphere is approximately [43.23]:

$$Q_{SS} = \frac{\lambda^2 D}{2\pi^2 \sigma^2 L_c} \tag{43.4}$$

where D is the microsphere diameter. Again, minimization of scattering losses is correlated with minimization of σ and L_c.

It is interesting to compare recent results for high index contrast structures in crystalline materials and glasses. Much of the work on very highly confining waveguides and photonic crystals has focused on semiconductor material systems, especially the silicon-on-insulator (SOI) system [43.29, 30] and III–V semiconductor systems [43.33]. Generally, structures are defined by electron-beam or photolithography, followed by a dry etching process. Each of these steps has limited precision and contributes to the overall roughness [43.31]. With use of sophisticated processes, roughness parameters are typically on the order of $\sigma \approx 5$ nm and $L_c \approx 50$ nm, resulting in losses on the order of 3–10 dB/cm for sub-micron strip waveguides [43.30]. Significant improvement in these numbers is mostly reliant on advancements in lithography or etching processes. These same issues limit the performance of other high index contrast structures formed in semiconductors. For example the Q factor of semiconductor microdisk and microring resonators typically does not exceed 10^5 [43.34], and the loss of photonic crystal defect waveguides in semiconductors remains impractically large [43.35].

By comparison, glasses (at least certain glasses) offer the potential for much smaller characteristic roughness. If a glass is formed from a supercooled liquid state, surface tension effects can produce glass surfaces with nearly atomic level smoothness. The roughness of a melt-formed or reflowed glass surface is determined by surface capillary waves, which are small amplitude fluctuations at a liquid surface that become frozen in place at T_g [43.25]. The resulting RMS roughness is approximately given by [43.36]:

$$\sigma \approx \sqrt{\frac{k_B T_g}{\gamma(T_g)}} \tag{43.5}$$

where $\gamma(T_g)$ is the surface tension of the supercooled liquid at T_g. For silica glass, this equation predicts rms roughness on the order of 0.1 nm or less, and the correlation length of this inherent roughness is estimated to be as small as 3 nm [43.23]. Note that glasses with lower T_g and/or higher surface tension can in theory exhibit even lower values of surface roughness. Interestingly, sub-nm surface roughness has been experimentally verified

Fig. 43.4 (a) A submicron silica glass wire with sub-nm surface roughness is shown [43.24]. (b) A silica glass toroid sitting atop a silicon post is shown. The atomic-level smoothness of the toroid surface was achieved using a selective reflow process. (After [43.32])

for both melt-formed and fractured surfaces of silica glass [43.36] and indirectly from loss measurements in photonic crystal fibers [43.25]. Sub-nm roughness has also been determined for wet-etched surfaces of SiO_2 glass [43.37] and is not untypical for glass films deposited by evaporation.

The smoothness of glass surfaces has important implications for realization of low loss microphotonic devices, as evidenced by several recently reported results. While little work has been conducted on small core, high index contrast planar waveguides in glass, a method for fabricating sub-micron silica wires was recently reported [43.24] (see Fig. 43.4a). The authors estimated the rms surface roughness of their wires to be less than 0.5 nm, and demonstrated losses on the order of 1 dB/cm at 633 nm for 400 nm diameter wires. Sub-nm surface roughness has also been measured in the case of sol-gel glass microlenses formed by a reflow process [43.27]. Silica glass based microsphere cavities, typically formed by melting the end of fused silica fiber, exhibit Q factors of nearly 10^{10} [43.22, 23], the highest for any solid state microcavity. Recently, this concept has been extended [43.21, 32] to the manufacture of silica toroid microresonators on a silicon substrate (see Fig. 43.4b). The latter devices exhibit Q factors exceeding 10^8. Finally, glass-based photonic crystal fibers have been demonstrated to exhibit scattering loss near fundamental limits set by surface capillary waves [43.25]. This last result contrasts with the situation mentioned above for semiconductor photonic crystals, which to date are significantly compromised by optical scattering.

In summary, processing enabled by the glass transition clearly offers unique benefits for manufacture of microphotonic structures. It is necessary to add a caveat to this discussion, however. Many amorphous materials are conditional glass formers, and do not offer the full range of processing advantages outlined above. As an extreme example, amorphous silicon films typically crystallize at temperatures well below their effective glass transition temperature [43.13], and are not amenable to reflow processes. To varying degrees, this is true of many 'glassy' films used in integrated optics, such as silicon oxynitride and alumina. Further, the tendency towards crystallization varies greatly even amongst the traditional glass formers, as discussed below. Thus, chalcogenide and fluoride glasses will not generally support the range of viscosity control that enabled many of the SiO_2-based devices mentioned. Having said that, novel techniques can sometimes circumvent crystallization problems. For example, pressure can be used as a tool for forming glass devices at temperatures below the onset of severe crystallization (but high enough to enable the glass to flow). Chalcogenide glass based lenses molded by the application of pressure and temperature are now commercially available [43.38]. Another interesting example is the recent work on extruded channel waveguides reported by *Mairaj* et al. [43.39]. Finally, much research has been directed at the fabrication of microphotonic elements in organic glasses by hot embossing [43.8]. In this technique, a hard master (such as a silicon wafer) is pre-patterned with the negative image of the desired photonic circuitry. The glass is heated above T_g and the master is pressed against the glass such that the image of its pattern is transferred to the softened glass. While studied mainly in polymer, this technique might also be applicable to low cost fabrication of photonic structures in 'soft' inorganic glasses.

43.1.2 Metastability

A glass is a metastable material, having been frozen as an amorphous solid possessing excess internal energy relative to the corresponding crystal. Metastability is a double-edged sword. On the one hand, it presents technological limitations with respect to the processing and use of a glass device. As an example, it is well known that commercial introduction of organic glass (polymer) based optical devices has been hindered by poor stability of many polymers at elevated temperatures or under exposure to high intensity light. On the other hand, metastability offers unique options for fabrication and optimization of photonic microstructures. Specifically, glass properties (optical, mechanical, chemical) can often be tailored or adjusted by the careful addition of energy, in the form of heat, light, electron beams, or ion beams. An example is provided by the commercially important fiber Bragg gratings (FBG), wherein UV light is used to induce a stable refractive index change in the Ge-doped core of a standard SiO_2-based fiber. In the following subsections, the implications of metastability for glass-based integrated photonics are discussed.

Devitrification

Well below its glass transition temperature, a glass sits in a local energy minimum and is impeded by its own viscosity from reaching the lower energy crystalline state. A frequently cited example is that of silicate glass windows in ancient buildings, which have not exhibited any significant crystallization. However, addition of sufficient energy (such as heating a glass to some temperature above T_g) can result in crystallization of a glass. Controlled crystallization is a technologically useful means of modifying the mechanical, thermal, or optical properties of a glass, used in the production of glass-ceramics. However, devitrification produces a polycrystalline material, increases optical scattering loss, and is usually a problem to be avoided in manufacture of photonic devices.

As mentioned in Sect. 43.1.1, crystallization tendencies vary widely between various glass systems. The difference between the onset temperature for crystallization and T_g, $\Delta T = T_x - T_g$, is one of the parameters that define the stability of a particular glass. The high stability against devitrification of silicate glasses is one of numerous reasons that they have traditionally dominated glass technology. This stability underlies the manufacture of the structures discussed in Sect. 43.1.1. Other glass forming systems have lower stability against devitrification, often making their technological application a greater challenge [43.39, 40]. For example, a major research thrust in the field of fluoride and chalcogenide glass fibers is the identification of new compositions that exhibit improved stability against crystallization during the fiber drawing process [43.41].

For integrated optics, this form of stability is not always so critical. Amorphous thin films created by chemical vapor deposition, sputtering, evaporation, etc. can often be processed at temperatures below the onset of any significant crystallization. This enables the use of conditional glass formers, such as Si_3N_4, Al_2O_3 and semiconductors, in their amorphous state. However, it is interesting to review the early development of integrated optics, much of it focused on the identification of suitable thin film materials. Studies [43.43–45] that compared waveguide fabrication in polycrystalline (ZnO, ZnS, Al_2O_3, etc.) and amorphous (Ta_2O_5, polymer, sputtered glass) thin films showed that losses were orders of magnitude higher for the polycrystalline films. The lowest losses were initially obtained for polymer films [43.44], attributed to their amorphous structure and the smoothness of surfaces arising from spin casting of films.

If present, uncontrolled crystallization within films can be a dominant source of propagation loss [43.46]. For conditional glass formers, such as Al_2O_3 and TiO_2, crystallite formation must be carefully avoided during the deposition process [43.47]. While less of an issue for natural glass formers, the possibility for crystallization must always be considered when a glass film is subjected to high temperature processing steps.

Structural Relaxation

Having been frozen in some metastable configuration at the time of formation, an amorphous solid will always be subject to some degree of aging (changes in the material properties on some time scale). At temperatures below T_g, the glass is usually inhibited from crystallization by a significant internal energy barrier. However, subtler changes in the network structure can occur, such as the transition from the initial state to a more stable (but still amorphous) second state. This so-called structural relaxation is manifested in many ways – experiments often probe the change in time of specific volume, structural signatures (such as by Raman spectroscopy), or optical properties of the glass. The rate of structural relaxation is highly dependent on temperature, being relatively fast at temperatures near T_g and significantly slower for temperatures far below T_g [43.13]. In fact, annealing of glasses at a temperature near T_g is a standard technique used to promote rapid structural relaxation and lessen the dependence of material properties on the processing history [43.6]. In many cases, this annealing step is not practical, and structural relaxation over the in-use lifetime of a glass-based device must be considered.

As the following discussion will illustrate, the rate of structural relaxation at temperature T_1 exhibits a logarithmic dependence on the temperature difference $T_g - T_1$. Thus, T_g is an important parameter in character-

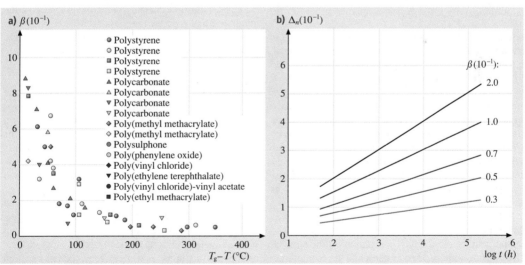

Fig. 43.5 (a) A plot of the structural relaxation rate parameter versus the supercooling temperature for a wide range of organic glasses is shown. (b) The predicted change in refractive index versus time for an organic glass, with the rate constant β as a parameter. (After [43.42])

izing the long-term stability of a glass component. For example, SiO_2 has a very high T_g ($\approx 1100\,°C$) and does not exhibit significant relaxation at room temperature, either on the time scale of typical experiments or over the typical lifetimes envisioned for glass devices. In fact, aging effects in silica fibers are predominately associated with the growth of defects formed in the glass at the time of manufacture. Environmental moisture is a particular contributor to this aging process. On the other extreme, structural relaxation in low T_g organic glasses is known to be an issue, and intrinsic aging effects in polymers have been widely studied [43.48]. In particular, low T_g glasses exhibit densification with time (relaxation of their specific volume) and a corresponding change in their refractive index. This is a particular concern for interferometric optical devices (gratings, resonators, Mach–Zehnder interferometers), which necessitate tight control of material indices.

Volume relaxation can be characterized by a rate parameter β, defined as:

$$\beta = \frac{1}{V}\left[\frac{\partial V}{\partial (\log t)}\right]_{P,T} \quad (43.6)$$

where V is volume, t is time, P is pressure, and T is temperature. Using the Lorentz–Lorenz expression, the following relationship has been derived from (43.6) [43.42]:

$$\left[\frac{\partial n}{\partial (\log t)}\right]_{P,T} = \frac{-\beta \left(\frac{\partial n}{\partial T}\right)_{P,t}}{\alpha} \quad (43.7)$$

where n is refractive index, $\partial n/\partial T$ is the thermo-optic coefficient, and α is the volume coefficient of thermal expansion. Zhang et al. [43.42] further showed that, for polymers, β is approximately a universal function of $T_g - T_1$ as shown in Fig. 43.5. This implies that the time rate of change of refractive index due to structural relaxation is strongly correlated with the parameter $\Delta T_{SC} = T_g - T_1$, termed the supercooling temperature. They also assessed the implications for polymer-based telecommunications devices, concluding that T_g needs to be higher than $300\,°C$ for many applications. This result provides a nice illustration of the impact of T_g on device stability.

While it is not necessarily reasonable to extend the foregoing results directly to inorganic glasses, the essential features of structural relaxation will be similar. The chalcogenide glasses in particular are sometimes called inorganic polymers, partly because of their softness and mechanical flexibility relative to silica-based glasses. Further, the range of T_g for chalcogenide glasses (≈ 50–$550\,°C$) is similar to that of organic polymers, and studies on chalcogenide glasses [43.14] support the conclusion that structural relaxation would present similar restrictions on inorganic and organic glasses. Thus, lifetime restrictions due to relaxation should be considered in the application of any glass having relatively low T_g, including chalcogenide glasses, fluoride glasses, and many non-silicate oxide glasses [43.6].

Both the processing temperatures (during manufacture) and the in-use temperatures are important considerations in assessing the stability of a glass. Telecommunication devices are typically designed to withstand in-use temperatures between $-40\,°C$ and $85\,°C$ [43.49]. As an example of a more demanding application, it has been proposed that glasses are good candidates for fabrication of integrated photonic elements on future integrated circuit chips [43.50]. In-use temperatures on modern microprocessors can reach 100–$200\,°C$.

Finally, it should be noted that any energy applied to the glass (light, electron beams, etc.) might induce structural relaxation. This has been most widely studied for the chalcogenide glasses, which exhibit a wide array of photoinduced structural changes [43.51]. A fascinating example is the so-called photoinduced fluidity phenomenon [43.52], in which intense sub-bandgap light can reduce the viscosity of a chalcogenide glass by several orders of magnitude (causing the glass to flow or melt). This effect is completely athermal, and in fact is often enhanced at low temperatures. While an extreme example, it illustrates the possibility of aging effects due to non-thermal processes in glass devices. For photonic devices, the main concern is usually the effect of light exposure over the in-use lifetime.

Metastability as Enabler

One highly desirable implication of metastability is the possibility for direct patterning of photonic structures in glasses using energetic beams (electron beams, ion beams, light). These effects can also be exploited for post-fabrication trimming of devices [43.53], which is important since many optical components have tolerances beyond the capabilities of practical fabrication processes.

Beam induced effects in glasses are generally linked to their random network structure. That is, unlike crystals, glasses contain a range of internal 'defects' – wrong bonds, dangling (missing) bonds, impurity atoms that act as network modifiers, as well as their inherent variation in bond angles. Addition of energy can cause a glass to undergo a transition from one metastable state to another. This transition is often accompanied by a change in the

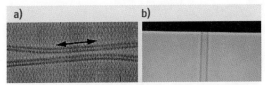

Fig. 43.6 (a) A directional coupler is shown, as written in a multicomponent silicate glass by an ultrafast Ti:sapphire laser [43.54]. The black arrow indicates the coupling region. (b) Microscope image of a rib waveguide formed by direct exposure of a chalcogenide glass (As_2Se_3) in a UV mask aligner, followed by wet etching [43.55]

density of defects or by a structural change – rearrangement or reorientation of bonds. Associated with these structural changes are changes in the physical, chemical, and optical properties of the glass. Most existing knowledge about these induced structural changes is of an empirical nature, whereas the underlying physical mechanisms are still the subject of research.

The ability to induce transitions between metastable states is probably a universal property of amorphous matter. Perhaps for practical reasons, such as the availability of light sources of suitable wavelength, metastability was initially studied in amorphous semiconductors and low band gap chalcogenide glasses. It should be added, though, that chalcogenide glasses seem to be unique with respect to the variety of induced structural changes they exhibit or at least the magnitude of those changes [43.51]. This might be related to the relatively weak, highly polarizable covalent bonds that are characteristic of chalcogenide glasses. Nevertheless, the later discovery [43.56] of photoinduced refractive index changes in Ge-doped silica glass prompted a huge research and development effort because of the technological importance of silicate glasses. The main outcome of this work was the fiber Bragg grating (FBG), which is a critical component in modern telecommunications and sensing networks. FBGs are manufactured by inducing a quasi-permanent index change within the Ge-doped core of a standard silica fiber, by exposure to an UV laser beam. Writing wavelengths are in the 150 to 350 nm range, corresponding to photon energies below the nominal bandgap of SiO_2. The mechanisms underlying the index increase are still debated [43.9, 56]. However, it is believed that contributing factors are absorption by defects in the glass (possibly generating further defects) and densification due to some sort of structural relaxation [43.57]. Typical index changes for standard fiber cores are in the 10^{-4} to 10^{-3} range, but several techniques (addition of other dopants, hydrogen-loading, etc.) have been developed to enable changes as high as 10^{-2}. Given the discussion in the preceding section, it is reasonable to ask whether photoinduced changes are themselves stable. The answer, in some cases, seems to be yes. Researchers have concluded that fiber Bragg gratings can have an operational lifetime of greater than 20 years, even at elevated temperatures of 80 °C [43.58].

Photosensitivity has been widely explored as a means to directly pattern integrated optics devices, especially in silica-based glasses [43.59]. In general, direct patterning is induced using photon energies close to the nominal bandgap energy of the glass or using intense light of lower photon energy [43.57]. In the latter case, photoinduced effects arise from nonlinear effects such as multi-photon absorption. In particular, the widely available Ti:sapphire laser has been applied to the writing of microstructures in a variety of glasses [43.54]. Because of the nonlinear nature of the writing process, writing with sub-band gap light allows the direct patterning of 3-dimensional structures (see Fig. 43.6).

Photosensitivity has been demonstrated within all of the standard glass families, and in many other amorphous thin films (see Sect. 43.2). Because of their large photosensitivity, direct patterning of integrated optics structures in chalcogenide glasses is possible using low power light sources [43.55, 60].

43.1.3 Glass as Host Material

For completeness, it should be mentioned that glasses are unique in their ability to incorporate a wide range of dopants, sometimes in very high concentration. This is related to the random network structure of a glass. Foreign species are much more likely to find a suitable bonding site, or simply space to reside, inside a glass versus a crystal. These foreign species can be rare-earth ions, transition metal ions, or semiconductor or metal nanoparticles, for example. As a result, a glass provides significant scope for active functionality. We can transform a passive glass into a laser glass, a nonlinear optical glass, or a magneto-optic glass, for example, by appropriate doping [43.3, 6].

43.2 Glasses for Integrated Optics

In general, thin films can be amorphous, polycrystalline, or crystalline. *Tien* and *Ballman* [43.45] provided an early review of waveguide results achieved for various materials lying within each of these categories. Unless the crystal grain size is sufficiently small relative to the wavelength of interest [43.47], polycrystalline films are too lossy for integrated optics. It should be noted that high attenuation might be tolerable if circuit length is sufficiently short. The present discussion is concerned with amorphous films, which Tien further subcategorized as low index ($n < 1.7$), medium index ($n < 2$), and high index ($n > 2$). For convenience, we will follow a similar approach in the following sections. This categorization is somewhat arbitrary, especially when considering material systems (such as the silicon oxynitride system) that enable a range of refractive index. It is interesting to note that 1.7 is the approximate upper limit for the refractive index of organic glasses [43.49].

As noted in Sect. 43.1, increased circuit density is one of the primary goals of integrated optics research [43.9]. Whether employing traditional total internal reflection effects or photonic band gap materials, increased density relies on high index contrast between at least two compatible materials. Note that the terminology 'high index' and 'high index contrast' are rather imprecise. For example, in silica PLC technologies core-cladding index differences of $\Delta n \approx 0.02$ have been labeled as 'superhigh' index contrast [43.10]. This is a very small value, however, relative to the index contrasts that characterize SOI photonic wire waveguides [43.30].

43.2.1 Low Index Glassy Films

SiO_2-Based Glasses

As mentioned in the introduction, the SiO_2 on silicon PLC system is the most widely used and developed glass-based integrated optics platform. PLC development (mainly in the 1990s) was driven by fiber optic long haul communication systems, especially the emergence of wavelength division multiplexing (WDM). Various devices, but especially arrayed waveguide grating (AWG) wavelength demultiplexers, were developed to a very high degree of sophistication by the telecommunications industry. Good early [43.18] and recent [43.9, 10] reviews are available in the literature. A brief overview of the technology is given below, as it illustrates some of the advantages and challenges associated with glass-based photonic integration.

As mentioned in Sect. 43.1.1, commercial PLC waveguides are fabricated primarily by FHD [43.10] or by CVD [43.9]. In addition, considerable research has been conducted on sol-gel synthesis [43.61], with the aim of reducing fabrication costs and providing greater flexibility over the choice of glass compositions. Typically, the undercladding is pure silica glass. To raise its refractive index, the core layer is doped with Ge or P. Ge-doped SiO_2 is known for its photosensitivity, as discussed in Sect. 43.1.2. Addition of P lowers the viscosity and characteristic reflow temperature, enabling the processing options discussed in Sect. 43.1.1. Typical relative index offset Δ between the core and cladding is in the 0.3 to 2% range, corresponding to minimum waveguide bend radius R in the 2 to 25 mm range [43.10]. After RIE to form an approximately square waveguide core, an upper cladding is deposited and subjected to heat treatment. The upper cladding is often a boro-phosphosilicate glass (BPSG), partly to enable reflow. Further, the B and P dopants have opposite effects on refractive index allowing a nearly symmetric waveguide structure to be obtained.

Given the target applications, it is not surprising that PLC technology placed a great emphasis on efficient coupling between the integrated waveguides and external fiber waveguides. Thus, waveguide cross-sectional dimensions and refractive index contrast between core and cladding layers were tailored to provide a good impedance match (low reflection and good modal overlap) to standard fiber. By employing essentially the same glass (doped SiO_2) as that used to construct fiber, it is even possible to achieve intimate reflowed (fusion type) coupling between the integrated and fiber guides [43.18]. The emphasis on impedance matching is due to the great importance of minimizing insertion loss in fiber systems, but presents some practical limitations:

1. The relatively low index contrast ($\approx 10^{-3}$) between core and cladding necessitates thick glass films. For example the undercladding or buffer layer, often a thermally grown SiO_2 layer, must typically exceed 12 μm in order to negate radiation losses into the high index silicon substrate. Further, high temperature anneals (typically 900–1150 °C) are required to drive out hydrogen impurities and to reflow the core and upper cladding layers. The combination of thick films, high temperature anneals, and thermal expansion mismatch can result in wafer bending and damage to the glass films. This is partly alleviated

in practice by depositing identical layers on both sides of the Si wafer [43.9], which adds cost and complexity.
2. Since modal area and minimum bend radius scale inversely with core-cladding index contrast, traditional PLC waveguides do not support high-density optical integration. As mentioned above, index contrasts in the 0.3 to 2% range correspond approximately to bending radii in the 25 to 2 mm range [43.10].

In recent work at Corning [43.62], very low loss (< 0.1 dB/cm) waveguides and ring resonators were realized in PECVD grown silica-germania waveguides having index contrast as high as 4%. Such high index contrasts can accommodate bending radius of less than 1 mm, which is comparable to the range explored recently by IBM researchers using silicon oxynitride materials [43.63].

Amorphous Aluminium Oxide

Sapphire (crystalline Al_2O_3) is amongst the most important solid-state laser hosts. Alumina (polycrystalline or amorphous Al_2O_3) is an important industrial material in its own right, possessing outstanding mechanical and thermal properties [43.15]. Amorphous Al_2O_3 films are of interest for integrated optics for several reasons [43.47]. First, the refractive index is relatively high (although falling within the low index range specified above), typically $n \approx 1.65$. Second, Al_2O_3 is an excellent host for rare earth and transition metal dopants. As discussed in Sect. 43.3, rare earth dopants of interest are typically trivalent, matching the valency of Al ions in Al_2O_3 [43.64]. As a result, rare earth ions can be incorporated easily into the alumina matrix. In short, Al_2O_3 can homogeneously dissolve large concentrations of rare earth ions and is therefore of interest for realization of integrated amplifiers and light sources. Finally, Al_2O_3 films typically have excellent transparency from the UV to mid-IR range.

Fluoride Glasses

Heavy-metal fluoride glasses (typically fluorozirconate glasses) such as ZBLAN have been widely studied since their discovery in 1974 [43.41]. One of their outstanding attributes is a wide transparency range, extending from the UV well into the mid-IR (see Fig. 43.1). They have received considerable attention as fiber optic materials, because theory predicts a minimum absorption well below that of silica glass. Further, rare earth ions exhibit the greatest number of useful radiative transitions when embedded in fluoride glasses [43.65]. This is due to their wide transparency window and low characteristic phonon energies (see Sect. 43.3).

Fluoride glasses typically have refractive indices in the 1.47–1.57 range, which is advantageous in terms of being well matched to silica glasses. Thin film deposition of these complex multicomponent glasses is difficult, and thermal expansion mismatches with standard substrates create further challenges. For these reasons, fluoride glasses have not been widely explored for applications in integrated optics. The so-called PZG fluoride glasses ($PbF_2 - ZnF_2 - GaF_3$) have been successfully deposited using straightforward evaporation techniques, enabling relatively low loss waveguides and erbium-doped amplifiers on a silicon platform [43.66].

43.2.2 Medium Index Glassy Films

Silicon Oxynitride

Silicon oxynitride (SiON) films are generally deposited by a CVD technique. The promise of this material system for integrated optics was identified in early work [43.45]. One of the main attributes of SiON is that it is a standard material system employed in silicon microelectronics, and the thin film technology has been developed accordingly [43.47, 63]. Because of this, SiON is currently being studied as potential material for on-chip interconnects [43.50]. Further, the system enables a continuous range of refractive index from approximately 1.45 (SiO_2) to 2 (Si_3N_4) at 1550 nm wavelength. This index range has been extended to approximately 2.2 by deposition of non-stoichiometric, silicon-rich nitride films [43.67]. SiON and SiN have been amongst the most explored materials for realization of microring resonator structures [43.53, 67].

As an optical material, SiON has some drawbacks. The main one is the presence of hydrogen impurities in films deposited by traditional techniques. Overtones due to hydrogen bonds (mainly N—H and O—H) can produce impractically large values of loss in the 1300 nm and 1550 nm telecommunications bands [43.62]. Long term, high temperature annealing (typically at $>1000\,°C$) is required to reduce this loss. Interestingly, structural relaxation on the time scale of hours and days has been observed in such annealed films [43.47]. Perhaps related to this metastability, UV light has been used as a means to trim the refractive index of SiN-based microring resonators [43.53]. Modified deposition processes that can produce SiON films having low stress and low hydrogen content (without requiring a high temperature annealing step) have been reported recently [43.63, 67]. However, inherent film stress can limit the maximum thickness to a few hundred nanometers in some cases [43.50].

43.2.3 High Index Glassy Films

Amorphous Metaloxides

Many amorphous metaloxides (Y_2O_3, Nb_2O_5, Ta_2O_5, etc.) have traditionally been used as high index layers in optical thin film stacks [43.49]. It is logical to look at these materials in the hunt for high index thin films for integrated optics. Often these oxides are miscible with SiO_2, making it possible to tune the refractive index and other material properties over a range of values (as in the case of silicon oxynitride) [43.69].

TiO_2 films are used as dielectric layers, optical coatings, and to protect underlying materials from mechanical damage or corrosion [43.47]. The high dielectric constant of TiO_2 is widely exploited in microelectronics applications. It also has useful optical properties, including a refractive index as high as 2.6 (depending on film deposition details) and good transparency from the UV to mid-IR. The TiO_2–SiO_2 material system has good properties for sol-gel synthesis of thin films [43.61]. Integrated waveguides based on TiO_2 have been studied [43.47, 69], and AWG devices in TiO_2-rich oxide glass with $n \approx 1.9$ were recently reported [43.70].

Ta_2O_5 (With $n \approx 2.1$) is another well-studied material, used for example as a high dielectric constant material in microelectronics. It has low absorption in the visible to near infrared wavelength range, and has been recently studied as a material for active integrated optics [43.71, 72]. Further, Ta_2O_5/SiO_2 has been applied to the study of compact microring resonators [43.73].

TiO_2 and Ta_2O_5 have been used in the fabrication of three-dimensional photonic crystals with fundamental bandgaps in the visible and near infrared [43.68, 74]. An example is shown in Fig. 43.7.

Heavy Metal Oxide Glasses

These glasses are based primarily on the oxides of bismuth (Bi_2O_3) and lead (PbO). They have relatively low phonon energies and transmit further into the infrared than most oxide glasses [43.75]. They also have interesting magneto-optic [43.6] and nonlinear optical [43.3] properties.

Chalcogenide Glasses

Many chalcogenide alloys, based on S, Se, or Te, are excellent glass formers. In fact, Se is the only element able to easily form a glass on its own [43.14]. These

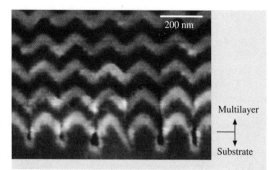

Fig. 43.7 A three-dimensional photonic crystal fabricated using the autocloning growth technique is shown. The *bright* and *dark* layers are TiO_2 and SiO_2 glass, respectively [43.68]

glasses are characterized by very low phonon energies, good transparency in the mid-infrared, and a wide array of beam-induced structural changes [43.51]. Further, chalcogenide glasses are essentially amorphous semiconductors, with electronic absorption edge typically in the visible to near infrared. Related to this, they have very high refractive indices, ranging from 2.2 for sulfide glasses to greater than 3 for some telluride glasses. They are currently of industrial interest for night-vision optics [43.38], fibers for infrared transmission [43.41], and optical and electronic memory elements [43.51].

Chalcogenide glasses have many appealing features for integrated optics. High quality thin films can often be deposited by straightforward techniques such as evaporation or sputtering. Their high indices make them suitable for realization of compact waveguides [43.55] and photonic crystals [43.76]. Their beam-induced properties enable many unique processing options. Some chalcogenide alloys have good potential as laser glasses [43.41], with the ability to uniformly dissolve large concentrations of rare-earth ions. Perhaps of greatest interest, the chalcogenide glasses are amongst the most promising materials for nonlinear integrated optics [43.77, 78]. On this last point, several chalcogenide glasses have ultrafast optical Kerr response 2 to 3 orders of magnitude higher than that of SiO_2, while satisfying basic figure of merit criteria for all-optical switching. Results for metal-doped chalcogenide glasses [43.78] suggest even higher nonlinearities might be attainable. Combined with the possibility of realizing highly confining waveguides, this makes chalcogenide glasses one of few material systems that can realistically satisfy long-standing goals for nonlinear processing in integrated optical circuits. For example, it has been estimated [43.77] that 1 pJ pulses of 1 ps duration (≈ 1 W peak power) could induce phase shifts (due to self phase modulation) greater than π in feasible chalcogenide glass waveguides with length on the order of a few centimeters. This is predicated on realization of low loss waveguides with modal area of $\approx 1~\mu m^2$. Recent experimental results suggest that this goal is within reach [43.55].

Tellurite Glasses

While TeO_2 is a conditional glass former, the addition of other oxides can result in stable glasses with many interesting properties [43.75]. They typically have refractive index greater than 2, low phonon energy, large optical nonlinearities, and a high acousto-optic figure of merit. Further, they have been widely studied as hosts for rare-earth ions. Of particular interest has been the wide bandwidth of the 1550 nm emission exhibited by erbium in tellurite glass [43.79]. Because of these numerous attractive properties, tellurite glasses and amorphous TeO_2 films ($n \approx 2.2$) have received some attention for applications in integrated optics [43.80].

43.3 Laser Glasses for Integrated Light Sources

Active functionality includes means to generate and detect light (especially stimulated emission and absorption) and means to control (switch, modulate, etc.) light signals. Numerous material properties are employed in active photonics, including thermo-optic, acousto-optic, magneto-optic, electro-optic, and nonlinear Kerr effects. Many glasses have attractive thermo-optic or acousto-optic properties, and a few examples were cited in Sect. 43.2. Also in Sect. 43.2, some glasses with promising nonlinear optical properties were discussed. Elsewhere in this volume, K. Tanaka has provided an excellent review of nonlinearities in photonic glasses. A recent, thorough review of magneto-optic glasses is also available [43.6].

Arguably, the most critical element for photonic integration is an integrated light source. Rare-earth doped glasses are well-established laser media, used especially for the realization of bulk and fiber lasers. Considerable effort has been directed towards development of integrated amplifiers and lasers based on such glasses. In the following, we attempt to highlight some ways in which light sources based on glasses are uniquely enabling, relative to those based on crystalline materials. The performance advantages discussed below combined with the properties discussed in earlier sections (fabrication options) make glasses particularly attractive.

43.3.1 Advantages of Glass-based Light Sources

Stimulated emission devices in glass are almost always based on trivalent rare-earth dopants [43.6]. Thus, the term laser glass can usually be equated with the term rare-earth doped glass. Rare-earth doped laser glasses have been widely studied and reviewed [43.64, 65]. Further, recent and comprehensive reviews on Nd- and Er-doped integrated glass amplifiers and lasers are available [43.6, 81]. Relative to single crystal hosts, glass hosts result in rare-earth ions exhibiting broadened luminescence lines and lower peak stimulated emission cross-sections [43.65]. This property is a result of the random network structure of glasses; embedded rare earth ions exist in a range of local environments. The broadened, weaker emission is of great importance for the realization of broadband, low noise fiber amplifiers.

It should be noted that the semiconductor injection laser is an extremely advanced technology, and is the dominant type of integrated light source at present and for the foreseeable future. Semiconductor lasers have important advantages over any glass-based device demonstrated to date. First, semiconductor gain media typically have gain coefficients of the order $\approx 100~cm^{-1}$ [43.17]. By comparison, glasses require high rare-earth dopant concentration to achieve gain coefficients exceeding $1~cm^{-1}$. Thus, semiconductor optical amplifiers (SOAs) and lasers have cavity lengths measured in tens to hundreds of μm while it is typical for glass waveguide amplifiers and lasers to be measured in cm. Second, semiconductor light sources are pumped electrically while glass devices are typically pumped optically. Electrical pumping is highly desirable for optoelectronic integration of photonic devices on electronic chips. However, as discussed below, integrated glass waveguide lasers have important advantages of their own [43.82].

The lifetime of the metastable lasing level in rare-earth doped glasses is usually on the order of ms, much longer than the ns lifetimes typical of semiconductor gain media. This long lifetime implies that the gain does not change rapidly with variations in input power (pump or signal). This is an essential feature of the commercially important erbium-doped fiber amplifier (EDFA); the long lifetime (≈ 10 ms) of the $^4I_{13/2}$ level of erbium in silicate glass contributes to low noise operation, high pump efficiency, and low crosstalk between wavelength channels in a WDM system [43.83]. Further, since the relaxation oscillations in glass waveguide lasers occur at relatively low frequency, glass lasers can be modelocked at correspondingly much lower repetition rates compared to semiconductor lasers [43.82]. This can enable much higher peak intensities from the glass laser.

Related to the discussion in Sect. 43.1, lower cavity loss (higher cavity Q) is generally possible for glass devices. Further, the long metastable lifetime of the rare-earth transition allows glass lasers to have linewidths approaching the Schawlow–Townes limit [43.84]:

$$\Delta \nu = \frac{2\pi h \nu_0 (\Delta \nu_c)^2}{P} \left(\frac{N_2}{N_2 - N_1} \right) \qquad (43.8)$$

where ν_0 is the laser center frequency, P is the laser output power, N_2 and N_1 are the population densities of the upper and lower lasing levels, and $\Delta \nu_c = (1/2\pi t_c)$ with t_c the photon lifetime in the cavity. The high Q of glass laser cavities coupled with relatively high output powers enables linewidths less than 10 kHz, orders of magnitude below that of semiconductor DFB lasers [43.82]. The high cavity Q and long metastable lifetime is also advantageous for achieving ultrastable passive mode locking, with low timing jitter and pulse-to-pulse power variation.

Finally, glasses offer the possibility of integration on various substrates. While semiconductor lasers are inherently integrated structures, they are not easily transportable between platforms. For example, III-V semiconductor lasers have not shown great promise (in spite of heroic efforts in some cases) to satisfy the desire for a compact, truly integrated light source on the silicon electronics platform. Glasses (perhaps doped with semiconductor nanocrystals) are increasingly viewed as the more promising route to achieving such a goal [43.64].

43.3.2 Alternative Glass Hosts

The theoretical maximum gain (cm^{-1}) of a rare-earth doped glass waveguide amplifier can be expressed [43.81] as $\gamma_p = \Gamma \sigma_p N_{RE}$, where σ_p is the peak (versus wavelength) stimulated emission cross-section (cm^2) for the transition of interest and N_{RE} is the volume density (cm^{-3}) of the rare-earth ions. Γ is a dimensionless factor (lying between 0 and 1) that accounts for the spatial overlap of the waveguide mode (at the wavelength to be amplified) and the active ions producing the gain. It can be optimized through waveguide design, irrespective of the glass host, and will not be considered further here. The expression for γ_p neglects all waveguide losses (due to scattering, etc.) and assumes that all of the rare-earth ions have been promoted to the desired lasing level; i.e. a complete population inversion. It is therefore an ideal and unattainable limit, but is useful for framing the following discussion.

Since compactness is a central goal of integrated waveguide lasers, alternative glass hosts can be compared on the basis of the maximum gain (γ_{max}) that they enable in practice (for a given transition of a given rare-earth ion). Further, since low noise operation relies on a near complete population inversion [43.81], it is desirable that $\gamma_{max} \approx \gamma_p$. In simple terms, the glass should dissolve a large concentration of the rare earth ion (high N_{RE}), should result in a large stimulated emission cross-section for the desired transition, and should enable the realization of a nearly complete population inversion. The importance of other practical considerations, such as stability, processing options, and physical properties of the glass, will depend on the intended application. Some hosts provide unique advantages, such as flexible pumping options, as discussed in Sect. 43.3.3.

As mentioned, laser transitions in glasses are generally provided by radiative decay between two energy levels of a trivalent rare-earth ion. Perhaps the most important example is the $^4I_{13/2}$ to $^4I_{15/2}$ transition of Er^{3+}, which produces luminescence in the 1500–1600 nm wavelength range. Once an ion has been promoted (by pumping) to the upper lasing level, it will eventually transition to another state by interactions with the glass, impurities in the glass, the photon fields (at the signal or pump wavelength), or with other rare-earth ions in its vicinity [43.65]. The metastable lifetime of the upper lasing level can be expressed [43.81]

$$\frac{1}{\tau} = A + W_{MP} + W_{ET} + W_{IMP} \qquad (43.9)$$

where A is the effective rate of spontaneous radiative decay to all lower lying levels, W_{MP} is the rate of non-radiative decay due to multi-phonon energy exchanges with the glass, W_{ET} is the rate of non-radiative energy transfer due to interactions between closely spaced rare-earth ions, and W_{IMP} is the rate of energy transfer to quenching impurity centers in the glass. The first three terms on the right will be discussed below. For the last term, a classic example is the quenching of the 1550 nm luminescence band of erbium due to resonant energy transfer to OH$^-$ impurities in silica glass [43.65].

The choice of a particular glass host will impact several important properties of a given transition: pumping efficiency, peak gain, linewidth, metastable lifetime, etc. Alternative glasses can be compared on the basis of a few key parameters, as discussed in the following sub-sections. Representative data for erbium in various glasses is given in Table 43.1.

Stimulated Emission Cross-Section

Stimulated transitions of rare-earth ions in glass tend to be predominately driven by electric dipole interactions [43.65]. For a given transition of interest, the spontaneous emission probability can be expressed in cgs units as [43.6, 85]:

$$A = \frac{64\pi^4 e^2 \chi}{3h(2J+1)\lambda_p^3} S \qquad (43.10)$$

where J is the total angular momentum of the upper lasing level, λ_p is the peak emission wavelength, and χ is the local field correction factor. For electric-dipole interactions of an ion in a dielectric medium, $\chi \approx n(n^2+2)^2/9$, with n the refractive index of the host glass. S is the quantum-mechanical line strength for the transition. Further, the peak stimulated emission cross-section can be expressed in terms of the spontaneous emission probability:

$$\sigma_p = \left(\frac{\lambda_p^4}{8\pi c n^2 \Delta\lambda_{\text{eff}}}\right) A \qquad (43.11)$$

where $\Delta\lambda_{\text{eff}}$ is the effective linewidth of the transition. From (43.10) and (43.11), the host-dependent factors that influence σ_p are the refractive index, the line strength, the effective linewidth, and to a lesser extent the peak emission wavelength (which typically varies only slightly between different hosts). The local field correction factor is significant in hosts with large refractive index, and can result in an enhancement of the stimulated emission cross-section and a reduction of the radiative lifetime. This is especially true for chalcogenide glass hosts, which typically have refractive index in the 2 to 3 range.

Metastable Lifetime

In the limit of low rare-earth dopant concentration, W_{ET} in (43.9) is zero because the ions are sufficiently well separated to negate their interaction. If the difference in energy between the upper lasing level and the adjacent state is several times the effective phonon energy, then W_{MP} can be neglected to first order. Further neglecting impurity quenching, we can then assert that $\tau_0 \approx 1/A$, where τ_0 is the metastable lifetime in the limit of low rare-earth concentration. From Einstein's relations, $A \approx \sigma_p$, so it follows that $\tau_0 \approx 1/\sigma_p$. In short, hosts that result in an enhancement of the peak stimulated emission cross-section (due to an enhancement of the electric-dipole interaction or because of a high local field correction factor) will also result in a reduction in metastable lifetime. In other words, both stimulated and spontaneous emission rates are enhanced.

The inverse scaling of stimulated emission cross-section and metastable lifetime represents a tradeoff, as it is generally desirable for the lifetime to be as large as possible. Some of the advantages of long lifetime were discussed in Sect. 43.3.1. In addition, the pumping efficiency (gain per applied pump power) of a waveguide amplifier scales directly with τ [43.6, 81] and, therefore, the threshold for CW lasing scales inversely with τ [43.86].

Concentration Quenching

Glasses differ greatly in the amount of a given rare-earth dopant that they are able to dissolve. To avoid problematic ion–ion interactions, the rare-earth ions should be

Table 43.1 Representative parameters for erbium ions embedded in various types of glass (after [43.61, 64, 65])

Glass host	Refractive index n	Peak stimulated emission cross-section σ_p (10^{-21} cm^2)	Metastable lifetime τ (ms)	Quenching concentration ρ_q (10^{20} cm^{-3})	Effective luminescence bandwidth (nm)
Silica	1.46	7	12	–	11
Amorphous Al_2O_3	1.64	6	7.8	–	55
Aluminosilicate	1.5	5.7	10	3.9–6.0	43
Phosphate	1.56	8	10	3.9–8.6	27
Fluoride	1.53	5	9	3.8–5.3	63
Tellurite	2.1	13	3.3	–	80
Sulfide	2.4	20	2.5	3.2	–

incorporated homogeneously into the glass structure. In the extreme case of high concentration, the rare-earth ions will form microscopic clusters (phase separation). Such clustering is highly detrimental, as typically all of the ions within a cluster are effectively removed from the desired stimulated emission process [43.81]. As is well known [43.64, 65], the onset of clustering occurs at quite small values (≈ 0.1 at%) in pure silica glass. The addition of Al_2O_3 to SiO_2 allows silica-based fibers to dissolve a significantly higher concentration. For example, a value of 10–20 for the Al ion to Er ion concentration ratio has been shown to greatly reduce clustering of Er ions in silica glass [43.86]. Interestingly, the addition of Ga (with similar ratio) to chalcogenide glasses has been shown to provide a similar reduction in rare-earth ion clustering [43.41]. These additives essentially modify the glass network, and create sites for isolated rare-earth ions to be incorporated. For similar reasons, multicomponent alumino-silicate and phosphosilicate glasses have been favored in the development of integrated waveguide amplifiers, where the rare-earth concentration must be orders of magnitude higher than in fibers [43.81].

Even in the absence of significant clustering, ion–ion interactions can occur at high concentrations. This is simply due to the reduction in inter-ion spacing, and is exacerbated by any non-uniform (non-homogeneous) distribution of the rare-earth ions in the glass host. These interactions are manifested by a reduction in the metastable lifetime, often well described by the semi-empirical expression [43.61, 65]:

$$\tau(\rho) = \frac{\tau_0}{1 + (\rho/\rho_Q)^p} \quad (43.12)$$

where ρ is the rare-earth ion concentration, ρ_Q is the so-called quenching concentration, and p is a fitting parameter ($p \approx 2$ for interactions between pairs of ions). The parameter ρ_Q is useful for comparing glasses in terms of their ability to uniformly dissolve a given ion.

Phonon Energies

The characteristic phonon energies of a glass depend on the weights of its constituent atoms and the strength and nature (ionic or covalent) of its bonds. Typical values are shown in Table 43.2. The rate of multi-phonon decay (WMP) between two energy levels depends exponentially on the number of phonons required to bridge the energy gap. Thus, the phonon energy has a great impact on the ultimate efficiency of a desired radiative transition. Low phonon energy can be a good or bad thing, depending on the transition of interest and the particular pumping scheme.

Mid- to far-infrared transitions of rare-earth ions can exhibit reasonably high quantum efficiency in low phonon energy hosts, such as fluoride, tellurite, and especially chalcogenide glasses. If the same ions are embedded in silicate or phosphate glasses, these transitions are completely quenched by non-radiative processes at room temperature. For this reason, rare-earth doped chalcogenide glasses are of interest for realization of long wavelength amplifiers and lasers [43.41, 64]. Also unique to low phonon energy hosts is the possibility of efficient upconversion lasers [43.87]. In simple terms, the long lifetimes of numerous energy levels allows processes such as ion–ion interactions and excited state absorption (ESA) to efficiently populate the higher energy levels. By contrast, these levels are rapidly depopulated by phonons in oxide glasses. On the other hand, population of the higher levels is highly detrimental if the desired transition is between two lower levels. For example, the efficient pumping of EDFAs using 980 nm wavelength sources relies on the rapid decay (via multiphonon processes) of ions from the $^4I_{11/2}$ pumping level to the $^4I_{13/2}$ lasing level. In fluoride and chalcogenide glass hosts, ions raised above the $^4I_{11/2}$ level tend to become trapped in higher levels (so-called 'population bottlenecking' [43.65]). Cerium co-doping has been shown to alleviate this problem [43.66]. Another approach is addition of light elements to the glass network, to increase the phonon energy [43.79].

43.3.3 Progress Towards Integrated Light Sources in Glass

Per the preceding discussion, important goals for glass-based lasers include size reduction and the need for simplified optical or (ideally) electrical pumping

Table 43.2 Characteristic maximum phonon energies for a variety of glass hosts (after [43.6, 65])

Glass host	Phonon energy (cm^{-1})
Borate	1400
Phosphate	1200
Silicate	1100
Germanate	900
Tellurite	700
Heavy-metal fluoride	500
Chalcogenide (sulfide)	450
Chalcogenide (selenide)	350

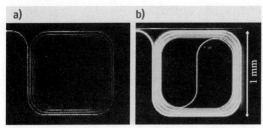

Fig. 43.8 Spiral geometry demonstrated for Al_2O_3-based waveguide amplifiers, where erbium was introduced by ion implantation (*left*) and co-sputtering (*right*). The bright green emission is a result of ion–ion interactions causing up-conversion for the co-sputtered sample [43.89]

schemes. It should be noted that integration brings its own advantages. As is well known, waveguide lasers have greatly enhanced pump efficiency (gain per applied pump power) relative to their bulk solid-state counterparts. This is due to the confinement of the pump beam and the stimulated emission beam within a small volume. As explored recently in [43.88], this benefit scales with confinement. Amplifiers based on very high index contrast waveguides can have extremely high pump efficiency, as well as requiring only a small amount of wafer real estate. Thus, while the gain per unit length is limited by concentration quenching in rare-earth doped glasses, devices can still be very compact through the use of folded spiral waveguide layouts as illustrated in Fig. 43.8 [43.89].

In terms of maximizing gain per unit length, phosphate glasses have produced the best results to date. This is due to their ability to uniformly dissolve large concentrations of rare-earth ions, and the high cross-sections exhibited by those ions. The high phonon energy can also be an advantage as discussed above. Phosphate glass disks (typically ≈ 1 mm thick) have been widely used for realization of compact microlasers [43.90] and ultrafast modelocked lasers with pulse length as short as 220 fs [43.91] and repetition rate exceeding 40 GHz [43.92]. Phosphate glass fiber lasers with a length of only 7 cm have recently generated 9.3 W of output power [43.93]. Sputtered phosphate glass waveguides have produced gains exceeding 4 dB/cm [43.94]. More recently, a phosphate glass co-doped with 8 at % erbium and 12 at % ytterbium was used to realize ion-exchange waveguides exhibiting 4.1 dB gain in only 3 mm length [43.95]. At such high concentrations, the rare-earth constituents can hardly be called dopants as they exceed the concentration of some of the host glass constituents.

On the theme of simplified pumping options, so-called broadband sensitizers [43.96] are of great interest. A broadband sensitizer is essentially a co-dopant that is able to efficiently absorb pump energy over a wide wavelength range and subsequently excite nearby rare-earth ions to the desired lasing level. This might enable planar waveguide amplifiers that are pumped transversely using low cost broadband light sources [43.64]. Silicon nanoclusters (embedded in oxide glass) are receiving the most attention at present. In addition to acting as a broadband sensitizer, there is evidence that these nanoclusters can enhance the stimulated emission cross-section of co-doped rare-earth ions by 1–2 orders of magnitude. Han et al. [43.97] have reported a waveguide amplifier (on a silicon platform) exhibiting signal enhancement of 7 dB/cm, in spite of a very low erbium concentration. The same group has also reported gain using transverse pumping by LED arrays [43.98]. Finally, the presence of a large concentration of silicon clusters makes electrical pumping a realistic possibility for some silicon-rich oxide glasses [43.64].

It seems very likely that chip-scale glass-based lasers are on the verge of playing a central role in photonics. These lasers will deliver moderate power in a very small package, and will offer extreme spectral purity or stable, high-repetion rate pulses. They might even satisfy the long-standing desire for a silicon-based laser [43.64, 91], facilitating the convergence of photonics and electronics.

43.4 Summary

Glasses have numerous advantages (ease of component fabrication, low scattering loss, flexible atomic structure) that have made them the workhorse material for traditional bulk optics applications. These properties can provide equal or greater advantage in the realization of microphotonic devices and circuits. The glass transition can enable manufacturing techniques unavailable to other materials, and realization of devices having surface roughness approaching fundamental limits. Owing to this, glasses underlie the highest Q microcavities and some of the lowest loss photonic crystal and microphotonic waveguides reported to date. The metastability of glass enables a rich array of processing and post-processing options. Direct patterning of waveguides, gratings, and other microphotonic elements by energetic beams is widely studied. These methods of material modification are also promising for post-fabrication trimming of devices. On the other hand, metastability has implications for processing and aging of glass devices.

Rare-earth doped glasses offer numerous advantages relative to other solid-state laser media, especially for realization of ultra-low noise single frequency and ultrafast lasers. Recent advances have greatly increased the maximum gain per unit length, and point to the potential for compact, on-chip glass-based light sources. Progress with respect to 2nd and 3rd order nonlinear effects in glasses is ongoing. It is expected that cm-scale switching or pulse shaping devices based on glasses will become feasible, at least for niche applications.

Overall, it is clear that glasses can contribute greatly to the development of compact, low-loss, multifunction optics integrated with electronics.

Defining Terms

Amorphous Metaloxides are glassy alloys of a transition metal with oxygen, typical examples being TiO_2, Ta_2O_5, Nb_2O_5, and Y_2O_3. In bulk form, these materials are typically polycrystalline or crystalline ceramics. However, amorphous thin films can be deposited with relative ease, and they have been widely used as high index layers in optical filter design and as dielectric layers in the microelectronics industry.

Broadband Sensitizer is typically some species co-doped along with rare-earth ions into a glass host, in order to increase the pumping efficiency or radiative efficiency of the rare-earth ions. Various sensitizers have been demonstrated, including silicon nanoclusters, silver ions, and other rare earth ions (such as in the sensitization of erbium by ytterbium).

Chalcogenide Glasses are amorphous alloys containing S, Se, and/or Te. Typical examples include Se, GeS_2, $GeSe_2$, As_2S_3, As_2Se_3, and As_2Te_3. By intermixing these and other binary chalcogenide glasses, a wide variety of multicomponent glasses can be formed. Further, a wide range of non-stoichiometric compositions is possible. Several compositions have become standard industrial materials, including $Ge_{33}As_{12}Se_{55}$ and $Ge_{28}Sb_{12}Se_{60}$. The chalcogenide glasses are characterized by narrow bandgaps and good transparency in the mid to far infrared wavelength range.

Concentration Quenching refers to the reduction in luminescence efficiency and luminescence lifetime of a laser glass when the rare-earth dopant concentration is high. Quenching is due to interactions between closely spaced rare-earth ions at high concentrations. These interactions create new pathways, other than the desired radiative decay, for the ions to relax to the ground state after they have been raised to a desired lasing level by pump energy.

Devitrification refers to the transition of a glassy material to its lower energy crystalline state. This process is usually driven by thermal energy, such as if the material is held at some characteristic temperature above its glass transition temperature. The difference between the crystallization temperature and the glass transition temperature for a particular glass is one measure of its stability.

Fluoride Glasses are multicomponent glasses, typically based on fluorides of zirconium, barium, lead, gallium, lanthanum, aluminium, and sodium. They have a wide transparency range, from ultraviolet to mid-infrared wavelengths. They also have low characteristic phonon energies and can dissolve large concentrations of rare-earth ions. For these reason, they are extremely popular as hosts for rare-earth doped amplifiers and lasers operating in the UV-vis and mid-infrared regions.

Glass Transition Temperature is the approximate temperature at which a material changes from a supercooled liquid to an amorphous solid, or vice versa. The transition is marked by an abrupt but continuous change in slope of the specific volume and enthalpy versus temperature curves. Viscosity varies rapidly near the glass transition temperature, which is also sometimes called the softening temperature.

High Index Contrast refers to waveguides or devices fabricated using two or more materials that have very different refractive index. High index contrast is the basis for the confinement of light to very small cross-sectional area waveguides or very small volume optical cavities, either using total internal reflection or photonic bandgap effects. High index contrast thus is the basis for increased density of optical integrated circuits.

Integrated Optics/Photonics refers to the manufacture of photonic elements and circuits on a planar substrate, typically using thin film deposition, lithography, and etching steps. Typically, the substrate is a glass or semiconductor wafer and the photonic elements are guided wave devices.

Metastability is a term that refers to the non-equilibrium nature of glasses or amorphous solids. Amorphous solids have excess internal energy relative to the corresponding crystalline state or states of the same material. The method of manufacture, such as melt quenching, inhibits a transition to the lowest energy crystalline state.

Microphotonics refers to the chip-scale manufacture of optical and photonic waveguide circuitry, using processing techniques borrowed from the microelectronics industry. Related to this is the need for high-density integrated optics, as facilitated by high index contrast waveguides and photonic crystals. By usual definition, microphotonics refers specifically to the monolithic manufacture of optical and photonic elements on silicon (CMOS) chips.

Photoinduced Effects are changes in the properties of a glass induced by light, involving transitions between metastable states of the glass or changes in defect sites within the glass. Typically, a laser beam is used to locally modify the refractive index, density, absorption coefficient, etc. of the glass. These processes are widely used to pattern photonic structures such as Bragg gratings, waveguides, and refractive lenses into glasses.

Planar Lightwave Circuit or PLC refers to the industrially established processes for manufacturing integrated optics devices in silica-based glasses deposited on silicon wafers. Typically, the glass layers are deposited by chemical vapor deposition or flame hydrolysis. These technologies were developed mainly for applications in fiber optics, and are widely used to manufacture wavelength multiplexers.

Reflow is the process of heating a glass above its glass transition temperature, to the point that its viscosity is sufficiently reduced to enable the material to flow. In combination with surface tension effects or other external forces, reflow is often exploited in the reshaping of optical devices.

Supercooling Temperature is the difference between the glass transition temperature and the in-use temperature for a glass-based device. For a large (small) supercooling temperature, the structural relaxation rate is low (high).

Structural Relaxation is essentially an aging effect associated with glasses. Because glasses are metastable materials with random network structures, they are inherently subject to short or long term changes in material properties. Often, structural relaxation is manifested by a change in specific volume (densification) at fixed temperature versus time. The rate of such changes is extremely sensitive to the difference between the glass transition temperature and the observation temperature. Structural relaxation can be induced rapidly by an annealing step, in which the glass is heated near its glass transition temperature for some period of time.

References

43.1 H. Rawson: Glass and its History of Service, Part A, IEE Proceedings **135**(6), 325–345 (1988)

43.2 W. J. Tropf, M. E. Thomas, T. J. Harris: *OSA Handbook of Optics, Vol. II*, 2nd edn. (McGraw-Hill, New York 1995)

43.3 K. Hirao, T. Mitsuyu, J. Si, J. Qiu: *Active Glass for Photonic Devices, Photoinduced Structures and Their Application* (Springer, Berlin, Heidelberg 2001)

43.4 Z. Yin, Y. Jaluria: Neck-down and thermally induced defects in high-speed optical fiber drawing, J. Heat Transfer **122**(2), 351–362 (2000)

43.5 W. A. Gambling: IEEE J. Sel. Top. Quant. Elec. **6**, 1084 (2000)

43.6 M. Yamane, Y. Asahara: *Glasses for Photonics* (Cambridge Univ. Press, Cambridge 2000)

43.7 J. Lucas: Curr. Op. Sol. St. Mat. Sci. **4**, 181 (1999)

43.8 H. Ma, A. K.-Y. Jen, L. R. Dalton: Adv. Mat. **14**, 1339 (2002)

43.9 M. R. Poulsen, P. I. Borel, J. Fage-Pederson, J. Hubner, M. Kristensen, J. H. Povlsen, K. Rottwitt, M. Svalgaard, W. Svendsen: Opt. Eng. **42**, 2821 (2003)

43.10 K. Okamoto: *Integrated Optical Circuits and Components, Design and Applications* (Dekker, New York 1999), Chapt. 4

43.11 S. Torquato: Nature **405**, 521 (2000)

43.12 P. G. Debenedetti, F. H. Stillinger: Nature **410**, 259 (2001)

43.13 C. A. Angell, K. L. Ngai, G. B. McKenna, P. F. McMillan, S. W. Martin: J. Appl. Phys. **88**, 3113 (2000)

43.14 J. M. Saiter, M. Arnoult, J. Grenet: Phys. B: Cond. Matt. **355**, 370 (2005)

43.15 J. C. Anderson, K. D. Leaver, R. D. Rawlings, J. M. Alexander: *Materials Science*, 4th edn. (Chapman Hall, London 1990)

43.16 B. Hendriks, S. Kuiper: IEEE Spectrum **41**, 32 (2004)

43.17 G. P. Agrawal: *Fiber-Optic Communication Systems*, 2nd edn. (Wiley, New York 1997)
43.18 Y. P. Li, C. H. Henry: IEE Proc.-Optoelectron **143**, 263 (1996)
43.19 R. R. A. Syms, W. Huang, V. M. Schneider: Elec. Lett. **32**, 1233 (1996)
43.20 R. R. A. Syms, A. S. Holmes: IEEE Phot. Tech. Lett. **5**, 1077 (1993)
43.21 T. J. Kippenberg, S. M. Spillane, B. Min, K. J. Vahala: IEEE J. Sel. Top. Quant. Elec. **10**, 1219 (2004)
43.22 D. W. Vernooy, V. S. Ilchenko, H. Mabuchi, E. W. Streed, H. J. Kimble: Opt. Lett. **23**, 247 (1998)
43.23 M. L. Gorodetsky, A. A. Savchenkov, V. S. Ilchenko: Opt. Lett. **21**, 453 (1996)
43.24 L. Tong, R. R. Gattass, J. B. Ashcom, S. He, J. Lou, M. Shen, I. Maxwell, E. Mazur: Nature **426**, 816 (2003)
43.25 P. J. Roberts, F. Couny, H. Sabert, B. J. Mangan, D. P. Williams, L. Farr, M. W. Mason, A. Tomlinson, T. A. Birks, J. C. Knight, P. St. J. Russell: Opt. Express **13**, 236 (2005)
43.26 C.-T. Pan, C.-H. Chien, C.-C. Hsieh: Appl. Opt. **43**, 5939 (2004)
43.27 M. He, X.-C. Yuan, N. Q. Ngo, J. Bu, V. Kudryashov: Opt. Lett. **28**, 731 (2003)
43.28 M. He, X.-C. Yuan, J. Bu: Opt. Lett. **29**, 2004 (2004)
43.29 Y. A. Vlasov, S. J. McNab: Opt. Express **12**, 1622 (2004)
43.30 F. Grillot, L. Vivien, S. Laval, D. Pascal, E. Cassan: IEEE Phot. Tech. Lett. **16**, 1661 (2004)
43.31 T. Barwicz, H. I. Smith: J. Vac. Sci. Tech. B **21**, 2892 (2003)
43.32 D. K. Armani, T. J. Kippenberg, S. M. Spillane, K. J. Vahala: Nature **421**, 925 (2003)
43.33 V. Van, P. P. Absil, J. V. Hryniewicz, P.-T. Ho: J. Light. Tech. **19**, 1734 (2001)
43.34 K. J. Vahala: Optical Microcavities, Nature **424**, 839–851 (August 2003)
43.35 S. J. McNab, N. Moll, Y. A. Vlasov: Opt. Express **11**, 2927 (2003)
43.36 P. K. Gupta, D. Inniss, C. R. Kurkijian, Q. Zhong: J. Non-Crystalline Sol. **262**, 200 (2000)
43.37 D. P. Bulla, W.-T. Li, C. Charles, R. Boswell, A. Ankiewicz, J. Love: Appl. Opt. **43**, 2978 (2004)
43.38 X. H. Zhang, Y. Guimond, Y. Bellec: J. Non-crystalline Sol. **326&327**, 519 (2003)
43.39 A. K. Mairaj, X. Feng, D. P. Shepherd, D. W. Hewak: Appl. Phys. Lett. **85**, 2727 (2004)
43.40 A. K. Mairaj, R. J. Curry, D. W. Hewak: Appl. Phys. Lett. **86**, 094102 (2005)
43.41 J. S. Sanghera, L. B. Shaw, I. D. Aggarwal: *Rare-Earth-Doped Fiber Lasers and Amplifiers*, 2nd edn. (Dekker, New York 2001), Chapter 9
43.42 Z. Zhang, G. Xiao, C. P. Grover: Appl. Opt. **43**, 2325 (2004)
43.43 P. K. Tien: Appl. Opt. **10**, 2395 (1971)
43.44 R. Ulrich: J. Vac. Sci. Tech. **11**, 156 (1974)
43.45 P. K. Tien, A. A. Ballman: J. Vac. Sci. Tech. **12**, 892 (1974)

43.46 J. M. Mir, J. A. Agostinelli: J. Vac. Sci. Tech. A **12**, 1439 (1994)
43.47 J. Mueller, M. Mahnke, G. Schoer, S. Wiechmann: AIP Conference Proceedings **709**, 268 (2004)
43.48 D. Cangialosi, M. Wubbenhorst, H. Schut, A. van Veen, S. J. Picken: Phys. Rev. B **69**, 134206–1 (2004)
43.49 M. B. J. Diemeer: AIP Conference Proceedings **709**, 252 (2004)
43.50 N. Daldosso, M. Melchiorri, F. Riboli, F. Sbrana, L. Pavesi, G. Pucker, C. Kompocholis, M. Crivellari, P. Bellutti, A. Lui: Mat. Sci. Semicond. Proc. **7**, 453 (2004)
43.51 A. V. Kolobov (Ed.): *Photo-induced Metastability in Amorphous Semiconductors* (Wiley-VCH, Weinheim 2003)
43.52 K. Tanaka: C.R. Chimie **5**, 805 (2002)
43.53 H. Haeiwa, T. Naganawa, Y. Kokubun: IEEE Phot. Tech. Lett. **16**, 135 (2004)
43.54 K. Minoshima, A. M. Kowalevicz, E. P. Ippen, J. G. Fujimoto: Opt. Express **10**, 645 (2002)
43.55 N. Ponnampalam, R. G. DeCorby, H. T. Nguyen, P. K. Dwivedi, C. J. Haugen, J. N. McMullin, S. O. Kasap: Opt. Express **12**, 6270 (2004)
43.56 K. O. Hill: IEEE J. Sel. Top. Quant. Elec. **6**, 1186 (2000)
43.57 P. R. Herman, R. S. Marjoribanks, A. Oettl, K. Chen, I. Konovalov, S. Ness: Appl. Surf. Sci. **154–155**, 577 (2000)
43.58 M. Aslund, J. Canning: Opt. Lett. **25**, 692 (2000)
43.59 C. Florea, K. A. Winick: J. Light. Tech. **21**, 246 (2003)
43.60 A. Zakery, Y. Ruan, A. V. Rode, M. Samoc, B. Luther-Davies: J. Opt. Soc. Am. B. **20**, 1844 (2003)
43.61 X. Orignac, D. Barbier, X. Min Du, R. M. Almeida, O. McCarthy, E. Yeatman: Opt. Mat. **12**, 1 (1999)
43.62 R. A. Bellman, G. Bourdon, G. Alibert, A. Beguin, E. Guiot, L. B. Simpson, P. Lehuede, L. Guiziou, E. LeGuen: J. Electrochem. Soc. **151**, 541 (2004)
43.63 G.-L. Bona, R. Germann, B. J. Offrein: IBM J. Res. & Dev. **47**, 239 (2003)
43.64 A. J. Kenyon: Prog. Quant. Elec. **26**, 225 (2002)
43.65 M. J. Miniscalco: *Rare-Earth-Doped Fiber Lasers and Amplifiers*, 2nd edn. (Dekker, New York 2001), Chapt. 2
43.66 Y. Gao, B. Boulard, M. Couchaud, I. Vasilief, S. Guy, C. Duverger, B. Jacquier: Opt. Mat. **27**, 195–199 (2005)
43.67 T. Barwicz, M. A. Popovic, P. T. Rakich, M. R. Watts, H. A. Haus, E. P. Ippen, H. I. Smith: Opt. Express **12**, 1437 (2004)
43.68 T. Kawashima, K. Miura, T. Sato, S. Kawakami: Appl. Phys. Lett. **77**, 2613 (2000)
43.69 R. Rabady, I. Avrutsky: Appl. Opt. **44**, 378 (2005)
43.70 H. Hirota, M. Itoh, M. Oguma, Y. Hibino: IEEE Phot. Tech. Lett. **17**, 375 (2005)
43.71 C.-Y. Tai, J. S. Wilkinson, N. M. B. Perney, M. Caterina Netti, F. Cattaneo, C. E. Finlayson, J. J. Baumberg: Opt. Express **12**, 5110 (2004)

43.72 B. Unal, C.-Y. Tai, D. P. Shepherd, J. S. Wilkinson, N. M. B. Perney, M. Caterina Netti, G. J. Parker: Nd:Ta2O5 rib waveguide lasers, Appl. Phys. Lett. **86**, 021110 (2005)

43.73 Y. Kokubun, Y. Hatakeyama, M. Ogata, S. Suzuki, N. Zaizen: IEEE J. Sel. Top. Quant. Elec. **11**, 4 (2005)

43.74 T. Sato, K. Miura, N. Ishino, Y. Ohtera, T. Tamamura, S. Kawakami: Opt. Quant. Elec. **34**, 63 (2002)

43.75 D. R. MacFarlane: Ceramics International **22**, 535 (1996)

43.76 A. Feigel, M. Veinger, B. Sfez, A. Arsh, M. Klebanov, V. Lyubin: Appl. Phys. Lett. **83**, 4480 (2003)

43.77 G. Lenz, S. Spalter: *Nonlinear Photonic Crystals* (Springer, Berlin, Heidelberg 2003), Chapt. 11

43.78 K. Ogusu, J. Yamasaki, S. Maeda, M. Kitao, M. Minakata: Opt. Lett. **29**, 265 (2004)

43.79 S. Hocde, S. Jiang, X. Peng, N. Peyghambarian, T. Luo, M. Morrell: Opt. Mat. **25**, 149 (2004)

43.80 R. Nayak, V. Gupta, A. L. Dawar, K. Sreenivas: Thin Sol. Films **445**, 118 (2003)

43.81 D. Barbier: *Integrated Optical Circuits and Components, Design and Applications* (Dekker, New York 1999), Chapt. 5

43.82 B. E. Callicoatt, J. B. Schlager, R. K. Hickernell, R. P. Mirin, N. A. Sanford: IEEE Circuits & Devices Mag. **19**, 18 (September 2003)

43.83 E. DeSurvire: *Rare-Earth-Doped Fiber Lasers and Amplifiers*, 2ed. (Dekker, New York 2001), Chapt. 10

43.84 A. Yariv: *Optical Electronics in Modern Communications*, 5th edn. (Oxford Univ. Press, New York 1997)

43.85 Q. Wang, N. K. Dutta: J. Appl. Phys. **95**, 4025 (2004)

43.86 M. J. F. Digonnet: *Rare-Earth-Doped Fiber Lasers and Amplifiers*, 2nd edn. (Dekker, New York 2001), Chapt. 3

43.87 D. S. Funk, J. G. Eden: *Rare-Earth-Doped Fiber Lasers and Amplifiers*, 2nd edn. (Dekker, New York 2001), Chapt. 4

43.88 S. Saini, J. Michel, L. C. Kimerling: J. Light. Tech. **21**, 2368 (2003)

43.89 P. G. Kik, A. Polman: J. Appl. Phys. **93**, 5008 (2003)

43.90 P. Laporta, S. Taccheo, S. Longhi, O. Svelto, C. Svelto: Opt. Mat. **11**, 269 (1999)

43.91 F. J. Grawert, J. T. Gopinath, F. O. Ilday, H. M. Shen, E. P. Ippen, F. X. Kartner, S. Akiyama, J. Liu, K. Wada, L. C. Kimerling: Opt. Lett. **30**, 329 (2005)

43.92 U. Keller: Nature **424**, 831 (2003)

43.93 T. Qiu, L. Li, A. Schulzgen, V. L. Temyanko, T. Luo, S. Jiang, A. Mafey, J. V. Moloney, N. Peyghambarian: IEEE Phot. Tech. Lett. **16**, 2592 (2004)

43.94 Y. C. Yan, A. J. Faber, H. de Waal, P. G. Kik, A. Polman: Appl. Phys. Lett. **71**, 2922 (1997)

43.95 F. D. Patel, S. DiCarolis, P. Lum, S. Venkatesh, J. N. Miller: IEEE Phot. Tech. Lett. **16**, 2607 (2004)

43.96 A. Polman, F. C. J. M. van Veggel: J. Opt. Soc. Am. B **21**, 871 (2004)

43.97 H.-S. Han, S.-Y. Seo, J. H. Shin, N. Park: Appl. Phys. Lett. **81**, 3720 (2002)

43.98 J. Lee, J. H. Shin, N. Park: J. Light. Tech. **23**, 19 (2005)

44. Optical Nonlinearity in Photonic Glasses

A brief review of optical nonlinearity in photonic glasses is given. For third-order nonlinearity, the relationship between two-photon absorption and nonlinear refractive index is considered using a formalism developed for crystalline semiconductors. Stimulated light scattering and supercontinuum generation in optical fibers are also introduced. Prominent resonant-type nonlinearity in particle-embedded glasses is described. For second-order nonlinearity, a variety of poling methods are summarized. Finally, it is pointed out that various photoinduced changes can appear when excited by linear and nonlinear optical processes, and this is related to glass structure.

44.1 Third-Order Nonlinearity in Homogeneous Glass 1064
    44.1.1 Experimental 1064
    44.1.2 Theoretical Treatment 1065
    44.1.3 Stimulated Light Scattering and Supercontinuum Generation 1068

44.2 Second-Order Nonlinearity in Poled Glass 1069

44.3 Particle-Embedded Systems 1070

44.4 Photoinduced Phenomena 1071

44.5 Summary 1072

References 1072

New developments in optical fibers and pulsed lasers have prompted increasing interest in optical nonlinearity in photonic glasses [44.1–3]. Third-order polarization yields several nonlinear phenomena, such as intensity-dependent absorption and intensity-dependent refractive index, which can be utilized in power stabilizers and all-optical switches. On the other hand, in high-power glass lasers, self-focusing effects arising from intensity-dependent increases in refractive index pose serious problems. Then again, the second-order polarization that occurs in poled glasses can be utilized in second harmonic generation for example. The present chapter provides a brief review of optical nonlinearity in *inorganic* glasses. At this point we should mention that, in many respects, organic polymers exhibit similar features to those of glass [44.4, 5]. In general, glass is more stable, while polymers can provide greater nonlinearity, so these two types of material are competitors in practical applications.

Glass also competes with crystals. A great advantage of glass is the ability to control its structure at three scales. First, the atomic composition of the glass can be tailored continuously. For instance, we can obtain nonlinear optical glass with any refractive index in the range of 1.4–3.2 at a wavelength of $\approx 1\,\mu$m. Second, the atomic structure can be modified reasonably easily using, e.g., light beams. Such modifications may be regarded as photoinduced phenomena, which can be employed in order to add second-order nonlinearity to selected regions and so forth. Lastly, macroscopic shapes can be changed into arbitrary bulk forms, fibers, thin layers and microparticles [44.6, 7].

Here, fiber and film waveguides may be very important for nonlinear applications for two reasons. One is that the light power density can be increased by reducing the lateral size, i.e. the film thickness or fiber diameter, to submicron scales. The other is, that fibers can provide long lengths for light–glass interactions that are not limited by diffraction. These scale factors produce apparent enhancements in nonlinear effects in glasses, although the intrinsic nonlinearity may be smaller than that in crystals.

Before proceeding further, it would be useful to introduce a nonlinear formula [44.4, 5]. For simplicity, we take the polarization P and the electric field E to be scalar quantities. Then, very simply, P can be written down in the cgs units as

$$P = \chi^{(1)}E + \chi^{(2)}EE + \chi^{(3)}EEE + \ldots, \qquad (44.1)$$

where the first term $\chi^{(1)}E$ depicts the conventional linear response, and $\chi^{(2)}$ and $\chi^{(3)}$ represent the second- and the third-order nonlinear susceptibilities. $\chi^{(1)}$ is related to the linear refractive index n_0 via $n_0 = \{1 + 4\pi\chi^{(1)}\}^{1/2}$. On the other hand, the second term provides such time dependence as $\sin\{(\omega \pm 2\omega)t\}$, so that it could produce dc and second-overtone (2ω) signals. In a similar way, the

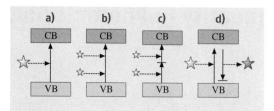

Fig. 44.1a–d Schematic illustrations of (**a**) one-photon absorption, (**b**) two-photon absorption, (**c**) two-step absorption via a mid-gap state, and (**d**) a Raman-scattering process in a semiconductor with a valence band (VB) and a conduction band (CB)

third could modify the fundamental (ω) signal and generate a third-overtone (3ω). Nevertheless, since a refractive index changes with frequency, it is difficult in the overtone generations to satisfy the so-called phase-matching condition [44.4,5], e.g. $\varphi_{3\omega} = 3\varphi_\omega$, where φ is the phase of electric fields in glasses. The, the fundamental-signal processing may be the most important. Microscopically, the nonlinear terms arise through several mechanisms, such as electronic, atomic (including molecular motions), electrostatic and thermal processes. Among these, electronic processes can provide the fastest responses, with fs–ps time scales, which will be needed for optical information technologies. Accordingly, we will focus on such processes from this point onwards.

The chapter is divided up as follows. In Sect. 44.1, we treat conventional glasses. These are optically isotropic, which gives $\chi^{(2)} = 0$, because the isotropic structure appears to be centrosymmetric. In this case, the lowest nonlinear term becomes $\chi^{(3)}EEE$. When $\chi^{(3)}EEE$ can be written as $\chi^{(3)}IE$, where I is the light intensity, we can define the refractive index n as $n = n_0 + n_2 I$, where n_2 is sometimes called the second-order index of refraction [44.5]. In nonlinear optics, this notation poses problems due to the definitions of several parameters. Therefore, the conversion $n_2(\text{cm}^2/\text{W}) \approx 0.04\chi^{(3)}(\text{esu})/n_0^2$ is frequently employed [44.5]. On the other hand, for light absorption, which is proportional to $\langle E\,\mathrm{d}P/\mathrm{d}t\rangle$, where $\langle \ldots \rangle$ represents a time average, we need to take the imaginary parts of $\chi^{(1)}$ and $\chi^{(3)}$, which cause one- and two-photon absorption, into account (Fig. 44.1). In such cases, an effective absorption coefficient can be written as $\alpha(\text{cm}^{-1}) + \beta(\text{cm/W})I(\text{W/cm}^2)$; n_2 and β will be connected through nonlinear Kramers–Kronig relations.

Section 44.2 focuses upon noncentrosymmetry. From the mid-1980s onwards, several kinds of poling treatments have been found to add $\chi^{(2)}$ to glasses [44.8]. For instance, Österberg and Margulis [44.9] demonstrated intense second harmonic generation in laser-irradiated glass fibers. Such work began with silica, and has more recently been directed towards more complicated glasses. The magnitude of $\chi^{(2)}$ seen in this case may be smaller than those observed in crystals, due to the disordered atomic structures in glasses, but the structural controllability may offer some advantages.

Section 44.3 introduces glasses incorporating fine semiconductor and metal particles. Such nanostructured glasses are known to exhibit large optical nonlinearities [44.4, 5, 10]. However, as will be described, their wider features remain to be studied.

In Sect. 44.4 we consider photoinduced phenomena. The intense light needed to produce optical nonlinearity is likely to modify the structure of glass. Light that is even more intense may cause damage. While this damage can be useful for optical engraving, the structural modification may ultimately be more important. Actually, fiber Bragg gratings [44.1, 11] have been produced commercially using excimer lasers, and nonlinear optical excitation may play an important role in these. Three-dimensional modifications are also produced through controlled light focusing [44.12]. Section 44.5 provides a summary of the chapter. This chapter is based on author's recent review [44.13].

44.1 Third-Order Nonlinearity in Homogeneous Glass

44.1.1 Experimental

Substantial data are available for n_2 at transparent wavelengths [44.4]. However, to the author's knowledge, all such experiments have utilized lasers, and no spectral dependence has been reported. We should also note that, in comparison with n_0 measurements, n_2 evaluations are much more difficult [44.4]. The biggest difficulty, which is common to all nonlinear experiments, is the light intensity. Normally, light is pulsed and focused, producing complex temporal and spatial profiles, which in turn means that intensity evaluations are not straightforward. For instance, if the profile is Gaussian, can we simply use the peak value to evaluate the nonlinearity? Because of such measuring difficulties and also some other specific problems inherent to

Table 44.1 Linear (E_g, n_0, α_0) and nonlinear (n_2, β, β_{max}) optical properties and figures of merit ($2\beta\lambda_0/n_2$, n_2/α_0) in some glasses. E_g is the optical bandgap energy [44.12], n_0 is the refractive index [44.12], α_0 is the attenuation coefficient [44.16], n_2 is the intensity-dependent refractive index [44.4, 5, 12], β is the two-photon absorption coefficient [44.16], and β_{max} is the maximum value. Except for E_g and β_{max}, the values are evaluated at a wavelength of 1–1.5 μm. BK-7 is a borosilicate glass and SF-59 indicates data for lead-silicate glasses with $\approx$ 57mol.% PbO

Glass	E_g (eV)	n_0	α_0 (cm^{-1})	n_2 (10^{-20} m^2/W)	β (cm/GW)	β_{max} (cm/GW)	$2\beta\lambda_0/n_2$	n_2/α_0 (cm^3/GW)
SiO$_2$	10	1.5	10^{-6}	2	< 10^{-2}	1	< 10	0.2
BK-7	4	1.5		3				
SF-59	3.8	2.0		30	< 10^{-1}	10		
As$_2$S$_3$	2.4	2.5	10^{-3}	200	10^{-2}	50	0.1	0.02
BeF$_2$	10	1.3		0.8				

glasses, reported n_2 values vary, even for a reference SiO$_2$ glass, over the range $1-3 \times 10^{-20}$ m^2/W [44.4]. In other glasses, values that differ by a factor of ≈ 10 have been reported [44.4, 14, 15].

However, we can see a general trend in n_2 in Table 44.1. Halide glasses (BeF$_2$) have the smallest values, light metal oxides (BK-7) and SiO$_2$ have similar values, heavy metal oxides such as PbO-SiO$_2$ (SF-59) have larger values, while the largest belong to the chalcogenides (As$_2$S$_3$) [44.4, 10]. The largest value reported so far may be $n_2 = 8 \times 10^{-16}$ m^2/W, seen in Ag$_{20}$As$_{32}$Se$_{48}$ at wavelength of 1.05 μm [44.21], which is $\approx 10^4$ times as large as that of SiO$_2$.

Figure 44.2 shows β spectra for SiO$_2$, As$_2$S$_3$ and two PbO-SiO$_2$ glasses [44.16]. Bi$_2$O$_3$-B$_2$O$_3$ glasses (*Imanishi* et al., [44.22]) exhibit similar features to those of lead glass. We can see that all of the β spectra seem to have maxima at midgap regions, $\hbar\omega \approx E_g/2$. However, this is not an absolute rule, since at $\hbar\omega \geq 2.0$ eV As$_2$S$_3$ provides two-step absorption (Fig. 44.1), which masks the two-photon absorption signal [44.20]. We also see that the maximal β ($\approx 10^0$ cm/GW) in SiO$_2$ is markedly smaller than those belonging to the other glasses. Note that such a marked difference does not exist in maximal α, which is $\approx 10^6$ cm^{-1} at the super-bandgap regions of SiO$_2$ and As$_2$S$_3$. It may be worth mentioning here that two-photon excitation produces photocurrents in some glasses [44.23, 24].

44.1.2 Theoretical Treatment

For $\chi^{(3)}$, theoretical treatments of glasses, liquids, and crystals are largely the same. Glasses and liquids can be treated similarly in the sense that both are optically isotropic. Hence, many ideas stemming from types of chemical bond pictures have been proposed [44.5]. On the other hand, the magnitude of $\chi^{(3)}$ in glass appears to be similar to that in the corresponding crystal, since their electronic structures are both governed by short-range atomic structures, within ≈ 0.5 nm [44.25], which is much shorter than the wavelength (≈ 500 nm) of visible light. Specifically, since n_2 is governed by integrated electronic absorption (44.8), the magnitudes are roughly the same in glass and crystal. For instance, in crystalline and glassy SiO$_2$ at near-infrared wavelengths, the difference in the n_2 values of both seems to be comparable with their experimental accuracy; i.e. the values for crystalline and glassy SiO$_2$ are $\approx 1.14 \times 10^{-13}$ and 0.85×10^{-13} esu respectively [44.4]. Then, using a formula derived for semiconductor crystals, we can apply

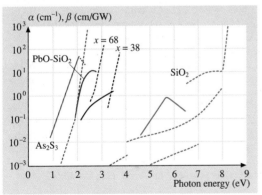

Fig. 44.2 One- (*dashed line*) and two-photon (*solid line*) absorption spectra, α and β, in SiO$_2$ [44.17, 18], 38PbO−62SiO$_2$, 68PbO−32SiO$_2$ [44.19], and As$_2$S$_3$ [44.20]. For SiO$_2$, in transparent regions ($\hbar\omega \leq 8.0$ eV) α should be regarded as an attenuation coefficient; these are not reproducible, probably due to impurities, defects and light scattering. The *dotted line* for As$_2$S$_3$ at $\hbar\omega \geq 2.0$ eV indicates two-step absorption

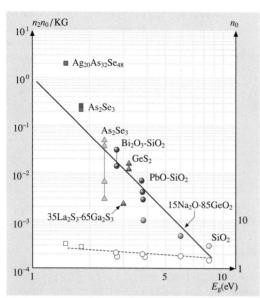

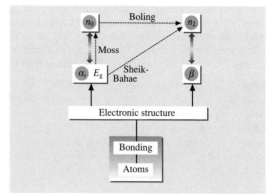

Fig. 44.3 The Sheik–Bahae relation $n_2 n_0 = KG(\hbar\omega/E_g)/E_g^4$ (*solid line*), related data (*solid symbols*), the Moss rule $n_0^4 E_g = 77$ (*dashed line*), and related data (*open symbols*) for an oxide (*circles*), sulfide (*triangles*) and selenide (*squares*). The four data for As_2S_3 are obtained from different publications, while those for $PbO-SiO_2$ with slightly different E_g correspond to different compositions. The illustration is modified from its previous form [44.16] due to the additions of $Ag_{20}As_{32}Se_{48}$ [44.21], Bi_2O_3-silica glasses [44.26], $35La_2S_3 - 65Ga_2S_3$ [44.27], and $15Na_2O - 85GeO_2$ [44.28]

Fig. 44.4 Relationship between atomic structure, electronic structure, linear absorption α including bandgap energy E_g, linear refractive index n_0, two-photon absorption β, and nonlinear refractive index n_2. Absorption and refraction are related by the linear and nonlinear Kramers–Kronig relations (*double arrows*). The Moss rule may be regarded as a simplified Kramers–Kronig relation. The Boling and Sheik–Bahae relations connect n_0 and E_g to n_2, respectively

a kind of band picture to glasses [44.16], as described below.

Estimation of Nonlinear Refractivity

Several semi-empirical relationships have been proposed for $\chi^{(3)}$ or n_2 [44.5]. Such relations are useful, since nonlinear optical constants are much more difficult to measure than linear constants. Naturally, since they are simplifications they are also less accurate. The simplest may be the one provided by *Wang*, $\chi^{(3)} \approx \{\chi^{(1)}\}^4$, which can be regarded as a generalized Miller's rule [44.5]. *Boling* et al. [44.29] have also derived some relations, among which the simplest may be

$$n_2(10^{-13} \text{ esu}) \approx 391(n_d - 1)/\nu_d^{5/4}, \quad (44.2)$$

where n_d is the refractive index at the d-line ($\lambda = 588$ nm) and ν_d is the Abbe number. This relation contains only two macroscopic parameters, which can be easily evaluated, and so it has often been used to estimate n_2. It actually provides a good approximation for small n_d glasses with $n_d \leq 1.7$ [44.29]. For high n_d glasses, *Lines* [44.14] proposes a relation that also contains the atomic distance, and others derive more complicated formula [44.30, 31]. Ab initio calculations for TeO_2-based glasses have also been reported recently [44.32]. However, these relations cannot predict wavelength dependence (Fig. 44.5). The estimated n_2 may be regarded as a long-wavelength limit. In addition, the material is regarded as transparent, so relations between n_2 and β cannot be found.

Nonlinear optical properties of semiconductor crystals have been studied fairly deeply [44.5], and applying the same concepts to glasses appears to be tempting. *Tanaka* [44.16] has adopted a band picture, which was developed by *Sheik-Bahae* et al. [44.33], to some glasses. Their concept gives a universal relationship of

$$n_2 = KG(\hbar\omega/E_g)/(n_0 E_g^4), \quad (44.3)$$

for crystals with energy gaps of $1-10$ eV, where K is a fixed constant and $G(\hbar\omega/E_g)$ is a spectral function (E_g is the bandgap energy). For E_g in glasses, we can take the so-called Tauc gap [44.25] if it is known, or otherwise the photon energy $\hbar\omega$ at $\alpha \approx 10^3$ cm^{-1}. As shown in Fig. 44.3, the universal line gives reasonable agreement with published data, while the agreement is

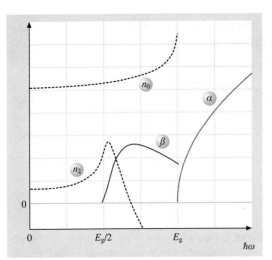

Fig. 44.5 Spectral dependence of linear absorption α, linear refractive index n_0, two-photon absorption β, and intensity-dependent refractive index n_2 in an ideal amorphous semiconductor with energy gap E_g. *Vertical scales are arbitrary*

not satisfactory as it is for crystals [44.33]. The poorer agreement for glasses may be partly due to experimental difficulties. In addition, band tail states probably cause larger deviations in smaller bandgap glasses such as $Ag_{20}As_{32}Se_{48}$. Incidentally, as shown by the dashed line in the figure, the Moss rule $n_0^4 E_g = 77$ [44.34], which was originally pointed out for crystalline semiconductors, also gives satisfactory fits for glasses.

Nonlinear Kramers–Kronig Relationship

We can take another approach in order to obtain unified insight into optical nonlinearity (n_2 and β). In this case, we start with absorption and then obtain the refractive index using Kramers–Kronig relationships. As illustrated in Fig. 44.4, this approach is able to provide a complete understanding of atomic structure and optical properties, since absorption spectra can be related to electronic structure more directly.

It is well-known that, for glasses, and neglecting momentum conservation rules, $\alpha(\hbar\omega)$ and $\beta(\hbar\omega)$ can be written as [44.16]

$$\alpha(\hbar\omega) \propto |\langle \phi_f | H | \phi_i \rangle|^2 \int D_f(E + \hbar\omega) D_i(E) \, dE \,, \tag{44.4}$$

$$\beta(\hbar\omega) \propto |\Sigma_n \langle \phi_f | H | \phi_n \rangle \langle \phi_n | H | \phi_i \rangle / (E_{ni} - \hbar\omega)|^2$$
$$\times \int D_f(E + 2\hbar\omega) D_i(E) \, dE \,, \tag{44.5}$$

where H is the perturbation Hamiltonian, ϕ is a related electron wavefunction, and D is the density of states, and the subscripts i, n, and f represent the initial, intermediate, and final states. α and β are linked to the electronic structure, i.e. the wavefunctions and densities of states, which is determined by the atomic species and bonding structures involved [44.25]. In this case, $n_0(\omega)$ is expressed using the conventional Kramers–Kronig relation as [44.5]

$$n_0(\omega) = 1 + (c/\pi)\wp \int \left[\alpha(\Omega)/(\Omega^2 - \omega^2)\right] d\Omega \,. \tag{44.6}$$

For a nonlinear response Δn, *Hutchings* et al. [44.35] have derived the relation

$$\Delta n(\omega; \zeta) = (c/\pi)\wp \int \left[\Delta\alpha(\Omega; \zeta)/(\Omega^2 - \omega^2)\right] d\Omega \,, \tag{44.7}$$

where $\Delta\alpha$ is the nonlinear absorption induced by an excitation at ζ and probed at Ω. Note that this relation is derived from the causality principle for nondegenerate cases, e.g., two-beam experiments with different photon energies, Ω and ζ. However, this relation may also be applied to degenerate cases ($\Omega = \zeta$) as a rough approximation as [44.36]

$$n_2(\omega) = (c/\pi)\wp \int \left[\beta(\Omega)/(\Omega^2 - \omega^2)\right] d\Omega \,. \tag{44.8}$$

Using these formulae, we can roughly predict how β and n_2 depend upon E_g. Equation (44.5) suggests that $\beta \propto 1/E_g^2$, provided that the E_g dependence is governed by $|\Sigma_n 1/(E_{ni} - \hbar\omega)|^2$. This relation may be consistent with the material dependence shown in Fig. 44.2. That is, $\beta_{max} \propto 1/E_g^2 \approx 1/E_g^3$. Note that this E_g dependence is comparable to $\beta \propto 1/E_g^3$, which is theoretically derived and experimentally confirmed for crystalline semiconductors [44.5, 33]. Then, putting $\beta \propto 1/E_g^2$ into (44.8), and assuming $\int [1/(\Omega^2 - \omega^2)] d\Omega \propto 1/E_g^2$, we obtain $n_2 \propto 1/E_g^4$, which is consistent with (44.3).

Under some plausible assumptions, we can also calculate the spectral dependence of β and n_2 for an ideal amorphous semiconductor, which contains no gap states [44.37]. Figure 44.5 shows that, at $\hbar\omega = E_g/2$, $\beta = 0$ and n_2 is maximal. This is similar to some degree to the feature at $\hbar\omega = E_g$ for α and n_0. The figure also shows that β becomes maximal at $E_g/2 - E_g$. Note, however, that it is not clear whether we can link this photon energy dependence to the experimental results shown in Fig. 44.2, since the observed spectra appear to have fairly sharp peaks.

Figure of Merit

Several figures of merit have been proposed for evaluating optical devices which utilize n_2. For instance, *Mizrahi* et al. [44.40] and others [44.15, 21, 41] emphasize the negative effects of two-photon absorption, and provide the criterion that $2\beta\lambda_0/n_2 < 1$. Then, using the above E_g dependence, we see that $\beta/n_2 \approx E_g^2$, so that small E_g materials are preferable. Actually (as also listed in Table 44.1), in this figure, As_2S_3 appears to be better than SiO_2. However, the criterion implicitly neglects α. Lines [44.14] uses n_2/α instead. Here, α arises from the so-called residual absorption, which is difficult to estimate quantitatively. In addition, it is the attenuation α_0 instead of the absorption α that could be decisive. Table 44.1 shows that, in this measure, SiO_2 behaves better than As_2S_3. These figures of merit, however, have presumed only nonresonant electronic contributions with sub-ps responses. More recently, *Jha* et al. [44.10] utilize $n_2/\tau\alpha$, where τ is the relaxation time. Nevertheless, theoretical predictions of τ remain to be studied.

The most appropriate definition of the figure of merit naturally depends upon the application of interest. For instance, for an optical fiber device, the maximum length might be ≈ 1 m [44.42], since the device must be compact and fast. In this case, the light absorption, which arises from $\alpha + \beta I$, should be suppressed to below 10^{-2} cm^{-1}, or the light propagation loss must be smaller than ≈ 1 dB/m. However, for optical integrated circuits, an effective propagation distance may be ≈ 1 cm, in which the attenuation could be as large as 10^0 cm^{-1}.

We can suggest another idea by taking the spectral dependence shown in Fig. 44.5 into account. That is, the best material that has n_2 at some $\hbar\omega$ is the one which satisfies a bandgap condition of $\hbar\omega = E_g/2$, where α and β are zero and n_2 becomes maximal, provided that there is no gap-state absorption. In practical glass samples, absorption due to impurities, dangling bonds such as E' centers in oxide glasses [44.11, 25], and wrong bonds in chalcogenide glasses [44.25, 37] cannot be neglected. Actually, we can see in Table 44.1 and Fig. 44.2 that residual attenuation exists in nominally transparent regions, parts of which are undoubtedly caused by absorption [44.43]. Note that these mid-gap states also give rise to two-step absorption (Fig. 44.1). We also cannot neglect photoinduced phenomena induced by these photoelectronic excitations (Sect. 44.4). Selecting an appropriate glass for a specific application is, therefore, not a straightforward process.

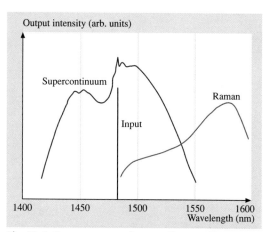

Fig. 44.6 Output spectra from a Raman fiber amplifier [44.38] and from a supercontinuum fiber generator [44.39] excited by 1483 nm light (schematic illustration)

44.1.3 Stimulated Light Scattering and Supercontinuum Generation

Recently, stimulated Raman scattering has attracted considerable interest, since it can be utilized for broadband light amplification [44.1, 5]. As illustrated in Fig. 44.1d, stimulated Raman scattering can be regarded as a kind of nondegenerate two-photon process. However, in contrast to two-photon absorption, one photon is absorbed by a virtual state at the same time as another photon is emitted in Raman scattering. As with conventional stimulated light emission, when the incident light is intense, the Stokes-shifted light can be amplified. The nonlinear polarization related to this is given by $P^{(3)}(\omega_s) \propto i\chi_R^{(3)} I(\omega_L) E(\omega_s)$, or the emitted light intensity is $\Delta I(\omega_s) \propto \chi_R^{(3)} I(\omega_L) I(\omega_s)$, where ω_L and $\omega_s (< \omega_L)$ are the exciting laser frequency and the Stokes-shifted frequency.

Most experiments have been performed using optical fibers [44.1, 42] in order to obtain long interaction lengths. For instance, as illustrated in Fig. 44.6, Raman fiber amplifiers comprising 5–10 km lengths of silica can provide broadband gains of ≈ 10 dB [44.38]. Raman lasers using silica microspheres of diameter $\approx 70\,\mu$m [44.7] are also interesting examples. On the other hand, stimulated Brillouin scattering in glasses has also been explored [44.1, 4].

Supercontinuum generation in silica fibers has now been studied intensively [44.4, 39, 44]. When pulsed or CW light propagates through a nonlinear medium, it may undergo spectral broadening [44.5].

For instance, as illustrated in Fig. 44.6, a 350 m single-mode fiber excited by a 2.22 W CW laser with a wavelength of 1.483 μm can emit 2.1 W over a broad spectrum of 1.43–1.53 μm [44.39]. Note that, unlike the stimulated light scattering described above, light of a shorter wavelength is also generated in this process. Such a spectral-conversion fiber could be utilized as a broadband optical amplifier. The phenomenon appears under strong and prolonged nonlinear interactions, and accordingly, several mechanisms such as intensity-dependent refractive index changes, third harmonic generation and stimulated Raman scattering could be responsible [44.39].

44.2 Second-Order Nonlinearity in Poled Glass

It has been discovered that several kinds of poling methods can add second-order nonlinearity to glasses [44.8]. At least five kinds have been demonstrated, which are listed in Table 44.2. Note that similar procedures are also employed for polymers. Most experiments utilize second harmonic signals to evaluate $\chi^{(2)}$ and, less commonly, electro-optical effects. Practical applications remain to be studied, while, for second-harmonic generations, the optical phase matching between exciting and nonlinearly-generated light is a prerequisite [44.45].

This so-called optical poling was demonstrated by Österberg and Margulis [44.9] using optical fibers. They found that exposing Ge-doped optical fibers of length ≈ 1 m to 70 kW Nd:YAG laser light for ≈ 1 h could increase the second-harmonic signal to 0.55 kW. Stolen and Tom [44.46] utilized two light beams (x1 and x2, of Nd:YAG laser light) for induction, which reduced the exposure time to ≈ 5 min. However, the method was only practical for optical fibers, since the nonlinearity induced was relatively small, $\chi^{(2)} \approx 10^{-4}$ pm/V. Second, so-called thermal poling, which was actually electrothermal in nature, was demonstrated in bulk SiO_2 samples by Myers et al. [44.47]. The nonlinearity induced, ≈ 1 pm/V, which was evaluated from the second harmonic signals of 1.06 μm laser light, is of a similar magnitude to that in quartz. Since this nonlinearity is reasonably large, this method has been widely applied to other glasses, such as $PbO\text{-}SiO_2$ [44.48], $Nb_2O_5\text{-}B_2O_3\text{-}P_2O_5\text{-}CaO$ [44.49], and $TeO_2\text{-}Bi_2O_3\text{-}ZnO$ [44.50]. Third, Okada et al. [44.51] demonstrated the corona-discharge poling at ≈ 200 °C of 7059 films deposited onto Pyrex glass substrates. This corona-poling procedure has often been employed for organic polymers. Fourth, electron-beam poling of PbO-silica glass was shown to produce ≈ 1 pm/V [44.52]. An advantage of this method is its high resolution, which may hold promise in the fabrication of optical integrated circuits, despite the fact that a vacuum is needed. Liu et al. (2001) [44.53] applied the method to chalcogenide glasses, which produced ≈ 1 pm/V. Proton implantation into silica can also add

Table 44.2 Reported poling methods, applied objects, typical procedures, and induced $\chi^{(2)}$ values in silica. For references, see the main text. The $\chi^{(2)}$ values listed are compared with $\chi^{(2)}_{11} = 1$ pm/V in crystalline SiO_2 and $\chi^{(2)}_{22} = 5$ pm/V in $LiNbO_3$

Method	Object	Procedure	$\chi^{(2)}$ (pm/V)
Optical	Fiber	Nd:YAG laser, 1 h	10^{-4}
Thermal	Bulk	4 kV, 300 °C, 1 h	1
Corona	Film	5 kV, 300 °C, 15 min	1
e-beam	Bulk	40 kV, 10 mA, 10 min	1
Proton	Bulk	500 kV, 1 mC, 100 s	1
UV	Bulk	ArF laser, 10 kV	3

a $\chi^{(2)}$ of ≈ 1 pm/V [44.54]. Sixth, Fujiwara et al. (1997) demonstrated UV poling in Ge-doped SiO_2 subjected to electric fields of ≈ 10^5 V/cm. The $\chi^{(2)}$ induced is reported to be ≈ 3 pm/V, comparable to that in $LiNbO_3$.

Two poling mechanisms of note have been proposed [44.8]. One is that space charge produces a built-in electric field of E_{DC} (≈ 10^6 V/cm), which induces an effective $\chi^{(2)}$ of $3E_{DC}\chi^{(3)}$ [44.46]. Here, E_{DC} is governed by the migration of ions such as Na^+ under applied or generated electric fields [44.55]. In agreement with this idea, $\chi^{(2)}$ decays with a time constant of 10^1–10^6 days at room temperature, which is connected to the alkali ion mobility [44.55]. The other idea is that oriented defects such as E' centers are responsible. It is reasonable to assume that UV excitation produces defective dipoles, which are oriented with the static electric field.

It may be reasonable to assume that the dominant mechanism depends upon the poling method. Actually, we can divide the procedures listed in Table 44.2 into two types, depending upon whether or not the glass is heated during the poling process. The heating tends to enhance macroscopic ion migration, while it can also relax microscopic defect orientations at the same time [44.25]. Therefore, it seems that the ion migration is responsible in thermal and coronal poling, while defect orientation dominates in the other methods. In this context, poling at low temperatures may be a promising way of enhancing defect orientation.

In so-called *glass ceramics*, embedded crystals seem to be responsible for prominent $\chi^{(2)}$ [44.50, 56, 57]. For instance, *Takahashi* et al. [44.57] demonstrated that oriented $Ba_2TiSi_2O_8$ crystals are produced in BaO-TiO_2-SiO_2 glass by heat treatment at 760 °C for 1 h, which gives a prominent $\chi^{(2)}$ of ≈ 10 pm/V. $\chi^{(2)}$ can also be generated at interfaces [44.58].

44.3 Particle-Embedded Systems

Glasses that contain nanoparticles of metals [44.65, 66] and semiconductors [44.67] have attracted considerable interest due to their unique third-order nonlinearities. Table 44.3 lists several recent results. Such glasses containing dispersed nanoparticles can be prepared by a variety of physical and chemical methods, e.g., vacuum deposition and sol-gel techniques [44.4, 65, 68]. For semiconductor systems, a lot of work has also been investigated in semiconductor-doped color glass filters [44.67], which are now commercially available.

These nanoparticle systems work efficiently at close to the resonant wavelengths of some electronic excitations. This feature produces at least three characteristics. First, the imaginary, not the real, part of $\chi^{(3)}$ may be more prominent. Accordingly, Table 44.3 compares absolute values. Second, the system exhibits a strong spectral dependence [44.4, 69]. For instance, Au-silica and CdSSe-silica are efficient at ≈ 580 nm and ≈ 800 nm [44.4]. Third, the response time τ and the linear absorption α tend to become longer and higher. Actually, a trade-off between $\chi^{(3)}$ and τ and α seems to exist. For instance, CdSSe-dispersed glasses show $\chi^{(3)}$ value of $\approx 10^{-9}$ esu with τ value of ≈ 20 ps, while Au-dispersed glasses give $\approx 10^{-11}$ esu and ≈ 1 ps, respectively [44.4]. Linear absorption can be as large as 10^4 cm^{-1} [44.65], so these systems can be utilized as small devices, not as fibers. Note that, in pure silica at nonresonant infrared wavelengths, $\chi^{(3)} \approx 10^{-13}$ esu, $\tau \approx 10$ fs, and $\alpha \leq 10^{-5}$ cm^{-1} [44.65]. As is suggested above, the particle-embedded system should surmount two problems for wide applications. One is the reduction of linear attenuation, including absorption and scattering, and the other is the shift of resonant wavelengths to the optical communication region, 1.3–1.5 μm. Is such a shift possible? What are the mechanisms that give rise to these prominent nonlinearities in particle-embedded systems?

When the particle is a semiconductor such as CuCl and CdSSe, excitons or confined electron-hole pairs are responsible [44.67]. Specifically, the excitons in semiconductor particles behave as two-level systems, and at the resonance frequency, $|\chi^{(3)}|$ is written as

$$|\chi^{(3)}| \approx \text{Im}\chi^{(3)} \propto |\mu|^4 NT_1T_2^2 , \qquad (44.9)$$

where μ is the dipole moment of the exciton, N is the particle number, $T_1 (\approx 100$ ps) is the lifetime of the exciton, and $T_2 (\approx$ fs) is the dephasing time. A quantitative estimation predicts that closely packed CuCl particles of radius 40 nm could provide a $|\chi^{(3)}|$ enhancement of a factor of $\approx 10^3$ when compared with the bulk value [44.70].

When metal particles such as spherical Au particles with diameters of 10–50 nm are used, we can envisage

Table 44.3 Several recently reported particle systems, along with their preparation methods, $|\chi^{(3)}|$ values, and response times τ at the measured wavelength λ, as well as references. PLD and VE depict pulsed laser deposition and vacuum evaporation

System	Preparation	$|\chi^{(3)}|$ (esu)	τ	λ (nm)	Reference
Au (15 nm)/ silica	Shell structure	10^{-9}	2 ps	550	[44.59]
Cu (2 nm)/Al_2O_3	PLD	10^{-7}	5–450 ps	600	[44.60]
Ag (20 nm)/BaO	VE	10^{-10}	0.2 ps	820	[44.61]
Fe (4 nm)/$BaTiO_3$	PLD	10^{-6}		532	[44.62]
SnO_2 (10 nm)/silica	Sol-gel	10^{-12}		1064	[44.63]
CdS (4 nm)/silica	Sol-gel	10^{-11}		500	[44.64]

local field enhancement and dynamic responses from conduction electrons, including plasmon effects [44.65, 66, 71]. The result is approximately written as

$$\chi^{(3)} \approx p_m \chi_m^{(3)} |3\varepsilon_h/(\varepsilon_m + 2\varepsilon_h)|^2 \{3\varepsilon_h/(\varepsilon_m + 2\varepsilon_h)\}^2 , \tag{44.10}$$

where p_m is the volume fraction of the metal particles, $\chi_m^{(3)}$ is the bulk nonlinearity of the metal, and ε_h and ε_m are the linear dielectric constants of the host (glass) and the metal. The nonlinearity of the host is neglected here for the sake of simplicity. Note that ε_h can be real, while ε_m is complex. We see that the metal's nonlinearity $\chi_m^{(3)}$ is decreased by the volume factor p_m, while $\chi_m^{(3)}$ may be enhanced by local fields if $\mathrm{Re}(\varepsilon_m + 2\varepsilon_h) \approx 0$, which determines the resonance wavelength. In agreement with this model, an Au-dispersed glass, for instance, gives a greater $\chi^{(3)}$ than that of an Au film [44.5].

However, since these systems are complex, consisting of particles and a glass matrix, a variety of situations arise. For the particle, as well as particle species, we need to consider size, size distribution, shape and concentration. Here, needless to say, a narrow size distribution is preferred for investigating fundamental mechanisms. The shape may be spherical, ellipsoidal, rod, and so forth. The concentration determines the mean separation distance between particles. As the concentration increases, electrical particle–particle interactions appear [44.59], and then the particles eventually percolate [44.60], which may produce dramatic changes in the nonlinear response. The matrix seems to be of secondary importance. Actually, liquids and organic materials have been employed as well as glasses [44.4]. However, some reports suggest that the nonlinearity is greatly affected by the surface states of the semiconductor particles [44.72] and by the dielectric constant of the matrix surrounding the metallic particles [44.4].

Lastly, it may be worth mentioning two recent results. One is that particles can be arrayed to produce photonic structures that exhibit novel nonlinear properties, such as light confinement [44.73]. The other is the generation of second harmonic signals from oriented ellipsoidal Ag nanoparticles [44.74]. Oriented particle structures have been produced in silica through tensile deformation and simultaneous heating, and this may be regarded as a kind of mechanical poling.

44.4 Photoinduced Phenomena

The intense pulsed light employed to produce nonlinear effects is also likely to produce a variety of transitory and (quasi-)stable optical changes [44.72, 75]. Photochromic effects induced by sub-gap light are the result of transitory electronic changes [44.76]. The increase in the refractive index of SiO_2 induced by fs–ns laser light

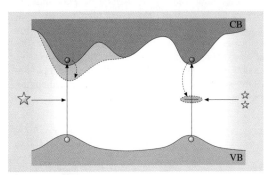

Fig. 44.7 Different relaxation paths from bandgap excitation due to one-photon (*left*) and two-photon (*right*) processes. For instance, the one-photon and two-photon excitations induce bonding strains and defect alterations

with $\hbar\omega \approx 1 \approx 5\,\mathrm{eV}$ is known to occur along with quasistable structural changes [44.11, 77]. The optical poling described in Sect. 44.2 can be regarded as a kind of photoinduced anisotropy. When the light is more intense than $1\,\mathrm{GW/cm^2}$–$1\,\mathrm{TW/cm^2}$, depending on the duration of the light pulse, permanent damage is likely to occur [44.5]. These photosensitive changes are induced by electronic excitations, which may be triggered by one-photon and multiphoton processes, depending upon the photon energy and the light intensity.

Are these photoinduced phenomena really excited by nonlinear processes? For instance, when a photoinduced phenomenon is induced by exposure to light of photonic energy $\hbar\omega \approx E_g/m$, where m is an integer, it is often asserted that m-photon absorption is responsible. Nevertheless, the phenomenon may be triggered by one-photon absorption by mid-gap states, which give weak absorption tails (Fig. 44.2) arising from defective structures in the glass [44.25, 78]. Otherwise, it may be triggered by m-step absorption processes [44.79], consisting of m one-photon absorptions in succession (Fig. 44.1c). These processes have probably been overlooked when analyzing experimental results.

In addition, as schematically illustrated in Fig. 44.7, the structural changes induced by linear and nonlinear excitation are not necessarily the same. *Tanaka* has demonstrated using As_2S_3 that the bandgap excitations produced by one-photon and two-photon absorption produce different changes [44.78]. One-photon excitation leads to photodarkening, while two-photon excitation results in an increase in the refractive index with no photodarkening. In these phenomena, temperature rises upon intense exposure can be neglected. Such changes related to excitation processes can appear in glasses because the *localized* atomic wavefunction plays an important role, in contrast to the extended Bloch wavefunctions in crystals. For instance, as we can see from (44.4), one-photon excitation occurs between wavefunctions with different parities, such as from s to p orbitals, since H is an odd function. In contrast, (44.5) shows that two-photon excitation can occur between states of the same parity states, such as from p to p. In addition, the $1/(E_{ni} - \hbar\omega)$ in the equation leads to the possibility of *resonant and localized* two-photon excitation if $E_{ni} - \hbar\omega \approx 0$ is satisfied for a mid-gap state. In such cases, the mid-gap state selectively absorbs the excitation energy, leading to an atomic change which shows itself as a macroscopic photoinduced phenomenon.

44.5 Summary

At present, there is a big gap between the science and the technology of glasses. From a physical point of view, glass science lags far behind crystal science, mainly due to experimental difficulties associated with atomic structure determination and the theoretical unavailability of Bloch-type wavefunctions. On the other hand, it is now difficult to imagine a world without optical fibers. These contrasting situations are promoting a deeper understanding of nonlinear photonic glasses and are leading to an increasingly wide range of applications for such glasses.

Third-order optical nonlinearity in homogeneous glasses has been studied in a fair amount of detail. Substantial experimental data have been obtained for n_2, which have been analyzed using empirical relations such as those from Boling. In contrast, less work has been done on the nonlinear absorption β. In the present article, therefore, we have tried to present, in a coherent way, the absorption and the refractive index using semiconductor terminology. This approach can be used to connect nonlinear properties to the energy gap. However, little work has been done on the dynamics associated with this field. For instance, it is difficult to theoretically predict the response time τ in a glass at a particular excitation energy.

Two inhomogeneous systems have aroused increasing interest. One is poled or crystallized glass. Enhanced second-order nonlinearity has been reported for such systems, and this can sometimes be added selectively to a region. The other is the particle-embedded system, which can give greater third-order nonlinearity, and in some cases enhanced second-order nonlinearity too. However, the mechanism associated with this system remains to be studied. One experimental problem is to find a method that can reproduce particles of a fixed size and shape in arbitrary concentrations. These two inhomogeneous systems can be combined using photonic structure concepts, which will be of interest for future applications.

Finally, nonlinear optical excitations appear to play important roles in a variety of photoinduced phenomena in glasses. However, fundamental studies are still lacking in this area. The phenomenon of interest may be nonlinear in nature, or the linear excitation of gap states may trigger successive changes. Attempting to understand the nonlinear photo–electro–structural process will provide a challenging problem.

References

44.1 G. P. Agrawal: *Nonlinear Fiber Optics*, 3rd edn. (Academic, San Diego 2001)
44.2 P. P. Mitra, J. B. Stark: Nature **441**, 1027 (2001)
44.3 L. F. Mollenauer: Science **302**, 996 (2003)
44.4 R. L. Sutherland: *Handbook of Nonlinear Optics*, 2nd edn. (Marcel Dekker, New York 2003)
44.5 R. W. Boyd: *Nonlinear Optics*, 2nd edn. (Academic, San Diego 2003)

44.6 M. H. Field, J. Popp, R. K. Chang: *Progress in Optics Vol. 41*, ed. by E. Wolf (North Holland, Amsterdam 2000) Chap. 1

44.7 S. M. Spillane, T. J. Kippenberg, K. J. Vahala: Nature **415**, 621 (2002)

44.8 Y. Quiquempois, P. Niay, M. Dounay, B. Poumellec: Curr. Opinion Solid State Mater. Sci. **7**, 89 (2003)

44.9 U. Österberg, W. Margulis: Opt. Lett. **11**, 516 (1986)

44.10 A. Jha, X. Liu, A. K. Kar, H. T. Bookey: Curr. Opinion Solid State Mater. Sci. **5**, 475 (2001)

44.11 G. Pacchioni, L. Skuja, D. L. Griscom: *Defects in SiO_2 and Related Dielectrics: Science and Technology* (Kluwer, Dortrecht 2000)

44.12 E. M. Vogel, M. J. Weber, D. M. Krol: Phys. Chem. Glasses **32**, 231 (1991)

44.13 K. Tanaka: J. Mater. Sci: Mater Electron. **16**, 633 (2005)

44.14 M. E. Lines: J. Appl. Phys. **69**, 6876 (1991)

44.15 A. Zakery, S. R. Elliott: J. Non-Cryst. Solids. **330**, 1 (2003)

44.16 K. Tanaka: J. Non-Cryst. Solids. **338–340**, 534 (2004)

44.17 T. Mizunami, K. Takagi: Opt. Lett. **19**, 463 (1994)

44.18 A. Dragonmir, J. G. McInerney, N. Nikogosyan: Appl. Opt. **41**, 4365 (2002)

44.19 K. Tanaka, N. Yamada, M. Oto: Appl. Phys. Lett. **83**, 3012 (2003)

44.20 K. Tanaka: Appl. Phys. Lett. **80**, 177 (2002)

44.21 K. Ogusu, J. Yamasaki, S. Maeda, M. Kitao, M. Minakata: Opt. Lett. **29**, 265 (2004)

44.22 K. Imanishi, Y. Watanabe, T. Watanabe, T. Tsuchiya: J. Non-Cryst. Solids **259**, 139 (1999)

44.23 R. C. Enck: Phys. Rev. Lett. **31**, 220 (1973)

44.24 P. S. Weitzman, U. Osterberg: J. Appl. Phys. **79**, 8648 (1996)

44.25 S. R. Elliott: *Physics of Amorphous Materials*, 2nd edn. (Longman Scientific, Essex 1990)

44.26 N. Sugimoto, H. Kanbara, S. Fujiwara, K. Tanaka, K. Hirao: Opt. Lett. **21**, 1637 (1996)

44.27 S. Smolorz, I. Kang, F. Wise, B. G. Aiken, N. F. Borrelli: J. Non-Cryst. Solids **256 and 257**, 310 (1999)

44.28 O. Sugimoto, H. Nasu, J. Matsuoka, K. Kamiya: J. Non-Cryst. Solids. **161**, 118 (1993)

44.29 N. L. Boling, A. J. Glass, A. Owyoung: IEEE Quant. Electron. **14**, 601 (1978)

44.30 V. Dimitrov, S. Sakka: J. Appl. Phys. **79**, 1741 (1996)

44.31 J. Qi and D. F. Xue, G. L. Ning: Phys. Chem. Glasses **45**, 362 (2004)

44.32 S. Suehara, P. Thomas andA. Mirgorodsky, T. Merle-Mejean, J. C. Champarnaud-Mesjard, T. Aizawa, S. Hishita, S. Todoroki, T. Konishi, S. Inoue: J. Non-Cryst. Solids **345&346**, 730 (2004)

44.33 M. Sheik-Bahae, D. J. Hagan, E. W. Van Stryland: Phys. Rev. Lett. **65**, 96 (1990)

44.34 T. S. Moss: *Optical Properties of Semiconductors* (Butterworths, London 1959) p. 48

44.35 D. C. Hutchings, M. Sheik-Bahae, D. J. Hagan, E. W. Van Stryland: Opt. Quantum. Electron. **24**, 1 (1992)

44.36 M. Sheik-Bahae, D. C. Hutchings, D. J. Hagan, E. W. Van Stryland: IEEE Quantum Electron. **27**, 1296 (1991)

44.37 K. Tanaka: *Optoelectronic Materials and Devices, Vol. 1*, ed. by G. Lucovsky, M. Popescu (INOE, Bucharest 2004) Chap. 3

44.38 J. Bromage: J. Lightwave Technol. **22**, 79 (2004)

44.39 A. Zheltikov: Appl. Phys. B **77**, 143 (2003)

44.40 V. Mizrahi, K. W. DeLong, G. I. Stegeman, M. A. Saifi, M. J. Andrejco: Opt. Lett. **14**, 1140 (1989)

44.41 M. Asobe: Opt. Fiber Technol. **3**, 142 (1997)

44.42 G. I. Stegeman, R. H. Stolen: J. Opt. Soc. Am. B **6**, 652 (1989)

44.43 K. Tanaka, T. Gotoh, N. Yoshida, S. Nonomura: J. Appl. Phys. **91**, 125 (2002)

44.44 A. K. Abeeluck, C. Headley: Appl. Phys. Lett. **85**, 4863 (2004)

44.45 H.-Y. Chen, C.-L. Lin, Y.-H. Yang, S. Chao, H. Niu, C. T. Shih: Appl. Phys. Lett. **86**, 81107 (2005)

44.46 R. H. Stolen, H. W. K. Tom: Opt. Lett. **12**, 585 (1987)

44.47 R. A. Myers, N. Mukherjee, S. R. J. Brueck: Opt. Lett. **22**, 1732 (1991)

44.48 Y. Luo, A. Biswas, A. Frauenglass, S. R. Brueck: Appl. Phys. Lett. **84**, 4935 (2004)

44.49 B. Ferreira, E. Fargin, J. P. Manaud, G. Le Flem, V. Rodriguez, T. Buffeteau: J. Non-Cryst. Solids **343**, 121 (2004)

44.50 G. S. Murugan, T. Suzuki, Y. Ohishi, Y. Takahashi, Y. Benino, T. Fujiwara, T. Komatsu: Appl. Phys. Lett. **85**, 3405 (2004)

44.51 A. Okada, K. Ishii, K. Mito, K. Sasaki: Appl. Phys. Lett. **60**, 2853 (1992)

44.52 P. G. Kazansky, A. Kamal, P. St. Russell: Opt. Lett. **18**, 683 (1993)

44.53 Q. M. Liu, F. X. Gan, X. J. Zhao, K. Tanaka, A. Narazaki, K. Hirao: Opt. Lett. **26**, 1347 (2001)

44.54 L. J. Henry, B. V. McGrath, T. G. Alley, J. J. Kester: J. Opt. Soc. Am. B **13**, 827 (1996)

44.55 O. Deparis, C. Corbari, G. Kazansky, K. Sakaguchi: Appl. Phys. Lett. **84**, 4857 (2004)

44.56 V. Pruneri, P. G. Kazansky, D. Hewak, J. Wang, H. Takebe, D. N. Payne: Appl. Phys. Lett. **70**, 155 (1997)

44.57 Y. Takahashi, Y. Benino, T. Fujiwara, T. Komatsu: Appl. Phys. Lett. **81**, 223 (2002)

44.58 R. T. Hart, K. M. Ok, P. S. Halasyamani, J. W. Zwanziger: Appl. Phys. Lett. **85**, 938 (2004)

44.59 Y. Hamanaka, K. Fukuta, A. Nakamura, L. M. Liz-Marzan, P. Mulvaney: Appl. Phys. Lett. **84**, 4938 (2004)

44.60 R. Del Coso, J. Requejo-Isidro, J. Solis, J. Gonzalo, C. N. Afonso: J. Appl. Phys. **95**, 2755 (2004)

44.61 Q. F. Zhang, W. M. Liu, Z. Q. Xue, J. L. Wu, S. F. Wang, D. L. Wang, Q. H. Gong: Appl. Phys. Lett. **82**, 958 (2003)

44.62 W. Wang, G. Yang, Z. Chen, Y. Zhou, H. Lu, G. Yang: J. Appl. Phys. **92**, 7242 (2002)

44.63 A. Clementi, N. Chiodini, A. Paleari: Appl. Phys. Lett. **84**, 960 (2004)

44.64 S. G. Lu, Y. J. Yu, C. L. Mak, K. H. Wong, L. Y. Zhang, X. Yao: Microelectronic Eng. **66**, 171 (2003)

44.65 F. Gonella, P. Mazzoldi: *Handbook of Nanostructured Materials and Nanotechnology*, Vol. 4, ed. by H. S. Nalwa (Academic, San Diego 2000) Chap. 2

44.66 V. M. Shalaev: *Nonlinear Optics of Random Media* (Springer, Berlin, Heidelberg 2000)

44.67 G. Banfi, V. Degiorgio, D. Ricard: Adv. Phys. **47**, 510 (1998)

44.68 M. Nogami, S. T. Selvan, H. Song: Photonic glasses: Nonlinear optical and spectral hole burning properties. In: *Handbook of Advanced Electronic and Photonic Materials and Devices*, Vol. 5, ed. by H. S. Nalwa (Academic, San Diego 2001) Chap. 5

44.69 J. He, W. Ji, G. H. Ma, S. H. Tang, H. I. Elim, W. X. Sun, Z. H. Zhang, W. S. Chin: J. Appl. Phys. **95**, 6381 (2004)

44.70 Y. Li, M. Takata, A. Nakamura: Phys. Rev. B **57**, 9193 (1998)

44.71 D. Stroud, P. M. Hui: Phys. Rev. B **37**, 8719 (1988)

44.72 A. Puzder, A. J. Williamson, F. Gygi, G. Galli: Phys. Rev. Lett. **92**, 217401 (2004)

44.73 M. Ajgaonkar, Y. Zhang, H. Grebel, C. W. White: Appl. Phys. Lett. **75**, 1532 (1999)

44.74 A. Podlipensky, J. Lange, G. Seifert, H. Graener, I. Cravetchi: Opt. Lett. **28**, 716 (2003)

44.75 R. C. Jin, Y. C. Cao, E. C. Hao, G. S. Metraux, G. C. Schatz, C. A. Mirkin: Nature **425**, 487 (2003)

44.76 Y. Watanabe, Y. Kikuchi, K. Imanishi, T. Tsuchiya: Mater. Sci. Eng. B **54**, 11 (1998)

44.77 J. S. Aitchison, J. D. Prohaska, E. M. Vogel: Met. Mater. Proc. **8**, 277 (1996)

44.78 K. Tanaka: Philos. Mag. Lett. **84**, 601 (2004)

44.79 K. Kajihara, Y. Ikuta, M. Hirano, H. Hosono: Appl. Phys. Lett. **81**, 3164 (2002)

45. Nonlinear Optoelectronic Materials

In a nonlinear optical material, intense light alters the real and imaginary components of the refractive index. The nonlinear response of the real part of refractive index modifies the phase of propagating light, while the imaginary part describes the change in absorption. These illumination-dependent properties of nonlinear materials provide the basis for all-optical switching—the ability to manipulate optical signals without the need for optical–electronic–optical conversion.

In this chapter we review the physical processes underlying the illumination-dependent refractive index. We review the real and imaginary nonlinear response of representative groups of materials: crystalline semiconductors, organic materials, and nanostructures, and we examine the practical applicability of these groups of materials to all-optical optical switching. We identify the spectral regions which offer the most favorable nonlinear response as characterized using engineering figures of merit.

45.1	**Background** 1075
	45.1.1 Signal Processing in Optical Networks 1075
	45.1.2 Optical Signal Processing Using Nonlinear Optics 1076
	45.1.3 The Approach Taken During this Survey of Nonlinear Optoelectronic Materials 1076
45.2	**Illumination-Dependent Refractive Index and Nonlinear Figures of Merit (FOM)** 1077
	45.2.1 Ultrafast Response..................... 1077
	45.2.2 Ultrafast Nonlinear Material Figures of Merit....................... 1078
	45.2.3 Resonant Response 1079
	45.2.4 Resonant Nonlinear Material Figures of Merit....................... 1079
45.3	**Bulk and Multi-Quantum-Well (MQW) Inorganic Crystalline Semiconductors** 1080
	45.3.1 Resonant Nonlinearities............. 1080
	45.3.2 Nonresonant Nonlinearities in Inorganic Crystalline Semiconductors 1083
45.4	**Organic Materials** 1084
	45.4.1 Resonant Nonlinear Response of Organic Materials................... 1085
	45.4.2 Nonresonant Nonlinear Response of Organic Materials................... 1086
45.5	**Nanocrystals** 1087
45.6	**Other Nonlinear Materials** 1088
45.7	**Conclusions**... 1089
References 1089

45.1 Background

Optical fiber provides a suitable medium in which it is possible to reach tremendous transmission rates over long distances [45.1]. The maximum information-carrying capacity has been estimated to be around 100 Tbps [45.2]. Very high data rates can be achieved using a combination of wavelength- and time-division multiplexing techniques (WDM and TDM). WDM involves sending many signals in parallel at closely spaced wavelengths along the same fiber, while TDM allows close spacing in time of bits in a single channel.

While there exist means to produce, transfer, and detect information at a very high bandwidth, there is a need for more agility in photonic networks.

The agility of present-day optical networks is limited by the electronic nature of a very important function: the processing of information-bearing signals. Signal processing is responsible for switching and routing traffic, establishing links, restoring broken links, and monitoring and managing the network.

45.1.1 Signal Processing in Optical Networks

At present, the important and functionally complex signal-processing operations of switching and routing are carried out electronically. Electronic signal processing imposes two significant limitations on the functionality of optical networks: cost and opacity. Signal switching and routing requires conversion of the optical information into electrical signals, processing in the electronic domain, and converting back to the optical domain before retransmission. Such an operation requires detection, retiming, reshaping, and regeneration at each switching and routing point. This necessitates complex and expensive electronic and electro-optical hardware at each routing and switching node. The use of electronic signal processing places strict requirements on the format of data streams transferred and processed, thus making the signal processing opaque. Repetition rates of optical signals, power levels, and packet lengths have to be standardized before they can be processed electronically.

The ability to perform signal processing operations entirely within the optical domain would eliminate the requirement for optical–electrical–optical conversions, while providing agility and speed inherent to optical elements. Optical signal processing, in contrast with electronics, may provide ultrafast sub-picosecond switching times [45.3].

45.1.2 Optical Signal Processing Using Nonlinear Optics

Nonlinear optics can potentially support ultrafast self-processing of signals.

A variety of nonlinear optical signal-processing functions can be realized with similar fundamental building blocks [45.4–6]. Nonlinear optical elements and devices can be either integrated in photonic circuits [45.7] or used in a free-standing configuration [45.8]. Nonlinear optics can enable signal processing without the requirement for external electrical, mechanical, or thermal control [45.9]. The response time of properly designed nonlinear optical devices is limited fundamentally only by the nonlinear response time of the constituent materials [45.3, 10–12].

Photons do not interact with each other in vacuo. In order to perform nonlinear optical signal-processing operations the properties of a medium through which the light travels must be modified by the light itself. Optical signals then propagate differently as a result of their influence on the medium. Nonlinear optical signal-processing elements utilize the illumination-dependent real and imaginary parts of the index of refraction [45.9]. Depending on the material and spectral position, the real part of the refractive index and absorption of a given nonlinear material can either increase or decrease with increasing illumination.

A wide range of broadband and wavelength-selective nonlinear optical signal-processing devices has been proposed and demonstrated.

The most commonly studied nonlinear optical switching elements are nonlinear Fabry–Perot interferometers, nonlinear Mach–Zehnder modulators, nonlinear directional couplers, optical limiters, and nonlinear periodic structures.

A nonlinear Fabry–Perot interferometer consists of two mirrors separated by a nonlinear material. As the refractive index of the nonlinear material changes with an increased level of illumination, the effective path length of the resonator is altered. A nonlinear Fabry–Perot interferometer can be tuned out of, or into, its transmission resonance. When illuminated with the continuous-wave light, a nonlinear Fabry–Perot interferometer can exhibit optical bistability. Optical bistability is a phenomenon in which the instantaneous transmittance of the device depends both on the level of incident illumination and on the prior transmittance of the device. Such an element enables all-optical memory.

In a nonlinear Mach–Zehnder modulator and a nonlinear directional coupler, a part of the waveguide is made out of a nonlinear material. Changing the intensity of the incident light changes the effective path length experienced by the light. This, in turn, through phase interference, results in an illumination-dependent transmittance in a Mach–Zehnder modulator, and an illumination-dependent coupling in a nonlinear directional coupler.

A number of techniques use nonlinear properties of materials to obtain power-limiting, and associated with it, on–off switching. Such devices are based on total internal reflection [45.13], self-focusing [45.14], self-defocusing, two-photon absorption [45.15], or photorefractive beam fanning [45.16].

Nonlinear periodic structures combine the phenomena of nonlinear index change and distributed Bragg reflection. The intensity-dependent transmission and reflection properties of nonlinear periodic structures can be harnessed to yield various signal-processing functions. It has been demonstrated that nonlinear periodic structures can support optical switching and limiting [45.4–6, 17–20], optical bistability [45.8, 21–25], solitonic propagation of pulses [45.26, 27], and pulse compression [45.28].

45.1.3 The Approach Taken During this Survey of Nonlinear Optoelectronic Materials

There exist excellent texts that describe nonlinear optical processes and review the published properties of nonlinear materials [45.29–35]. This chapter will discuss the applicability of different material groups to nonlinear optical switching. Following the introduction of the concept of nonlinear refractive index and figures of merit in Sect. 45.2, the nonlinear properties of inorganic crystalline semiconductors (Sect. 45.3), organic materials (Sect. 45.4), nanocrystals (Sect. 45.5), and selected other materials (Sect. 45.6) will be reviewed and summarized. A critical review is given with a focus on figures of merit and processability.

45.2 Illumination-Dependent Refractive Index and Nonlinear Figures of Merit (FOM)

In a nonlinear optical medium intense light alters the real and imaginary components of the refractive index. The nonlinear response of the real part of the refractive index modifies the phase of propagating light, while the imaginary part describes the change in absorption.

This subsection will present the formalism used to describe how light affects the ultrafast and resonant changes in the nonlinear refractive index. The ultrafast nonlinear index changes take place in the spectral region where the material is nonabsorbing, while the resonant nonlinear index changes take place in the absorbing spectral region.

45.2.1 Ultrafast Response

Ultrafast nonlinear response is characterized by the instantaneous response, weak nonlinear index changes, and weak nonlinear absorption. The formalism that describes the ultrafast changes in the real and imaginary parts of the refractive index can be derived from the theory of nonlinear polarization.

The polarization $\boldsymbol{P}(r, \omega)$ of a material in the presence of an electric field $\boldsymbol{E}(r, \omega)$ at a frequency ω and position r is defined as

$$\boldsymbol{P}(r, \omega) = \epsilon_0 \chi(r, \omega) \boldsymbol{E}(r, \omega) \,, \quad (45.1)$$

where ϵ_0 is the permittivity of free space and $\chi(r, \omega)$ is the dielectric susceptibility tensor. $\chi(r, \omega)$ is related to the index of refraction $n(\omega)$ by

$$\chi(r, \omega) = n^2(r, \omega) - 1 \,. \quad (45.2)$$

In a homogeneous nonlinear material $\chi(r, \omega) = \chi(\omega)$ but $\chi(\omega)$ is not constant with electric field and the influence of $\boldsymbol{E}(r, \omega)$ on $\boldsymbol{P}(r, \omega)$ is not linear. In this case it is customary to expand $\boldsymbol{P}(r, \omega)$ in a power series of $\boldsymbol{E}(r, \omega)$ [45.34]:

$$\begin{aligned}
\boldsymbol{P}(r, \omega) = & \epsilon_0 \chi^{(1)}(\omega) \boldsymbol{E}(\omega) \\
& + \epsilon_0 \left[D^{(2)} \sum_{j,k} \chi^{(2)}_{ijk}(-\omega_3; \omega_1, \omega_2) \right. \\
& \times \boldsymbol{E}_j(\omega_1) \boldsymbol{E}_k(\omega_2) \\
& + D^{(3)} \sum_{jkl} \chi^{(3)}_{ijkl}(-\omega_4; \omega_1, \omega_2, \omega_3) \\
& \times \boldsymbol{E}_j(\omega_1) \boldsymbol{E}_k(\omega_2) \boldsymbol{E}_l(\omega_3) \\
& \left. + \text{higher-order terms} \right] \,, \quad (45.3)
\end{aligned}$$

where $\chi^{(1)}$ is the linear susceptibility, while $\chi^{(2)}$ and $\chi^{(3)}$ are the coefficients of the second- and third-order nonlinear susceptibility. The coefficients $D^{(2)}$ and $D^{(3)}$ are defined as:

$$D^{(2)} = \begin{cases} 1, & \text{for indistinguishable fields} \\ 2, & \text{for distinguishable fields} \end{cases} \quad (45.4)$$

and

$$D^{(3)} = \begin{cases} 1, & \text{for all fields indistinguishable} \\ 2, & \text{for two fields indistinguishable} \\ 3, & \text{for all fields distinguishable} \end{cases} \quad (45.5)$$

In all known materials the higher-order components of the effective nonlinear susceptibility tensor $\chi(\omega)$ yield smaller contributions to the effective polarization than the preceding terms of the same parity. On the other hand, in the presence of strong electric field the terms designated as the *higher-order terms* in (45.3) [i.e. terms proportional to the powers of $\boldsymbol{E}(r, \omega)$ higher than four], can be larger than the first three terms. However, the assumption of moderate intensities and the aim to illustrate the concept of nonlinear refractive index justifies retaining only the first three terms of (45.3) in the derivation that follows.

Nonlinear optical switching relies on nonlinear effects in which intense light changes the refractive index. Under such conditions there are no direct-current (DC) or low-frequency electro-optic effects present and the second term in (45.3) can be neglected. In addition all values of ω are degenerate. $P(r, \omega)$ reduces to

$$P(r, \omega) = \epsilon_0[\chi^{(1)}(\omega) + \chi^{(3)}(\omega)E(\omega)E(\omega)]E(\omega)$$

$$= \epsilon_0 \left[\chi^{(1)}(\omega) + \frac{2\chi^{(3)}(\omega)I}{\epsilon_0 n_0 c}\right] E(\omega), \quad (45.6)$$

where I is the local intensity

$$I = \frac{\epsilon_0}{2} n_0 c |E(\omega)|^2, \quad (45.7)$$

and c is the speed of light in vacuum.

The first term in (45.6) represents the linear contribution to the polarization and the second term represents the nonlinear, intensity-dependent part. This intensity-dependent part gives rise to the nonlinear index of refraction fundamental to this work.

To obtain the direct expression for the nonlinear refractive index the effective susceptibility from (45.6) is substituted into (45.2) to yield

$$n^2 = 1 + \chi^{(1)} + \frac{2\chi^{(3)}(\omega)I}{\epsilon_0 n_0 c}. \quad (45.8)$$

In order to relate directly this nonlinear part of polarization to the intensity-dependent part of the refractive index – a macroscopic measurable quantity – the effective index of refraction is expressed as

$$n = n_0 + n_2 I. \quad (45.9)$$

Taking the square of (45.9) and neglecting the terms proportional to I^2 under the assumption of weak relative nonlinearity $\left(n_2^2 I^2 \ll n_0 n_2 I \text{ and } n_2^2 I^2 \ll n_0^2\right)$ gives

$$n^2 = n_0^2 + 2n_0 n_2 I. \quad (45.10)$$

Equating (45.8) and (45.10) gives an expression for n_2

$$n_2 = \frac{\chi^{(3)}}{\epsilon_0 n_0^2 c}, \quad (45.11)$$

where all the factors are in SI units.

In general, n_2 can have real (Re) and imaginary (Im) parts with $n_{2\,\text{Re}}$ responsible for the nonlinear refraction and $n_{2\,\text{Im}}$ responsible for the nonlinear absorption or gain. There are many conventions used to express the real and imaginary parts of the nonlinear refractive index. The approach used by researchers must always be determined prior to comparison with absolute numbers. However, in general it is safe to write

$$n_{2\,\text{Re}} = \frac{K}{n_0^2} \text{Re}\left(\chi^{(3)}\right) \quad (45.12)$$

and

$$n_{2\,\text{Im}} = \frac{K}{n_0^2} \text{Im}\left(\chi^{(3)}\right), \quad (45.13)$$

where the constant K depends on the convention and units used [45.34].

In the rest of this chapter n_2 will be used to express the real part of the ultrafast nonlinear index of refraction, i.e. n_2 will be as used in (45.12).

In order to account for the imaginary component of the ultrafast nonlinear response in a commonly used way the following relationship is defined

$$\alpha = \alpha_0 + \beta I. \quad (45.14)$$

Equation (45.14) expresses the total absorption (α) in terms of its linear (α_0) and nonlinear (βI) contributions. β is the measurable, macroscopic quantity that will be used throughout this chapter to quantify the effects of the ultrafast imaginary nonlinear response.

45.2.2 Ultrafast Nonlinear Material Figures of Merit

A nonlinear material useful in a nonlinear optical signal-processing device must simultaneously satisfy the following conditions:

- The excitation time of the nonlinear effect must be less than the pulse width.
- The sum of the excitation and the relaxation times must be shorter than the pulse spacing.

In addition, an ultrafast nonlinear material must satisfy the following requirements:

- The effect of linear absorption must be weak compared to the effect of nonlinear refraction. *Stegeman* quantifies this condition in terms of the unitless figure of merit W [45.12]

$$W = \frac{|\Delta n|}{\alpha_0 \lambda} > 1, \quad (45.15)$$

where Δn is the induced change in the real part of the refractive index, α_0 is the linear absorption (expressed in units of inverse length) and λ

is the wavelength of light (with units of length). To facilitate consistent comparison between different nonlinear materials, Δn in (45.15) was assumed to be evaluated as the intensity approaches the saturation intensity, at which the rate of change of the refractive index drops noticeably below a linear dependence on intensity [45.12]. In general (45.15) can be used to quantify the nonlinear quality of a given material at any intensity, not only at the saturation.

- The effect of two-photon absorption must be weak compared to the effect of nonlinear refraction. This condition is quantified using the figure of merit T [45.12]

$$T = \frac{\beta_2 \lambda}{n_2} < 1, \quad (45.16)$$

where β is the two-photon absorption coefficient from (45.14) (expressed in units of length/power).

The conditions (45.15) and (45.16) can be combined in terms of a single figure of merit F

$$F = \frac{|\Delta n|}{\alpha_{\text{eff}} \lambda} > 1, \quad (45.17)$$

where α_{eff} is the effective absorption experienced by the sample at a given intensity. F can be used to quantify the quality of materials for signal processing with respect to nonlinear processes of any order rather than with respect to only third-order processes as in (45.16).

Condition (45.17) ensures that the nonlinear phase shift $\Delta \phi^{\text{NL}} = 2\pi \Delta n L / \lambda$, where L is the length of the material, reaches 2π before the intensity decays to $1/\text{e}$ of its input value as a result of the effective absorption. Phase shifts between 0.5π and 3.5π are required for most optical switching devices [45.12].

45.2.3 Resonant Response

The resonant response of a nonlinear material is the dominant nonlinear effect in the linearly absorbing spectral region. A different formalism than that presented in Sects. 45.2.1 and 45.2.2 is used to describe the resonant changes in the real and imaginary parts of the refractive index.

Illumination with light which is resonant with the material results in the direct absorption of the incoming photons, generating excited states and giving rise to a decrease in the effective absorption. If the relaxation time of the excited states is longer than the length of the pulse, the resonant effect is proportional to the fluence, rather than to the intensity of the incident ultrafast pulse. This saturation of the absorption is described by the following expression for the effective absorption α_{eff} [45.35]

$$\alpha_{\text{eff}} = \frac{\alpha_0'}{1 + \frac{P}{P_{\text{sat}}}} + \alpha_{\text{u}}. \quad (45.18)$$

where α_{u} is the unsaturable absorption, $P = \int_0^t I(t') \, \mathrm{d}t'$ is the incident fluence and P_{sat} is the saturation fluence at which the effective absorption decreases to half of its initial value $\alpha_0 = \alpha_0' + \alpha_{\text{u}}$. P accounts for the cumulative (up to the duration of the pulse) character of the resonant nonlinear response.

The saturation of absorption is accompanied by a change in the real part of the refractive index [45.35]

$$\Delta n = \frac{n_2' P}{1 + \frac{P}{P_{\text{sat}}}}. \quad (45.19)$$

n_2' describes the strength of the real part of the resonant nonlinear refractive index.

In this chapter, nonresonant and resonant phenomena are considered. The parameters n_2 and β from Sect. 45.2.1 are used to quantify the ultrafast response and Δn and $\Delta \alpha$ from Sect. 45.2.1 are used to describe the resonant response.

45.2.4 Resonant Nonlinear Material Figures of Merit

Figures of merit for the nonresonant response have been defined in Sect. 45.2.2. This section will introduce resonant figures of merit that account for the nonlinear phase shift that accumulates over the duration of a pulse.

For illustrative purposes, first-order approximations to (45.18) and (45.19) of the form $\Delta n(t) = \int_0^t n_2' I(t') \, \mathrm{d}t'$ and $\alpha_{\text{eff}}(t) = \alpha_0$ are considered under the assumption $P \ll P_{\text{sat}}$. A resonant nonlinear material is assumed to be illuminated with a square pulse of the form:

$$I(t) = \begin{cases} I_0, & \text{if } 0 < t < \tau_{\text{p}} \\ 0, & \text{if } t < \tau_{\text{p}}. \end{cases} \quad (45.20)$$

In analogy to (45.17) a time-averaged nonlinear figure of merit is defined for the resonant response

$$\langle F \rangle = \frac{|\langle \Delta n \rangle|}{\langle \alpha \rangle \lambda}. \quad (45.21)$$

The time-averaged nonlinear index change is

$$\langle \Delta n \rangle = \frac{1}{\tau_p} \int_0^{\tau_p} \left[\int_0^t n_2' I(t') \, dt' \right] dt$$

$$= \frac{n_2' P_{\text{total}}}{2} = \frac{|\Delta n_{\text{peak}}^{\text{ultrafast}}|}{2}, \quad (45.22)$$

where P_{total} is the total fluence of the pulse $P_{\text{total}} = \int_0^{\tau_p} I(t') \, dt' = I_0 \tau_p$. The time-averaged absorption is $\langle \alpha \rangle = \alpha_0$. For the case considered, the figure of merit (45.21) becomes:

$$\langle F \rangle = \frac{|\Delta n_{\text{peak}}|}{2 \alpha_0 \lambda}, \quad (45.23)$$

which is half of the ultrafast figure of merit. For simplicity, (45.17) will be used throughout this chapter for both resonant and ultrafast response.

45.3 Bulk and Multi-Quantum-Well (MQW) Inorganic Crystalline Semiconductors

The illumination-dependent refractive and absorptive nonlinear properties of inorganic crystalline semiconductors have been studied comprehensively. Since semiconductors are at the heart of the electronics industry, semiconductor micro- and nanofabrication techniques are well established. This enables the preparation of high-quality nonlinear samples and devices. The ability to change the composition of semiconductor compounds allows the tuning of the electronic band gap over the visible and infrared spectral ranges. The spectral position of the band gap, in turn, tunes the nonlinear properties.

When a semiconductor is illuminated with light at a frequency within the absorbing region, the dominant nonlinear effect relies on the presence of linear absorption. Upon absorption of the incident light, the electrons undergo a transition from the valence band to the conduction band, saturating the absorption. This band-filling effect is accompanied by a very large change in the real part of the refractive index.

In a nonresonant nonlinear process no single-photon absorption takes place. Under illumination with intense light the electronic clouds of the constituent atoms are distorted, changing the refractive index of the material. Associated with this is a multiphoton absorption process which takes place when the sum of the photon energies is larger than the band-gap energy. This effect changes the absorption characteristics of the material. Both the real and imaginary parts of ultrafast nonlinear index change, given their connection through the nonlinear Kramers–Kronig relations.

In addition, when subjected to an intense continuous wave or a high-repetition-rate pulsed illumination, the temperature of the absorbing materials, including semiconductors, increases. This in turn changes the refractive index. Thermal effects have relaxation times as long as

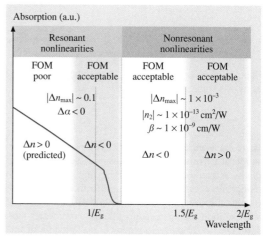

Fig. 45.1 Trends in the nonlinear response of bulk semiconductors

milliseconds and are not useful in processing trains of closely spaced pulses.

Figure 45.1 shows typical trends in the nonthermal nonlinear response of bulk inorganic crystalline semiconductor material under picosecond, low-repetition-rate illumination.

45.3.1 Resonant Nonlinearities

The two most important characteristics of the resonant nonlinear response are saturation of absorption and large nonlinear index change [45.36–38]. The relaxation of resonant nonlinear effects in semiconductors is not instantaneous. As long as the duration of the incident pulse is shorter than the relaxation time of the material, the magnitude of a nonlinear resonant response is proportional to the fluence, rather than to the intensity of the incident pulse. The relatively long relaxation time of nonlinear effects in bulk and multi-quantum-well (MQW) inorganic crystalline semiconductors (from hundreds of picoseconds to tens of nanoseconds) is often used as an argument against using resonant nonlinearities. However, established techniques such as low-temperature growth and doping can reduce the relaxation time down to tens of picosecond [45.39, 40].

The phenomenon of saturation of absorption ($\Delta\alpha < 0$) translates into absorption that decreases with increasing incident fluence. Resonant figures of merit of semiconductors are acceptable near the band edge and become worse at lower wavelengths. Such behavior is due to a stronger, lower-threshold saturation of absorption around the band edge.

Nonlinear index change is negative around the band edge and has been predicted to be positive at wavelengths lower than those corresponding to the first heavy-hole and light-hole excitonic peaks.

Resonant Nonlinearities in Bulk Semiconductors

In 1991 *Gupta* et al. measured the time of nonlinear response of GaAs grown at low temperatures (LT-GaAs) [45.42]. Changes in reflectivity were monitored during a pump-probe experiment at 620 nm. Relaxation times of several 2-μm-thick samples grown at temperatures ranging between 190 °C and 400 °C were measured. A decrease in the decay time to 0.4 ps was recorded with decreasing growth temperatures. This short relaxation time is drastically lower than the typical value of nanoseconds for unannealed GaAs [45.42].

In 1993 *Harmon* et al. studied the dependence of the nonlinear relaxation time in LT-GaAs on annealing temperatures. A decrease in the relaxation time down to sub-picosecond values was observed with decreasing annealing temperatures [45.43].

In a number of papers published between 1994 and 1998, the group of Smith, Othonos, Benjamin, and Loka reported on a series of comprehensive experiments carried out on various LT-GaAs samples grown using molecular beam epitaxy (MBE). The dependence of the magnitude and the response time of nonlinear effects on the growth and annealing temperatures was studied. Very large negative nonlinear index changes were measured ($\Delta n_{\max} = -0.13$) accompanied by a strong saturation of absorption [45.44]. The relaxation time was measured to decrease to a few picoseconds for samples grown at 500 °C [45.45]. The pump-probe measurements were carried out in the band-edge region at

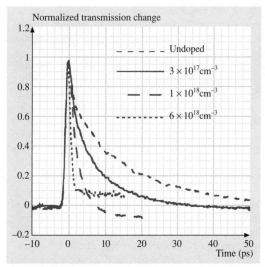

Fig. 45.2 Results of the pump-probe measurements illustrating the time-resolved change in transmission for He-InGaAsP samples with different Be doping concentrations. (After [45.41] with permission)

wavelengths of 870–890 nm. The decreased response time was attributed to the fast decay of excited carriers to mid-gap states. These states are an effect of the LT growth [45.40].

In subsequent years another group of researchers studied the strength and dynamics of intensity-dependent response in InGaAsP doped with Be grown with He-plasma-assisted MBE. As illustrated in Fig. 45.2, sub-picosecond relaxation times were obtained [45.41]. The rapid decay was explained by a short lifetime of excited states due to the existence of mid-gap He and He–Be trap states. Again, large negative changes in the real part of the refractive index and strong saturation of absorption were observed [45.39].

Resonant Nonlinearities in Semiconductor Multi-Quantum-Wells

The nonlinear properties of semiconductor multi-quantum-wells (MQWs) are similar to those of bulk semiconductors [45.38, 46, 47]. The nonlinear response in MQWs around the band edge is stronger and begins at lower fluences than in bulk materials.

Compared with bulk semiconductors, semiconductor MQWs offer an additional degree of freedom in selecting their nonlinear properties. The effective electronic band gap of a given semiconductor MQW structure, and hence the dispersion of real and imaginary parts of its linear and nonlinear refractive index, are influenced by two factors: the choice of the compositions of the constituent compounds and the well-to-barrier thickness ratio.

In 1982 *Miller* et al. reported on the measurements of resonant nonlinear properties of semiconductor MQWs. A very strong absorption saturation was noticed around the first excitonic peak in GaAs/AlGaAs MQWs. Based on these results a large refractive nonlinearity was deduced from the nonlinear Kramers–Kronig relation [45.48]. A theoretical paper followed, explaining the dynamics of transient excitonic nonlinearities [45.49]. A 20-ns excited-carrier relaxation time was predicted.

In 1986 *Lee* et al. measured the nonlinear saturation of the absorption of bulk GaAs and 29.9-nm GaAs/AlGaAs wells grown by molecular beam epitaxy. The measurement was performed using a monochromatic pump and a broadband probe over a 40-nm spectral range near the MQW band edge. Using the nonlinear Kramers–Kronig relation, large index changes of both signs were predicted. In MQWs, absorptive and refractive nonlinearities were enhanced compared to bulk GaAs. Index changes ranging from $\Delta n = -0.06$ to $\Delta n = 0.03$ were predicted in the samples analyzed [45.50].

This report was followed in 1988 by a study of nonlinearities around the band edge carried out by the same research group [45.38]. The response of bulk GaAs was compared with that of three sets of GaAs/AlGaAs MQWs, with well thicknesses of 7.6 nm, 15.2 nm and 29.9 nm. Again, a strong saturation of absorption was measured and nonlinear index changes of both signs were predicted from the nonlinear Kramers–Kronig relation [45.51]. The magnitude of the change in the real part of the refractive index was predicted to increase with decreasing well size. The sign of the refractive nonlinearity changed at wavelengths slightly shorter than that corresponding to the first excitonic peak [45.38].

Since 1988 many results of research on the nonlinear properties of GaAs/AlGaAs MQWs have been reported by *Garmire* et al. In a series of papers, the saturation of absorption was studied in GaAs/AlGaAs MQWs grown by metalorganic chemical vapor deposition epitaxy. The nonlinear Kramers–Kronig relation was used to predict the associated change in the real part of the refractive index. Figure 45.3 shows the predicted enhancement of nonlinearity with decreasing well size and the change of sign near the excitonic peak. Attempts were made to use the illumination-dependent shift of Fabry–Perot fringes to estimate directly the negative nonlinear index change along the band edge. However, this approach was admitted to yield significant errors, with the Fabry–Perot technique sometimes giving a value of Δn at twice the magnitude predicted

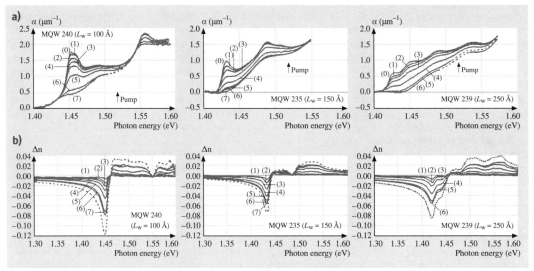

Fig. 45.3a,b Spectra of GaAs/AlGaAs MQWs of three different well widths, measured at various incident intensities by pulsed pump at 1.52 eV: (**a**) absorption coefficient α; (**b**) the change in the real part of the refractive index Δn. (After [45.46] with permission)

from the Kramers–Kronig relation [45.46]. In 1987 *Fox et al.* reported nonlinear measurements around the band edge of bulk GaInAs [45.52] and GaInAs/InP MQWs near wavelengths of 1.6 μm [45.53]. Full saturation of absorption was observed. The nonlinear index changes deduced from the nonlinear Kramers–Kronig relation were slightly larger than that observed in GaAs [45.35].

Recently *Brzozowski* et al. published results of direct picosecond measurements of nonlinear refractive-index change and nonlinear absorption in $In_{0.530}Al_{0.141}Ga_{0.329}As/In_{0.530}Ga_{0.470}As$ multi-quantum-wells in the range 1480–1550 nm. Large low-threshold nonlinear index changes were found: Δn of up to 0.14 with a figure of merit of 1.38 at a fluence of 116 μJ/cm². The figure of merit F was greater than unity over much of the spectrum. The results are summarized in Fig. 45.4.

In 1996 *Judawlikis* et al. reported the decreased nonlinear relaxation time in LT-grown Be-doped InGaAs/InAlAs MQWs. Nonlinear relaxation times of a few tens of picoseconds were observed in a pump-probe experiment near the band edge. The nonlinear change in the real part of the refractive index was not reported [45.56].

A different approach to decrease the response times of band-edge nonlinearities of semiconductor MQWs was taken by the groups of White, Sibbet, and Adams. An electric current was applied to active InGaAsP/InP waveguides and the nonlinear optical response under electrical bias was studied. It was found that under a forward bias the refractive nonlinear response was quenched. Under a reverse bias the nonlinear response was slightly reduced, but the initially long recovery time was reduced to 50 ps [45.57] and 18 ps in subsequent experiments [45.58]. Further, it was found that, when the waveguide was biased at transparency, the nonlinear coefficients of the semiconductor MQW waveguides were $n_2 = 4 \times 10^{-11}$ cm²/W and $\beta = 4 \times 10^{-9}$ cm/W, giving a combined figure of merit of $F = 7$ [45.59]. In all measurements the negative nonlinear index changes were measured to have magnitude smaller than $|\Delta n| < 0.001$ [45.60].

45.3.2 Nonresonant Nonlinearities in Inorganic Crystalline Semiconductors

Nonresonant nonlinearities are not triggered by direct electronic transitions due to single photons. Much weaker effects of distortion of electronic clouds and multi-photon absorption are responsible for nonresonant nonlinear response. Maximum nonresonant nonlinear index changes are of the order $|\Delta n_{max}| \approx 1 \times 10^{-3}$. Since in certain spectral regions a typical nonresonant Kerr coefficient is $n_2 \approx 1 \times 10^{-13}$ cm²/W, the linear absorption is around 5 cm⁻¹, and the corresponding two-photon absorption coefficient is $\beta \approx 1 \times 10^{-9}$ cm/W, the figures of merit associated with nonresonant semiconductor nonlinearities can be acceptable.

The biggest advantage of nonresonant semiconductor nonlinearities is their sub-picosecond response time. The sum of the rise and relaxation times of nonresonant nonlinearity has been argued to be comparable to the orbital period of an electron in its motion about the nucleus, estimated to be around 10^{-16} s [45.29].

Depending on the spectral region, bulk and MQW inorganic crystalline semiconductors may exhibit either positive or negative refractive nonresonant nonlinearities. Under illumination with sub-nanosecond pulses at low repetition rates, the nonlinear index change is negative for wavelengths up to $1.5 ch/E_g$, where ch/E_g is the wavelength corresponding to the band gap, and h is Planck's constant. Δn is positive for wavelengths longer than $1.5 ch/E_g$ [45.37, 61]. In MQWs the spectral position of the sign change in Δn depends on the nanostructure of MQWs [45.62]. In 1993 *Shaw* and *Jaros* predicted through theory the dispersion of refractive nonlinearity in semiconductor MQWs and superlattices. They found that in MQWs the proximity of the spectral position of the Δn sign change to the band edge increases with increasing quantum confinement [45.62].

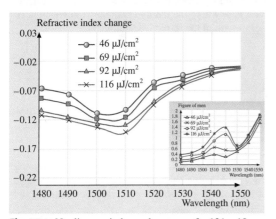

Fig. 45.4 Nonlinear index changes of 121 10-nm $In_{0.530}Al_{0.141}Ga_{0.329}As$ barriers and 120 5-nm $In_{0.530}Ga_{0.470}As$ wells grown on S-doped 001 InP wafer substrate. The *inset* shows the corresponding figures of merit. (After [45.54, 55])

Under nonresonant illumination with pulses longer than one nanosecond, there is no sign change in the refractive nonlinearity. The negative nonlinearity originating from two-photon absorption-induced free-carrier effects is much stronger than any positive third-order refractive effects at moderate and high intensities for $hc/E_g < \lambda < 2hc/E_g$. Consequently, the measured Δn is always negative in this spectral range [45.63].

The group of Sheik-Bahae and Van Stryland has authored several reports on the prediction of the spectral dependence of nonresonant nonlinearities in semiconductors. In 1985 *Van Stryland* et al. predicted trends in the absorptive ultrafast nonlinear response of semiconductors. An equation for the two-photon absorption below the band gap was derived and compared with experimental values. Dispersion of two-photon absorption is expected to mimic the dispersion of linear absorption; i.e. two-photon absorption is strong and relatively flat from the band gap to almost the midpoint of the band gap, at which point it goes to zero. Good agreement was obtained between experiment and theory for photon energies not in the vicinity of the band gap, with two-photon absorption coefficients of various semiconductors ranging from $\beta = 3 \times 10^{-9}$ cm/W to $\beta = 25 \times 10^{-9}$ cm/W [45.36].

In the ensuing years the same research group reported theory describing the spectral dependence of the real part of the ultrafast nonlinearity and compared it with experiments. The results are shown in Fig. 45.5. The magnitude of n_2 is largest near the photon energy corresponding to half of the band gap. Since, for wavelengths longer than that corresponding to half the band gap, two-

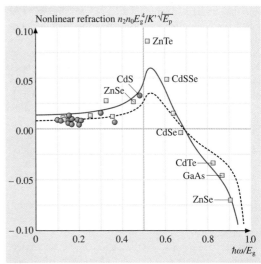

Fig. 45.5 Dispersion of the refractive nonlinearity of inorganic crystalline semiconductors in the transparent region. The *points* correspond to the experimental data explained in [45.37], while the *lines* are a fit to the theory. (After [45.37])

photon absorption vanishes, large figures of merit can be expected in these spectral region. In addition, n_2 was predicted to be positive for wavelengths longer than that corresponding to $0.75E_g$, and negative between $0.75E_g$ and E_g [45.37, 64]. A large discrepancy between theoretical and experimental results was observed near the band gap, where the theory drastically underestimated the strength of the refractive nonlinearity.

45.4 Organic Materials

Organic materials constitute another class of promising nonlinear materials. Organic materials exhibit significant nonlinearities across the visible and infrared spectral regions [45.65]. They are readily processable into thin-film waveguide structures [45.33, 66] and in general do not rely on a high degree of perfection in ordering or purity to manifest their desired properties. The molecules that make up organic materials provide a tremendous range of structural, conformational, and orientational degrees of freedom for exploration with the aid of novel synthetic chemistry. This permits flexible modification and optimization of linear and nonlinear properties [45.33].

As is the case with semiconductor nonlinearities, the nonlinear response of organic materials can be divided into resonant and nonresonant parts, occurring in the absorbing and transparent regions, respectively. The resonant nonlinearities are a result of a single-photon absorption, while the nonresonant nonlinearities arise as a result of perturbations of electronic clouds and multi-photon absorption.

Depending on the structure of the constituent molecules, organic materials may exhibit many absorption resonances and hence many spectral areas of different strength and sign of nonlinear response. Phenomena such as molecular reorientation and pho-

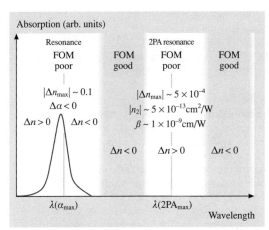

Fig. 45.6 Typical trends in the nonlinear response of organic materials with one absorption resonance

toisomerization, which are often found in organic materials, make the picture even more complex.

Some organic materials, such as most nonlinear dyes, have only one absorption resonance, which permits the qualitative prediction of their nonlinear response in the visible and near-infrared regions. Fig. 45.6 shows the nonlinear response of a typical nonlinear organic material with one absorption resonance.

In general, the figures of merit of organic materials in the absorbing region are poor. However, molecular effects such as trans–cis photoisomerization strongly increase the nonlinear index change along the absorption edge. The magnitudes of ultrafast nonlinearities and associated figures of merit of organic materials are comparable to those of inorganic crystalline semiconductors.

45.4.1 Resonant Nonlinear Response of Organic Materials

Since 1998 *Rangel* and *Rojo* et al. have reported resonant nonlinear properties of various organic compounds. The nonlinear refraction and absorption effects were reported in and near the absorbing regions of: solid-state PMMA samples doped with nonlinear azobenzene dye Disperse Red 1 [45.67], polydiacetylene microcrystals in aqueous suspension [45.68], cyclohexane suspensions of vanadium-oxide phthalocyanine microcrystals [45.69, 70], solid-state samples of plythiophene/selenophene copolymer [45.71], and a chloroform solution of triazole-quinone derivation [45.72].

PMMA films doped with Disperse Red 1 (10% molar concentration) have been shown to exhibit large, low-threshold nonlinear index changes and saturation of absorption as a result of optically induced structural changes in the middle and near the edge of the absorption resonance at 490 nm [45.67]. The nonlinear index changes associated with this photochemical phenomenon, called *trans–cis photoisomerization* have exceeded 0.1 in the spectral region $\lambda < 590$ nm. As illustrated in Fig. 45.7 [45.67], light near the main absorption resonance causes the azobenzene molecule to change from the *trans* to the *cis* configuration. During this process, the distance between the two carbons from which the acceptor and donor groups extend reduces from about 9.0 Å to 5.5 Å. This results in a drastic reduction in the molecule's dipole moment, which reduces the material's polarizability, providing a large negative nonlinearity with nonlinear index changes reaching $\Delta n_{\max} = 0.12$ at 560 nm under illumination with 20-ps pulses at a repetition rate of 10 Hz [45.67]. The figures of merit calculated according to (45.17) did not exceed 0.42 over the spectral range studied.

The refractive and absorptive nonlinear properties of polydiacetylene microcrystals in aqueous solution have been analyzed across their absorbing and near resonance regions 500–800 nm [45.68] with the same system as in [45.67]. As shown in Fig. 45.8, both negative and pos-

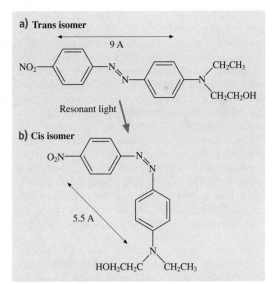

Fig. 45.7 Chemical structure of the azobenzene dye Disperse Red 1 undergoing trans–cis photoisomerization. Following resonant absorption, the azobenzene molecule changes its configuration, resulting in a decreased dipole moment. (After [45.67] with permission)

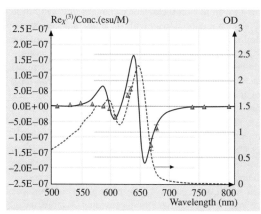

Fig. 45.8 $\chi^{(3)}$ and linear absorbance as a function of wavelength for 100-nm polydiacetylene microcrystals. (After [45.68])

itive nonlinear index changes were observed, depending on the spectral position relative to one of the two absorption resonances in the visible region. Saturation of absorption was observed across the entire absorbing region. A three-level model was developed to explain the data. Because of the low concentration of microcrystals in water, the absolute values of the nonlinear index change did not exceed 2×10^{-4}, while the figures of merit stayed below 0.26.

Cyclohexane suspensions of vanadium-oxide phthalocyanine microcrystals have also been characterized around its main absorption resonance peak in the spectral range 600–680 nm [45.69, 70]. Similarly to the nonlinear response of polydiacetylene microcrystals, both signs of refractive nonlinearity were observed, with the sign of the nonlinearity tracing the derivative of the linear absorption. Solid-state samples of plythiophene/selenophene copolymer were analyzed near the absorbing region and have shown only negative nonlinearity [45.71]. These results contradicted the predictions from nonlinear Kramers–Kronig transformation, which predicts positive nonlinearity in the spectral region where absorption increases with wavelength. The origins of this discrepancy were not understood.

In 1997 *Demenicis* et al. reported on the measurements of nonlinear properties of poly(3-hexadecylthiophene) in a chloroform solution around the absorption edge of 532 nm with 70-ps pulses at a repetition rate of 5 Hz [45.73]. Saturation of absorption and negative nonlinearity was observed. The saturation intensity decreased with increasing concentration, while the nonlinear absorption and nonlinear refraction at low intensity increased linearly. Figures of merit were not given.

45.4.2 Nonresonant Nonlinear Response of Organic Materials

In the 1970s it was predicted from theory and experimentally demonstrated that the conjugation of organic molecules results in a strong electronic delocalization and an associated large third-order nonlinearity in the transparent spectral region. This region extends between the main absorption peak due to the band-to-band transition and the first of the vibronic modes of the conjugated chain [45.74, 75]. It was demonstrated that the solid-state polymerization of organic monomers results in $\chi^{(3)}$ values comparable to these observed in inorganic crystalline semiconductors. Numerous studies followed that concentrated on the determination of the length and bond order of the conjugation chain, and acceptor and donor strengths that yield the strongest nonlinear response. *Marder* et al. provide an excellent summary of the field of nonresonant nonlinear response of organic molecules in [45.76].

Single-crystal polydiacetylene para-toluene sulphonates (PTS) have received special attention among the organic nonlinear compounds. In 1994 *Lawrence* et al. reported that at 1600 nm PTS has a large nonlinear refractive index of $n_2 = 2.2 \times 10^{-12}$ cm^2/W [45.77]. Since linear absorption of PTS is very low at this wavelength, and two-photon absorption was below experimental sensitivity ($\beta < 0.5$ cm/GW), and very large figures of merit were predicted with W exceeding unity for incident intensities as low as 20 MW/cm^2, and T never exceeding 0.1. A report followed in which *Lawrence* et al. showed measured two-photon absorption and nonlinear refraction values of PTS in the spectral range 800–1600 nm [45.78]. The two-photon-absorption coefficient varied in the range 0–700 cm/GW while the measured n_2 coefficient was between $n_2 = -2.2 \times 10^{-12}$ cm^2/W and $n_2 = 4.3 \times 10^{-12}$ cm^2/W. In 2000 and 2003 *Yoshino* et al. published two reports where the influence of three- and four-photon absorption on the nonlinear response was studied in the near-infrared region [45.79, 80]. For wavelengths of 1600–2200 nm the nonlinear refraction coefficient n_2 was around $n_2 = 5 \times 10^{-13}$ cm^2/W while nonlinear absorption was dominated by three- and four-photon effects.

Although PTS exhibits nonlinear refractive-index changes and figures of merit that are very large for nonlinear material in the transparent region, PTS suffers from low processability – as a single-crystal it cannot be easily processed into desired shapes.

Third-order nonresonant nonlinear properties of fullerene organic compounds have also attracted significant attention [45.81, 82]. Recently *Chen* et al. reported nonlinear coefficients and figures of merit of high-quality polyurethane films heavily loaded with (60)fullerene (C_{60}) [45.83]. Nonlinear refractive coefficients in excess of 10^{-12} cm^2/W were reported in the wavelength range 1150–1600 nm with very good figures of merit.

45.5 Nanocrystals

Nanoscale quantum-confined inorganic crystalline semiconductors represent an interesting group of nonlinear materials [45.84]. The size of such quantum dots is less than the bulk radii of excitons, holes, and electrons in a given semiconductor. As in the case of semiconductor MQWs, this results in quantum confinement of carriers. In a nanocrystal, this takes place in all three dimensions [45.85, 86]. Quantized energy levels make nanocrystals an artificial analogue of noninteracting atoms in a gas, raising the possibility of explaining the nonlinear processes by adopting the models of atomic physics.

To allow processability nanocrystals are usually embedded in either solid or liquid, organic or glass, optically linear hosts. Nanocrystal material systems are thus hybrids of semiconducting and insulating materials and combine interesting properties from both material groups. As in the case of semiconductor MQWs, the composition and size of quantum dots determines the energy of the electronic transitions. This allows spectral tunability of absorption features and nonlinear properties over the entire visible and infrared spectrum. On the other hand, the organic or glass host permits flexible fabrication of samples, waveguides, and other integrated components using polymer photonics technologies [45.87].

Figure 45.9 shows the properties of a typical resonant and nonresonant nonlinear response of strongly confined semiconductor nanocrystal composites. The data presented in this figure are based on the published theoretical predictions and experimental reports.

The finite number of allowed lower electronic levels leads to more pronounced excitonic features and resonant nonlinearities, which take place at lower fluences than in bulk or MQW inorganic crystalline semiconductors [45.88].

Similarly to the nonlinear response of bulk and MQW semiconductors, the resonant nonlinear response of nanocrystals is characterized by the saturation of absorption and an associated large change in the real part of the refractive index.

Saturation of absorption in strongly confined PbS quantum-dot glasses was measured in the spectral range 1.2–1.3 µm [45.89], covering the spectral position of the valley between the first and second excitonic peaks in the 6.6-nm-diameter sample studied. This material system was used as a passive saturable absorber in the production of 4.6-ps pulses via mode-locking around a wavelength of 1.3 µm [45.89]. The report was followed by studies of saturation of absorption dynamics in quantum dots of various sizes at a wavelength of 1.3 µm. This wavelength covered spectral regions ranging from the first to second electronic transitions, depending on the size of nanocrystals. The saturation energy and nonlinear decay times at a given wavelength were found to decrease with increasing size of nanocrystals [45.90]. Values for the refractive nonlinearity were not reported.

Saturation of absorption in PbS quantum-dot-doped glasses was also studied under illumination with 70-ps and 15-ns pulses 1.06 µm [45.91]. The

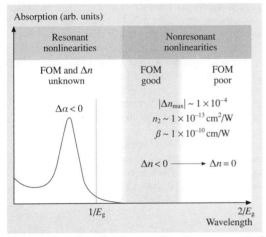

Fig. 45.9 Trends in the picosecond nonlinear response of inorganic semiconductor nanocrystals

nanocrystals analyzed had mean radii of 1.7–2.7 nm, resulting in excitonic peaks at wavelengths in the range 1.0–1.5 μm. The saturation intensity of the for the samples analyzed was found to be 2.3 MW/cm^2 and the relaxation time was measured to be 23 ± 2 ps.

Lu et al. measured the resonant nonlinear refractive properties of strongly confined 3.3-nm-diameter PbS nanocrystals in polymeric coatings over the spectral range 580–630 nm. This spectral range corresponds to the wavelengths around the first excitonic peak. The degenerate four-wave-mixing technique was used to measure the nonlinear susceptibility of nanocrystals near the photon energy of the first electronic transition at 595 nm. The values were found to fall in the range $\chi^{(3)} = 1 \times 10^{-6}$–$1 \times 10^{-5}$ esu. Combined with the 50 kW/cm^2 intensity used in the experiment these values of $\chi^{(3)}$ suggest maximum nonlinear index changes of $\Delta n \approx 0.013$. No data on saturation of the absorption were reported.

The nonresonant nonlinear response of nanocrystals shows different dispersion characteristics than that of any other material group. Under illumination with picosecond pulses, the nonresonant third-order nonlinearity is negative for photon energies between half the band gap and the band gap [45.92]. The magnitude of the third-order nonlinear response increases with proximity to the band gap, and disappears entirely near the half-band-gap energy. Such a response is in contrast to the Δn sign change between absorption and two-photon absorption resonances as observed in bulk semiconductors and organic materials. However, similarly to the nonresonant response of bulk semiconductors, the sign of the nonresonant refractive nonlinearity in the semiconductor-doped glasses in some spectral regions depends on the duration of the pulses used to measure the nonlinear effect. For pulses shorter than 1 ps the contribution of third-order positive refractive effects is comparable to that of the negative free-carrier absorption nonlinearities of the fifth order. Consequently, the measured Δn can be positive [45.63].

In 2000 and 2001 *Liu* et al. published several reports on measurements of the ultrafast nonlinearities of PbS nanoparticles, and PbS-coated CdS nanocomposites [45.93–95]. Surprisingly large refractive nonlinear indices of refraction up to -5×10^{-12} cm^2/W and unmeasurable nonlinear absorption were observed in surface modified polymer–PbS composites at moderate concentrations of 1.9×10^{-3} mol/l. This large refractive nonlinearities were attributed to the surface recombination owing to the high surface-to-volume ratio of PbS nanoparticles.

45.6 Other Nonlinear Materials

Metallic nanocomposites and cascaded second-order materials are two other promising groups of nonlinear materials. Metalorganic nanocomposites are made out of metallic quantum dots embedded in organic or glass hosts. Resonant nonlinear properties of copper [45.96] and silver nanoparticles [45.97, 98] embedded in a glass host were measured using degenerate four-wave-mixing experiments at visible wavelengths. The nonlinear coefficients and figures of merit of the metallic nanocrystals characterized were similar to those of semiconductor nanocrystals and reached a maximum near the plasma-frequency absorption peak.

It has been argued that metallic nanoparticles can potentially exhibit stronger nonlinear effects than other material systems [45.99]. This is associated with local field effects that enhance nonlinear response of the composite systems if the refractive index of the nonlinear constituent is lower than that of the linear host. Such a scenario can be realized in metallic nanoparticle–glass composites, since around the spectral positions of the plasma resonance the refractive index of metals can be lower than 1.

Cascaded nonlinear materials are made out of materials with second-order nonlinear properties. An appropriate design of such structures design results in a net accumulated phase shift for the illumination at a fundamental optical frequency at the end of a cascaded system. Cascaded material system acts as an effective third-order nonlinear material [45.100–102].

45.7 Conclusions

Following the preceding review, this section will summarize the major conclusions, as well as the missing pieces, of the published literature on nonlinear optical materials.

Bulk and MQW semiconductors have been demonstrated to exhibit low-threshold saturation of absorption near the band edge. The spectral position of the band edge can be tuned over the entire visible and near-infrared spectrum. It has been predicted from the nonlinear Kramers–Kronig relation, and has been measured directly in isolated cases, that the band-edge saturation of absorption results in large changes of the real part of refractive index.

In the regions of transparency, semiconductors exhibit weak nonlinear refractive effects of both signs. The nonresonant effects can be accompanied by two-photon absorption. Depending on the spectral position, the nonresonant nonlinear response of semiconductors can be characterized by good figures of merit.

Semiconductor nanocrystals also permit spectral tunability of their linear and nonlinear optical properties over the entire visible and near-infrared regions. Semiconductor nanocrystals have been demonstrated to exhibit strong saturation of absorption near the excitonic peak associated with the first allowed electronic transition. The nonresonant nonlinear response of nanocrystals is of a similar magnitude as in bulk and in MQW inorganic crystalline semiconductors.

The figures of merit for organic materials in the absorbing region are in general poor. In the transparent region the Kerr and two-photon absorption coefficients of organic materials are of magnitudes comparable to these of inorganic crystalline semiconductors. The sign of the refractive nonlinearity varies across the spectrum depending on the proximity to various absorption resonances.

Although the nonlinear properties of many materials systems have been reported, further characterization is needed to assess the applicability of various nonlinear material systems to optical signal processing. In contrast to previously reported measurements carried out at isolated wavelengths, measurements of the refractive and absorptive nonlinear response over wide spectral ranges, which would permit determination of figures of merit, need to be carried out. In particular, the refractive and absorptive nonlinear response in the most-promising absorption-edge regions of MQW semiconductors and semiconductor nanocrystals should be examined comprehensively and the applicability of these material systems to optical signal processing should be determined.

References

45.1 E. Cotter, J. K. Lucek, D. D. Marcenac: IEEE Commun. Mag. **34**, 90–95 (1997)
45.2 P. P. Mitra, J. B. Stark: Nature **411**, 1027 (2001)
45.3 P. W. Smith: The Bell Syst. Tech. J. **61**, 1975–1983 (1982)
45.4 L. Brzozowski, E. H. Sargent: J. Opt. Soc. Am. B **17**, 1360–1365 (2000)
45.5 L. Brzozowski, E. H. Sargent: IEEE J. Quantum Electron. **36**, 550–555 (2000)
45.6 L. Brzozowski, E. H. Sargent: IEEE J. Quantum Electron. **36**, 1237–1242 (2000)
45.7 P. W. Smith, I. P. Kaminov, P. J. Maloney, L. W. Stulz: Appl. Phys. Lett. **34**, 62–65 (1979)
45.8 P. W. Smith, E. H. Turner: Appl. Phys. Lett. **30**, 280–281 (1977)
45.9 B. E. A. Saleh, M. C. Teich: *Fundamentals of Photonics* (Wiley, New York 1991)
45.10 P. W. E. Smith, L. Qian: IEEE Circuits Dev. Mag. **15**, 28–33 (1999)
45.11 P. W. E. Smith: All-optical devices: materials requirements. In: *Nonlinear Optical Properties of Advanced Materials*, Vol. 1852 (SPIE, Los Angeles, CA 1993) pp. 2–9
45.12 G. I. Stegeman: All-optical devices: materials requirements. In: *Nonlinear Optical Properties of Advanced Materials*, Vol. 1852 (SPIE, Los Angeles, CA 1993) pp. 75–89
45.13 I. C. Khoo, M. Wood, B. D. Guenther: Nonlinear liquid crystal optical fiber array for all-optical switching/limiting, In: LEOS 96 9th Annual Meeting, Vol. 2, pp. 211–212, IEEE, (Bellingham, 1996)
45.14 G. L. Wood, W. W. II. I. Clark, M. J. Miller, G. J. Salamo, E. J. Sharp: Evaluation of passive optical limiters and switches. In: *Materials for Optical Switches, Isolators, and Limiters*, Vol. 1105 (SPIE, Orlando, FL 1989) pp. 154–181
45.15 R. Bozio, M. Meneghetti, R. Signorini, M. Maggini, G. Scorrano, M. Prato, G. Brusatin, M. Guglielmi: Optical limiting of fullerene derivatives embedded in sol–gel materials, In: Photoactive Organic Materials. Science and Applications, Proc. NATO Adv.

45.16 J. A. Hermann, P. B. Chapple, J. Staromlynska, P. Wilson: Design criteria for optical power limiters. In: *Nonlinear Optical Materials for Switching and Limiting*, Vol. 2229, ed. by M. J. Soileau (SPIE, Orlando, FL 1994) pp. 167–178

45.17 N. G. R. Broderick, D. Taverner, D. J. Richardson: Opt. Express **3**, 447–453 (1998)

45.18 N. D. Sankey, D. F. Prelewitz, T. G. Brown: Appl. Phys. Lett. **60**, 1427–1429 (1992).

45.19 L. Brzozowski, E. H. Sargent: IEEE J. Lightwave Technol. **19**, 114–119 (2000)

45.20 L. Brzozowski, V. Sukhovatkin, E. H. Sargent, A. SpringThorpe, M. Extavour: IEEE J. Quantum Electron. **39**, 924–930 (2003)

45.21 H. M. Gibbs, S. L. McCall, T. N. C. Venkatesan, A. C. Gossard, A. Passner, W. Wiegmann: Appl. Phys. Lett. **35**, 451–453 (1979)

45.22 H. M. Gibbs, S. S. Tang, J. L. Jewell, D. A. Winberger, K. Tai, A. C. Gossard, S. L. McCall, A. Passner: Appl. Phys. Lett. **41**, 221–222 (1982)

45.23 H. G. Winful, J. H. Marburger, E. Garmire: Appl. Phys. Lett. **35**, 379–381 (1979)

45.24 D. Pelinovsky, L. Brzozowski, E. H. Sargent: Phys. Rev. E **60**, R4536–R4539 (2000)

45.25 D. Pelinovsky, J. Sears, L. Brzozowski, E. H. Sargent: J. Opt. Soc. Am. B **19**, 45–53 (2002)

45.26 C. M. de Sterke, J. E. Sipe: Progress Opt. **33**, 203–260 (1994)

45.27 W. Chen, D. L. Mills: Phys. Rev. Lett. **58**, 160–163 (1987)

45.28 W. N. Ye, L. Brzozowski, E. H. Sargent, D. Pelinovsky: J. Opt. Soc. Am. B **20**, 695–705 (2003)

45.29 R. W. Boyd: *Nonlinear Optics* (Academic, New York 1992)

45.30 M. G. Kuzyk, C. W. Dirk: *Characterization Techniques and Tabulations for Organic Nonlinear Optical Materials* (Dekker, New York, N.Y. 1998)

45.31 D. L. Mills: *Nonlinear Optics: Basic Concepts* (Springer, Berlin, Heidelberg 1998)

45.32 P. Gunter: *Nonlinear Optical Effects and Materials* (Springer, Berlin, Heidelberg 2000)

45.33 P. N. Prasad, D. J. Williams: *Introduction to Nonlinear Optical Effects in Molecules and Polymers* (Wiley, New York 1991)

45.34 R. L. Sutherland: *Handbook of Nonlinear Optics* (Dekker, New York 1996)

45.35 E. Garmire: IEEE J. Selected Topics Quantum Electron. **6**, 1094–1110 (2000)

45.36 E. W. Van Stryland, M. A. Woodall, H. Vanherzeele, M. J. Soileau: Opt. Lett. **10**, 490–492 (1985)

45.37 M. Sheik-Bahae, D. C. Hutchings, D. J. Hagan, E. W. Van Stryland: IEEE J. Quantum Electron. **27**, 1296–1309 (1991)

45.38 S. H. Park, J. F. Morhange, A. D. Jeffery, R. A. Morgan, A. Chevez-Pirson, H. M. Gibbs, S. W. Koch, N. Peyghambarian, M. Derstine, A. C. Gossard, J. H. English, W. Weigmann: Appl. Phys. Lett. **52**, 1201–1203 (1988)

45.39 L. Qian, S. D. Benjamin, P. W. E. Smith, H. Pinkney, B. J. Robinson, D. A. Thompson: Opt. Lett. **22**, 108–110 (1997)

45.40 H. S. Loka, S. D. Benjamin, P. W. E. Smith: IEEE J. Quantum Electron. **34**, 1426–1437 (1998)

45.41 H. Pinkney, D. A. Thompson, B. J. Robinson, L. Qian, S. D. Benjamin, P. W. E. Smith: J. Cryst. Growth **209**, 237–241 (2000)

45.42 S. Gupta, M. Y. Frankel, J. A. Valdmanis, J. F. Whitaker, G. A. Mourou, F. W. Smith, A. R. Calawa: Appl. Phys. Lett. **59**, 3276–3278 (1991)

45.43 E. S. Harmon, M. R. Melloch, J. W. Woodall, D. D. Nolte, N. Olsuka, C. L. Chang: Appl. Phys. Lett. **63**, 2248–2250 (1993)

45.44 S. D. Benjamin, A. Othonos, P. W. E. Smith: Electron. Lett. **30**, 1704–1706 (1994)

45.45 P. W. E. Smith, S. D. Benjamin, H. S. Loka: Appl. Phys. Lett. **71**, 1156–1158 (1997)

45.46 M. Kawase, E. Garmire, H. C. Lee, P. D. Dapkus: IEEE J. Quantum Electron. **30**, 981–988 (1994)

45.47 L. Brzozowski, E. H. Sargent, A. SpringThorpe, M. Extavour: Appl. Phys. Lett. **82**, 4429–4431 (2003)

45.48 D. A. B. Miller, D. S. Chemla, D. J. Eilenbergeer, P. W. Smith, A. C. Gossard, W. T. Tsang: Appl. Phys. Lett. **41**, 679–681 (1982)

45.49 S. Schmitt-Rink, D. S. Chemla, D. A. B. Miller: Phys. Rev. B **32**, 6601–6609 (1985)

45.50 Y. H. Lee, A. Chavez-Pirson, S. W. Koch, H. M. Gibbs, S. H. Park, J. Morchange, A. Jeffery, N. Peyghambrian, J. Banyai, A. C. Gossard, W. Wiegmann: Phys. Rev. Lett. **57**, 2446–2449 (1986)

45.51 F. Stern: Phys. Rev. **133**, A1653–A1664 (1964)

45.52 A. M. Fox, A. C. Maciel, J. F. Ryan, M. D. Scott: Nonlinear excitonic optical absorption in GaInAs/InP quantum wells

45.53 A. M. Fox, A. C. Maciel, M. G. Shorthose, J. F. Ryan, M. D. Scott, J. I. Davies, J. R. Riffat: Nonlinear excitonic optical absorption in GaInAs/InP quantum wells

45.54 L. Brzozowski, E. H. Sargent, A. SpringThorpe, M. Extavour: Appl. Phys. Lett., 4429–4431 (2003)

45.55 L. Brzozowski, E. H. Sargent, A. SpringThorpe, M. Extavour: Virtual J. Ultrafast Sci. **2**(7) (2003)

45.56 P. W. Joudawlkis, D. T. McInturff, S. E. Ralph: Appl. Phys. Lett. **69**, 4062–4064 (1996)

45.57 R. V. Penty, H. K. Tsang, I. H. White, R. S. Grant, W. Sibert, J. E. A. Whiteaway: Electron. Lett. **27**, 1447–1449 (1991)

45.58 I. E. Day, P. A. Snow, R. V. Penty, I. H. White, R. S. Grant, G. T. Kennedy, W. Sibbett, D. A. O. Davies, M. A. Fisher, M. J. Adams: Appl. Phys. Lett. **65**, 2657–2659 (1994)

45.59 M. A. Fisher, H. Wickes, G. T. Kennedy, R. S. Grant, W. Sibbett: Electron. Lett. **29**, 1185–1186 (1993)

45.60 D. A. O. Davies, M. A. Fisher, D. J. Elton, S. D. Perrin, M. J. Adams, G. T. Kennedy, R. S. Grant,

P. D. Roberts, W. Sibbett: Electron. Lett. **29**, 1710–1711 (1993)

45.61 C. Aversa, J. E. Sipe, M. Sheik-Bahae, E. W. V. Stryland: Phys. Rev. B **24**, 18073–18082 (1994)

45.62 M. J. Shaw, M. Jaros: Phys. Rev. B **47**, 1620–1623 (1993)

45.63 K. S. Bindra, A. K. Kar: Appl. Phys. Lett. **79**, 3761–3763 (2001)

45.64 M. Sheik-Bahae, D. J. Hagan, E. W. Van Stryland: Phys. Rev. Lett. **65**, 96–99 (1990)

45.65 N. J. Long: Angew. Chem., Int. Ed. **34**, 21–38 (1995)

45.66 I. Liakatas, C. Cai, M. Bösch, C. B. M. Jäger, P. Günter: Appl. Phys. Lett. **76**, 1368–1370 (2000)

45.67 R. Rangel-Rojo, S. Yamada, H. Matsuda, D. Yankelevicg: Appl. Phys. Lett. **72**, 1021–1023 (1998)

45.68 R. Rangel-Rojo, S. Yamada, H. Matsuda, H. Kasai, H. Nakanishi, A. K. Kar, B. S. Wherrett: J. Opt. Soc. Am. B **203**, 2937–2945 (1998)

45.69 R. Rangel-Rojo, S. Yamada, H. Matsuda, H. Kasai, Y. Komai, S. Okada, H. Oikava, H. Nakanishi: Jap. J. Appl. Phys. **38**, 69–73 (1999)

45.70 R. Rangel-Rojo, H. Matsuda, H. Kasai, H. Nakanishi: J. Opt. Soc. Am. **17**, 1376–1382 (2000)

45.71 E. Van Keuren, T. Wakebe, R. Andreaus, H. Möhwald, W. Schrof, V. Belov, H. Matsuda, R. Rangle-Rojo: J. Opt. Soc. Am. B **203**, 2937–2945 (1998)

45.72 R. Rangel-Rojo, L. Stranges, A. K. Kar, M. A. Mendez-Rojas, W. H. Watson: Opt. Commun. **203**, 385–391 (2002)

45.73 L. Demenicis, A. S. L. Gomes, D. V. Petrov, C. B. de Araújo, C. P. de Molo, C. G. dos Santos, R. Souto-Major: J. Opt. Soc. Am. B **14**, 609–614 (1997)

45.74 C. Sauteret, J. P. Hermann, R. Frey, F. Pradère, J. Ducling, R. H. Baughman, R. R. Chance: Phys. Rev. Lett. **36**, 956–959 (1976)

45.75 G. P. Agrawal, C. Cojan, C. Flytzanis: Phys. Rev. B **17**, 776–789 (1978)

45.76 S. R. Marder, B. Kippelen, A. Y. Jan, N. Peyghambarian: Nature **388**, 845–951 (1997)

45.77 B. L. Lawrence, M. Cha, J. U. Kang, W. Torruellas, G. Stegeman, G. Baker, J. Meth, S. Etemad: Electron. Lett. **30**, 447–448 (1994)

45.78 B. L. Lawrence, W. Torruellas, M. C. G. Stegeman, J. Meth, S. Etemad, G. Baker: Electron. Lett. **30**, 447–448 (1994)

45.79 F. Yoshino, S. Polyakov, L. Friedrich, M. Liu, H. Shim, G. I. Stegeman: J. Nonlinear Opt. Phys. Mater. **9**, 95–104 (2000)

45.80 F. Yoshino, S. Polyakov, M. Liu, G. Stegeman: Phys. Rev. Lett. **91**, 063901-1–063901-4 (2003)

45.81 S. Wang, W. Huang, R. Liang, Q. Gong, H. Li, H. Chen, D. Qiang: Phys. Rev. B **63**, 153408(1–4) (2001)

45.82 B. L. Yu, H. P. Xia, C. S. Zhu, F. X. Gan: Appl. Phys. Lett. **81**, 2701–2703 (2002)

45.83 Q. Chen, L. Kuang, E. H. Sargent, Z. Y. Wang: Appl. Phys. Lett. **83**, 2115–2117 (2003)

45.84 F. W. Wise: Accounts Chem. Res. **33**, 773–780 (2000)

45.85 L. Banyai, S. W. Koch: Phys. Rev. Lett. **57**, 2722–2724 (1986)

45.86 S. Schmitt-Rink, D. A. B. Miller, D. S. Chemla: Phys. Rev. B **35**, 8113–8125 (1987)

45.87 M. A. Hines, G. D. Scholes: Synthesis of colloidal PbS nanocrystals with size-tunable NIR emissions, submitted

45.88 G. Wang, K. Guo: Physica B **315**, 234–239 (2001)

45.89 P. T. Guerreiro, S. Ten, N. F. Borrelli, J. Butty, G. E. Jabbour, N. Peyghambarian: Appl. Phys. Lett. **71**, 1595–1597 (1997)

45.90 K. Wundke, S. Pötting, J. Auxier, A. Schülzgen, N. Pegyghambarian, N. F. Borrelli: Appl. Phys. Lett. **76**, 10–12 (2000)

45.91 A. M. Malyarevich, V. G. Savitski, P. V. Prokoshin, N. N. Posonov, K. V. Yumashev, E. Raaben, A. A. Zhilin: J. Chem. Phys. **78**, 1543–1551 (1983)

45.92 D. Cotter, M. C. Burt, R. J. Manning: Phys. Rev. Lett. **68**, 1200–1203 (1992)

45.93 H. P. Li, B. Liu, C. H. Kam, Y. L. Lam, W. X. Que, L. M. Gan, C. H. Chew, G. Q. Xu: Opt. Mater. **14**, 321–327 (2000)

45.94 B. Liu, H. Li, C. H. Chew, W. Que, Y. L. Lam, C. H. Kam, L. M. Gan, G. Q. Xu: Mater. Lett. **51**, 461–469 (2001)

45.95 B. Liu, C. H. Chew, L. M. Gan, G. Q. Xu, H. Li, Y. L. Lam, C. H. Kam, W. X. Que: Mater. Lett. **51**, 461–469 (2001)

45.96 L. Yang, K. Becker, F. M. Smith, R. H. Magruder III, R. F. Haglund Jr., L. Yang, R. Dorsinville, R. R. Alfano, R. A. Zuhr: Size dependence of the third-order susceptibility of copper nanocrystals investigated by four-wave mixing

45.97 K. Uchida, S. Kaneko, S. Omi, C. Hata, H. Tanji, Y. Asahara, A. J. Ikushima, T. Tokizaki, A. Nakamura: J. Opt. Soc. Am. **11**, 1236–1243 (1994)

45.98 H. Inouye, K. Tanaka, I. Tanahashi, T. Hattori, H. Nakatsuka: Jap. J. Appl. Phys. **39**, 5132–5133 (2000)

45.99 D. Ricard, P. Roussignol, C. Flytanis: Opt. Lett. **10**, 511–513 (1985)

45.100 D. V. Petrov: Opt. Commun. **13**, 102–106 (1996)

45.101 C. Bosshard: Adv. Mater. **5**, 385–397 (1996)

45.102 G. I. Stegeman, D. J. Hagan, L. Torner: Opt. Quantum Electron. **28**, 1691–1740 (1996)